T0234127

Structural Mechanics in Lightweight Engineering

Christian Mittelstedt

Structural Mechanics
in Lightweight Engineering

 Springer

Christian Mittelstedt
Department of Mechanical Engineering
Institute for Lightweight Construction and Design
Technical University Darmstadt
Darmstadt, Germany

ISBN 978-3-030-75195-1   ISBN 978-3-030-75193-7   (eBook)
https://doi.org/10.1007/978-3-030-75193-7

Translation from the German language edition: *Rechenmethoden des Leichtbaus* by Christian Mittelstedt, © Studienbereich Mechanik der TU Darmstadt 2017. Published by Studienbereich Mechanik der TU Darmstadt. All Rights Reserved.

This Springer imprint is published by the registered company Springer Nature Switzerland AG
The registered company address is: Gewerbestrasse 11, 6330 Cham, Switzerland

For Siham

Preface

This book is the result of lectures that I gave in various positions at universities and technical colleges over the past years as well as during my time in the aerospace industry. It is the first volume of a book series dealing with structural mechanics and analysis methods in lightweight engineering and deals with the static analysis of beam and bar structures. Further volumes of the planned book series will deal with the theory of stability of lightweight structures as well as with the statics of flat and curved two-dimensional load-bearing structures, i.e. plates and shells. I found it to be a very important task during the creation of the manuscript to not only to present the contents of this book in a scientifically correct and stringent way but also to keep an eye on the clear and understandable presentation and practical feasibility of the contents. I sincerely hope that this undertaking has been successful.

Of course one may ask why such a detailed description of the contents presented in this book is necessary. My years of industrial experience have shown me that engineers in lightweight engineering practice have to deal with the statics and stability of lightweight structures on a daily basis where quick solutions for given problems must be found, often under immense time pressure. In order to find answers to such lightweight engineering problems, a particularly solid basic knowledge of the special features of lightweight structures is necessary. The application of numerical methods, above all the finite element method in the form of powerful commercial program systems, has become the industry standard, with the consequence that such methods are used even if the purpose does not justify their application and more convenient methods are available. The principles and methods presented in this book are also, as will be shown in detail, the basis for all kinds of approximation methods in lightweight engineering. Consequently, the error-free operation of commercial program systems, which are often used like a 'black box', requires a profound basic knowledge of the methods and principles detailed in this book. This book is therefore to be understood as a work for this fundamental area, in which the basics of structural mechanics are substantially and profoundly expanded and specialized for the analysis of lightweight constructions. For this reason, this book focuses especially on the closed-analytical treatment of bar and beam structures. From these fundamentals, numerical approximation methods, which have become an indispensable part of daily engineering work, follow quite naturally as will be shown in detail. Numerical methods, however, are not the main focus of this book, for they generally contribute little to the fundamental understanding of the problems that are dealt with here.

This book is aimed at students of aerospace engineering, mechanical engineering, civil engineering, aircraft and vehicle construction, and related fields of study at technical colleges and universities, who already have some fundamental knowledge of general structural mechanics. I do hope though that this book and all following volumes of this book series will be interesting and useful for engineers with an industrial background as well as for researchers and scientists at universities for whom this book series may serve as basic introductory literature or as a reference for their own work and research.

Questions, suggestions, and any other kind of feedback are of course welcome at any time.

Darmstadt, Germany   Christian Mittelstedt
Spring 2021

Contents

Chapter 1
Introduction

1.1 Definition and Tasks of Lightweight Construction and Engineering

1.1.1 Introduction

In many technical applications, the weight of a construction plays an important role. This special field of knowledge, which is generally referred to as lightweight construction or lightweight engineering and which is particularly dominant everywhere where engineers deal with moving structures, has been pioneered by the aerospace industry. Aircrafts as an example cannot be realized without special attention to lightweight construction and design. But also technical disciplines such as automotive engineering, rail vehicle construction, structural engineering and general mechanical engineering bring about lightweight construction tasks in many regards that need to be addressed and solved. In all that follows we will use the terminology lightweight engineering, realizing that lightweight engineering always encompasses construction and design tasks. The definition and main task of lightweight engineering can be summarized with the following quotation from Schapitz[1] (1963):

> *Lightweight engineering has the task of reducing the weight of technical constructions.*

The overall goal of lightweight engineering can also be formulated in such a way that structures that are generally subject to the extreme stress states need to be realized with the lowest possible weight without compromising load-bearing capacity, stiffness, operational strength or other necessary functions. Consequently, lightweight engineering includes aspects such as material science and selection, manufacturing

---

[1]The original German quotation reads: *Der Leichtbau hat die Aufgabe, das Gewicht technischer Konstruktionen zu verringern.*

technology, research and development technology, design and construction aspects, analysis methods, and many more. This book is mainly dedicated to the implementation of lightweight engineering tasks from the viewpoint of structural mechanics and analysis. It is not so much involved with looking at concrete tasks and structures, but rather deals with providing general and universally applicable analysis methods and rules that can be used to analyze and evaluate lightweight constructions, not only in the above-mentioned technical areas. The correct handling of lightweight structures in terms of structural mechanics and analysis is essential for a safe but also economical design of structures with minimal weight. The planned loose book series on the structural analysis of lightweight structures is specifically dedicated to those aspects in the field of strength of materials which are recurrent in the context of lightweight construction tasks. This includes analysis tasks from the field of statics and, because lightweight structures are mostly slender and thin-walled structures, the theory of stability. The present book is dedicated to the statics of beams and bars in the context of lightweight engineering. The treatment of plane and curved two-dimensional load bearing structures (i.e. plates and shells) and the consideration of the stability of lightweight structures is reserved for further volumes of the planned book series.

1.1.2 What Is Lightweight Engineering?

Lightweight engineering becomes relevant in all technical areas where minimizing the weight of a structure plays a decisive role. This may be due to the fact that a structure at rest under static load (e.g. a hall structure or a bridge) naturally has to bear its own weight in addition to other loads and therefore one should always try to reduce the dead weight as much as possible. However, lightweight engineering is particularly important when components or structures have to be moved. In order to minimize the necessary power for propulsion, it may therefore be advisable to minimize the weight of the structure to be moved to such an extent that propulsion effort and the resulting benefits are balanced. This is an important aspect in automotive engineering, for example, but especially in aircraft construction, where weight savings in the kilogram range can also result in significant savings in the operating costs of an aircraft. Especially aircraft construction, but also aerospace engineering in general, is only made possible by a consistent lightweight construction and design in the first place—an aircraft that is too heavy can neither be operated economically nor will it be possible to find buyers for it. The financial savings resulting from weight reduction are particularly dramatic in the aerospace industry, so that in these areas no one is shying away from significant efforts in order to extract the last weight reserves from aircraft or from spacecraft. Finally, it should also be mentioned that especially in times of scarce resources, materials of all kinds should be used responsibly, so that weight-saving designs of all kinds are fundamentally desirable.

A consistently pursued lightweight construction is in any case desirable. However, lightweight construction and design also means additional work and additional costs compared to traditional or classical design philosophies. Often, converting an existing classical process to lightweight construction and design also means using new and potentially more expensive materials. If, for instance, one is familiar with the design and verification of metallic structures, then the changeover to composite materials as an example will involve considerable additional expenditure in all technically relevant areas. In addition to material science aspects, this also concerns suitable manufacturing and research techniques, the general design possibilities, the mathematical treatment and the verification procedure. The consistent application of lightweight construction and design principles is also reflected in significantly longer product development times compared to simpler design approaches and, due to the typically high degree of utilization of lightweight and thus potentially filigree structures, entails a high experimental effort, on the one hand to prove the safety of structures by means of tests, but also to qualify new materials for certain applications or to validate new analysis methods experimentally and thus make them applicable for analysis and design work.

In all, this shows that lightweight construction and design is generally very desirable, but is also associated with additional efforts and costs. Whenever it is the case that the additional benefit outweighs the additional expenses, lightweight construction and design will prevail. In all cases where this is not the case, however, the application of lightweight construction principles will not be feasible. Planning engineers must always be aware of the tension between technical fascination on the one hand, and economic constraints on the other.

Irrespective of the specific application, lightweight construction can be divided into various approaches that become relevant at various stages of the product development process and that address issues such as the expected force flow within the structure, the required material properties, operating and environmental conditions, structural safety requirements, available manufacturing possibilities, individual or series production, assembly and disassembly, recycling and similar aspects. Classically, a distinction is made between lightweight constructions and designs as follows:

- Differential construction and design
- Integral construction and design

In addition, there is the so-called composite or hybrid design, which in turn can be divided as follows:

- Monolithic structures: Typically single-shell structures (e.g. monocoque or wing cover shells) made of a uniform material in one piece with closed outer skin,
- Sandwich structures: Multilayer composite of uusually two thin and very stiff outer layers, which are joined together by a typically quite soft so-called core,
- Hybrid structures: This refers to the combination of different materials within a structure, with the aim of achieving optimum material utilization through targeted positioning and thus also a significant reduction in weight.

Another important concept is lightweight construction by adequate choice of materials, which we will discuss later on. There is a multitude of other classifications which will not be discussed here. Structural analysis and structural optimization are of course important aspects in lightweight engineering and will be discussed later on.

Differential construction and design

Differential construction (Fig. 1.1, upper) means that a structure consists of several subcomponents, which are put together by means of suitable joining procedures to form the desired structure. Usually, the subcomponents to be joined are easy to manufacture so that the structure to be created can be produced with little effort and at low cost. In general, all joining and manufacturing processes which are suitable for the desired construction method and which are able to meet the requirements of the lightweight structure can be considered. A structure that is representative for aircraft construction and which is manufactured in a typical differential design is the fuselage (see Fig. 1.4), which we will discuss in detail later on.

Differential constructions have a number of advantages. First of all, they are generally quite easy to establish, both with regard to the analysis methods (individual subcomponents are generally much easier to analyze and design than very complex integral structures) and with regard to the manufacturing process. In addition, a differential construction method makes it possible to easily process several different materials in a lightweight structure, e.g., high-strength materials in areas of high stresses and less performant and thus usually also less expensive materials in areas of lower loads. A further point that cannot be neglected in times of scarce resources is the recycling capability: The structure can be completely dismantled again at the end of its lifetime, and the individual subcomponents can either be reused or sent for

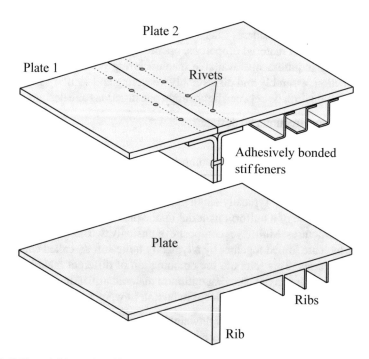

Fig. 1.1 Differential (upper) and integral construction and design (lower)

professional recycling. In the event of damage, all individual components can also be repaired or fully replaced. Tolerances are also easier to maintain with this kind of design than when a complex lightweight structure is manufactured in one piece. Lastly, the damage behavior must also be mentioned: The failure of one component does not necessarily result in the total failure of the entire structure. Cracks that form in one component cannot easily propagate to the adjacent components.

The disadvantage of a differential construction is that the high number of connections generally significantly increases the overall structural weight, especially due to the required overlaps. In addition, due to the notch effect at holes of bolted joints such connections are stress concentrators, which in many cases can also be decisive for the design. The assembly of the individual sub-components to form the desired lightweight structure also results in a very high number of work steps, which can increase not only the construction time but also the construction costs.

Integral construction and design

The integral construction (Fig. 1.1, lower, and Fig. 1.2) is the exact opposite of the differential construction. Its goal is to produce a lightweight structure from one piece. The manufacturing can be done with classical methods like turning or milling, but also with newer methods like additive manufacturing. Usually, a structure produced in this way is made of a uniform material which usually has isotropic properties.

The advantage of integral construction is that the geometry and wall thickness of the resulting components can be optimally adapted to the occurring stress distributions, and material is only arranged where it is actually needed. This allows for optimal lightweight construction and design. The integral design also eliminates the

Fig. 1.2 An example for an integral construction: cable bracket of the AIRBUS A350XWB. Shown here is the comparison between the original differential design (top and right) and an integral design (bottom left) for additive manufacturing (*Source* Schürg et al. (2016))

need for connections of any kind within the structure which is a significant advantage not only from the point of view of mechanical performance, but also with regard to the resulting weight savings. Furthermore, integral design allows the number of process steps to be reduced to a minimum, which is an important factor in reducing manufacturing costs.

A disadvantage is that the manufacturing processes used here can result in high material and tool costs. The demands on the manufacturing process, especially in terms of accuracy, are enormous. In addition, integral structures are susceptible to damage, cracks can migrate unhindered through the structure and, in the worst case, cause catastrophic failure. Furthermore, such components are very sensitive to tolerance deviations, which is an important point especially in highly sensitive areas such as aerospace engineering. Due to the nature of the required manufacturing processes, integrally manufactured structures are usually subject to clear limits, even in their component dimensions.

Lightweight construction and design by adequate choice of materials

Material lightweight construction, i.e. the lightweight construction and design by adequate choice of materials, means that the material previously used in a specific application is replaced by another, lightweight construction-specific and more suitable material. Prominent examples of the recent past for material lightweight construction are the development of the AIRBUS A380 and the AIRBUS A350 (Fig. 1.3). The extensive use of fiber-reinforced plastics was promoted on both air-

Fig. 1.3 The AIRBUS A350XWB landing in Hamburg-Finkenwerder, Germany (courtesy of AIRBUS Operations GmbH, Hamburg, Germany)

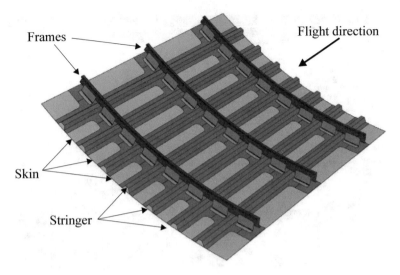

Frames

Flight direction

Skin

Stringer

Fig. 1.4 Generic representation of a typical segment of a fuselage shell of the AIRBUS A350XWB, consisting of the main components skin/stringer/frame

craft. For example, the upper fuselage shells of the A380 consist of hybrid laminates, the so-called GLARE® laminates (see e.g. Beumler (2004), GLARE is an abbreviation for Glass Laminate Aluminium Reinforced Epoxy). These are thin layered structures made of aluminum and glass fiber reinforced plastics (GFRP). The use of fiber reinforced materials was taken even further with the AIRBUS A350XWB, with large parts of the primary structure in the fuselage, wings and vertical and horizontal tail plains made of laminates of carbon fiber reinforced plastics (CFRP). A typical segment from the fuselage of the AIRBUS A350XWB is shown in Fig. 1.4, a view of the side shell of the AIRBUS A350XWB is given in Fig. 1.5. Figure 1.6 shows a view into the fuselage of the AIRBUS A350XWB. Laminates (Fig. 1.7) are thin layered structures consisting of a number of individual layers that can be oriented at any fiber angle (Mittelstedt and Becker (2016)). The main reason for the use of such laminate structures against metals in lightweight construction is the particularly advantageous mechanical behavior of such materials, in particular the extremely positive relationship between stiffness and density as well as strength and density. However, the automotive industry is also increasingly engaged in lightweight construction and is relying more and more on fiber composite components, as demonstrated by the fairly new BMWi3 and BMWi8 vehicles. The BMWi3 is an electric vehicle from Bayrische Motorenwerke (BMW) whose passenger cell is made of CFRP. The BMW hybrid model i8 was constructed in a similar manner.

The use of special lightweight construction materials is usually associated with significantly higher costs than with conventional construction methods, so that the top priority must be not only to make an intelligent choice of materials appropriate

Fig. 1.5 Side shell of the AIRBUS A350XWB (courtesy of AIRBUS Operations GmbH, Hamburg, Germany)

Fig. 1.6 View inside the fuselage of the AIRBUS A350XWB, Sects. 13 and 14 (courtesy of AIR-BUS Operations GmbH, Hamburg, Germany)

to the application, but also to use the material to the fullest possible extent. It should be noted that the suitability of a material for lightweight construction is not only determined by its mechanical properties such as the modulus of elasticity E or its strength properties R (this can be both a strength value for brittle materials, or a yield point for ductile materials), but that these values are rather related to the density ρ.

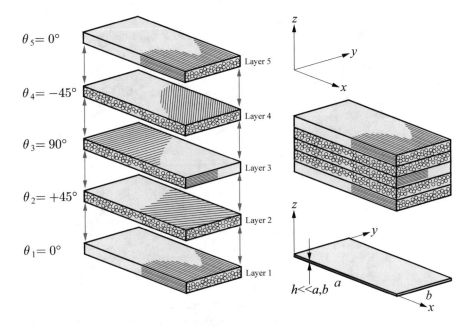

$\theta_5 = 0°$

$\theta_4 = -45°$

$\theta_3 = 90°$

$\theta_2 = +45°$

$\theta_1 = 0°$

Layer 5

Layer 4

Layer 3

Layer 2

Layer 1

$h \ll a,b$

Fig. 1.7 Joining individual layers of fiber-reinforced plastic (left) to form a laminate (right). It should be noted that the illustration is strongly exaggerated, laminates are usually very thin layered structures

The specific stiffness E/ρ and the specific strength R/ρ are therefore of primary importance.

Hybrid lightweight construction

Hybrid lightweight construction combines several materials suitable for lightweight construction in such a way that the properties of the individual materials involved are optimally matched to the intended purpose. One example is the already mentioned GLARE® laminates in which aluminum layers are combined with GFRP layers to form an optimal hybrid laminate. Although the GFRP layers are comparable in density to the aluminum alloys typically used in aircraft construction, they play the important role of a crack stopper so that any cracks in the aluminum layers cannot spread through the entire thickness of the component. Furthermore, GLARE® has more advantageous properties than pure aluminum. However, the disadvantage here is the significantly higher material costs compared to aluminum.

Examples of hybrid bar- or beam-type components are shown in Fig. 1.8. The beam shown in Fig. 1.8, left, consists of different materials in its flanges and its web and thus represents an example of a hybrid or composite beam. A special case of such a hybrid beam is shown in Fig. 1.8, middle. Here, reinforced concrete plate is supported by a steel beam. This method of construction which is very common in civil engineering, is also called composite or a hybrid construction, but is of course not the focus of this book. Steel-reinforced concrete is nevertheless a very

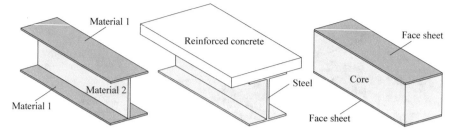

Fig. 1.8 Hybrid structures

Fig. 1.9 Reinforced
concrete beam

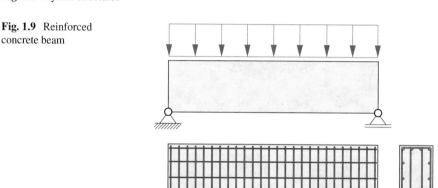

interesting example of a hybrid construction method (see also Fig. 1.9). The non-
reinforced concrete is a composite material consisting of cement, water and the
so-called aggregate (usually in the form of pebble stones of various diameters).
However, since concrete is extremely brittle and has only very low strength properties,
especially under tensile stresses, a reinforcement adapted to the respective stress state
is applied in the form of steel reinforcing bars. The reinforced concrete beam shown
in Fig. 1.9 thus has a pronounced steel reinforcement in both the upper and lower area
and is therefore suitable for alternating moment loads which cause tensile stresses in
both the upper and lower area. In contrast, the steel elements that are perpendicular
to the reinforcements in the upper and lower area have the task of relieving shear
stresses and preventing according crack formations.

   A special case of a hybrid beam or bar structure which is especially relevant
for lightweight construction is the so-called sandwich construction, as shown in
Fig. 1.8, right. In a sandwich structure, a core material is provided with two so-called
facesheets. Depending on the given specific application, a variety of materials are
used for core and facesheets. Cover layers can be made of wood, aluminum or fiber
reinforced composite materials. An equally wide range of materials is available for
the core. In Fig. 1.10 a so-called honeycomb core is shown which is typically made
of aluminum. Equally common are e.g. foams, metallic folded core structures or
paper-based materials.

Fig. 1.10 Sandwich beam
with honeycomb core

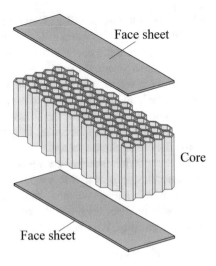

Face sheet

Core

Face sheet

A sandwich structure is particularly suitable for bending load cases. The surface layers usually have significantly higher strength and stiffness properties than the core, so that they are essentially responsible for absorbing the bending moments. The simultaneously occurring transverse shear forces are absorbed primarily by the core. This design principle corresponds to the load-bearing effect of an I-beam, in which the flanges contribute significantly to the bending stiffness of such beams due to the outer position and thus also carry the highest bending stresses. Applications of sandwich structures are again very prominent in aircraft engineering, here e.g. as floor plates in the fuselage, but also, for example, in civil engineering as facade elements.

Ultimately, the aim of hybrid lightweight construction is to achieve optimum material utilization with the best possible stiffness and strength of the lightweight structure and thus to achieve the best possible material utilization with optimum weight and material savings. Naturally, these advantages are accompanied by the problem that the costs can be higher than those of classic construction methods, that the design of components made of different materials can be a challenge and that attention must be paid to the different mechanical, thermal and hygroscopic properties of the individual components.

1.2 Structural Analysis in Lightweight Engineering

The conception, construction and design of any engineering structure requires first of all statements about the requirements that apply for the planned structure. This includes, for example, the required loads and strengths, the desired type and duration of use, general conditions such as environmental effects, but also the expected static and dynamic loadings. This is usually summarized in a list of requirements, without which a design is not possible. However, if the list of requirements is only rough in its specifications or if there are inaccurate specifications or if there is initially insufficient knowledge about the mechanical behavior when a new material is introduced into a lightweight structure, the resulting structure will contain unnecessary conservatisms and safety margins which can destroy the potential for lightweight construction. This not only affects the later performance of the lightweight structure, but also the chances and possibilities of the structural analysis to which this section is dedicated. The structural analysis becomes less precise the more imprecise the specifications of the list of requirements are.

Once the requirements for a lightweight structure have been clarified and an initial idea for a design is available, then it is possible to proceed to the structural analysis. By structural analysis we mean the treatment of the state of a structure with the help of suitable analysis methods as well as the proof of the load-bearing capacity and the serviceability. We structure these verifications as follows:

- State of stress: The verification that the stresses in any point of the lightweight structure under consideration remain below a defined limit value, for example a strength value or a yield strength. A distinction is made between static strength and fatigue strength when the structure is subject to dynamic loads. This is a form of the proof of load-bearing capacity.
- Deformation state: In addition to strength requirements, many structures are also subject to requirements regarding the maximum permissible deformations. However, since exceeding a limit deformation is often not synonymous with the loss of load-bearing capacity but rather restricts the general serviceability, we also refer to this as the proof of serviceability.
- Structural stability: Lightweight structures are typically slender and thin-walled. Therefore, in addition to the types of proofs already mentioned it must also be ensured that structural stability is not affected by any of the following buckling modes: column buckling (Fig. 1.11), torsional buckling (Fig. 1.12), flexural torsional buckling (Fig. 1.13), lateral stability (Fig. 1.14) and local buckling (Fig. 1.15). Further, buckling in thin-walled two-dimensional structures (i.e. plates and shells, see Figs. 1.16, 1.17 and 1.18) must be considered as well. The mentioned stability cases can lead to a catastrophic failure of a structure which must be avoided in any case.
- Dynamic verification: If the structure to be verified is subjected to vibrational loads, it must also have natural frequencies that are sufficiently far from the frequencies under service conditions. This is verified by a so-called modal analysis.
- Detailed design and verification: Usually, the previously mentioned types of verification are performed globally without capturing deep details of the design. Once the global verification process is complete, it is possible to proceed to the verification of the structure in all its details. We then speak of refined or local verifications. This applies, for example, to the introduction of forces, which is often a challenge in thin-walled structures, but also to verifications in the vicinity of openings which are often introduced to reduce weight, and so on.

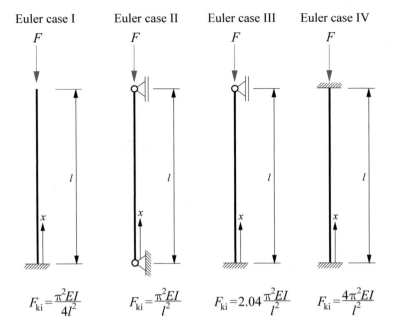

Euler case I	Euler case II	Euler case III	Euler case IV
$F_{ki}=\dfrac{\pi^2 EI}{4l^2}$	$F_{ki}=\dfrac{\pi^2 EI}{l^2}$	$F_{ki}=2.04\,\dfrac{\pi^2 EI}{l^2}$	$F_{ki}=\dfrac{4\pi^2 EI}{l^2}$

Fig. 1.11 The four Euler cases as basic examples of column buckling

Fig. 1.12 Torsional buckling of a bar with doubly symmetric cross-section under centric axial compressive load

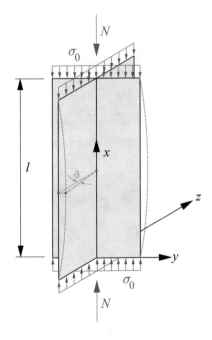

Fig. 1.13 Column buckling and torsional flexural buckling of a singly symmetric C-section with the excentricity e_y of the shear center M, measured from the center of gravity S of the section

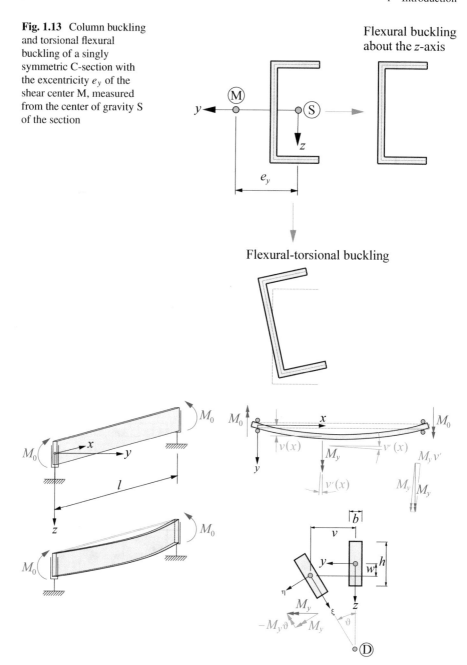

Flexural buckling about the z-axis

Flexural-torsional buckling

Fig. 1.14 Lateral stability of a beam under edge moments: This loss of stability is caused by a lateral deformation of the compressed area of the beam (left), which additionally causes a twisting of the cross section (right). In aircraft construction, this case becomes relevant when considering the fuselage, for example, in the design of cross beams or frames

Fig. 1.15 Local buckling (left), column buckling (middle) and interactive buckling (right) of a bar under centric compressive load

Fig. 1.16 Experimental buckling investigations at the DLR—Deutsches Zentrum für Luft- und Raumfahrt in Braunschweig (German Aerospace Center); cylindircal shell under axial compression (left), combined compression-shear-load (right) (*Source* Degenhardt et al. (2014))

The consideration of global and local verifications shows that structural analysis does not consist of a single step, but is performed iteratively several times in a row. In this way, various stages of structural analysis can be identified whose level of detail and thus the effort to be expended is constantly increasing.

Every project of lightweight engineering and every new development begins at the stage of the so-called preliminary design. If it is necessary to first get an idea of possible structural designs without a high degree of accuracy, then usually quite simple engineering models with correspondingly rough idealizations will be sufficient. It is important that these still very rough models take into account all essential influencing parameters, so that the basic structural behavior can be estimated through

Fig. 1.17 Buckling of liquid containers (*QSource* shellbuckling.com)

Fig. 1.18 Local buckling of the side shells of a B-52 Superfortress (*Source* shellbuckling.com)

dedicated parameter studies and first systematic optimizations. At this stage, simple closed-form analytical methods or, if not applicable, coarse finite element models are sufficient in many cases. If not explicitly necessary, non-linearities of any kind (usually geometrical and/or material nonlinearities) will be omitted at first and calculations based on simple structural elements will suffice.

Fig. 1.19 Static test pyramid

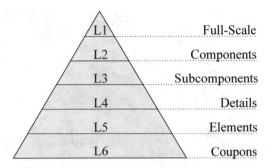

As soon as the preliminary design stage is finalized, the level of detail of the structural analysis will also increase and thus also the computational effort. Thus, the stage of the detailed analysis is reached. The structure to be analyzed is now usually refined within the framework of numerical models, boundary conditions and load scenarios are represented more and more precisely, and any non-linearities that may occur are included in the calculation as well. After the rather rough optimizations of the preliminary design, increasingly more accurate optimization models and thus results will typically be developed. At the end of the detailed analysis the design will be finalized.

In safety critical applications such as aerospace engineering, every structural analysis and thus every structural design is validated experimentally. Basically, safety-relevant lightweight structures are not released for operation without experimental proof of the load-bearing capacity. The test pyramid regarding the static strength, which is used e.g. for the AIRBUS A350XWB, is shown in Fig. 1.19. The degree of complexity increases the further one moves up the pyramid. For example, the lowest level of experimental verification consists of the so-called coupon tests. This refers to small sample tests under simple load conditions, such as the tensile test. Coupon tests are used to determine material properties such as stiffness and strength. At the top of the pyramid, however, is the experimental investigation of the entire aircraft. The effort required for such test campaigns is immense (see Fig. 1.20).

1.3   Structural Optimization in Lightweight Engineering

Lightweight engineering is by definition always an optimization task. Optimization can be performed on different levels, whereby the application purpose, the assumed lightweight construction potential, the complexity of the task, but also the time constraints determine the concrete optimization procedure. The term structural optimization is used for different approaches. Optimization can refer both to simple parameter studies with the subsequent selection of the best resulting structure design, but also to the consistent application of numerical-iterative optimization methods.

Fig. 1.20 Front view of the test aircraft in the so-called full scale test of the AIRBUS A350XWB (courtesy of AIRBUS Operations GmbH, Hamburg, Germany)

Structural optimization in lightweight construction and design is usually applied in all design stages and can be divided into the following types, among other criteria:

- Parameter studies: Targeted variation of the essential descriptive parameters of the structure to be optimized (such as dimensions, material parameters, etc.) and based on the results, the selection of a design according to defined criteria.
- Shape optimization: Targeted variation of the shape of a structure with the aim of obtaining an improved or, in the best case, optimized structure.
- Topology optimization: Determination of those areas of the design space (i.e. the maximum volume available for the structure to be designed) which are to be filled with material.
- material optimization: Selection of the most suitable material(s) for the given purposes.

We consider the situation shown in Fig. 1.21, in which a maximum available volume (the so-called design space) is supported as indicated. The load case here is a centric single force acting at the lower edge of the design space. It could be an optimization goal to determine the structural design that has the lowest weight while fulfilling certain given requirements. The design space can be freely filled with any material. The two points of support and the point of load introduction, on the other hand, cannot be changed during the concept development and subsequent optimization.

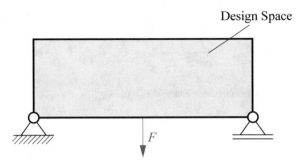

Design Space

Fig. 1.21 Given model situation

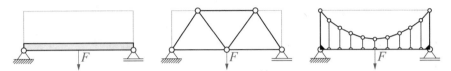

Fig. 1.22 Various structural concepts: Straight bending beam (left), truss (middle), Hanging construction (right)

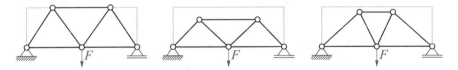

Fig. 1.23 Shape optimization at the example of a truss

The first step would be the finding of a basic concept (Fig. 1.22) and the selection of a material suitable for the given purpose.

We now assume that the given problem can be solved advantageously by the design of a truss structure which is shown in Fig. 1.22, middle, and which represents the starting point for the following optimization procedures. A shape optimization would mean, for example, that the positions of the nodes of the truss are varied to find an improved structural design based on the initial design (Fig. 1.23). Topology optimization, on the other hand, attempts to find the best possible configuration for the truss by adding or removing specific elements (Fig. 1.24). When carrying out a parameter study, an attempt would be made to determine the best possible design (Fig. 1.25) on the basis of the initial design by targeted variation of the member and cross-section parameters (material, cross-section, represented by the tensile stiffness EA), whereby not only the actual cross-section shape but also the individual wall thicknesses of the (potentially thin-walled) cross-sections could be taken as variable parameters.

An example for the consistent application of topology optimization of a structure relevant to lightweight construction and design, here for the production by means

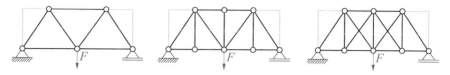

Fig. 1.24 Topology optimization at the example of a truss

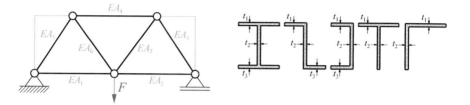

Fig. 1.25 Parameter variation at the example of a truss

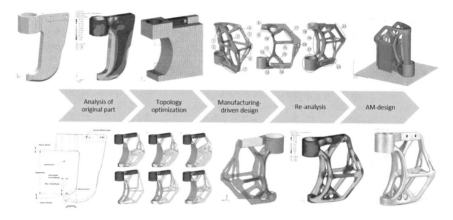

Fig. 1.26 Product development process for the optimized and production-oriented design of a riveting tool for additive manufacturing (Großmann et al. (2020))

of additive manufacturing (short: AM), is the riveting tool for self-pierce riveting in automotive engineering (Fig. 1.26, see also Großmann et al. (2020)). A numerical analysis of the initial design (here a full-walled tool) is followed by a dedicated topology optimization with the goal of minimizing the structural weight while adhering to limit values for stresses and deformations. The results of the topology optimization always require an engineer's interpretation, whereby in this case the production-oriented design is the main focus. The iteratively determined final design is finally subjected to a new analysis and a verification procedure is carried out. In many cases the final AM design has the shape of a truss, as shown in Figs. 1.26 and 1.27.

The optimization of a structure requires a clear and unambiguous mathematical formulation. Specifically, the following quantities need to be determined respectively defined:

Fig. 1.27 Prototype of an AM-design of a riveting tool for additive manufacturing (Großmann et al. (2020))

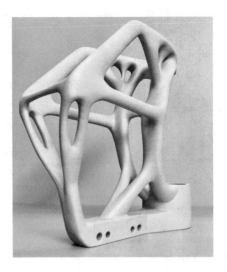

- Objective function: A property of the structure under consideration which usually required minimization, e.g. the structural weight.
- Restrictions: Conditions that must be met to ensure the load-bearing capacity and/or serviceability of the structure. Restrictions can be formulated e.g. for the stresses occurring in the structure which must remain below a certain limit value (e.g. a strength value).
- Analysis model: The analysis model with which the structure is analyzed. These can be simple hand calculations as well as high-resolution numerical models.
- Design variables: Those parameters that are available for optimization and can be used to find the optimal design. In the case of a truss as shown in Fig. 1.25, these could be, for example, the wall thicknesses of the member sections or the section dimensions.
- Optimization algorithm: The mathematical methods used to minimize the objective function. This concerns the field of knowledge of mathematical programming respectively nonlinear optimization.

As far as structural optimization in the framework of lightweight engineering is concerned, such questions have been dealt with intensively for several decades. While initially the main problems were in lightweight construction and design in the aerospace industry, structural optimization is now being used more and more widely, e.g. in automotive engineering, mechanical engineering and civil engineering. Structural optimization draws on the methods of mathematical programming and offers a wide range of different optimization techniques (see e.g. Vanderplaats (1984), Eschenauer et al. (1990), Eschenauer and Schnell (1993), Kamat (1993), Baier et al. (1994), Harzheim (2019)). Structural optimization requires appropriate structural modeling, the formation of a suitable optimization model, the dedicated use of optimization algorithms and the integration of these subtasks into an effective optimization procedure.

Fig. 1.28 Ideal planes (*Source* aviatormag.com.au)

The general mathematical form of an optimization problem with the objective function $f(\underline{x})$ to be minimized is:

$$\min. f(\underline{x}), \quad \underline{x} = (x_1, x_2, ..., x_n)^T, \tag{1.1}$$

where $\underline{x}$ is the vector of the n design variable $x_1, x_2, ..., x_n$. The problem is also determined by the following m inequality restrictions:

$$g_j(\underline{x}) \le 0, \quad j = 1, 2, ...m. \tag{1.2}$$

Further, let there be q equality restrictions as follows:

$$h_k(\underline{x}) = 0, \quad k = 1, 2, ...q. \tag{1.3}$$

Finally, for certain design variables, for example the design variable x_i, certain limits can be imposed:

$$a_i \le x_i \le b_i. \tag{1.4}$$

The application of structural optimization may also hold a certain potential for conflict. Usually, experts from various disciplines work on a sophisticated lightweight structure, so that, depending on the demands placed on the structure by the individual experts, very different and usually contradictory optimal designs of a structure can result. This is shown in a humorous way in the following cartoon, which not only shows in a very vivid way the potentially opposing structural designs, but also represents the field of tension in which design engineers of different disciplines operate (Fig. 1.28).

The term lightweight construction can be defined more precisely at this point. Lightweight construction and design does not only try to design a structure in the most weight-optimized way possible by using the means of structural optimization and a most detailed and high-resolution structural analysis. Rather, the aim of lightweight construction and design is to create a weight-optimized structure with the best possible performance by applying universal design solutions specialized for the given application. It is clear from the term design, which is used prominently here, that the consistent implementation of lightweight construction principles requires a certain level of experience of the involved engineers. Lightweight construction and design often leads to very individual solutions: The best design solution for a certain structure with specified requirements cannot necessarily be transferred to similar structures in every case. Nevertheless, certain general design guidelines can be formulated for many different problem areas, which we will deal with in the course of this book series if they can be derived from the calculation methods and structural-mechanical considerations. Such design principles must always be taken into account in the design stage, be it in the preliminary design phase or during detailed analysis, from the first step on. Subsequent or late fundamental changes in the design process cause costs, disrupt schedules, and are often not manageable when the project or development process is subject to strict time constraints.

1.4 Structural Elements in Lightweight Engineering

The structural analysis of a lightweight design always requires, independent of the concrete application, a conversion of the given situation into an idealization, which can be evaluated computationally by means of an adequate analysis method. This idealization requires knowledge of the given loads and boundary conditions as well as the selection of a suitable computational model. Analysis methods are based on common structural elements, which we will introduce in this section. For this purpose we consider the model situation of the Fig. 1.29. The available design space has the dimensions l, b, h and is loaded at its upper surface by the constant line load f. The reference axes x, y, z are defined as shown. The design space is supported at its two lower transverse edges in such a way that no displacements in the $z-$ direction can occur at those locations. The boundary conditions are not specified further at this point, but are assumed to be such that no rigid body degrees of freedom occur. We will discuss the given situation for different ratios between l, b, h and draw conclusions on the best possible structural elements for the individual situation. We assume isotropic material with the material properties E (modulus of elasticity) and ν (Poisson's ratio). In addition to the stress constraint $\sigma \leq R$ (where σ refers to each occurring stress component, and R refers to the respective strength property), the displacement constraints $u \leq u_{max}$, $v \leq v_{max}$ and $w \leq w_{max}$ also apply, where u, v, w are the displacements in the direction of the coordinate axes x, y, z.

Case 1: $b, h << l$

First, the case is considered that the dimensions b and h are significantly smaller than the dimension l: $b, h << l$. This is shown in Fig. 1.30, left. The resulting structural element is the beam, whose load-bearing effect is achieved primarily by bending (Fig. 1.30, right). The line load f can then be simplified and represented as a single

Fig. 1.29 Model situation, consisting of design space, supports and load

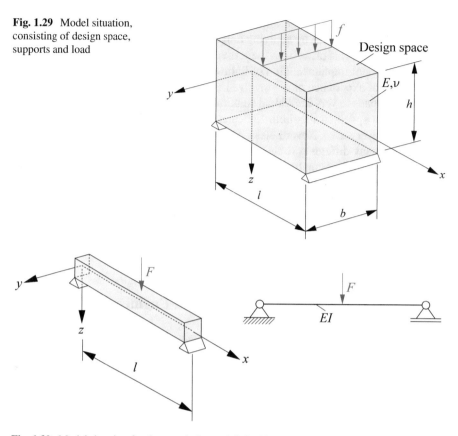

Fig. 1.30 Model situation for the case $b, h << l$ (left), idealization as a beam (right)

force F. A beam structure for the simple case of uniaxial bending (as in the present case) is described essentially by its length l and its bending stiffness EI, where I is the corresponding second-order moment of inertia. The main task of the designer is to find the best possible cross-sectional shape for the application (in Fig. 1.31 a few exemplary standard thin-walled sections are shown) and to perform all required verifications of the beam structure. In the case of the cross-section shapes that are common in lightweight construction, the considerations usually refer to the so-called skeleton line, which halves the wall thickness of the cross-section at each point and is shown in Fig. 1.31 as a dashed line.

In this elementary example, the static justification procedure will be limited to the verification of the maximum deflection w_{max} in the center of the field at the point of force application as well as to the verification of the stresses that occur (the normal stress σ and the shear stress τ) in the beam caused by the bending moment M and the transverse shear force Q (Fig. 1.32).

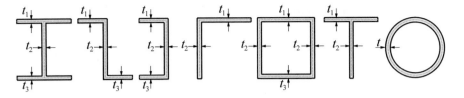

Fig. 1.31 Typical thin-walled cross-sections in lightweight engineering

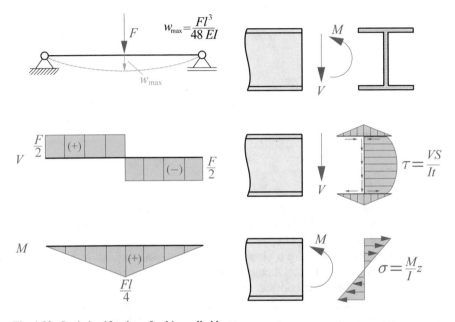

Fig. 1.32 Static justification of a thin-walled beam

The maximum deflection w_{max} of the beam can be determined using the means of elementary strength of elastostatics:

$$w_{max} = \frac{Fl^3}{48EI}. \tag{1.5}$$

We assume here that the most suitable cross-sectional shape is an I-beam. The shear stress τ caused by the transverse shear force Q is calculated at the position of the maximum shear force and then compared to its allowable value. It is calculated as follows:

$$\tau = \frac{QS}{It}, \tag{1.6}$$

wherein S is the first-order moment of inertia of the considered cross-section area, and t is the wall thickness at the point under consideration. The shear stress shows a

Fig. 1.33 Bending moments
and transverse shear forces
in the case of biaxial bending

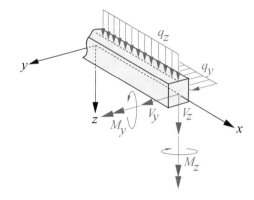

linear distribution in the flanges and a parabolic distribution in the web, as shown in
Fig. 1.32, middle. Furthermore, the normal stress σ, caused by the bending moment
M, occurs. It can be calculated as

$$\sigma = \frac{M}{I} z \tag{1.7}$$

and expands linearly over the thickness of the cross section, as shown in Fig. 1.32,
lower. It is verified at the location of the maximum bending moment, here in the
center point of the beam.

The simple load case with a single force in the middle of the beam which is
considered here first is a special case. A more general load case representation is
shown in Fig. 1.33. Here we consider the line loads q_y and q_z. Analogous to uniaxial
bending the given loads cause transverse shear forces and bending moments, where
two bending moments must be considered, namely the bending moment M_y (with
direction of rotation about the $y-$ axis) and the bending moment M_z (rotating about
the $z-$ axis). Likewise, two transverse shear forces occur in this case, namely the
transverse shear force Q_y (acting in the $y-$ direction) and the transverse shear force
Q_z (acting in the z direction). These forces and moments cause stresses which must
be taken into account in a design procedure. This general case is also called biaxial
bending as it contains bending about both axes y and z.

Up to now it was assumed that the considered beam is subjected to pure bending.
In many technically relevant situations, however, torsional stresses can occur as
well, e.g. as the result of an eccentric load introduction. The basic idea is shown in
Fig. 1.34. We consider a beam loaded eccentrically by the line load q_y and the single
force F_y (Fig. 1.34, left), where q_y and F_y act eccentrically to the longitudinal axis
x (eccentricity e_y). This load situation can be decomposed into the two load cases
of bending and torsion. Bending results when the two acting loads are moved to the
center of gravity of the beam (Fig. 1.34, top right). In addition, the eccentricity of the
two loads must be taken into account by calculating the equivalent torsional moment
$M_T = F_z e_y$ and the distributed torsional moment flux $m_T = q_z e_y$. Torsion means, in

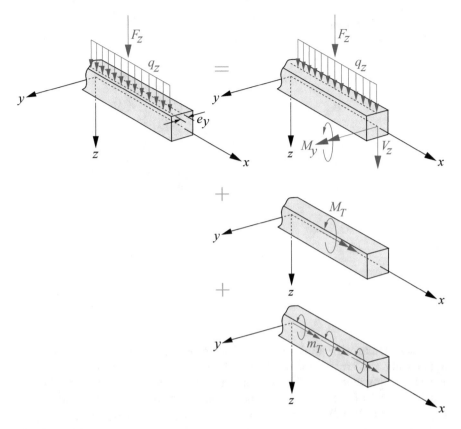

Fig. 1.34 Beam eccentrically loaded by the line load q_z and the single force F_z (left), decomposition into the load cases bending and torsion (right)

contrast to bending with the displacements v and w in the $y-$ and the $z-$-direction, a rotation ϑ of the cross-section about the $x-$ axis and, depending on the shape of the cross-section, also longitudinal displacements u, the so-called warping.

In addition to the static strength verification, stability considerations must always be made for beam structures. In addition to lateral stability of the beam, local buckling can also occur in a thin-walled beam which must be analyzed and verified accordingly.

The statics of beam structures are described in detail in a number of textbooks. A selection is given with the works of Gross et al. (1995, 2013, 2014) or Hirschfeld (2006) and Meskouris and Hake (1999).

Case 2: $b << h, l$

The next case is given with a design space whose dimension b is significantly smaller than h and l (Fig. 1.35). For the moment we use the coordinate system as shown in Fig. 1.35. If the design space has to be fully utilized due to constructive reasons or other requirements, then we speak of a so-called disk in the current case. A disk

Fig. 1.35 Model situation
for the case $b << h, l$

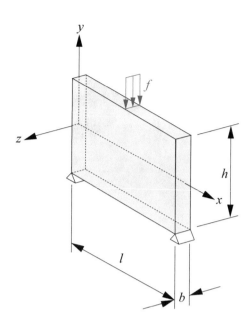

is a flat structure with a thickness (i.e. its dimension in the $z-$-direction) that is
significantly smaller than the dimensions in the $xy-$-plane. A disk is only loaded
by forces acting in its plane that are constant over z, as shown in Fig. 1.35. The disk
is halved at each (x, y) point by its middle plane, which is spanned by the $x-$ and
$y-$ axes. In a disk, only the stress components $\sigma_{xx}, \sigma_{yy}, \tau_{xy} = \tau_{yx}$ occur in the $xy-$
plane (Fig. 1.36) and, under certain circumstances which we will discuss later, the
normal stress σ_{zz} in the thickness direction z. The two shear stresses τ_{xy} and τ_{yx} are
always identical, so that $\tau_{xy} = \tau_{yx}$ applies. The plane stress components σ_{xx}, σ_{yy}
and τ_{xy} are only functions of x and y due to the applied loads that are assumed to be
constant over the thickness: $\sigma_{xx} = \sigma_{xx}(x, y), \sigma_{yy} = \sigma_{yy}(x, y)$ und $\tau_{xy} = \tau_{xy}(x, y)$.

The load-bearing effect of a disk is thus via tension, compression and shear in the
disk plane. It should also be noted that due to their thin-walled nature, depending on
the applied load case disks are generally susceptible to buckling failure. Thus, the
static properties of disks have to be verified with respect to the stress state and, if
necessary, their buckling behavior. The load-bearing behavior of disks as well as the
analysis methods used in this context are described e.g. in Altenbach et al. (1998),
Becker and Gross (2002), Eschenauer and Schnell (1993), Eschenauer et al. (1997),
Girkmann (1974), Göldner et al. (1979), Göldner et al. (1985), Gross et al. (1995).

If there is not the necessity to fill the design space of Fig. 1.35 with material
completely, a flat truss structure can be developed as an alternative to the solid
construction of a disk (Fig. 1.37, left). The applied load can be idealized as a single
force F which is applied in the upper truss node. A truss is defined as an assembly of
straight bars that are assumed to be without any dead weight and which meet in ideal
frictionless joints (so-called nodes) and where the external load occurs only in the

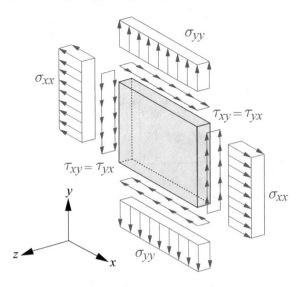

Fig. 1.36 Infinitesimal element of a disk with the stress components σ_{xx}, σ_{yy}, τ_{xy}

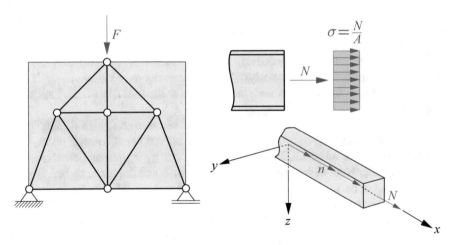

Fig. 1.37 Truss structure

form of forces in the truss nodes. The load-bearing effect of a truss structure is then exclusively via tensile and compressive normal forces N in the individual members (Fig. 1.37, top right). Consequently, the only stress component that occurs in this case is the normal stress σ, which results from the normal force N of the respective member according to

$$\sigma = \frac{N}{A}. \tag{1.8}$$

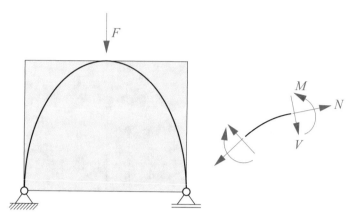

Fig. 1.38 Curved beam

The normal stress σ is assumed to be constant over cross-section A of the bar.

The more general situation of a bar is shown in Fig. 1.37, bottom right. It is possible that member situations may occur where the external loads consist of linearly distributed loads n which, like single loads, for example in a truss, result in normal forces N in the member.

It should be noted that compressively loaded bars are also fundamentally susceptible to buckling so that a static analysis must include not only a stress analysis and the investigation of the displacement state, but also an analysis of sufficient buckling resistance.

A further design solution is to adapt a curved beam to the given design space (see Fig. 1.38). The load-bearing effect of such a curved beam consists on the one hand of an effect as a bar, i.e. the transfer of the loads by tension and compression, and on the other hand of a beam effect, i.e. the transfer of the loads by bending. A clear separation between bar and beam action is therefore not possible in the case of a curved beam, both load bearing effects usually occur simultaneously. It should also be noted that in addition to the static strength of a curved beam, the stability behavior can also play a significant role. Thus, column buckling, torsional flexural buckling and lateral stability as well as local buckling can become significant and must therefore be taken into account in a suitable design procedure.

Case 3: $h \ll b, l$

We consider the situation that the dimensions b and l are significantly higher than h, i.e. $h \ll b, l$. If it is necessary to fill the entire design space with material for constructive reasons, we refer to the resulting structure as a so-called plate. (Fig. 1.39). A plate, like a disk, is a flat thin-walled structure whose thickness is significantly smaller than its other two dimensions. In fact, disks and plates are completely identical in terms of their geometric properties. They differ only in the applied load. While the disk is only loaded in its plane, the plate is only loaded perpendicular to its plane (compare the two Figs. 1.35 and 1.39). The disk and plate behavior can be

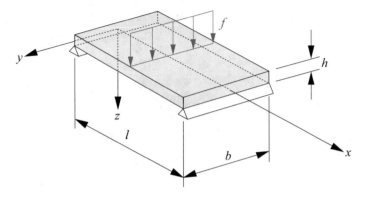

Fig. 1.39 Plate

considered separately, provided that small deformations occur (i.e. a geometrically linear problem), even if inplane and out-of-plane loads are present simultaneously.

As in the case of a disk, the plane stress components σ_{xx}, σ_{yy} and τ_{xy} occur in the plate as well. However, these are no longer constant over the thickness, but rather are linear functions of the thickness coordinate z. In addition, the two shear stresses τ_{yz} and τ_{xz} also occur in the thickness direction z for reasons of equilibrium and will be distributed parabolically across the thickness of the plate. The stress components are shown in Fig. 1.40.

Plate structures are discussed e.g. in Altenbach et al. (1998), Ambartsumyan (1970), Ashton and Whitney (1970), Becker and Gross (2002), Eschenauer and Schnell (1993), Eschenauer et al. (1997), Girkmann (1974), Göldner et al. (1979), Göldner et al. (1985), Gross et al. (1995), Lekhnitskii (1968), Mansfield (1989), Marguerre and Woernle (1975), Mittelstedt and Becker (2016), Reddy (2004, 2006), Turner (1965), Ugural (1981), Vinson (1974).

Further structural elements

If, for example, the general case exists that the given design space of Fig. 1.29 does not fit into any of the previously mentioned cases of structural elements, another important structural element of lightweight engineering, the so-called shell, can be discussed. An example of a situation that would occur, for example, if the middle area of the design space was not usable, is shown in Fig. 1.41.

A shell is a load bearing structure which, in contrast to the disk and plate which are flat structures, can be arbitrarily curved in space as is necessitated by the technical application. Technically relevant examples of shell structures are shown in Figs. 1.4 or 1.17. The analysis of a shell structure is complex insofar as a pure disk effect and a pure plate effect cannot be identified—both occur simultaneously due to the spatial curvature of the shell. The state of stress of the shell and thus the resulting section forces and moments correspond—in simple terms—to a superposition of the stress state of a disk and a plate.

Fig. 1.40 Stresses in a plate

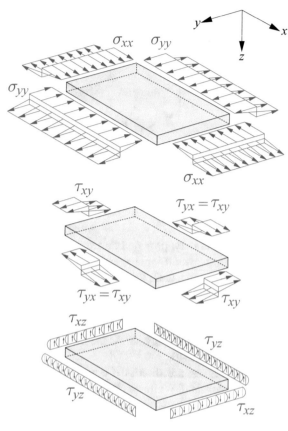

Fig. 1.41 Shell

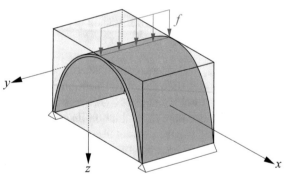

Shell structures are discussed e.g. in Becker and Gross (2002), Dym (1990), Eschenauer and Schnell (1993), Eschenauer et al. (1997), Flügge (1981),Girkmann (1974), Göldner et al. (1979), Göldner et al. (1985), Mazurkiewicz and Nagorski (1991), Møllmann (1981), Pflüger (1967), Reddy (2004, 2006), Turner (1965), Ugural (1981), Vinson (1974), Wlassow (1958).

1.5 About the Functionality of an Aircraft Fuselage

The contents of this book can be applied in many places e.g. to load-bearing aircraft structures, which justifies a short and very concentrated overview of aircraft structures in general and a summary of fuselage statics in particular. Interested readers will find more detailed information e.g. in Bruhn (1973), Donaldson (1993), Ermanni (2007), Megson (1999), Niu (2011), Niu and Niu (2011a, b), Torenbeek (1982), among others.

1.5.1 Main Components of an Aircraft Fuselage

The load-bearing structure of a modern passenger aircraft consists of wings, tail planes and the fuselage and is shown in Fig. 1.42 which shows the typical main components. The fuselage is divided into the cabin area (passenger area as well as on-board kitchen and other installations), the baggage area and the cockpit. In addition, there is the rear fuselage section below the vertical tail plane which is separated from the cabin by the rear pressure bulkhead. The cockpit is also closed in the front section by a pressure bulkhead.

The control of the aircraft is done with the help of the tail planes. These are the so-called vertical tail plane (VTP) including its rudder as well as the two horizontal tail planes (HTP) including their rudders. In addition to the ailerons and the spoilers which are both located on the wings they are used for navigation of the aircraft, which is described by three axes:

- Longitudinal axis
- Vertical axis
- Transverse axis

The position change of an airplane is then described by rotations about these three axes:

- Yaw: Describes a rotation about vertical axis (z-axis)
- Pitch: Describes a rotation about the transverse axis (y-axis)
- Rolling: Describes a rotation about the longitudinal axis (x-axis)

With the help of the VTP rudder a rotation about the vertical axis is initiated, and the VTP serves to control the flight direction with regard to such a rotation. The rudders of the horizontal stabilizers, on the other hand, initiate the pitching, and the control of the flight direction with respect to the pitching is accomplished by the two horizontal stabilizers. A rolling motion is initiated by the ailerons and supported by the spoilers.

The landing flaps, also located on the rear wing, are movable and are used to control or increase the lift of an aircraft at certain speeds. They are retracted when they are not needed to minimize the drag during flight. A similar task is performed by the so-called leading edge wings which are located in the front part of the wing.

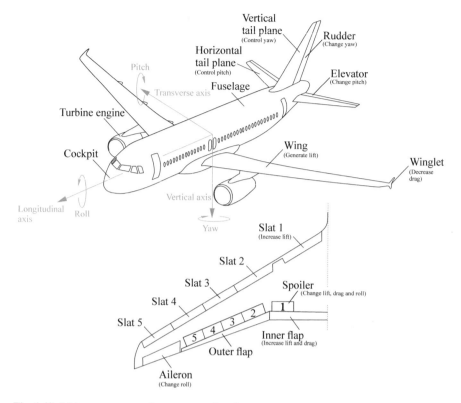

Fig. 1.42 Main components of a passenger aircraft

In this section we will focus exclusively on the primary structure of the aircraft. By primary structure we mean all those components for which the loss of load-bearing capacity can potentially pose a danger to the crew and passengers and whose load-bearing capacity must be proven under load, whereby lightweight construction naturally plays a very prominent role at this point. Specifically, we will refer here mainly to the fuselage and briefly describe its essential components.

The fuselage of a passenger aircraft is designed to carry both crew and payloads and to protect them against external influences (low temperatures and pressures, extremely high air speeds, significant aircraft noise and a hostile atmosphere). Passengers are accommodated in the cabin, and cargo is stored in the cargo hold which is usually filled with standardized containers. The aircraft is controlled by the crew from the cockpit which is located in the bow of the fuselage. The wings as well as the vertical and horizontal stabilizers are connected to the fuselage. In addition, landing gears are usually integrated into the fuselage. Further, the fuselage contains all systems necessary for flight operations, such as air conditioning, oxygen systems, hydraulics, etc. The fuselage is thus the link between all the aircraft's components, carries considerable loads and must therefore meet the most stringent stiffness and

Fig. 1.43 Section of the aluminum fuselage of a Boeing 747 (*Source* en.wikipedia.org). Skin, stringer and frames are clearly visible here

strength requirements and be designed and verified accordingly, always with extreme lightweight construction in mind.

In the following we will assume that the fuselage has already been configured in all its basic configurations and that the actual design has already been carried out so that we can concentrate on the characteristic components and their verification. The occurring flight loads result from the pressure difference between the pressurized cabin and the external area under environmental conditions, but also from the forces introduced by the wings, tail planes and landing gear, from the floor and cargo floor as well as from the multitude of systems in the aircraft.

The fuselage of modern passenger airplanes is usually built in a so-called shell construction (Fig. 1.43).

The main load-bearing components of an aircraft fuselage are the skin, the stringers in the flight direction, and the frames running in the circumferential direction, where the stringers and frames are stiffening elements. They have the following main static tasks:

- The skin gives the fuselage its shape. In addition to aerodynamic requirements and the protection of the structure, cargo and passengers from external influences, its function is to support the membrane stresses resulting from internal pressure in a pressurized cabin. Furthermore, the skin takes over the absorption of the shear stresses resulting from shear forces and torsional loads. To a certain extent, the skin also absorbs the longitudinal forces or normal stresses resulting from the bending of the fuselage under flight conditions. In unperturbed areas, i.e. at locations where

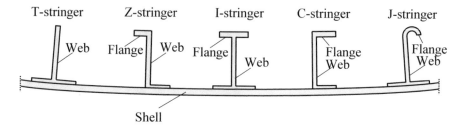

Fig. 1.44 Stringer cross-sections

no concentrated loads are applied and no openings such as doors or windows occur, the skin thickness is only a few millimeters.

- The stringers which usually run through the entire fuselage in the flight direction have the essential task of supporting the airframe forces resulting from the bending of the fuselage. In addition, they serve to stiffen the fuselage skin, which is highly susceptible to stability failure in the form of buckling under shear and compressive stresses.
- The frames ensure that the shape of the fuselage is maintained during flight. They also limit the buckling length of the stringers. They are therefore decisive for the safety and integrity of the fuselage and especially for the buckling resistance. At the same time they stiffen the skin with respect to shear buckling and introduce concentrated forces into the fuselage shell in the tail plane and landing gear areas. The frames also transfer loads from the wings, tail planes, passenger and cargo areas into the fuselage shell and are much more massive in areas of high loads than shown in Fig. 1.43. The shape of the frames is largely determined by the shape of the fuselage. The more the fuselage cross section deviates from the (ideal) circular cross section, the more massive the frames will become.

Skin, stringers and frames are connected to each other by the so-called clips. Clips are required to ensure the flow of forces between frames and skin, but they also stiffen the stringers and thus prevent stability failure.

The stringers are available in a variety of different cross-sectional shapes. In the AIRBUS A350XWB for example, both open-profile sections and closed sections are used. These are the so-called T-stringers (Fig. 1.44) and the so-called omega-stringers (see the fuselage section in Fig. 1.4). However, many other cross-sectional shapes are also possible (Fig. 1.44). There can be many reasons for or against the choice of a specific cross-sectional shape which can be found in aspects such as manufacturability, static requirements, requirements on the damage tolerance behavior, inspectability, manufacturing and assembly possibilities, and the like.

With regard to the frame construction, several variants are available which are determined by given requirements. The Figs. 1.45, 1.46 and 1.47 show a number of variants for metal fuselage structures as they are installed in unperturbed areas without high point load transfer (the necessary rivets between the individual components are not shown here). Figure 1.45 shows a differential frame with a Z-section which

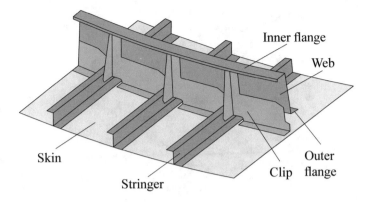

Fig. 1.45 Differential frame with clips

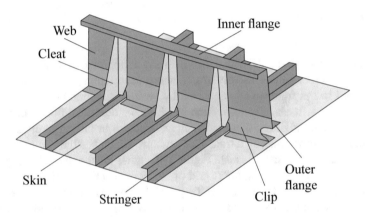

Fig. 1.46 Differential frame with clips and cleats

is connected to the fuselage skin and the stringers by the so-called clips (shown here with integrated flanges, the so-called cleats). This design is very popular in the fuselage segments of modern passenger aircraft due to its ease of fabrication, excellent lightweight construction quality and performance. Figure 1.46 shows a similar construction where the cleats are not integrated into the clips (designed as shear plates). An integral frame, however, where the shear plates are already integrated is shown in Fig. 1.47.

The connection of the frame with the skin and the stringers is done via the so-called clips, often also supported by the so-called cleats. The clips have the essential task of introducing loads into the fuselage shell and at the same time stabilizing the frame against lateral displacements. The cleats, in turn, support the frame against lateral movements and also support the stringers, thus limiting their buckling length. The Figs. 1.45, 1.46 and 1.47 show some details.

Another important part of the fuselage is the floor structure. Its essential components are the so-called transverse beams (often also called cross-beams) and the

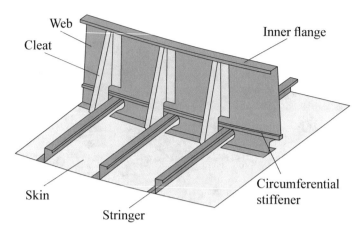

Fig. 1.47 Integral frame with cleats

longitudinal beams/seat rails, which together form a support grid. This support grid is covered by the floor panels, usually of sandwich construction. The transverse beams are attached to the frames and supported by vertical supports at the bottom. The Fig. 1.48 shows a fuselage cross-section in which the described floor structures are clearly visible. In addition, there is the floor structure in the cargo area, which is supported in its lower part by a truss-like structure which has the additional task of absorbing energy in case of a crash. Figure 1.49 shows the the lower fuselage shells and the side shells with the support grid of the floor structure during the assembly process. The cross beams can be seen very clearly here. The floor structure absorbs loads due to passengers, cargo and parts of the interior equipment. In addition, various systems are attached to it, which must be supported by the floor structure. However, the transverse beams not only bear bending loads, but are also subject to tensile loads when the fuselage is subjected to internal pressure. Thus the cross beams and the frames also contribute to maintaining the hull shape under operating conditions. The floor structure also divides the fuselage cross-section into several so-called cells, which benefits the torsional stiffness under torsional loading.

Finally, the pressure dome/rear pressure bulkhead must be considered (Fig. 1.48). The pressure bulkhead is designed to seal off the rear end of the cabin and cargo area from the outside atmosphere. It therefore essentially carries the internal pressure and must be designed and constructed accordingly. Since the dimensions of the pressure dome correspond to those of the fuselage section in which it is installed, it is a comparatively large component which is usually reinforced by stiffeners.

Further components of the primary structure of the fuselage will not be discussed further at this point.

Fig. 1.48 View into the fuselage of the AIRBUS A350-1000, showing the floor panels and the rear pressure bulkhead (courtesy of AIRBUS Operations Gmbh, Hamburg, Germany)

Fig. 1.49 Assembly of the lower fuselage shells and the side shells of the AIRBUS A350XWB with the support grid of the floor structure (courtesy of AIRBUS Operations Gmbh, Hamburg, Germany)

1.5.2 Loads and Classification into Structural Elements

The fuselage structure of a passenger aircraft is exposed to the following significant loads, among others:

- Internal pressure: In order to enable the passengers and crew of a passenger aircraft to fly at high altitudes in a potentially hostile environment, a certain amount of air pressure is maintained in the cabin, cockpit and cargo compartment, which is higher than the external conditions and which is a function of the altitude. This causes the fuselage to expand and leads not only to significant membrane forces in the skin, but also stresses on the frames.
- Horizontal tail planes: The maneuvering with the help of the tail planes causes bending of the fuselage, which leads to tensile and compressive forces in the stringers. Stability considerations in the skin and in the stringers under compressive stresses and forces also play a role here.
- Vertical tail plane: As with the horizontal tail planes, maneuvering with the help of the vertical tail plane also leads to a fuselage bending. There are also torsional moments which result from maneuvering with the aid of the vertical stabilizer and which are transmitted as shear stresses in the skin.
- Wings: The entire lift from the wings is transmitted into the fuselage by transverse shear forces, bending and torsional moments.
- Dead weight: The dead weight of the fuselage causes bending which is carried by the skin and the stringers through tension and compression.
- Landing gear and engines: In these areas significant local loads occur as a result of landing gear, engine forces and the like.
- Wind loads: During flight operations, considerable loads occur due to wind and wind pressure, which must be borne by the individual fuselage elements.
- Crash: Crash loads must be borne by the crash elements on the underside of the fuselage.

The design and dimensioning of the individual components is often done with closed-form analytical methods or, in areas with complex conditions, with the help of specially adapted finite element models. This concerns, among other things, door, window and cargo door areas, which are not considered here and are subject to special requirements.

The following division of the components of an aircraft fuselage into structural elements can be made:

- Stringers are loaded in tension or compression and can therefore be regarded as straight bar structures. This also applies to the vertical supports of the floor structure.
- Frames are subject to normal forces as well as transverse sehar forces and bending moments. They are therefore to be considered as beams. The same applies to the beams of the floor structures. Regarding the frame, a distinction must be made as to whether its curvature is to be considered in the analysis or not, i.e. if it is to be considered as straight or curved beam.

- The skin supports both the internal pressure and the membrane force flows resulting from the operation of the aircraft. It thus acts both as a plate and a disk.
- The floor plates act as plate structures.
- The pressure bulk heads act as curved components like shell structures.

We concretize the necessary stress analysis, verification and proof by means of the frame cross section a little closer. Since this is a component which is considered as a beam, the following static verifications have to be performed:

- Material strength: The occurring stress state as a result of N (normal force), Q (transverse shear force) and M (bending moment) must be verified at the points of maximum stress with respect to the material strength. Depending on the cross-section of the frame, biaxial bending must be taken into account. If it is a metallic frame made of aluminum, the check is carried out using equivalent stresses against the yield point. In the case of a composite laminated frame, however, corresponding strength criteria will be used.
- Global Stability: A frame is a thin-walled and very slender component that is subjected to normal forces and bending moments. Therefore, both the resistance to lateral stability as well as torsional flexural buckling must be verified. The clips and cleats support the frame laterally which should be taken into account in the calculation.
- Local stability: Due to the thin-walled nature of its components, the frame is also endangered with respect to local buckling. This must be verified accordingly, both for compressive stresses and for shear stresses as well as for their interaction.
- Local minimum stiffnesses: The stiffening elements such as the flanges must have certain minimum stiffnesses to exclude local buckling failure.
- Fasteners: The fasteners must be verified. This applies not only to the connection between frame and clips / cleats, but also to the connections of the individual frame sections at frame couplings.
- Damage tolerance: It must be demonstrated that the frame can continue to fulfil its function even in the case of certain defined damages and does not lose its load-bearing capacity.
- Crippling: This is a very local form of failure which can be interpreted as local buckling due to compressive stresses and where the plastic behavior of the cross-section is taken into account. The treatment of crippling is usually based on empirical methods derived from experimental results.
- Global minimum stiffness: The frame cross section must meet certain minimum requirements in order to maintain the fuselage shape in the operating condition and to allow the fuselage to perform its function.
- Stability in the area of the fasteners: The resistance against so-called inter-rivet buckling, i.e. the very local buckling of a thin-walled structure between two fasteners (here: rivets), must also be suitably demonstrated.

Similar conclusions can be drawn for all other parts of the primary structure.

1.6   Aim and Structure of This Book

The present book is the first volume of a planned series of books on the subject of structural analysis in lightweight engineering. It belongs to the field of strength of materials and applied structural mechanics. The reader is assumed to have a basic understanding of mathematics and engineering mechanics as they are usually taught in the first two semesters of a course of study in engineering design. The present book is therefore intended as an extensive consolidation and, to a large extent, an extension of the contents of these mechanics courses, but also as a partial revision of these fundamentals which are so immensely important for lightweight construction. This first volume of a book series deals exclusively with the static analysis of bars and beams and is divided into four main parts consisting of a series of chapters:

- Part I—Fundamentals
- Part II—Thin-walled beam structures
- Part III—Energy methods
- Part IV—Further beam models

Part I—Fundamentals—is divided into the Chaps. 2–4. Chapter 2 (Theory of elasticity) offers a concise summary of the basics of the theory of elasticity, which are essential for the understanding of the present book, but also of the later following planned volumes. Chapter 3 (Plane problems) specializes the considerations on the analysis of plane problems of theory of elasticity, before Chap. 4 (Strength criteria for isotropic materials) concludes part I of this book with an overview of the most important strength hypotheses.

Part II—Thin-walled beam structures—consists of the Chaps. 5–8 and introduces the basic relationships of beam and bar statics. Chapter 5 (Beams under normal forces and bending) sheds light on the determination of cross-sectional values and the calculation of normal stresses for arbitrary thin-walled beams within the framework of the Euler-Bernoulli-Beam Theory, whereby the considerations are initially referred to an arbitrary coordinate system. The necessary cross-sectional normalizations for the determination of the main axes of the cross-section and their effects on the calculations are explained in detail, whereby integral tables are advantageously used for the determination of the necessary cross-sectional values. The Chap. 5 concludes with the determination of displacement of beam and bar structures. However, since the Euler-Bernoulli beam theory excludes transversal shear strains and thus no constitutive relations can be used to determine shear stresses in the cross-sectional plane, Chap. 6 (Beams under transverse shear) deals with the question how this shear stress distribution can be obtained from the normal stresses. In this chapter, the considerations are also limited to the analysis of thin-walled cross-sections. In addition to the general procedure for the consideration of open cross-sections, the question of shear stress determination on closed cross-sections is also dealt with. Another important geometric parameter is the so-called shear center, i.e. the point at which transverse beam loads have to be applied in order to avoid (typically unwanted) torsional effects. For the determination of the position of the shear center a general calculation method for open and closed cross-sections is discussed. Chapter 7 (St. Venant torsion) is then devoted to torsion of bars in the context of the so-called St. Venant torsion. This chapter includes the determination of the shear stresses as well as the torsional moment of inertia at different cross-sectional shapes, whereby both open and closed cross-sections are considered. For applications such as truss segments and stiffened walls, calculation rules for equivalent wall thicknesses are also derived before this chapter concludes with the determination of the internal forces in a bar under torsion. The considerations concerning torsion of thin-walled rods are extended in Chap. 8 (Warping torsion) to include warping of cross-sections, i.e. those cross-sections which under torsional load show deformations in axial direction in addition to a pure rotation of the cross-section. In this chapter, the calculation of the warping deformation at open and closed cross-sections is dealt with after a detailed illustrative motivation of the warping torsion and its significance for the design of thin-walled components. Furthermore, the constitutive law of the so-called first-order bending torsion is considered and the importance of the cross-sectional normalizations to be performed here is discussed. In addition, the conditions under which the torsion problem is decoupled from the bar and beam problems are examined. Following this, some standard sections are considered with respect to their warping behaviour and respective stiffnesses before the chapter ends with the determination of inner moments in the context of warping torsion and a comparative study between an open and a closed section.

Part III—Energy methods—is divided into the Chaps. 9–14. Chapter 9 (Work and energy) fundamentally motivates the concepts of work and energy and provides explanations of the strain energy and complementary strain energy of bars and beams before making a generalization for the elastic continuum. From this, we can derive the principle of work and energy, which is used in this chapter for the determination of displacements in beam and bar systems. Chapter 9 ends with the presentation of the general principle of work and energy of elastostatics, which is formulated for a three-dimensional continuum and from which a number of important principles can be derived. Chapter 10 (Principle of virtual displacements) motivates the fundamental and important principle of virtual displacements. The application to the determination of force and moments as well as influence lines for force and moments in statically determinate beam and bar structures is discussed, whereby for more complex systems the so-called kinematic method for the determination of kinematic figures is introduced. Finally, this chapter shows in detail how to derive differential equations and boundary conditions for a given static problem with the help of the principle of virtual displacements. In Chapter 11 (Principle of stationary value of the total elastic potential) a very basic elementary introduction into the calculus of variations is given before the so-called principle of the stationary value of the total elastic potential is derived from the principle of virtual displacements. This principle later forms the basis for energy-based lightweight construction methods (see Chaps. 13 and 14) and therefore has a special significance. Two interesting principles can be derived from this, namely the first theorem of Castigliano and the theorem of Clapeyron. Chapter 12 (Principle of virtual forces) includes the static analysis of structures by means of the consideration of virtual forces. Besides the analysis of displacements of statical bar and beam systems, the calculation of statically indeterminate systems is also discussed. Connected to this is the term of the so-called force method, a standard method of statics for the analysis of beam and bar structures. From the principle of the stationary value of the elastic complementary potential which is also discussed here, the second theorem of Castigliano and the theorem of Menabrea can be derived which are very useful for the analysis of beam and bar structures. Reciprocity theorems in the form of Betti's theorem and Maxwell's theorem and their applications are also discussed here, before Chap. 12 ends with explanations of the reduction theorem of statics and the analysis of continuous beams. Chapter 13 (Energy-based approximation methods) discusses in detail the derivation and application of the Ritz method to the static analysis of beam and bar structures, before the Galerkin method is briefly discussed at the end. Both methods can be derived from variational or energy statements and thus represent a consistent further development of the previous contents of this book section. Part III of this book ends with Chap. 14 (The finite element method), in which this discretizing method for bars and beams is developed from the Ritz method. In addition to the consideration of truss structures, this chapter also deals in detail with beam and bar systems and includes different types of element configurations.

The last part of this book, *Part IV—Further beam models*—contains the Chaps. 15–18. Chapter 15 (Shear wall girders) deals with the development of an easy to handle analysis model for stiffened beam-like structures, the so-called shear wall girders. These are plane disk-like beams which are reinforced by stiffeners and where the calculation methodology assumes a clear division of tasks between the individual components. In addition to the static analysis of shear wall girders with rectangular skin fields, parallelogram fields and trapezoidal fields are also considered and a number of applications of this rather easy to handle computational model are presented. Chapter 16 (The Timoshenko beam) deals with an improved beam theory which extends the limitations of the Euler-Bernoulli beam theory. Specifically, the so-called Timoshenko beam theory is discussed here which does not take the so-called normal hypothesis into account and thus allows to determine transverse shear stresses from constitutive relations. Chapter 16 contains the derivation of the constitutive law of the Timoshenko beam theory as well as explanations for the determination of stresses and displacements. A special feature of Timoshenko's theory is the need to take into account the so-called shear correction factor the determination of which is discussed in detail in this chapter. This is followed by an energetic consideration of the Timoshenko beam and the application of the force method. Chapter 16 is completed by a general description of the Ritz method for the Timoshenko beam and a finite element formulation for an arbitrary number of element nodes. The consideration of hybrid bars and beams is then described in Chap. 17. This chapter contains the static analysis of bars and beams with thin-walled lightweight cross-sections in which the individual segments may have different elastic properties. The theoretical foundation here is again the Euler-Bernoulli beam theory, and besides providing effective cross-section and stiffness values, the analysis of normal and shear stresses in open and closed cross-sections is discussed. This chapter is rounded off by a brief consideration of the torsional stiffnesses of such hybrid cross-sections. Part IV of this book is concluded by Chap. 18 (Laminate and sandwich beams), in which the basics of the analysis of laminate and sandwich bars and beams are discussed.

1.7 Notes on Relevant Literature

Readers are referred to regular journals published worldwide which are dedicated to a variety of topics in applied mechanics and are therefore also relevant for engineers in lightweight construction and design. A selection is given with the following journals:

- *Applied Mechanics Reviews,*
- *Archive of Applied Mechanics,*
- *European Journal of Mechanics A—Solids,*
- *Engineering Structures,*
- *International Journal of Engineering Science,*
- *International Journal of Mechanical Sciences,*
- *International Journal of Non-Linear Mechanics,*
- *International Journal of Solids and Structures,*
- *International Journal of Structural Stability and Dynamics,*
- *Journal of Applied Mechanics,*
- *Journal of Engineering Mechanics,*
- *Journal of Mechanics of Materials and Structures,*
- *Journal of the Mechanics and Physics of Solids,*
- *Journal of Structural Engineering,*
- *Mechanics of Advanced Materials and Structures,*
- *Zeitschrift für Angewandte Mathematik und Mechanik (ZAMM).*

Especially for engineers in lightweight construction and design practice there are technical journals of interest which mainly deal with the mechanics and the analysis of thin-walled structures:

- *Aerospace Science and Technology,*
- *AIAA Journal,*
- *Journal of Aircraft,*
- *Marine Structures,*
- *Thin-Walled Structures.*

In addition, publications focusing on the field of numerical analysis of solid bodies and here also lightweight structures should be mentioned:

- *Computational Materials Science,*
- *Computational Mechanics,*
- *Computer Methods in Applied Mechanics and Engineering,*
- *Computers and Structures,*
- *International Journal for Numerical Methods in Engineering.*

Composites science and technology and here especially the analysis of fiber composite laminates is also of great interest in the context of this book series. Without claiming completeness, we refer to the following international journals:

- *Composites Part A: Applied Science and Manufacturing,*
- *Composites Part B: Engineering,*
- *Composites Science and Technology,*
- *Composite Structures,*
- *Journal of Composite Materials.*

Finally, a few standard works on lightweight construction and design and also on analysis methods in the framework of lightweight engineering should be mentioned. These include the works of Bauchau and Craig (2009), Czerwenka and Schnell (1970a, b), Hertel (1960), Klein (2007), Kossira (1996), Linke and Nast (2015), Megson (1999), Rammerstorfer (1992), Schapitz (1963) and Wiedemann (2007a, b).

1.8   A Few Notes on Nomenclature

This book contains many mathematical quantities in the form of tensors, vectors and matrices. We want to agree on the following notation. Vectors / first-order tensors are underlined. The vector $\underline{a}$ with its components a_x, a_y, a_z then reads:

$$\underline{a} = \begin{pmatrix} a_x \\ a_y \\ a_z \end{pmatrix}. \tag{1.9}$$

Matrices / second-order tensors are assigned with two underlines:

$$\underline{\underline{A}} = \begin{bmatrix} A_{11} & A_{12} & A_{13} \\ A_{21} & A_{22} & A_{23} \\ A_{31} & A_{32} & A_{33} \end{bmatrix}. \tag{1.10}$$

A compact representation of formulaic relations is very helpful in some places. Here the so-called Einstein's summation convention has proven to be very suitable. It consists of simply omitting the sum signs for sums and summing over identical indices. The sum

$$a_{11} + a_{22} + a_{33} = \sum_{i=1}^{i=3} a_{ii} \tag{1.11}$$

then reads when using Einstein's summation convention:

$$a_{11} + a_{22} + a_{33} \equiv a_{ii}, \tag{1.12}$$

where the summation is performed over the double appearing index i from 1 to 3. For the following sum we have:

$$a_1 b_1 + a_2 b_2 + a_3 b_3 = \sum_{i=1}^{i=3} a_i b_i \equiv a_i b_i. \tag{1.13}$$

The notation $a_i b_i$ is also denoted as the so-called index notation, wherein it is always assumed that the summation is to be performed over identical indices (here the index i).

If a summation is only to be performed from 1 to 2, then instead of Latin letters, Greek letters are usually used for the indices:

$$a_{11} + a_{22} = \sum_{\alpha=1}^{\alpha=2} a_{\alpha\alpha} \equiv a_{\alpha\alpha}. \tag{1.14}$$

The equilibrium condition

$$\frac{\partial \sigma_{11}}{\partial x_1} + \frac{\partial \sigma_{12}}{\partial x_2} + f_1 = 0 \tag{1.15}$$

reads in index notation:

$$\frac{\partial \sigma_{1\alpha}}{\partial x_\alpha} + f_1 = 0. \tag{1.16}$$

References

Altenbach H, Altenbach J, Naumenko K (1998) Ebene Flächentragwerke. Springer Berlin et al, Germany

Ambartsumyan SA (1970) Theory of anisotropic plates. Technomic Publishing Co. Inc, Stamford, USA

Ashton JE, Whitney JM (1970) Theory of laminated plates. Technomic Publishing Co. Inc, Stamford, USA

Baier H, Seeßelberg C, Specht B (1994) Optimierung in der Strukturmechanik. Vieweg, Braunschweig, Germany

Bauchau OA, Craig JI (2009) Structural analysis: with applications to aerospace structures. Springer Dordrecht et al, The Netherlands

Becker W, Gross D (2002) Mechanik elastischer Körper und Strukturen. Springer Berlin et al, Germany

Beumler T (2004) Flying GLARE—a contribution to aircraft certification issues on strength properties in non-damaged and fatigue damaged GLARE structures, Dissertation thesis, Delft University Press, Delft, The Netherlands

Bruhn EF (1973) Analysis and design of flight vehicle structures. S.R. Jacobs & Associates, Carmel, USA

Czerwenka G, Schnell W (1970a) Einführung in die Rechenmethoden des Leichtbaus, Band 1. Bibliographisches Institut, Mannheim et al., Germany

Czerwenka G, Schnell W (1970b) Einführung in die Rechenmethoden des Leichtbaus, Band 2, Bibliographisches Institut, Mannheim et al., Germany

Degenhardt R, Castro S, Arbelo M, Zimmerman R, Kling A, Khakimova R (2014) Future structural stability design for composite space and airframe structures. Thin-Walled Struct 81:29–38

Donaldson BK (1993) Analysis of aircraft structures: an introduction. McGraw-Hill Inc., New York, USA

Dym CL (1990) Introduction to the theory of shells. Hemisphere Publishing Co., New York et al., USA

Ermanni P (2007) Aerospace structures. Lecture Notes, Eidgenössische Technische Hochschule Zürich, Switzerland

Eschenauer H, Koski J, Osyczka A (1990) Multicriteria design optimization. Springer, Berlin, Germany

Eschenauer H, Schnell W (1993) Elastizitätstheorie, Bibliographisches Institut. Mannheim et al, Germany

Eschenauer H, Olhoff N, Schnell W (1997) Applied structural mechanics. Springer Berlin et al, Germany

Flügge W (1981) Statik und Dynamik der Schalen, 3rd edn. Springer Berlin et al, Germany

Girkmann K (1974) Flächentragwerke, 6th edn. Springer Vienna et al, Austria

Göldner H, Altenbach J, Eschke K, Garz KF, Sähn S (1979) Lehrbuch Höhere Festigkeitslehre, vol 1. Physik Verlag, Weinheim, Germany

Göldner H, Altenbach J, Eschke K, Garz KF, Sähn S (1985) Lehrbuch Höhere Festigkeitslehre, vol 2. Physik Verlag, Weinheim, Germany

Gross D, Hauger W, Schnell W, Wriggers P (1995) Technische Mechanik 4: Hydromechanik, Elemente der Höheren Mechanik, Numerische Methoden, 2nd edn. Springer Berlin et al, Germany

Gross D, Hauger W, Schröder J, Wall WA (2013) Technische Mechanik 1: Statik, 12th edn. Springer Berlin et al, Germany

Gross D, Hauger W, Schröder J (2014) Technische Mechanik 2: Elastostatik, 12th edn. Springer Berlin et al, Germany

Großmann A, Weis P, Clemen C, Mittelstedt C (2020) Optimization and re-design of a metallic riveting tool for additive manufacturing—a case study. Additive Manuf 31:100892

Harzheim L (2019) Strukturoptimierung: Grundlagen und Anwendungen, 3rd edn. Europa-Lehrmittel, Haan-Gruyten, Germany

Hertel H (1960) Leichtbau. Springer Berlin et al, Germany

Hirschfeld K (2006) Baustatik, 5th edn. Springer Berlin et al, Germany

Kamat MP (1993) Structural optimization: status and promise, AIAA, Progress in astronautics and aeronautics, Vol 150

Klein B (2007) Leichtbau-Konstruktion. Vieweg, Braunschweig, Germany

Kossira H (1996) Grundlagen des Leichtbaus. Springer Berlin et al, Germany

Lekhnitskii SG (1968) Anisotropic plates, Gordon and Breach. New York et al, USA

Linke M, Nast E (2015) Festigkeitslehre für den Leichtbau. Springer Berlin et al, Germany

Mansfield EH (1989) The bending and stretching of plates. Cambridge University Press, Cambridge et al, USA

Marguerre K, Woernle HT (1975) Elastische Platten, Bibliographisches Institut. Mannheim et al, Germany

Mazurkiewicz EM, Nagorski RT (1991) Shells of revolution. Elsevier Science Publication, Amsterdam, The Netherlands

Megson THG (1999) Aircraft structures for engineering students, 3rd edn. Arnold, London, UK

Meskouris K, Hake E (1999) Statik der Stabtragwerke: Einführung in die Tragwerkslehre. Springer Berlin et al, Germany

Mittelstedt C, Becker W (2016) Strukturmechanik ebener Laminate. Technische Universität Darmstadt, Germany, Studienbereich Mechanik

Møllmann H (1981) Introduction to the theory of thin shells. Wiley Chichester et al, UK

Niu MCY (2011) Airframe stress analysis and sizing, 3rd edn. Adaso/Adastra Engineering Center, Northridge, USA

Niu MCY, Niu M (2011a) Airframe structural design: practical design information and data on aircraft structures, 2nd edn. Adaso/Adastra Engineering Center, Northridge, USA

Niu MCY, Niu M (2011b) Composite airframe structures, 3rd edn. Hong Kong Conmilit Press Ltd., Hong Kong, China

Pflüger A (1967) Elementare Schalenstatik, 4th edn. Springer Berlin et al, Germany

Reddy JN (2004) Mechanics of laminated composite plates and shells, 2nd edn. CRC Press Boca Raton et al, USA

Reddy JN (2006) Theory and analysis of elastic plates and shells. CRC Press Boca Raton et al, USA

Rammerstorfer FG (1992) Repetitorium Leichtbau, Oldenbourg Verlag. Wien et al, Austria

Schapitz E (1963) Festigkeitslehre für den Leichtbau, 2nd edn. VDI-Verlag, Düsseldorf, Germany

Schürg M, Scheibe M, Mittelstedt C (2016) Anwendungsmöglichkeiten der additiven Fertigung - Entwicklung eines Kabelführungshalters im Seitenleitwerk des Airbus A350, 3. NAFEMS DACH Regionalkonferenz, 25th-27th April 2016, Bamberg, Germany

Torenbeek E (1982) Airplane design. Delft University Press, Delft, The Netherlands

Turner CE (1965) Introduction to plate and shell theory. Longmans Green and Co., Ltd., London, UK

Ugural AC (1981) Stresses in plates and shells. McGraw Hill New York et al, USA

Vanderplaats GN (1984) Numerical optimization techniques for engineering design. McGraw-Hill Inc., New York, USA

Vinson JR (1974) The behaviour of plates and shells. Wiley New York et al, USA

Wiedemann J (2007a) Leichtbau 1: Elemente. Springer Berlin et al, Germany

Wiedemann J (2007b) Leichtbau 2: Konstruktion. Springer Berlin et al, Germany

Wlassow WS (1958) Allgemeine Schalentheorie und ihre Anwendung in der Technik. Akademie-Verlag, Berlin, Germany

Part I
Fundamentals

Chapter 2
Theory of Elasticity

2.1 Introduction

This chapter contains the basics of the theory of elasticity of three-dimensional anisotropic bodies which are essential for the understanding of the further contents of this book as well as of all further planned volumes of this book series. Even though this chapter is quite extensive we will limit ourselves to absolutely essential and indispensable contents. The reader is therefore recommended to work through this chapter thoroughly in order to be able to process the contents of the other chapters of this book.

A three-dimensional structure (a so-called continuum) which is loaded by applied outer forces will develop forces in its interior, namely the so-called stresses. In addition, there are deformations of the considered structure, i.e. the displacements of each body point. This is accompanied by the so-called strains. In order to describe the state of a structure under a given load and under given boundary conditions, the following set of equations is generally required:

- Kinematic equations: The kinematic equations represent a relationship between the displacements of a solid and the resulting strains.
- Constitutive equations: The strains are coupled with the stresses by the so-called constitutive equations. In the case of linear elasticity the constitutive equations are described by Hooke's law.
- Equilibrium conditions: At each point of the structure under consideration the equilibrium of forces and moments must be ensured. This is described by the equilibrium conditions.

C. Mittelstedt, *Structural Mechanics in Lightweight Engineering*,
https://doi.org/10.1007/978-3-030-75193-7_2

2.2   State of Stress

2.2.1   Stress Vector and Stress Tensor

We start the considerations with an arbitrary solid body (Fig. 2.1) which is not specified further at this point. The body under consideration is exposed to a combination
of loads and is subject to certain boundary conditions. For the description of the
occurring state variables we introduce the spatial Cartesian reference system x, y,
z with the associated displacements u, v and w, where the displacements can be
functions of all three spatial directions:

$$u = u(x, y, z), \quad v = v(x, y, z), \quad w = w(x, y, z). \tag{2.1}$$

Various types of loads can occur. First the so-called volume forces have to be
mentioned, among which all types of load can be summarized which are distributed
in the volume of the body under consideration, such as the dead weight. The unit of
the volume forces is that of a force per volume unit, e.g. [N/m^3]. Volume forces can
occur in all three spatial directions. The volume force components are denoted f_x,
f_y, f_z and are combined in the vector $\underline{f}$:

$$\underline{f} = \begin{pmatrix} f_x \\ f_y \\ f_z \end{pmatrix}. \tag{2.2}$$

In addition, loads can also occur in the form of surface loads which have the unit
of a force per area unit, e.g. [N/m^2]. Surface loads can occur on the entire surface
of the structure under consideration or only on parts of it. We want to denote the
components of a surface load as t_x, t_y, t_z and summarize them in the vector

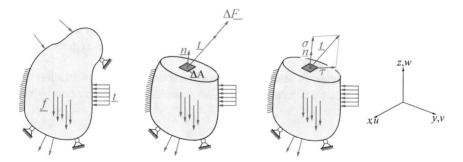

Fig. 2.1   Three-dimensional solid body under load (left), section through an arbitrary body point
(middle), decomposition of the stress vector into normal and shear stress (right)

$$t = \begin{pmatrix} t_x \\ t_y \\ t_z \end{pmatrix}. \tag{2.3}$$

In addition, loads can also occur in the form of line loads (in the unit force per length unit, e.g. [N/m]) and as point loads (unit [N]), which are concentrated along a line or concentrated in a point. However, it should always be noted that both line and point loads represent only idealized scenarios.

If a given body is under load, then forces are generated inside this body which are usually related to an imaginary cutting surface. We commonly refer to these forces as the so-called stresses. Since stresses correspond to a force acting on a surface, they are given in a corresponding unit, e.g. [N/m²]. To define stresses, we cut through any point of the considered solid body and consider an infinitesimal sectional area ΔA, as shown in Fig. 2.1, middle. The spatial orientation of this sectional area is indicated by the normal vector $\underline{n}$:

$$\underline{n} = \begin{pmatrix} n_x \\ n_y \\ n_z \end{pmatrix}. \tag{2.4}$$

The force $\Delta \underline{F}$ acts on this sectional area, and the corresponding stress vector is then defined as follows:

$$\underline{t} = \lim_{\Delta A \to 0} \frac{\Delta \underline{F}}{\Delta A}. \tag{2.5}$$

It is obvious that the stress vector depends on the orientation of the section and the surface under consideration. Thus the stress vector $\underline{t}$ is a function of the normal vector $\underline{n}$, i.e.: $\underline{t} = \underline{t}\left(\underline{n}\right)$.

The stress vector is usually decomposed into a component parallel to the normal vector and a component tangential to the considered surface. These components are called normal stress σ and shear stress τ. The stress components are shown in Fig. 2.1, right. They result as:

$$\sigma = \underline{t} \cdot \underline{n}, \quad \tau = \sqrt{\underline{t} \cdot \underline{t} - \sigma^2}. \tag{2.6}$$

The stress state in any point of the considered solid is clearly defined when the stress vector is given in three independent sections through this point. The normal vectors of these three sections must be linearly independent. Usually, the sections through the considered body point are placed in such a way that they are perpendicular to the three coordinate axes x, y, z. Fig. 2.2 shows an infinitesimally small element that has been cut out of the considered solid body. Once these three stress vectors are available, the stress vectors under any other orientation can be determined from them as will be shown later on.

Fig. 2.2 Infinitesimal
volume element and stress
components

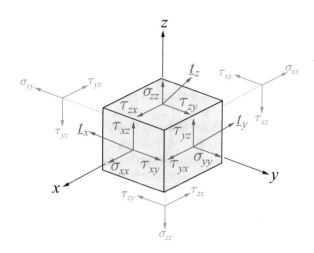

Let the three cutting surfaces be defined in such a way that their normal vectors coincide exactly with the coordinate axes x, y, z. Then the three stress vectors $\underline{t}_x$, $\underline{t}_y$, $\underline{t}_z$ can be written as:

$$
\underline{t}_x = \begin{pmatrix} \sigma_{xx} \\ \tau_{xy} \\ \tau_{xz} \end{pmatrix}, \quad
\underline{t}_y = \begin{pmatrix} \tau_{yx} \\ \sigma_{yy} \\ \tau_{yz} \end{pmatrix}, \quad
\underline{t}_z = \begin{pmatrix} \tau_{zx} \\ \tau_{zy} \\ \sigma_{zz} \end{pmatrix}. \tag{2.7}
$$

The indices of the stresses are used as follows. The first index indicates the direction of the surface normal of the considered section, while the second index indicates the direction of action of the considered stress component. Thus, stresses with identical indices are always normal stresses, they are perpendicular to the considered section. Stresses with two different indices, on the other hand, are shear stresses, and they are tangential to the considered section surface. Obviously, nine stress components must be known for the complete description of the stress state in one point of a solid, including the three normal stresses σ_{xx}, σ_{yy} and σ_{zz} as well as the six shear stresses τ_{xy}, τ_{yx}, τ_{xz}, τ_{zx}, τ_{yz} and τ_{zy}. A stress is always assumed to be positive if it points in positive coordinate direction at a positive cutting surface. A cutting surface is considered positive if the normal vector of the considered surface points in positive coordinate direction.

In many cases, an alternative notation is used such that the reference axes are named x_1, x_2, x_3 and the stress components are indexed accordingly. When using this notation all stress components are usually designated σ so that the normal stresses σ_{11}, σ_{22}, σ_{33} and the shear stresses σ_{12}, σ_{21}, σ_{13}, σ_{31}, σ_{23} and σ_{32} occur. This is shown in Fig. 2.3. In this case the corresponding displacement components would be called u_1, u_2, u_3 instead of u, v, w. Analogously, we then refer to the stress vectors as $\underline{t}_1$, $\underline{t}_2$, $\underline{t}_3$, and they are assigned as follows:

Fig. 2.3 Infinitesimal volume element and stress components, use of alternative notations

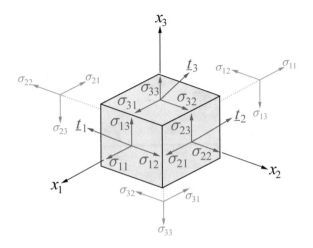

$$t_1 = \begin{pmatrix} \sigma_{11} \\ \sigma_{12} \\ \sigma_{13} \end{pmatrix}, \quad t_2 = \begin{pmatrix} \sigma_{21} \\ \sigma_{22} \\ \sigma_{23} \end{pmatrix}, \quad t_3 = \begin{pmatrix} \sigma_{31} \\ \sigma_{32} \\ \sigma_{33} \end{pmatrix}. \tag{2.8}$$

Using Figs. 2.2 and 2.3, by formulating the moment equilibrium around the center of gravity of the cube about all three reference axes it becomes clear that shear stresses with identical but interchanged indices must be identical in order to maintain equilibrium. Thus it follows:

$$\tau_{xy} = \tau_{yx}, \quad \tau_{xz} = \tau_{zx}, \quad \tau_{yz} = \tau_{zy}, \tag{2.9}$$

and

$$\sigma_{12} = \sigma_{21}, \quad \sigma_{13} = \sigma_{31}, \quad \sigma_{23} = \sigma_{32}. \tag{2.10}$$

This reduces the number of independent stress components to six, namely the three normal stresses σ_{xx}, σ_{yy}, σ_{zz} or σ_{11}, σ_{22}, σ_{33} and the three shear stresses $\tau_{xy} = \tau_{yx}$, $\tau_{xz} = \tau_{zx}$, $\tau_{yz} = \tau_{zy}$ or $\tau_{12} = \tau_{21}$, $\tau_{13} = \tau_{31}$, $\tau_{23} = \tau_{32}$.

The stress components are typically summarized in a symmetrical matrix $\underline{\underline{\sigma}}$ as follows:

$$\underline{\underline{\sigma}} = \begin{bmatrix} \sigma_{xx} & \tau_{xy} & \tau_{xz} \\ \tau_{xy} & \sigma_{yy} & \tau_{yz} \\ \tau_{xz} & \tau_{yz} & \sigma_{zz} \end{bmatrix} = \begin{bmatrix} \sigma_{11} & \sigma_{12} & \sigma_{13} \\ \sigma_{12} & \sigma_{22} & \sigma_{23} \\ \sigma_{13} & \sigma_{23} & \sigma_{33} \end{bmatrix}. \tag{2.11}$$

Therein, $\underline{\underline{\sigma}}$ is Cauchy's[1] stress tensor in which the stress vectors t_x, t_y, t_z respectively t_1, t_2, t_3 are arranged column by column. The tensor $\underline{\underline{\sigma}}$ is a symmetric tensor of second order.

---

[1] Augustin-Louis Cauchy, 1789–1857, French mathematician.

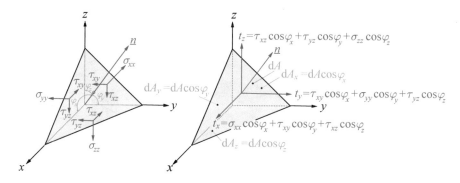

Fig. 2.4 Forces acting on an infinitesimal tetrahedron element

Once the Cauchy stress tensor $\underline{\underline{\sigma}}$ is known in a specific body point, the stress state for any other section through the considered point can be determined by elementary transformation rules. For this purpose, we consider the tetrahedron cut from the infinitesimal element as shown in Fig. 2.4. The tetrahedron is defined by the three surface areas dA_x, dA_y, dA_z and the arbitrarily oriented tetrahedron surface dA. The orientation of the normal vector $\underline{n}$ of the tetrahedral surface with respect to the reference system x, y, z is given by the angles φ_x, φ_y, φ_z. Furthermore, $\underline{t} = \left(t_x, t_y, t_z\right)^T$ is the vector of the stresses with respect to the tetrahedral surface dA. The normal vector $\underline{n}$ is then defined as $\underline{n} = \left(n_x, n_y, n_z\right)^T = \left(\cos \varphi_x, \cos \varphi_y, \cos \varphi_z\right)^T$. Equilibrium of forces in the $x-$, y- and z-direction then yields:

$$t_x dA = \sigma_{xx} dA_x + \tau_{xy} dA_y + \tau_{xz} dA_z,$$
$$t_y dA = \tau_{xy} dA_x + \sigma_{yy} dA_y + \tau_{yz} dA_z,$$
$$t_z dA = \tau_{xz} dA_x + \tau_{yz} dA_y + \sigma_{zz} dA_z. \tag{2.12}$$

With $dA_x = dA n_x$, $dA_y = dA n_y$, $dA_z = dA n_z$ we have:

$$t_x = \sigma_{xx} n_x + \tau_{xy} n_y + \tau_{xz} n_z,$$
$$t_y = \tau_{xy} n_x + \sigma_{yy} n_y + \tau_{yz} n_z,$$
$$t_z = \tau_{xz} n_x + \tau_{yz} n_y + \sigma_{zz} n_z. \tag{2.13}$$

In a vector-matrix notation this reads:

$$\underline{t} = \underline{\underline{\sigma}} \underline{n}. \tag{2.14}$$

Using an index notation with the axes x_1, x_2, x_3 this can be written as follows:

$$t_i = \sigma_{ij} n_j. \tag{2.15}$$

This is the so-called Cauchy formula or the Cauchy theorem. It states that if the stress tensor $\underline{\underline{\sigma}}$ in a body point is known, the stress tensor for any other section through the considered body point can be easily determined as defined.

2.2.2 Stress Transformation

We have so far assumed that all stresses are given with respect to the spatial axes x, y, z respectively their base vectors $\underline{e}_x, \underline{e}_y, \underline{e}_z$. We will now consider the effects on the stress state if we convert the orthonormal basis $\underline{e}_x, \underline{e}_y, \underline{e}_z$ into an arbitrary basis $\bar{\underline{e}}_x, \bar{\underline{e}}_y, \bar{\underline{e}}_z$ with the reference system $\bar{x}, \bar{y}, \bar{z}$. This is shown in Fig. 2.5, left. Both the original orthonormal basis $\underline{e}_x, \underline{e}_y, \underline{e}_z$ and the basis $\bar{\underline{e}}_x, \bar{\underline{e}}_y, \bar{\underline{e}}_z$ have the same origin, and the base vectors $\underline{e}_x$ and $\bar{\underline{e}}_y$ include the direction angle $\angle\left(\bar{\underline{e}}_y, \underline{e}_x\right)$. The cosine of this angle is denoted as the directional cosine R_{yx}, whereby this applies analogously to all other R_{ij}. Here, the first index indicates the base vector from the transformed coordinate system $\bar{x}, \bar{y}, \bar{z}$, and the second index indicates the one from the original coordinate system x, y, z ($i, j = x, y, z$):

$$R_{ij} = \cos\left(\angle\bar{\underline{e}}_i, \underline{e}_j\right). \tag{2.16}$$

We obtain the directional cosine R_{ij} from the scalar product of the two considered base vectors:

$$R_{ij} = \frac{\bar{\underline{e}}_i \cdot \underline{e}_j}{\left|\bar{\underline{e}}_i\right|\left|\underline{e}_j\right|}. \tag{2.17}$$

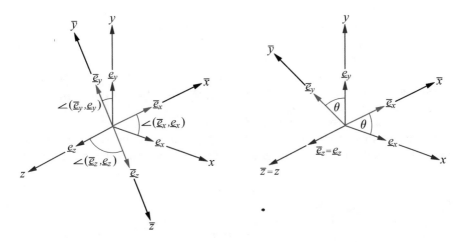

Fig. 2.5 Rotation of an orthonormal basis $\underline{e}_x, \underline{e}_y, \underline{e}_z$ into an arbitrary basis $\bar{\underline{e}}_x, \bar{\underline{e}}_y, \bar{\underline{e}}_z$ (left), rotation of an orthonormal basis $\underline{e}_x, \underline{e}_y, \underline{e}_z$ into an orthonormal basis $\bar{\underline{e}}_x, \bar{\underline{e}}_y, \bar{\underline{e}}_z$ with fixed z−axis (right)

We now arrange the stresses in the column vector $\underline{\sigma} = (\sigma_{xx}, \sigma_{yy}, \sigma_{zz}, \tau_{yz}, \tau_{xz}, \tau_{xy})^T$. The transformation of the stresses from $\underline{e}_x, \underline{e}_y, \underline{e}_z$ to $\bar{\underline{e}}_x, \bar{\underline{e}}_y, \bar{\underline{e}}_z$ can then simply be concluded from equilibrium at the tetrahedron and results as:

$$\bar{\underline{\sigma}} = \underline{\underline{T}}_\sigma \, \underline{\sigma}. \tag{2.18}$$

The matrix $\underline{\underline{T}}_\sigma$ is denoted as transformation matrix and reads:

$$\underline{\underline{T}}_\sigma = \begin{bmatrix} R_{11}^2 & R_{12}^2 & R_{13}^2 & 2R_{12}R_{13} & 2R_{11}R_{13} & 2R_{11}R_{12} \\ R_{21}^2 & R_{22}^2 & R_{23}^2 & 2R_{22}R_{23} & 2R_{21}R_{23} & 2R_{21}R_{22} \\ R_{31}^2 & R_{32}^2 & R_{33}^2 & 2R_{32}R_{33} & 2R_{31}R_{33} & 2R_{31}R_{32} \\ R_{21}R_{31} & R_{22}R_{32} & R_{23}R_{33} & R_{22}R_{33}+R_{23}R_{32} & R_{21}R_{33}+R_{23}R_{31} & R_{21}R_{32}+R_{22}R_{31} \\ R_{11}R_{31} & R_{12}R_{32} & R_{13}R_{33} & R_{12}R_{33}+R_{13}R_{32} & R_{11}R_{33}+R_{13}R_{31} & R_{11}R_{32}+R_{12}R_{31} \\ R_{11}R_{21} & R_{12}R_{22} & R_{13}R_{23} & R_{12}R_{23}+R_{13}R_{22} & R_{11}R_{23}+R_{13}R_{21} & R_{11}R_{22}+R_{12}R_{21} \end{bmatrix}. \tag{2.19}$$

In the special case that the vectors $\underline{e}_x, \underline{e}_y, \underline{e}_z$ form an orthonormal basis and are transformed into the orthonormal basis $\bar{\underline{e}}_x, \bar{\underline{e}}_y, \bar{\underline{e}}_z$ by a rotation around the fixed $z-$axis by the angle θ (see Fig. 2.5, right), then the above relations simplify significantly and we obtain:

$$\begin{pmatrix} \bar{\sigma}_{xx} \\ \bar{\sigma}_{yy} \\ \bar{\sigma}_{zz} \\ \bar{\tau}_{yz} \\ \bar{\tau}_{xz} \\ \bar{\tau}_{xy} \end{pmatrix} = \begin{bmatrix} \cos^2\theta & \sin^2\theta & 0 & 0 & 0 & 2\cos\theta\sin\theta \\ \sin^2\theta & \cos^2\theta & 0 & 0 & 0 & -2\cos\theta\sin\theta \\ 0 & 0 & 1 & 0 & 0 & 0 \\ 0 & 0 & 0 & \cos\theta & -\sin\theta & 0 \\ 0 & 0 & 0 & \sin\theta & \cos\theta & 0 \\ -\cos\theta\sin\theta & \cos\theta\sin\theta & 0 & 0 & 0 & \cos^2\theta-\sin^2\theta \end{bmatrix} \begin{pmatrix} \sigma_{xx} \\ \sigma_{yy} \\ \sigma_{zz} \\ \tau_{yz} \\ \tau_{xz} \\ \tau_{xy} \end{pmatrix}. \tag{2.20}$$

2.2.3 Principal Stresses, Invariants, Mohr's Circles

There is a particular reference system x_p, y_p, z_p (the so-called principal axes) where the stress vectors all run parallel to the associated unit vectors $\underline{e}_{x,p}, \underline{e}_{y,p}, \underline{e}_{z,p}$ or to the normal vectors of the section surfaces. This means that in this particular reference system the shear stresses disappear, i.e. $\tau_{yz} = \tau_{xz} = \tau_{xy} = 0$. The normal stress components $\sigma_{xx}, \sigma_{yy}, \sigma_{zz}$ remaining in this state are called principal stresses. Most often, the symbolic designation $\sigma_1, \sigma_2, \sigma_3$ is used for the principal stresses. If the principal stresses are the only components of the stress vectors and thus are perpendicular to the section surface, then the components of $\underline{t}$ are each a multiple of the corresponding normal vector multiplied by the corresponding principal stress σ:

$$\underline{t} = \sigma\underline{n}. \tag{2.21}$$

Using (2.14) yields:

$$\underline{\underline{\sigma}}\underline{n} = \sigma\underline{n}. \tag{2.22}$$

This leads to:

$$\left[\underline{\underline{\sigma}} - \sigma\underline{\underline{I}}\right]\underline{n} = 0. \tag{2.23}$$

This represents an eigenvalue problem where σ plays the role of the eigenvalue and $\underline{n}$ is the assigned eigenvector. The matrix $\underline{\underline{I}}$ is the unit matrix. For a non-trivial solution it is required that the coefficient determinant in (2.23) disappears:

$$\det\left[\underline{\underline{\sigma}} - \sigma\underline{\underline{I}}\right] = 0, \tag{2.24}$$

respectively:

$$\det\begin{bmatrix} \sigma_{xx} - \sigma & \tau_{xy} & \tau_{xz} \\ \tau_{xy} & \sigma_{yy} - \sigma & \tau_{yz} \\ \tau_{xz} & \tau_{yz} & \sigma_{zz} - \sigma \end{bmatrix} = 0. \tag{2.25}$$

This leads to the following third-order polynomial the roots of which constitute the principal stresses $\sigma_1, \sigma_2, \sigma_3$:

$$\sigma^3 - I_1\sigma^2 - I_2\sigma - I_3 = 0, \tag{2.26}$$

with:

$$I_1 = \sigma_{xx} + \sigma_{yy} + \sigma_{zz} = \text{spur}\left[\underline{\underline{\sigma}}\right],$$
$$I_2 = \tau_{xy}^2 + \tau_{yz}^2 + \tau_{xz}^2 - \left(\sigma_{xx}\sigma_{yy} + \sigma_{yy}\sigma_{zz} + \sigma_{xx}\sigma_{zz}\right),$$
$$I_3 = \det\underline{\underline{\sigma}}. \tag{2.27}$$

Equation (2.26) always leads to real values for the principal stresses $\sigma_1, \sigma_2, \sigma_3$ and is independent of the reference system. The quantities I_1, I_2, I_3 are therefore also called the invariants of the stress state.

The result (2.26) can be interpreted as follows. One of the stresses $\sigma_1, \sigma_2, \sigma_3$ is always the maximum normal stress occurring in the body point under consideration. We will refer to this stress as σ_1. Another of the three normal stresses (σ_3) is the smallest occurring normal stress, and the remaining stress σ_2 has a value between σ_1 and σ_3. The unit vectors of the principal axis system x_p, y_p, z_p are the eigenvectors of $\underline{\underline{\sigma}}$ to the respective eigenvalues. They always form an orthogonal basis. The calculation of the principal stresses and the corresponding principal axis system is called principal axis transformation. In the principal axis system, the stress tensor is:

$$\underline{\underline{\sigma}} = \begin{bmatrix} \sigma_1 & 0 & 0 \\ 0 & \sigma_2 & 0 \\ 0 & 0 & \sigma_3 \end{bmatrix}. \tag{2.28}$$

The invariants I_1, I_2, I_3 of the stress tensor then read:

Fig. 2.6 Mohr's circles for a spatial stress state with the principal normal stresses σ_1, σ_2, σ_3 and the principal shear stresses τ_1, τ_2, τ_3

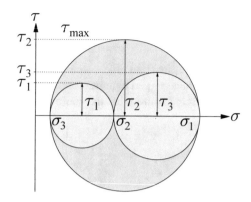

$$I_1 = \sigma_1 + \sigma_2 + \sigma_3,$$
$$I_2 = \sigma_1\sigma_2 + \sigma_2\sigma_3 + \sigma_1\sigma_3,$$
$$I_3 = \sigma_1\sigma_2\sigma_3. \tag{2.29}$$

In addition, a section orientation can be identified where the maximum shear stresses (the so-called principal shear stresses) τ_1, τ_2, τ_3 result. They occur in sections whose normals are perpendicular to a principal axis and form an angle of 45° with the remaining axes. Note that if the principal shear stresses are present, the associated normal stresses are not necessarily zero. For the principal shear stresses we obtain:

$$\tau_1 = \frac{1}{2}\left(\sigma_2 - \sigma_3\right), \quad \tau_2 = \frac{1}{2}\left(\sigma_1 - \sigma_3\right), \quad \tau_3 = \frac{1}{2}\left(\sigma_1 - \sigma_2\right). \tag{2.30}$$

Therein, τ_2 is the maximum shear stress.

The spatial state of stress can also be interpreted with the help of the so-called Mohr's[2] circles (see Fig. 2.6). The circles shown there each indicate the stress state in one of the three sections which are identified by the unit vectors $\underline{e}_{x,p}, \underline{e}_{y,p}, \underline{e}_{z,p}$ of the principal axis system. The tangent points of the circles contain the three principal stresses. Every other stress state which differs from the principal axes with the base vectors $\underline{e}_{x,p}, \underline{e}_{y,p}, \underline{e}_{z,p}$ is located in the area highlighted in dark gray. The circle which is described by σ_1 and σ_3 and has its center at the position $\sigma = \frac{1}{2}\left(\sigma_1 - \sigma_3\right)$ limits the stress state on both diagram axes.

Formally Mohr's circle can be derived as follows. We start from the principal stress state with the principal stresses $\sigma_1, \sigma_2, \sigma_3$, for which the stress vector $\underline{t}$ can be represented as:

$$\underline{t} = \underline{\underline{\sigma}}\underline{n} = \begin{pmatrix} \sigma_1 n_1 \\ \sigma_2 n_2 \\ \sigma_3 n_3 \end{pmatrix}, \tag{2.31}$$

---

[2]Christian Otto Mohr, 1835–1918, German civil engineer.

with $\underline{n} = (n_1, n_2, n_3)^T$. Then with $\sigma^2 + \tau^2 = t^2$ we obtain:

$$\sigma^2 + \tau^2 = t^2 = (\sigma_1 n_1)^2 + (\sigma_2 n_2)^2 + (\sigma_3 n_3)^2. \tag{2.32}$$

In addition we have:

$$\sigma = \underline{t}\underline{n} = \sigma_1 n_1^2 + \sigma_2 n_2^2 + \sigma_3 n_3^2. \tag{2.33}$$

The identity $n_1^2 + n_2^2 + n_3^2 = 1$ holds at this point. Using

$$\left(\sigma - \frac{\sigma_2 + \sigma_3}{2}\right)^2 + \tau^2 = -\sigma\left(\sigma_2 + \sigma_3\right) + \left(\frac{\sigma_2 + \sigma_3}{2}\right)^2 + \sigma^2 + \tau^2, \tag{2.34}$$

the expression

$$\left(\sigma - \frac{\sigma_2 + \sigma_3}{2}\right)^2 + \tau^2 = n_1^2\left(\sigma_1 - \sigma_2\right)\left(\sigma_1 - \sigma_3\right) + \left(\frac{\sigma_2 - \sigma_3}{2}\right)^2 \tag{2.35}$$

results. This a circle equation in the $\sigma\tau$−plane where the center of the circle is located on the σ−axis at the position $(\sigma_2 + \sigma_3)/2$. The circle radius R depends on n_1:

$$R^2 = n_1^2\left(\sigma_1 - \sigma_2\right)\left(\sigma_1 - \sigma_3\right) + \left(\frac{\sigma_2 - \sigma_3}{2}\right)^2. \tag{2.36}$$

The radius R assumes a minimum for $n_1 = 0$ and a maximum for $|n| = 1$ so that $R_{min} = (\sigma_2 - \sigma_3)/2$ and $R_{max} = \sigma_1 - (\sigma_2 + \sigma_3)/2$. Consequently, all possible value pairs for σ and τ for a given stress state can be found within the annulus with R_{min} and R_{max}. The two further circle equations as already shown in Fig. 2.6 are obtained by cyclically swapping indices. As a result, all possible value pairs for σ and τ can be found at the intersection of the three resulting circles.

2.2.4 Decomposition of the Stress Tensor

In some cases it may be useful to split the stress tensor into two parts, namely the so-called hydrostatic stress state or the so-called spherical stress tensor on the one hand, and the so-called stress deviator on the other hand. For this purpose, we first introduce the mean normal stress σ_0 which is the mean value of the principal stresses σ_1, σ_2 and σ_3 as follows

$$\sigma_0 = \frac{1}{3}\left(\sigma_1 + \sigma_2 + \sigma_3\right). \tag{2.37}$$

The stress tensor $\underline{\underline{\sigma}}$ is then decomposed as follows:

$$
\underline{\underline{\sigma}} = \begin{bmatrix} \sigma_0 & 0 & 0 \\ 0 & \sigma_0 & 0 \\ 0 & 0 & \sigma_0 \end{bmatrix} + \begin{bmatrix} \sigma_{xx} - \sigma_0 & \tau_{xy} & \tau_{xz} \\ \tau_{xy} & \sigma_{yy} - \sigma_0 & \tau_{yz} \\ \tau_{xz} & \tau_{yz} & \sigma_{zz} - \sigma_0 \end{bmatrix}
$$
$$
= \begin{bmatrix} \sigma_0 & 0 & 0 \\ 0 & \sigma_0 & 0 \\ 0 & 0 & \sigma_0 \end{bmatrix} + \begin{bmatrix} s_{11} & s_{12} & s_{13} \\ s_{12} & s_{22} & s_{23} \\ s_{13} & s_{23} & s_{33} \end{bmatrix}. \tag{2.38}
$$

The first component is the so-called spherical tensor $\sigma_0\underline{\underline{I}}$. The second component $\underline{\underline{s}}$ is the so-called stress deviator. In symbolic form (2.38) can also be written as:

$$
\underline{\underline{\sigma}} = \sigma_0\underline{\underline{I}} + \underline{\underline{s}}. \tag{2.39}
$$

The hydrostatic component $\sigma_0\underline{\underline{I}}$ describes the state of stress that is also found in a liquid at rest. This means that the same pure normal stress state σ_0 always occurs under any section with any normal vector $\underline{n}$ and that no shear stresses are present. With respect to the deviator $\underline{\underline{s}}$, the mean normal stress σ_0 disappears. The stress deviator thus describes the deviation of the given stress state from the hydrostatic state.

The invariants I_1^0, I_2^0 and I_3^0 of the hydrostatic stress state are

$$
I_1^0 = 3\sigma_0, \quad I_2^0 = 3\sigma_0^2, \quad I_3^0 = \sigma_0^3. \tag{2.40}
$$

For the invariants I_1^D, I_2^D and I_3^D of the stress deviator we have in index notation:

$$
I_1^D = 0, \quad I_2^D = -\frac{1}{2}s_{ij}s_{ij}, \quad I_3^D = \frac{1}{3}s_{ij}s_{jk}s_{ki}. \tag{2.41}
$$

It is worth mentioning that the principal axes of the stress tensor $\underline{\underline{\sigma}}$ and the stress deviator $\underline{\underline{s}}$ are always identical.

2.2.5   Equilibrium Conditions

After we have gained clarity about the state of stress in any point of a three-dimensional solid, we now want consider how the states of stress of two points of the body infinitely distant from each other are connected. Obviously, equilibrium must be ensured between these two points, i.e. the so-called local equilibrium conditions must be fulfilled. We now consider an infinitesimally small volume element with the edge dimensions dx, dy, dz which we separate from the solid and attach the respective stress components to the cut areas (Fig. 2.7). For reasons of clarity, the volume forces f_x, f_y, f_z are not shown here. We now assume that the stress components at the respective positive and negative cut areas (which could be representative

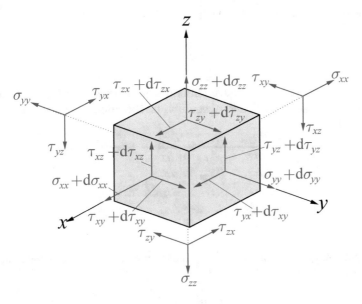

Fig. 2.7 Local equilibrium at an infitesimal volume element

of different infinitesimally adjacent body points) differ by an infinitesimal increment. We will refer to these increments as $d\sigma_{xx}$, $d\sigma_{yy}$, $d\sigma_{zz}$ and $d\tau_{xy}$, $d\tau_{xz}$, $d\tau_{yz}$ respectively. As an example, the normal stress $\sigma_{xx}(x)$ appears on the negative sectional plane with respect to the x-direction, whereas the stress $\sigma_{xx}(x + dx) = \sigma_{xx} + d\sigma_{xx}$ appears on the opposite positive sectional plane. We can develop these infinitesimal increments as a Taylor series which we terminate after the first term:

$$d\sigma_{xx} = \frac{\partial \sigma_{xx}}{\partial x}dx, \qquad d\sigma_{yy} = \frac{\partial \sigma_{yy}}{\partial y}dy, \qquad d\sigma_{zz} = \frac{\partial \sigma_{zz}}{\partial z}dz,$$

$$d\tau_{xy} = \frac{\partial \tau_{xy}}{\partial x}dx, \qquad d\tau_{yx} = \frac{\partial \tau_{yx}}{\partial y}dy, \qquad d\tau_{xz} = \frac{\partial \tau_{xz}}{\partial x}dx,$$

$$d\tau_{zx} = \frac{\partial \tau_{zx}}{\partial z}dz, \qquad d\tau_{yz} = \frac{\partial \tau_{yz}}{\partial y}dy, \qquad d\tau_{zy} = \frac{\partial \tau_{zy}}{\partial z}dz. \qquad (2.42)$$

We now formulate the equilibrium of forces in the x-direction and obtain:

$$\left(\sigma_{xx} + \frac{\partial \sigma_{xx}}{\partial x}dx\right)dydz - \sigma_{xx}dydz$$

$$+ \left(\tau_{yx} + \frac{\partial \tau_{yx}}{\partial y}dy\right)dxdz - \tau_{yx}dxdz$$

$$+ \left(\tau_{zx} + \frac{\partial \tau_{zx}}{\partial z}dz\right)dxdy - \tau_{zx}dxdy + f_x dxdydz = 0. \qquad (2.43)$$

Using $\tau_{ij} = \tau_{ji}$ we eventually achieve:

$$\frac{\partial \sigma_{xx}}{\partial x} + \frac{\partial \tau_{xy}}{\partial y} + \frac{\partial \tau_{xz}}{\partial z} + f_x = 0. \tag{2.44}$$

Similar expressions result from equilibrium of forces with respect to the $y-$ and the $z-$direction. The so-called local equilibrium conditions eventually result as:

$$\frac{\partial \sigma_{xx}}{\partial x} + \frac{\partial \tau_{xy}}{\partial y} + \frac{\partial \tau_{xz}}{\partial z} + f_x = 0,$$

$$\frac{\partial \tau_{xy}}{\partial x} + \frac{\partial \sigma_{yy}}{\partial y} + \frac{\partial \tau_{yz}}{\partial z} + f_y = 0,$$

$$\frac{\partial \tau_{xz}}{\partial x} + \frac{\partial \tau_{yz}}{\partial y} + \frac{\partial \sigma_{zz}}{\partial z} + f_z = 0. \tag{2.45}$$

Using the reference system x_1, x_2, x_3 and the notation $\tau_{ij} = \sigma_{ij}$ in index notation, we can formulate this as follows

$$\frac{\partial \sigma_{ij}}{\partial x_j} + f_i = 0, \tag{2.46}$$

where in i and j can have values between 1 and 3.

The equations (2.45) represent three coupled partial differential equations for six unknown stress components. Obviously, a three-dimensional problem in elastostatics is statically indeterminate in itself since only three equilibrium conditions are available to determine the six independent stress components. Therefore, the three local equilibrium conditions alone are not sufficient for the unambiguous determination of the state of stress so that we have to use further equations.

It should be emphasized that the equilibrium conditions (2.45) and (2.46) are local statements for a body point. If one is interested in the equilibrium conditions for a solid body with the volume V, then an integral formulation offers itself whereby by means of the Gaussian[3] integral theorem and the Cauchy formulas the following formulation can be obtained in index notation:

$$\int_V \left(\frac{\partial \sigma_{ij}}{\partial x_j} + f_i \right) \mathrm{d}V = \int_V \frac{\partial \sigma_{ij}}{\partial x_j} \mathrm{d}V + \int_V f_i \mathrm{d}V$$

$$= \int_{\partial V} \sigma_{ij} n_j \mathrm{d}A + \int_V f_i \mathrm{d}V$$

$$= \int_{\partial V} t_i \mathrm{d}A + \int_V f_i \mathrm{d}V = 0. \tag{2.47}$$

It follows that the total surface load and the total volume forces must be in equilibrium with each other.

---

[3] Johann Carl Friedrich Gauss, 1777–1855, German mathematician.

2.3 Deformations and Strains

2.3.1 Introduction

The points of a solid under load will undergo displacements u, v, w, which in turn will lead to the so-called local strains. Both the displacements and the strains are usually summarized under the term deformation state. To describe the deformation state it is necessary to refer to a certain configuration or to a certain deformation state. We want to distinguish here between the undeformed initial state C_0 (also called reference configuration) and the current deformed state or momentary configuration C_1. Both states are shown in Fig. 2.8. We define a spatially fixed Cartesian coordinate system X, Y, Z in which we can describe the position of each body point P in the initial configuration C_0 by the position vector $\underline{X}$ with the components X, Y, Z. The components X, Y, Z are also called material coordinates or Lagrangian[4] coordinates. The set V_0 of all body points P or the set of all position vectors $\underline{X}$ in the state C_0 then characterizes the undeformed body. We refer to its surface as ∂V_0.

We now consider the body in its deformed state C_1. Each body point P has now shifted by a measure $\underline{u}$, and the corresponding position vector in the state C_1 is $\underline{x}$. We refer to its components as x, y, z. Apparently, the following applies:

$$\underline{x} = \begin{pmatrix} x \\ y \\ z \end{pmatrix} = \begin{pmatrix} X \\ Y \\ Z \end{pmatrix} + \begin{pmatrix} u \\ v \\ w \end{pmatrix} = \underline{X} + \underline{u}. \tag{2.48}$$

Thus, the deformation of a body is characterized by the displacement vector $\underline{u}$:

$$\underline{u} = \underline{x} - \underline{X}. \tag{2.49}$$

The coordinates x, y, z are also known as the so-called Eulerian[5] coordinates. The set of all body points $\bar{P}$ in the state C_1 then constitutes the deformed body. We call this set V, its surface is denoted as ∂V.

If one now wants to describe the deformation process in a solid body, two approaches are possible. On the one hand the so-called Lagrangian approach can be chosen, where all state variables are understood as functions of the coordinates X, Y, Z:

$$u = u\,(X, Y, Z)\,, \quad v = v\,(X, Y, Z)\,, \quad w = w\,(X, Y, Z)\,, \tag{2.50}$$

or:

$$\underline{u} = \underline{u}\,(X, Y, Z)\,. \tag{2.51}$$

---

[4]Joseph-Louis de Lagrange, 1736–1813, Italian mathematician.
[5]Leonhard Euler, 1707–1783, Swiss mathematician.

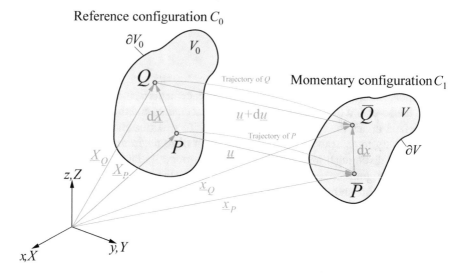

Fig. 2.8 Deformation of a solid body

It is assumed that the coordinate system X, Y, Z remains at the same position during the deformation process and thus the deformation process is related to the original coordinates.

On the other hand one can choose the so-called Eulerian approach where all state variables like e.g. the displacements u, v, w are described as functions of the local coordinates x, y, z:

$$u = u(x, y, z), \quad v = v(x, y, z), \quad w = w(x, y, z), \tag{2.52}$$

or

$$\underline{u} = \underline{u}(x, y, z). \tag{2.53}$$

Usually the Lagrangian approach is preferred in the theory of elasticity. We assume that the location $\underline{x}$ of a body point as well as its displacements $\underline{u}$ are described in dependence of the material coordinates X, Y, Z, thus $\underline{u} = \underline{u}(X, Y, Z)$ and $\underline{x} = \underline{x}(X, Y, Z)$.

We refer to Fig. 2.8 again and examine the deformation of the two body points P and Q. In the original state C_0 these are infinitesimally adjacent from each other and exhibit the two position vectors $\underline{X}_P = (X, Y, Z)^T$ and $\underline{X}_Q = (X + \mathrm{d}X, Y + \mathrm{d}Y, Z + \mathrm{d}Z)^T$. The difference vector $\mathrm{d}\underline{X}$ can then be calculated as:

$$\mathrm{d}\underline{X} = \begin{pmatrix} \mathrm{d}X \\ \mathrm{d}Y \\ \mathrm{d}Z \end{pmatrix} = \underline{X}_Q - \underline{X}_P, \tag{2.54}$$

with

$$|d\underline{X}| = dS = \sqrt{dX^2 + dY^2 + dZ^2}. \tag{2.55}$$

When the momentary configuration C_1 is reached, the two body points P and Q are denoted as $\bar{P}$ and $\bar{Q}$ to indicate the deformed configuration. Their position vectors are $\underline{x}_P = (x, y, z)^T$ and $\underline{x}_Q = (x + dx, y + dy, z + dz)^T$. The difference vector then results as:

$$d\underline{x} = \begin{pmatrix} dx \\ dy \\ dz \end{pmatrix} = \underline{x}_Q - \underline{x}_P, \tag{2.56}$$

with

$$|d\underline{x}| = ds = \sqrt{dx^2 + dy^2 + dz^2}. \tag{2.57}$$

From Fig. 2.8 we get:

$$d\underline{X} + \underline{u} + d\underline{u} = \underline{u} + d\underline{x}, \tag{2.58}$$

or

$$d\underline{u} = d\underline{x} - d\underline{X}. \tag{2.59}$$

Written down in its components this relation reads:

$$\begin{pmatrix} du \\ dv \\ dw \end{pmatrix} = \begin{pmatrix} dx \\ dy \\ dz \end{pmatrix} - \begin{pmatrix} dX \\ dY \\ dZ \end{pmatrix}. \tag{2.60}$$

The components du, dv, dw of $d\underline{u}$ are the total differentials of the displacement u, v, w:

$$
\begin{aligned}
du &= \frac{\partial u}{\partial X}dX + \frac{\partial u}{\partial Y}dY + \frac{\partial u}{\partial Z}dZ, \\
dv &= \frac{\partial v}{\partial X}dX + \frac{\partial v}{\partial Y}dY + \frac{\partial v}{\partial Z}dZ, \\
dw &= \frac{\partial w}{\partial X}dX + \frac{\partial w}{\partial Y}dY + \frac{\partial w}{\partial Z}dZ.
\end{aligned}
\tag{2.61}
$$

The partial derivatives of the displacements u, v, w with respect to the material coordinates X, Y, Z are summarized in a second order tensor $\underline{\underline{H}}$, the so-called displacement gradient:

$$\underline{\underline{H}} = \begin{bmatrix} H_{xx} & H_{xy} & H_{xz} \\ H_{yx} & H_{yy} & H_{yz} \\ H_{zx} & H_{zy} & H_{zz} \end{bmatrix} = \begin{bmatrix} \dfrac{\partial u}{\partial X} & \dfrac{\partial u}{\partial Y} & \dfrac{\partial u}{\partial Z} \\ \dfrac{\partial v}{\partial X} & \dfrac{\partial v}{\partial Y} & \dfrac{\partial v}{\partial Z} \\ \dfrac{\partial w}{\partial X} & \dfrac{\partial w}{\partial Y} & \dfrac{\partial w}{\partial Z} \end{bmatrix}. \tag{2.62}$$

2.3.2 Green-Lagrange Strain Tensor

The so-called Green[6]-Lagrange strain tensor results from the difference of the length squares ds^2 and dS^2:

$$
\begin{aligned}
ds^2 - dS^2 &= dx^2 + dy^2 + dz^2 - \left(dX^2 + dY^2 + dZ^2\right) \\
&= (du + dX)^2 + (dv + dY)^2 + (dw + dZ)^2 - \left(dX^2 + dY^2 + dZ^2\right) \\
&= \left(\frac{\partial u}{\partial X}dX + \frac{\partial u}{\partial Y}dY + \frac{\partial u}{\partial Z}dZ + dX\right)^2 \\
&\quad + \left(\frac{\partial v}{\partial X}dX + \frac{\partial v}{\partial Y}dY + \frac{\partial v}{\partial Z}dZ + dY\right)^2 \\
&\quad + \left(\frac{\partial w}{\partial X}dX + \frac{\partial w}{\partial Y}dY + \frac{\partial w}{\partial Z}dZ + dZ\right)^2 - \left(dX^2 + dY^2 + dZ^2\right)
\end{aligned}
\tag{2.63}
$$

This leads to:

$$
\begin{aligned}
ds^2 - dS^2 &= 2E_{xx}dX^2 + 2E_{yy}dY^2 + 2E_{zz}dZ^2 \\
&\quad + 4E_{xy}dXdY + 4E_{xz}dXdZ + 4E_{yz}dYdZ.
\end{aligned}
\tag{2.64}
$$

In index notation we have:

$$
ds^2 - dS^2 = 2E_{ij}dX_idX_j.
\tag{2.65}
$$

The quantities E_{xx}, E_{yy}, E_{zz}, E_{xy}, E_{xz}, E_{yz} are the components of the symmetric Green-Lagrange strain tensor $\underline{\underline{E}}$ and are defined as:

$$
\begin{aligned}
E_{xx} &= \frac{\partial u}{\partial X} + \frac{1}{2}\left(\frac{\partial u}{\partial X}\right)^2 + \frac{1}{2}\left(\frac{\partial v}{\partial X}\right)^2 + \frac{1}{2}\left(\frac{\partial w}{\partial X}\right)^2, \\
E_{yy} &= \frac{\partial v}{\partial Y} + \frac{1}{2}\left(\frac{\partial u}{\partial Y}\right)^2 + \frac{1}{2}\left(\frac{\partial v}{\partial Y}\right)^2 + \frac{1}{2}\left(\frac{\partial w}{\partial Y}\right)^2, \\
E_{zz} &= \frac{\partial w}{\partial Z} + \frac{1}{2}\left(\frac{\partial u}{\partial Z}\right)^2 + \frac{1}{2}\left(\frac{\partial v}{\partial Z}\right)^2 + \frac{1}{2}\left(\frac{\partial w}{\partial Z}\right)^2, \\
E_{xy} &= \frac{1}{2}\left(\frac{\partial u}{\partial Y} + \frac{\partial v}{\partial X} + \frac{\partial u}{\partial X}\frac{\partial u}{\partial Y} + \frac{\partial v}{\partial X}\frac{\partial v}{\partial Y} + \frac{\partial w}{\partial X}\frac{\partial w}{\partial Y}\right), \\
E_{xz} &= \frac{1}{2}\left(\frac{\partial u}{\partial Z} + \frac{\partial w}{\partial X} + \frac{\partial u}{\partial X}\frac{\partial u}{\partial Z} + \frac{\partial v}{\partial X}\frac{\partial v}{\partial Z} + \frac{\partial w}{\partial X}\frac{\partial w}{\partial Z}\right), \\
E_{yz} &= \frac{1}{2}\left(\frac{\partial v}{\partial Z} + \frac{\partial w}{\partial Y} + \frac{\partial u}{\partial Y}\frac{\partial u}{\partial Z} + \frac{\partial v}{\partial Y}\frac{\partial v}{\partial Z} + \frac{\partial w}{\partial Y}\frac{\partial w}{\partial Z}\right).
\end{aligned}
\tag{2.66}
$$

---

[6]George Green, 1793–1841, British mathematician and physicist.

The Green-Lagrange strain tensor $\underline{\underline{E}}$ describes large (finite) deformations and can be represented as:

$$\underline{\underline{E}} = \begin{bmatrix} E_{xx} & E_{xy} & E_{xz} \\ E_{xy} & E_{yy} & E_{yz} \\ E_{xz} & E_{yz} & E_{zz} \end{bmatrix}. \tag{2.67}$$

In index notation (2.66) reads:

$$E_{ij} = \frac{1}{2}\left(\frac{\partial u_i}{\partial X_j} + \frac{\partial u_j}{\partial X_i} + \frac{\partial u_k}{\partial X_i}\frac{\partial u_k}{\partial X_j}\right). \tag{2.68}$$

The components E_{xx}, E_{yy}, E_{zz} of the Green-Lagrange strain tensor must not be confused with the moduli of elasticity of an orthotropic material which will be introduced at a later point.

2.3.3 Von-Kármán Strains

Many lightweight structures can be classified to be thin-walled. Often the so-called von-Kármán[7] theory using the associated von-Kármán strains is applied. These are based on the assumption that all three displacement components u, v, w can occur in a thin-walled structure, but that the inplane displacements u and v are relatively small compared to the deflection w. Therefore, it can be assumed that in the Green-Lagrange strain tensor all higher-order terms in u and v can be omitted, thus:

$$\begin{aligned}
E_{xx} &= \frac{\partial u}{\partial x} + \frac{1}{2}\left(\frac{\partial w}{\partial x}\right)^2, \\
E_{yy} &= \frac{\partial v}{\partial y} + \frac{1}{2}\left(\frac{\partial w}{\partial y}\right)^2, \\
E_{zz} &= \frac{\partial w}{\partial z} + \frac{1}{2}\left(\frac{\partial w}{\partial z}\right)^2, \\
E_{xy} &= \frac{1}{2}\left(\frac{\partial u}{\partial y} + \frac{\partial v}{\partial x} + \frac{\partial w}{\partial x}\frac{\partial w}{\partial y}\right), \\
E_{xz} &= \frac{1}{2}\left(\frac{\partial u}{\partial z} + \frac{\partial w}{\partial x} + \frac{\partial w}{\partial x}\frac{\partial w}{\partial z}\right), \\
E_{yz} &= \frac{1}{2}\left(\frac{\partial v}{\partial z} + \frac{\partial w}{\partial y} + \frac{\partial w}{\partial y}\frac{\partial w}{\partial z}\right).
\end{aligned} \tag{2.69}$$

There in usually the distinction between material and Eulerian coordinates is no longer made and the coordinates X, Y, Z are replaced by x, y, z.

---

[7]Theodore von Kármán, 1881–1963, Austro-Hungarian engineer and physicist.

2.3.4 Infinitesimal Strain Tensor

If problems are considered where the components of the displacement gradient are very small, i.e. $H_{ij} \ll 1$ applies, then all product terms and all quadratic terms can be neglected in the Green-Lagrange strain tensor. The distinction between derivatives with respect to material coordinates X, Y, Z and spatial coordinates x, y, z can then also be omitted, and the Eulerian and Lagrangian views then coincide, so that $x_i \rightarrow X_i$. The displacements u, v, w can then be interpreted as functions of the coordinates x, y, z: $u = u(x, y, z), v = v(x, y, z), w = w(x, y, z)$. The components of the infinitesimal strain tensor $\underline{\underline{\varepsilon}}$ are then obtained as follows:

$$\varepsilon_{xx} = \frac{\partial u}{\partial x}, \quad \varepsilon_{yy} = \frac{\partial v}{\partial y}, \quad \varepsilon_{zz} = \frac{\partial w}{\partial z},$$

$$\gamma_{xy} = 2\varepsilon_{xy} = \frac{\partial u}{\partial y} + \frac{\partial v}{\partial x},$$

$$\gamma_{xz} = 2\varepsilon_{xz} = \frac{\partial u}{\partial z} + \frac{\partial w}{\partial x},$$

$$\gamma_{yz} = 2\varepsilon_{yz} = \frac{\partial v}{\partial z} + \frac{\partial w}{\partial y}. \tag{2.70}$$

Here in the quantities $\gamma_{xy}, \gamma_{xz}, \gamma_{yz}$ are the so-called technical shear strains. We can represent the symmetrical infinitesimal strain tensor $\underline{\underline{\varepsilon}}$ as follows:

$$\underline{\underline{\varepsilon}} = \begin{bmatrix} \varepsilon_{xx} & \varepsilon_{xy} & \varepsilon_{xz} \\ \varepsilon_{xy} & \varepsilon_{yy} & \varepsilon_{yz} \\ \varepsilon_{xz} & \varepsilon_{yz} & \varepsilon_{zz} \end{bmatrix}. \tag{2.71}$$

The diagonal elements $\varepsilon_{xx}, \varepsilon_{yy}, \varepsilon_{zz}$ are called strains or stretches. The remaining entries $\gamma_{xy} = 2\varepsilon_{xy}, \gamma_{xz} = 2\varepsilon_{xz}, \gamma_{yz} = 2\varepsilon_{yz}$ are called shear strains. In index notation (2.70) reads:

$$\varepsilon_{ij} = \frac{1}{2}\left(\frac{\partial u_i}{\partial x_j} + \frac{\partial u_j}{\partial x_i}\right). \tag{2.72}$$

The entries of the infinitesimal strain tensor can be interpreted geometrically which we will examine more closely below. We consider the infinitesimal element of Fig. 2.9 and examine the strains ε_{xx} and ε_{yy} in the $xy-$ plane more closely. From elementary geometric relations we get:

$$[(1 + \varepsilon_{xx})\,dx]^2 = \left[\left(1 + \frac{\partial u}{\partial x}\right)dx\right]^2 + \left(\frac{\partial v}{\partial x}dx\right)^2,$$

$$[(1 + \varepsilon_{yy})\,dy]^2 = \left[\left(1 + \frac{\partial v}{\partial y}\right)dy\right]^2 + \left(\frac{\partial u}{\partial y}dy\right)^2. \tag{2.73}$$

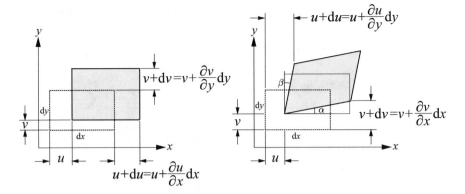

Fig. 2.9 Definition of the infinitesimal strains ε_{xx} and ε_{yy} (left) and the infinitesimal shear strain γ_{xy} (right)

Dividing by dx respectively dy and neglect of quadratic terms yields:

$$\varepsilon_{xx} = \frac{\partial u}{\partial x} + \frac{1}{2}\left(\frac{\partial u}{\partial x}\right)^2 + \frac{1}{2}\left(\frac{\partial v}{\partial x}\right)^2,$$

$$\varepsilon_{yy} = \frac{\partial v}{\partial y} + \frac{1}{2}\left(\frac{\partial u}{\partial y}\right)^2 + \frac{1}{2}\left(\frac{\partial v}{\partial y}\right)^2. \tag{2.74}$$

This corresponds to the components of the Green-Lagrange strain tensor assuming $\underline{X} = \underline{x}$. If we also neglect the square terms in u and v, we get the following result

$$\varepsilon_{xx} = \frac{\partial u}{\partial x}, \quad \varepsilon_{yy} = \frac{\partial v}{\partial y}. \tag{2.75}$$

This corresponds to the expressions already derived with (2.70). The corresponding expression for ε_{zz} can be derived analogously.

The strain ε_{xx} can also be derived by assuming small displacements from the beginning so that:

$$\varepsilon_{xx} = \frac{\Delta l}{l} = \frac{\left[u\,(x) + \dfrac{\partial u}{\partial x}dx\right] - u\,(x)}{dx} = \frac{\partial u}{\partial x}. \tag{2.76}$$

Obviously this corresponds exactly to the formulation already derived with (2.70). A similar procedure can be used for the two strains ε_{yy} and ε_{zz}.

For the shear strain γ_{xy} in the $xy-$ plane, on the basis of Fig. 2.9 we get:

$$(1 + \varepsilon_{xx})\, dx \left(1 + \varepsilon_{yy}\right) dy \cos\left(\frac{\pi}{2} - \gamma_{xy}\right) = \left(1 + \frac{\partial u}{\partial x}\right) dx \frac{\partial u}{\partial y} dy + \left(1 + \frac{\partial v}{\partial y}\right) dy \frac{\partial v}{\partial x} dx.$$
(2.77)

with $\cos\left(\frac{\pi}{2} - \gamma_{xy}\right) = \sin\gamma_{xy}$ this results in:

$$\sin\gamma_{xy} = \frac{\dfrac{\partial u}{\partial y} + \dfrac{\partial v}{\partial x} + \dfrac{\partial u}{\partial x}\dfrac{\partial u}{\partial x} + \dfrac{\partial v}{\partial x} + \dfrac{\partial v}{\partial y}}{(1 + \varepsilon_{xx})\left(1 + \varepsilon_{yy}\right)}.$$
(2.78)

Assuming that the strains ε_{xx} and ε_{yy} can be neglected in the denominator of (2.78) and that we can also apply the small angle approximation $\sin\gamma_{xy} \simeq \gamma_{xy}$, then we get:

$$\gamma_{xy} = \frac{\partial u}{\partial y} + \frac{\partial v}{\partial x} + \frac{\partial u}{\partial x}\frac{\partial u}{\partial y} + \frac{\partial v}{\partial x}\frac{\partial v}{\partial y}.$$
(2.79)

This expression has also been derived in the Green-Lagrange strain tensor. If in this expression the terms of higher order are deleted, the result is:

$$\gamma_{xy} = \frac{\partial u}{\partial y} + \frac{\partial v}{\partial x}.$$
(2.80)

This result can also be achieved if we assume from the beginning that the change of length of the edges dx and dy is negligible:

$$\tan\alpha \simeq \alpha = \frac{\dfrac{\partial v}{\partial x} dx}{dx} = \frac{\partial v}{\partial x}, \quad \tan\beta \simeq \beta = \frac{\dfrac{\partial u}{\partial y} dy}{dy} = \frac{\partial u}{\partial y}.$$
(2.81)

The shear strain γ_{xy} then results as:

$$\gamma_{xy} = \alpha + \beta = \frac{\partial u}{\partial y} + \frac{\partial v}{\partial x}.$$
(2.82)

We can determine corresponding expressions for γ_{xz} and γ_{yz} in a similar manner.

The equations (2.70) establish relations between the displacements u, v, w on the one hand and the strains ε_{xx}, ε_{yy}, ε_{zz}, γ_{xy}, γ_{xz}, γ_{yz} on the other hand. This set of equations is therefore also referred to as the so-called kinematic equations.

Similar to the transformation of the stress components, analogous transformation relations can be derived for the strain components. However, in the case of the strains the resultant transformation expressions are slightly different due to the use of technical shear strains. The corresponding transformation matrix $\underline{\underline{T}}_\varepsilon$ which is used in the transformation relation $\underline{\bar{\varepsilon}} = \underline{\underline{T}}_\varepsilon \underline{\varepsilon}$ with $\underline{\varepsilon} = \left(\varepsilon_{xx}, \varepsilon_{yy}, \varepsilon_{zz}, \gamma_{yz}, \gamma_{xz}, \gamma_{xy}\right)^T$ results as:

$$
\underline{\underline{T}}_\varepsilon =
\begin{bmatrix}
R_{11}^2 & R_{12}^2 & R_{13}^2 & R_{12}R_{13} & R_{11}R_{13} & R_{11}R_{12} \\
R_{21}^2 & R_{22}^2 & R_{23}^2 & R_{22}R_{23} & R_{21}R_{23} & R_{21}R_{22} \\
R_{31}^2 & R_{32}^2 & R_{33}^2 & R_{32}R_{33} & R_{31}R_{33} & R_{31}R_{32} \\
2R_{21}R_{31} & 2R_{22}R_{32} & 2R_{23}R_{33} & R_{22}R_{33} + R_{23}R_{32} & R_{21}R_{33} + R_{23}R_{31} & R_{21}R_{32} + R_{22}R_{31} \\
2R_{11}R_{31} & 2R_{12}R_{32} & 2R_{13}R_{33} & R_{12}R_{33} + R_{13}R_{32} & R_{11}R_{33} + R_{13}R_{31} & R_{11}R_{32} + R_{12}R_{31} \\
2R_{11}R_{21} & 2R_{12}R_{22} & 2R_{13}R_{23} & R_{12}R_{23} + R_{13}R_{22} & R_{11}R_{23} + R_{13}R_{21} & R_{11}R_{22} + R_{12}R_{21}
\end{bmatrix}.
$$
$$(2.83)$$

In the case of a coordinate rotation by the angle θ with the z−axis fixed, we get:

$$
\begin{pmatrix}
\bar{\varepsilon}_{xx} \\
\bar{\varepsilon}_{yy} \\
\bar{\varepsilon}_{zz} \\
\bar{\gamma}_{yz} \\
\bar{\gamma}_{xz} \\
\bar{\gamma}_{xy}
\end{pmatrix} =
\begin{bmatrix}
\cos^2\theta & \sin^2\theta & 0 & 0 & 0 & \cos\theta\sin\theta \\
\sin^2\theta & \cos^2\theta & 0 & 0 & 0 & -\cos\theta\sin\theta \\
0 & 0 & 1 & 0 & 0 & 0 \\
0 & 0 & 0 & \cos\theta & -\sin\theta & 0 \\
0 & 0 & 0 & \sin\theta & \cos\theta & 0 \\
-2\cos\theta\sin\theta & 2\cos\theta\sin\theta & 0 & 0 & 0 & \cos^2\theta - \sin^2\theta
\end{bmatrix}
\begin{pmatrix}
\varepsilon_{xx} \\
\varepsilon_{yy} \\
\varepsilon_{zz} \\
\gamma_{yz} \\
\gamma_{xz} \\
\gamma_{xy}
\end{pmatrix}.
$$
$$(2.84)$$

The following relations exist between the transformation matrices $\underline{\underline{T}}_\sigma$ and $\underline{\underline{T}}_\varepsilon$:

$$
\underline{\underline{T}}_\sigma^{-1} = \underline{\underline{T}}_\varepsilon^T, \quad \underline{\underline{T}}_\varepsilon^{-1} = \underline{\underline{T}}_\sigma^T.
$$
$$(2.85)$$

As a final remark it should be noted that in the context of continuum mechanics a number of other strain tensors exist which, however, are not relevant for the contents of this book and therefore will not be treated further. The interested reader is referred to the corresponding literature a selection of which is cited at the end of this chapter.

2.3.5 Compatibility Conditions

If the three displacements u, v, w are known, then the six components ε_{xx}, ε_{yy}, ε_{zz}, γ_{yz}, γ_{xz}, γ_{xy} of the infinitesimal strain tensor $\underline{\varepsilon}$ can be derived by means of the kinematic relations (2.70). In turn, this also means that in the case that the six strain components are known, six kinematic equations are available to calculate the three displacements. Thus, the system of equations is kinematically overdetermined and as a consequence the components in $\underline{\varepsilon}$ cannot be independent of each other. The infinitesimal strains ε_{xx}, ε_{yy}, ε_{zz}, γ_{yz}, γ_{xz}, γ_{xy} must rather satisfy the so-called compatibility conditions, which represent necessary and sufficient conditions for a unique displacement field. We obtain the compatibility conditions by eliminating the displacements in the kinematic equations. We have:

$$
\frac{\partial^2 \varepsilon_{ij}}{\partial x_k \partial x_l} + \frac{\partial^2 \varepsilon_{kl}}{\partial x_i \partial x_j} - \frac{\partial^2 \varepsilon_{ik}}{\partial x_j \partial x_l} - \frac{\partial^2 \varepsilon_{jl}}{\partial x_i \partial x_k} = 0,
$$
$$(2.86)$$

wherein we use the notation $\varepsilon_{ij} = \frac{1}{2}\gamma_{ij}$ for the shear strains in the reference system x_1, x_2, x_3. For the index pairs $(i, j), (k, l) = (1, 1), (2, 2), (3, 3), (1, 2), (1, 3), (2, 3)$ we get a total of six different compatibility conditions, three of which are independent from each other. When using the reference system x, y, z and the technical shear strains they read:

$$\frac{\partial^2 \varepsilon_{xx}}{\partial y^2} + \frac{\partial^2 \varepsilon_{yy}}{\partial x^2} - \frac{\partial^2 \gamma_{xy}}{\partial x \partial y} = 0,$$

$$\frac{\partial^2 \varepsilon_{xx}}{\partial z^2} + \frac{\partial^2 \varepsilon_{zz}}{\partial x^2} - \frac{\partial^2 \gamma_{xz}}{\partial x \partial z} = 0,$$

$$\frac{\partial^2 \varepsilon_{yy}}{\partial z^2} + \frac{\partial^2 \varepsilon_{zz}}{\partial y^2} - \frac{\partial^2 \gamma_{yz}}{\partial y \partial z} = 0,$$

$$2\frac{\partial^2 \varepsilon_{xx}}{\partial y \partial z} + \frac{\partial^2 \gamma_{yz}}{\partial x^2} - \frac{\partial^2 \gamma_{xz}}{\partial x \partial y} - \frac{\partial^2 \gamma_{xy}}{\partial x \partial z} = 0,$$

$$2\frac{\partial^2 \varepsilon_{yy}}{\partial x \partial z} + \frac{\partial^2 \gamma_{xz}}{\partial y^2} - \frac{\partial^2 \gamma_{xy}}{\partial y \partial z} - \frac{\partial^2 \gamma_{yz}}{\partial x \partial y} = 0,$$

$$2\frac{\partial^2 \varepsilon_{zz}}{\partial x \partial y} + \frac{\partial^2 \gamma_{xy}}{\partial z^2} - \frac{\partial^2 \gamma_{yz}}{\partial x \partial z} - \frac{\partial^2 \gamma_{xz}}{\partial y \partial z} = 0. \qquad (2.87)$$

2.3.6   Volume Strain

It is convenient to introduce the so-called volume strain ε_V which results as the volume increase ΔdV of an infinitesimal volume element under a given strain state, relative to its initial volume $dV = dxdydz$:

$$\varepsilon_V = \frac{\Delta dV}{dV}. \qquad (2.88)$$

The volume element has the edge lengths dx, dy, dz. Then the following applies:

$$\Delta dV = (1 + \varepsilon_{xx})(1 + \varepsilon_{yy})(1 + \varepsilon_{zz}) \, dxdydz - dxdydz. \qquad (2.89)$$

If square and cubic elements in the strains are neglected, the result is

$$\Delta dV = (\varepsilon_{xx} + \varepsilon_{yy} + \varepsilon_{zz}) \, dV. \qquad (2.90)$$

The volume strain ε_V then results in:

$$\varepsilon_V = \varepsilon_{xx} + \varepsilon_{yy} + \varepsilon_{zz}. \qquad (2.91)$$

Obviously the volume strain is described by the diagonal elements of the infinitesimal strain tensor $\underset{=}{\varepsilon}$ and is identical to its first invariant I_1. The mean strain ε_0 is thus one third of the volume strain:

$$\varepsilon_0 = \frac{1}{3} \left(\varepsilon_{xx} + \varepsilon_{yy} + \varepsilon_{zz} \right) = \frac{1}{3} \varepsilon_V. \tag{2.92}$$

2.3.7 Decomposition of the Infinitesimal Strain Tensor

Analogous to Cauchy's stress tensor $\underset{=}{\sigma}$ the infinitesimal strain tensor $\underset{=}{\varepsilon}$ can be decomposed into a spherical tensor and a deviator:

$$
\begin{aligned}
\underset{=}{\varepsilon} &=
\begin{bmatrix}
\varepsilon_0 & 0 & 0 \\
0 & \varepsilon_0 & 0 \\
0 & 0 & \varepsilon_0
\end{bmatrix}
+
\begin{bmatrix}
\varepsilon_{11} - \varepsilon_0 & \varepsilon_{12} & \varepsilon_{13} \\
\varepsilon_{12} & \varepsilon_{22} - \varepsilon_0 & \varepsilon_{23} \\
\varepsilon_{13} & \varepsilon_{23} & \varepsilon_{33} - \varepsilon_0
\end{bmatrix} \\
&=
\begin{bmatrix}
\varepsilon_0 & 0 & 0 \\
0 & \varepsilon_0 & 0 \\
0 & 0 & \varepsilon_0
\end{bmatrix}
+
\begin{bmatrix}
e_{11} & e_{12} & e_{13} \\
e_{12} & e_{22} & e_{23} \\
e_{13} & e_{23} & e_{33}
\end{bmatrix} .
\end{aligned}
\tag{2.93}
$$

The first part in (2.93) describes a pure volume expansion whereas the second part describes the deviating part of the strain state and thus a pure shape change state with constant volume. In symbolic notation (2.93) reads:

$$\underset{=}{\varepsilon} = \varepsilon_0 \underset{=}{I} + \underset{=}{e}. \tag{2.94}$$

2.4 Constitutive Equations

2.4.1 Introduction

All relations considered so far—i.e. equilibrium conditions and kinematic equations - are material-independent. Thus, the relations established so far are obviously not sufficient to describe the mechanical behavior of a given solid completely, but material-specific equations have to be taken into account as well which characterize the given material behavior appropriately. Such material-specific equations establish a relation between the stress components on the one hand and the strain components on the other hand and are addressed as constitutive equations or material law.

In the course of this book we will always assume ideal elastic material behavior unless otherwise stated. This means that after a complete unloading of the solid under consideration, no permanent deformations remain and thus the deformation state is completely reversible. Strain energy stored in the solid due to the deformations can be fully recovered and converted into work. In addition, the stress state is independent of time and also independent of the load history and is uniquely defined by the momentary deformation state. If we now take a closer look at the case that we are dealing with small deformations and thus with a geometrically linear problem which is described by the infinitesimal strain tensor $\underline{\underline{\varepsilon}}$, then we can express a corresponding material law in general terms as follows:

$$\underline{\underline{\sigma}} = \underline{\underline{\sigma}}\left(\underline{\underline{\varepsilon}}\right). \tag{2.95}$$

The concrete form of (2.95) is of course dependent on the material under consideration. In the following we will take a closer look at this in the case of linear elasticity, i.e. there is a linear relationship between stresses and strains.

2.4.2 Hooke's Generalized Law

If the considered material is not only elastic but if there is even a linear relationship between the stresses and the strains, then we speak of a so-called linear elastic material, and the material law is given in the form of the so-called generalized Hooke's[8] law. We distinguish between isotropic materials (i.e. materials that exhibit a direction-independent behavior) on the one hand, and anisotropic materials on the other hand (i.e. materials that can exhibit different behavior in different directions in space). In the simplest case of isotropy in the one-dimensional case, Hooke's law results as follows:

$$\sigma = E\varepsilon, \quad \tau = G\gamma, \quad \varepsilon_t = -\nu\varepsilon. \tag{2.96}$$

Therein, σ is the normal stress and τ is the shear stress. The linear elastic material behavior is described by the so-called modulus of elasticity E, the shear modulus G as well as the Poisson's[9] number ν. The corresponding strain quantities are the normal strain ε, the transverse strain ε_t and the shear strain γ. However, if we focus our attention again on a three-dimensional solid body which may also exhibit anisotropic but linearly elastic material behavior, then the representation (2.96) is not sufficient and we formulate the following general representation in index notation (with the reference system x_1, x_2, x_3 and the designations σ_{ij} for the shear stresses and $2\varepsilon_{ij} = \gamma_{ij}$ for the shear strains) which is the generalized Hooke's Law:

$$\sigma_{ij} = C_{ijkl}\varepsilon_{kl}. \tag{2.97}$$

---

[8]Robert Hooke, 1635–1703, English physicist.
[9]Siméon Denis Poisson, 1781–1840, French physicist and mathematician.

The indices i, j, k, l take on values from 1 to 3. The quantity C_{ijkl} is the so-called elasticity tensor. It is a fourth-order tensor and has a total of $3^4 = 81$ components which describe the material behavior. In expanded form (2.97) reads:

$$\begin{aligned}
\sigma_{11} &= C_{1111}\varepsilon_{11} + C_{1112}\varepsilon_{12} + C_{1113}\varepsilon_{13} + C_{1121}\varepsilon_{21} + C_{1122}\varepsilon_{22} + C_{1123}\varepsilon_{23} \\
&\quad + C_{1131}\varepsilon_{31} + C_{1132}\varepsilon_{32} + C_{1133}\varepsilon_{33}, \\
\sigma_{22} &= C_{2211}\varepsilon_{11} + C_{2212}\varepsilon_{12} + C_{2213}\varepsilon_{13} + C_{2221}\varepsilon_{21} + C_{2222}\varepsilon_{22} + C_{2223}\varepsilon_{23} \\
&\quad + C_{2231}\varepsilon_{31} + C_{2232}\varepsilon_{32} + C_{2233}\varepsilon_{33}, \\
\sigma_{33} &= C_{3311}\varepsilon_{11} + C_{3312}\varepsilon_{12} + C_{3313}\varepsilon_{13} + C_{3321}\varepsilon_{21} + C_{3322}\varepsilon_{22} + C_{3323}\varepsilon_{23} \\
&\quad + C_{3331}\varepsilon_{31} + C_{3332}\varepsilon_{32} + C_{3333}\varepsilon_{33}, \\
\sigma_{12} &= C_{1211}\varepsilon_{11} + C_{1212}\varepsilon_{12} + C_{1213}\varepsilon_{13} + C_{1221}\varepsilon_{21} + C_{1222}\varepsilon_{22} + C_{1223}\varepsilon_{23} \\
&\quad + C_{1231}\varepsilon_{31} + C_{1232}\varepsilon_{32} + C_{1233}\varepsilon_{33}, \\
\sigma_{13} &= C_{1311}\varepsilon_{11} + C_{1312}\varepsilon_{12} + C_{1313}\varepsilon_{13} + C_{1321}\varepsilon_{21} + C_{1322}\varepsilon_{22} + C_{1323}\varepsilon_{23} \\
&\quad + C_{1331}\varepsilon_{31} + C_{1332}\varepsilon_{32} + C_{1333}\varepsilon_{33}, \\
\sigma_{21} &= C_{2111}\varepsilon_{11} + C_{2112}\varepsilon_{12} + C_{2113}\varepsilon_{13} + C_{2121}\varepsilon_{21} + C_{2122}\varepsilon_{22} + C_{2123}\varepsilon_{23} \\
&\quad + C_{2131}\varepsilon_{31} + C_{2132}\varepsilon_{32} + C_{2133}\varepsilon_{33}, \\
\sigma_{23} &= C_{2311}\varepsilon_{11} + C_{2312}\varepsilon_{12} + C_{2313}\varepsilon_{13} + C_{2321}\varepsilon_{21} + C_{2322}\varepsilon_{22} + C_{2323}\varepsilon_{23} \\
&\quad + C_{2331}\varepsilon_{31} + C_{2332}\varepsilon_{32} + C_{2333}\varepsilon_{33}, \\
\sigma_{31} &= C_{3111}\varepsilon_{11} + C_{3112}\varepsilon_{12} + C_{3113}\varepsilon_{13} + C_{3121}\varepsilon_{21} + C_{3122}\varepsilon_{22} + C_{3123}\varepsilon_{23} \\
&\quad + C_{3131}\varepsilon_{31} + C_{3132}\varepsilon_{32} + C_{3133}\varepsilon_{33}, \\
\sigma_{32} &= C_{3211}\varepsilon_{11} + C_{3212}\varepsilon_{12} + C_{3213}\varepsilon_{13} + C_{3221}\varepsilon_{21} + C_{3222}\varepsilon_{22} + C_{3223}\varepsilon_{23} \\
&\quad + C_{3231}\varepsilon_{31} + C_{3232}\varepsilon_{32} + C_{3233}\varepsilon_{33}.
\end{aligned} \tag{2.98}$$

It is apparent that of the 81 components of the elasticity tensor C_{ijkl} not all can be independent of each other. Rather, C_{ijkl} must exhibit certain symmetry properties which we will clarify, for example, by means of the stress components σ_{12} and σ_{21} (which, as we already know, must be identical). From $\sigma_{12} = \sigma_{21}$ we can immediately conclude $C_{12kl} = C_{21kl}$ which must also apply analogously to the other stress components. We can conclude:

$$C_{ijkl} = C_{jikl}. \tag{2.99}$$

In addition, due to $\varepsilon_{ij} = \varepsilon_{ji}$ the infinitesimal strain tensor is also symmetrical so that in addition, if the stress component σ_{12} is considered further, $C_{12kl} = C_{12lk}$ must also apply. In general this leads to the conclusion

$$C_{ijkl} = C_{ijlk}. \tag{2.100}$$

Apparently only 36 independent components of the 81 components of the elasticity tensor C_{ijkl} remain. For this reason, the following compact representation of

Hooke's Law in a vector-matrix notation has proven to be convenient in engineering practice as follows (so-called Voigt notation)[10]:

$$
\begin{pmatrix}
\sigma_{11} \\
\sigma_{22} \\
\sigma_{33} \\
\sigma_{23} \\
\sigma_{13} \\
\sigma_{12}
\end{pmatrix}
=
\begin{bmatrix}
C_{11} & C_{12} & C_{13} & C_{14} & C_{15} & C_{16} \\
C_{21} & C_{22} & C_{23} & C_{24} & C_{25} & C_{26} \\
C_{31} & C_{32} & C_{33} & C_{34} & C_{35} & C_{36} \\
C_{41} & C_{42} & C_{43} & C_{44} & C_{45} & C_{46} \\
C_{51} & C_{52} & C_{53} & C_{54} & C_{55} & C_{56} \\
C_{61} & C_{62} & C_{63} & C_{64} & C_{65} & C_{66}
\end{bmatrix}
\begin{pmatrix}
\varepsilon_{11} \\
\varepsilon_{22} \\
\varepsilon_{33} \\
2\varepsilon_{23} \\
2\varepsilon_{13} \\
2\varepsilon_{12}
\end{pmatrix}.
\tag{2.101}
$$

The connection between the double indexed stiffness values C_{ij} and the fourth-order tensor components C_{ijkl} can be easily established by direct comparison of (2.98) with (2.101). We recognize directly that, for example, the relationships $C_{11} = C_{1111}$, $C_{12} = C_{1122}$, $C_{16} = \frac{1}{2}(C_{1112} + C_{1121}) = C_{1112}$ etc. must apply.

In symbolic notation (2.101) reads:

$$
\underline{\sigma} = \underline{\underline{C}}\,\underline{\varepsilon}.
\tag{2.102}
$$

The matrix $\underline{\underline{C}}$ is the so-called stiffness matrix of the material under consideration. Its components C_{ij} ($i, j = 1, 2, 3, 4, 5, 6$) are referred to as stiffnesses. The elasticity tensor C_{ijkl} and also the stiffness matrix $\underline{\underline{C}}$ are both always symmetrical as will be proven later. Therefore, the following applies:

$$
C_{ijkl} = C_{klij} \quad \text{bzw.} \quad C_{ij} = C_{ji}.
\tag{2.103}
$$

Arbitrarily anisotropic and linear-elastic material behavior is therefore always described by 21 independent material constants.

At this point it should already be noted that a completely anisotropic material behavior (characterized by the fact that all tensor components C_{ijkl} and stiffnesses C_{ij} are not equal to zero and are generally different from each other) is characterized by the so-called coupling effects. From (2.101) we can conclude that any strain component ε_{ij} causes all stress components σ_{ij} which not only makes the analysis of such materials complicated, but in many cases is also an undesirable side effect of anisotropy. We will discuss this in more detail at a later stage.

In inverted form the generalized Hooke's Law can be represented as follows:

$$
\varepsilon_{ij} = S_{ijkl}\sigma_{kl}.
\tag{2.104}
$$

Here S_{ijkl} is the so-called compliance tensor. It has the same symmetry properties as the elasticity tensor C_{ijkl} and thus also has a maximum of 21 independent components:

$$
S_{ijkl} = S_{jikl} = S_{ijlk} = S_{klij}.
\tag{2.105}
$$

---

[10]Woldemar Voigt, 1850–1919, German physicist.

In a vector-matrix notation we have:

$$
\begin{pmatrix}
\varepsilon_{11} \\
\varepsilon_{22} \\
\varepsilon_{33} \\
2\varepsilon_{23} \\
2\varepsilon_{13} \\
2\varepsilon_{12}
\end{pmatrix}
=
\begin{bmatrix}
S_{11} & S_{12} & S_{13} & S_{14} & S_{15} & S_{16} \\
S_{21} & S_{22} & S_{23} & S_{24} & S_{25} & S_{26} \\
S_{31} & S_{32} & S_{33} & S_{34} & S_{35} & S_{36} \\
S_{41} & S_{42} & S_{43} & S_{44} & S_{45} & S_{46} \\
S_{51} & S_{52} & S_{53} & S_{54} & S_{55} & S_{56} \\
S_{61} & S_{62} & S_{63} & S_{64} & S_{65} & S_{66}
\end{bmatrix}
\begin{pmatrix}
\sigma_{11} \\
\sigma_{22} \\
\sigma_{33} \\
\sigma_{23} \\
\sigma_{13} \\
\sigma_{12}
\end{pmatrix},
\tag{2.106}
$$

and in symbolic form:

$$
\underline{\varepsilon} = \underline{\underline{S}}\,\underline{\sigma}.
\tag{2.107}
$$

Herein $\underline{\underline{S}}$ is the so-called (symmetric, $S_{ij} = S_{ji}$) compliance matrix. The compliance matrix $\underline{\underline{S}}$ is the inverse of the stiffness matrix $\underline{\underline{C}}$, i.e. $\underline{\underline{S}} = \underline{\underline{C}}^{-1}$.

2.4.3 Strain Energy

If an elastic solid body is subjected to an external load, this results in a state of deformation and strain. The stresses caused inside the body then perform work along the associated displacements, and in the case of an elastic body the work thus performed is stored as energy that can be fully recovered after the load is removed. This is the so-called strain energy which we will explain in more detail below. We consider an infinitesimal volume element cut out of an elastic body with the edge lengths dx, dy, dz, as shown in Fig. 2.10. Let us first consider only the case where the solid element is loaded exclusively by a normal stress σ_{xx}. The resultant force is $\sigma_{xx}dydz$. In this case an infinitesimal strain $d\varepsilon_{xx}$ occurs from which the change in length $d\varepsilon_{xx}dx$ results. The resultant force then performs the work increment $\sigma_{xx}d\varepsilon_{xx}dxdydz$. If one relates this work increment to the volume $dV = dxdydz$, then one receives as work increment per volume unit:

Fig. 2.10 Infinitesimal volume element under normal stress σ_{xx}

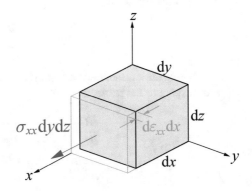

$$dU_0 = \sigma_{xx} d\varepsilon_{xx}. \tag{2.108}$$

If not only the normal stress σ_{xx} is present but all other stress components also occur, then these also perform work along the displacements they cause. The total work increment per volume unit is then:

$$dU_0 = \sigma_{xx} d\varepsilon_{xx} + \sigma_{yy} d\varepsilon_{yy} + \sigma_{zz} d\varepsilon_{zz} + \tau_{yz} d\gamma_{yz} + \tau_{xz} d\gamma_{xz} + \tau_{xy} d\gamma_{xy}. \tag{2.109}$$

In index notation we have:

$$dU_0 = \sigma_{ij} d\varepsilon_{ij}. \tag{2.110}$$

In order to calculate the work starting from the strain-free state and ending in the current deformed state with the strains $\varepsilon_{xx}, \varepsilon_{yy}, \varepsilon_{zz}, \gamma_{yz}, \gamma_{xz}, \gamma_{xy}$, we integrate over the total work increment dU_0 from the strain-free initial state to the deformed state with the currently present strains:

$$U_0 = \int_0^{\varepsilon_{xx}} \sigma_{xx} d\hat{\varepsilon}_{xx} + \int_0^{\varepsilon_{yy}} \sigma_{yy} d\hat{\varepsilon}_{yy} + \int_0^{\varepsilon_{zz}} \sigma_{zz} d\hat{\varepsilon}_{zz} + \int_0^{\gamma_{yz}} \tau_{yz} d\hat{\gamma}_{yz}$$
$$+ \int_0^{\gamma_{xz}} \tau_{xz} d\hat{\gamma}_{xz} + \int_0^{\gamma_{xy}} \tau_{xy} d\hat{\gamma}_{xy}. \tag{2.111}$$

The quantity U_0 is the so-called specific deformation work or strain energy density. The term 'specific' indicates that the work is related to the volume. For an elastic material, the strain energy density U_0 does not depend on the deformation history but depends exclusively on the current strains $\varepsilon_{xx}, \varepsilon_{yy}, \varepsilon_{zz}, \gamma_{yz}, \gamma_{xz}, \gamma_{xy}$. In this case the work increment dU_0 is the total differential of the strain energy density U_0:

$$\sigma_{xx} d\varepsilon_{xx} + \sigma_{yy} d\varepsilon_{yy} + \sigma_{zz} d\varepsilon_{zz} + \tau_{yz} d\gamma_{yz} + \tau_{xz} d\gamma_{xz} + \tau_{xy} d\gamma_{xy}$$
$$= \frac{\partial U_0}{\partial \varepsilon_{xx}} d\varepsilon_{xx} + \frac{\partial U_0}{\partial \varepsilon_{yy}} d\varepsilon_{yy} + \frac{\partial U_0}{\partial \varepsilon_{zz}} d\varepsilon_{zz} + \frac{\partial U_0}{\partial \gamma_{yz}} d\gamma_{yz} + \frac{\partial U_0}{\partial \gamma_{xz}} d\gamma_{xz} + \frac{\partial U_0}{\partial \gamma_{xy}} d\gamma_{xy}. \tag{2.112}$$

In index notation we have:

$$\sigma_{ij} d\varepsilon_{ij} = \frac{\partial U_0}{\partial \varepsilon_{ij}} d\varepsilon_{ij}. \tag{2.113}$$

Apparently, the stresses can be determined as the partial derivatives of the potential U_0 with respect to the respective strains:

$$\sigma_{xx} = \frac{\partial U_0}{\partial \varepsilon_{xx}}, \quad \sigma_{yy} = \frac{\partial U_0}{\partial \varepsilon_{yy}}, \quad \sigma_{zz} = \frac{\partial U_0}{\partial \varepsilon_{zz}},$$
$$\tau_{yz} = \frac{\partial U_0}{\partial \gamma_{yz}}, \quad \tau_{xz} = \frac{\partial U_0}{\partial \gamma_{xz}}, \quad \tau_{xy} = \frac{\partial U_0}{\partial \gamma_{xy}}, \tag{2.114}$$

or in a general form:

$$\sigma_{ij} = \frac{\partial U_0}{\partial \varepsilon_{ij}}. \tag{2.115}$$

In the case of a linear elastic material the stiffnesses C_{ijkl} can then be calculated as partial derivatives of the stresses σ_{ij} with respect to the strains ε_{kl}:

$$C_{ijkl} = \frac{\partial \sigma_{ij}}{\partial \varepsilon_{kl}}. \tag{2.116}$$

Since according to (2.115) the stresses σ_{ij} result from the first partial derivatives of the potential U_0, the stiffnesses C_{ijkl} are obtained as second partial derivatives of U_0:

$$C_{ijkl} = \frac{\partial^2 U_0}{\partial \varepsilon_{ij} \partial \varepsilon_{kl}}. \tag{2.117}$$

The order of the differentiations is arbitrary so that the following applies

$$C_{ijkl} = \frac{\partial \sigma_{ij}}{\partial \varepsilon_{kl}} = \frac{\partial^2 U_0}{\partial \varepsilon_{ij} \partial \varepsilon_{kl}} = \frac{\partial^2 U_0}{\partial \varepsilon_{kl} \partial \varepsilon_{ij}} = \frac{\partial \sigma_{kl}}{\partial \varepsilon_{ij}} = C_{klij}. \tag{2.118}$$

This proves the already assumed symmetry property $C_{ijkl} = C_{klij}$ of the elasticity tensor and thus also the symmetry of the stiffness matrix $\underline{C}$.

In the case of linear elasticity it can be shown that the specific strain energy U_0 is a square function of the strains. In index notation we have:

$$U_0 = \int_0^\varepsilon \sigma_{ij} d\hat{\varepsilon}_{ij} = C_{ijkl} \int_0^\varepsilon \varepsilon_{kl} d\hat{\varepsilon}_{ij} = \frac{1}{2} C_{ijkl} \varepsilon_{ij} \varepsilon_{kl} = \frac{1}{2} \sigma_{ij} \varepsilon_{ij}, \tag{2.119}$$

or:

$$U_0 = \frac{1}{2} \left(\sigma_{xx} \varepsilon_{xx} + \sigma_{yy} \varepsilon_{yy} + \sigma_{zz} \varepsilon_{zz} + \tau_{yz} \gamma_{yz} + \tau_{xz} \gamma_{xz} + \tau_{xy} \gamma_{xy} \right). \tag{2.120}$$

2.4.4 Complementary Strain Energy

We want to assume that in the case of elastic material behavior as considered here there is a clearly reversible relationship between the stresses σ_{ij} and the strains ε_{ij}. At this point, we introduce the specific complementary energy or complementary strain energy density $\bar{U}_0$ in index notation as follows:

$$\bar{U}_0 = \sigma_{ij} \varepsilon_{ij} - U_0 = \int_0^\sigma \varepsilon_{ij} d\hat{\sigma}_{ij}. \tag{2.121}$$

As in the case of the strain energy density U_0, $\bar{U}_0$ is independent of the load history so that the increment $\varepsilon_{ij} d\sigma_{ij}$ represents a complete differential:

$$\varepsilon_{ij} d\sigma_{ij} = d\bar{U}_0 = \frac{\partial \bar{U}_0}{\partial \sigma_{ij}} d\sigma_{ij}. \tag{2.122}$$

Thus, the strains ε_{ij} are obtained as the partial derivatives of the specific complementary energy $\bar{U}_0$ with respect to the stress components:

$$\varepsilon_{ij} = \frac{\partial \bar{U}_0}{\partial \sigma_{ij}}. \tag{2.123}$$

It can also easily be shown that in the case of linear elasticity the following holds:

$$\bar{U}_0 = U_0 = \frac{1}{2} \left(\sigma_{xx}\varepsilon_{xx} + \sigma_{yy}\varepsilon_{yy} + \sigma_{zz}\varepsilon_{zz} + \tau_{yz}\gamma_{yz} + \tau_{xz}\gamma_{xz} + \tau_{xy}\gamma_{xy} \right). \tag{2.124}$$

2.5 Boundary Value Problems

In the previous sections we have provided all equations necessary for describing an arbitrary problem of three-dimensional elasticity theory. If the case in question is linear elasticity (which is commonly addressed as material linearity), and if the considerations are limited to the case of small changes of shape and here especially small strains (which is commonly summarized as geometric linearity), then the basic equations of such a problem can be summarized as follows. First the local equilibrium conditions have to be fulfilled:

$$\frac{\partial \sigma_{xx}}{\partial x} + \frac{\partial \tau_{xy}}{\partial y} + \frac{\partial \tau_{xz}}{\partial z} + f_x = 0,$$

$$\frac{\partial \tau_{xy}}{\partial x} + \frac{\partial \sigma_{yy}}{\partial y} + \frac{\partial \tau_{yz}}{\partial z} + f_y = 0,$$

$$\frac{\partial \tau_{xz}}{\partial x} + \frac{\partial \tau_{yz}}{\partial y} + \frac{\partial \sigma_{zz}}{\partial z} + f_z = 0. \tag{2.125}$$

In index notation we have:

$$\frac{\partial \sigma_{ij}}{\partial x_j} + f_i = 0. \tag{2.126}$$

Further, the kinematic equations hold which establish relations between displacements and strains:

$$\varepsilon_{xx} = \frac{\partial u}{\partial x}, \quad \varepsilon_{yy} = \frac{\partial v}{\partial y}, \quad \varepsilon_{zz} = \frac{\partial w}{\partial z},$$

$$\gamma_{xy} = \frac{\partial u}{\partial y} + \frac{\partial v}{\partial x}, \quad \gamma_{xz} = \frac{\partial u}{\partial z} + \frac{\partial w}{\partial x}, \quad \gamma_{yz} = \frac{\partial v}{\partial z} + \frac{\partial w}{\partial y}, \qquad (2.127)$$

respectively in index notation:

$$\varepsilon_{ij} = \frac{1}{2} \left(\frac{\partial u_i}{\partial x_j} + \frac{\partial u_j}{\partial x_i} \right). \qquad (2.128)$$

Linear elastic material behavior is described by the generalized Hooke's Law:

$$\begin{pmatrix} \sigma_{xx} \\ \sigma_{yy} \\ \sigma_{zz} \\ \sigma_{yz} \\ \sigma_{xz} \\ \sigma_{xy} \end{pmatrix} = \begin{bmatrix} C_{11} & C_{12} & C_{13} & C_{14} & C_{15} & C_{16} \\ C_{21} & C_{22} & C_{23} & C_{24} & C_{25} & C_{26} \\ C_{31} & C_{32} & C_{33} & C_{34} & C_{35} & C_{36} \\ C_{41} & C_{42} & C_{43} & C_{44} & C_{45} & C_{46} \\ C_{51} & C_{52} & C_{53} & C_{54} & C_{55} & C_{56} \\ C_{61} & C_{62} & C_{63} & C_{64} & C_{65} & C_{66} \end{bmatrix} \begin{pmatrix} \varepsilon_{xx} \\ \varepsilon_{yy} \\ \varepsilon_{zz} \\ \gamma_{yz} \\ \gamma_{xz} \\ \gamma_{xy} \end{pmatrix}. \qquad (2.129)$$

This can be represented in index notation as follows:

$$\sigma_{ij} = C_{ijkl} \varepsilon_{kl}. \qquad (2.130)$$

This set of equations represents the governing equations of a given problem of three-dimensional linear theory of elasticity and describes the distribution of the state variables or field variables - i.e. the displacements u_i, the strains ε_{ij} and the stresses σ_{ij} - inside the solid under consideration. The governing equations thus represent 15 equations for the 15 unknown quantities u_i, ε_{ij} and σ_{ij}.

However, the above set of equations is not sufficient to describe a given problem of three-dimensional elasticity theory completely. Rather, statements about given boundary values for the state variables in question must be given and suitably considered in the analysis. These are the so-called boundary conditions of a given elasticity problem. If we designate the entire surface of the considered solid as ∂V, then boundary conditions are usually formulated in such a way that on a part ∂V_t of the body surface ∂V loads are given in the form of the stress vector. On ∂V_t $\underline{t} = \underline{t}_0$ applies if $\underline{t}_0$ is the stress vector given on ∂V_t. On the remaining surface ∂V_u values for the displacements are then given, thus $u = u_0$, $v = v_0$, $w = w_0$, where u_0, v_0, w_0 are prescribed displacements. The sum of the two partial surfaces ∂V_t and ∂V_u again results in the total body surface ∂V, i.e. $\partial V_t \cup \partial V_u = \partial V$. A boundary value problem can then be specified as follows. On the partial surface ∂V_t the following applies:

$$\sigma_{xx} n_x + \tau_{xy} n_y + \tau_{xz} n_z = t_{x0},$$
$$\tau_{xy} n_x + \sigma_{yy} n_y + \tau_{yz} n_z = t_{y0},$$
$$\tau_{xz} n_x + \tau_{yz} n_y + \sigma_{zz} n_z = t_{z0}, \qquad (2.131)$$

or in index notation:

$$\sigma_{ij} n_j = t_{i0}. \tag{2.132}$$

On the partial surface ∂V_u we have:

$$u = u_0, \quad v = v_0, \quad w = w_0. \tag{2.133}$$

In addition to the kinematic relations, the compatibility equations (2.86) or (2.87) apply, depending on the requirements.

Boundary value problems can be divided into three different classes. The so-called first boundary value problem describes a problem with stress boundary conditions in the form of surface loads t_{x0}, t_{y0}, t_{z0} on the entire surface of the body. In this case $\partial V_t = \partial V$ applies. With the second boundary value problem, however, displacement boundary conditions in the form of the displacements u_0, v_0, w_0 are present on the entire body surface, thus $\partial V_u = \partial V$. The third boundary value problem describes a combination of the two aforementioned classes, i.e., if stress boundary conditions are prescribed on a partial surface ∂V_t and displacement boundary conditions on the partial surface ∂V_u. This is also called a mixed boundary value problem.

If a linear-elastic material and a geometrically linear problem is given, the governing equations given in this section are linear equations with constant coefficients. Thus the superposition principle applies, meaning that with two solutions $\sigma_{ij}^{(1)}, \varepsilon_{ij}^{(1)}$, $u_i^{(1)}$ and $\sigma_{ij}^{(2)}, \varepsilon_{ij}^{(2)}, u_i^{(2)}$ of a given boundary value problem, the linear combination $C_1 \sigma_{ij}^{(1)} + C_2 \sigma_{ij}^{(2)}, C_1 \varepsilon_{ij}^{(1)} + C_2 \varepsilon_{ij}^{(2)}, C_1 u_i^{(1)} + C_2 u_i^{(2)}$ is also a solution of the given problem.

2.6 Material Symmetries

A linear-elastic material whose stiffness matrix $\underline{\underline{C}}$ is fully populated (see Eq. (2.101) exhibits pronounced coupling effects. Accordingly, any strain component can cause any stress component. The same is true in reverse if we consider the inverted form of the material law (2.101) and find a fully occupied compliance matrix $\underline{\underline{S}}$. On the other hand, many materials relevant for lightweight engineering show significantly less complex properties, and we can use certain advantageous properties of common materials, namely the so-called material symmetries where it can be shown that in many cases certain entries in $\underline{\underline{C}}$ and $\underline{\underline{S}}$ can be omitted. Nevertheless, anisotropy properties remain for many materials, as we will show in the further course of this section.

In the following we will consider a material that has three distinct principal directions (the so-called material principal axes), e.g. a fiber reinforced plastic, as shown in Fig. 2.11. Such a material obviously exhibits certain anisotropic properties. In this section, we will always designate the principal material axes as x_1, x_2, x_3 unless otherwise mentioned. In the case of a material that exhibits particularly pronounced

Fig. 2.11 Fiber reinforced
plastic

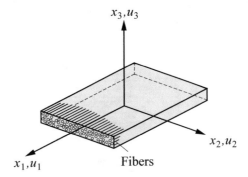

Fibers

properties in one direction (such as unidirectionally reinforced fiber composite materials that exhibit particularly advantageous properties in the fiber direction), this particular preferred direction is defined as the x_1−axis. The displacements associated with x_1, x_2, x_3 are u_1, u_2, u_3.

2.6.1 Full Anisotropy

In the case that both the stiffness matrix $\underline{\underline{C}}$ and the compliance matrix $\underline{\underline{S}}$ are fully populated, we speak of the so-called full anisotropy. All stiffnesses C_{ij} and flexibilities S_{ij} occur, none of those quantities assume zero values in this case. The generalized Hooke's law then reads:

$$
\begin{pmatrix} \sigma_{11} \\ \sigma_{22} \\ \sigma_{33} \\ \tau_{23} \\ \tau_{13} \\ \tau_{12} \end{pmatrix} = \begin{bmatrix} C_{11} & C_{12} & C_{13} & C_{14} & C_{15} & C_{16} \\ C_{12} & C_{22} & C_{23} & C_{24} & C_{25} & C_{26} \\ C_{13} & C_{23} & C_{33} & C_{34} & C_{35} & C_{36} \\ C_{14} & C_{24} & C_{34} & C_{44} & C_{45} & C_{46} \\ C_{15} & C_{25} & C_{35} & C_{45} & C_{55} & C_{56} \\ C_{16} & C_{26} & C_{36} & C_{46} & C_{56} & C_{66} \end{bmatrix} \begin{pmatrix} \varepsilon_{11} \\ \varepsilon_{22} \\ \varepsilon_{33} \\ \gamma_{23} \\ \gamma_{13} \\ \gamma_{12} \end{pmatrix} , \qquad (2.134)
$$

and

$$
\begin{pmatrix} \varepsilon_{11} \\ \varepsilon_{22} \\ \varepsilon_{33} \\ \gamma_{23} \\ \gamma_{13} \\ \gamma_{12} \end{pmatrix} = \begin{bmatrix} S_{11} & S_{12} & S_{13} & S_{14} & S_{15} & S_{16} \\ S_{12} & S_{22} & S_{23} & S_{24} & S_{25} & S_{26} \\ S_{13} & S_{23} & S_{33} & S_{34} & S_{35} & S_{36} \\ S_{14} & S_{24} & S_{34} & S_{44} & S_{45} & S_{46} \\ S_{15} & S_{25} & S_{35} & S_{45} & S_{55} & S_{56} \\ S_{16} & S_{26} & S_{36} & S_{46} & S_{56} & S_{66} \end{bmatrix} \begin{pmatrix} \sigma_{11} \\ \sigma_{22} \\ \sigma_{33} \\ \tau_{23} \\ \tau_{13} \\ \tau_{12} \end{pmatrix} . \qquad (2.135)
$$

In this very general case there are 21 independent material constants in total. For the shear stresses and the shear strains we have used the designations τ_{ij} and γ_{ij} respectively. Apparently, complete anisotropy entails an extremely complex relation

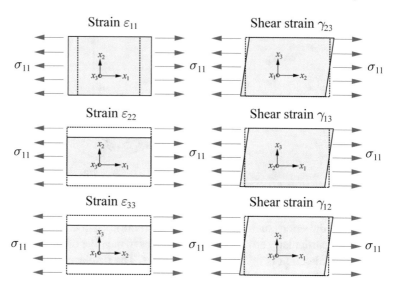

Fig. 2.12 Coupling effects in a fully anisotropic material

between stresses and strains in the form of the so-called coupling effects. This is shown in Fig. 2.12. The figure shows a cube separated from a completely anisotropic material under the normal stress σ_{11}. Obviously, according to the material law (2.135) the applied stress component σ_{11} causes all strain components ε_{11}, ε_{22}, ε_{33}, γ_{23}, γ_{13} and γ_{12}. However, many lightweight materials show a much less complex behavior due to certain symmetry properties which we will discuss in more detail below. Symmetry properties refer to rotations of the reference system where no changes in material behavior occur which eventually leads to the disappearance of certain stiffnesses C_{ij} and compliances S_{ij}.

2.6.2 Monotropy

A monoclinic or monotropic material is a material that has a single plane of symmetry. As an example, let the $x_1 x_2$−plane be a symmetry plane of a given material. Then the material properties are mirror-symmetrical with respect to the plane $x_3 = 0$. This is shown in Fig. 2.13. In contrast to the previous explanations, for the moment the axes x_1, x_2, and x_3 are not the principal axes of the material but are global reference axes that do not necessarily coincide with the principal axes of the material. If, for example, a fiber-reinforced plastic is given as shown in Fig. 2.13, the fibers can enclose any arbitrary angle with the x_1−axis. The presence of a single plane of symmetry means that the material properties are invariant with respect to a mirroring of the x_3−axis on the $x_1 x_2$−plane, i.e. a transformation of the type $x_1 = \bar{x}_1$, $x_2 = \bar{x}_2$,

Fig. 2.13 Monotropic
material

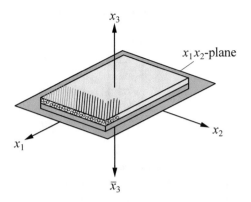

$x_3 = -\bar{x}_3$, and thus do not change due to this mirroring. However, the signs of the shear strains γ_{23} and γ_{13} and the shear stresses τ_{23} and τ_{13} are reversed, so that $\bar{\gamma}_{23} = -\gamma_{23}$, $\bar{\gamma}_{13} = -\gamma_{13}$ and $\bar{\tau}_{23} = -\tau_{23}$, $\bar{\tau}_{13} = -\tau_{13}$. All other stress and strain components remain unchanged by this transformation. From the fourth and the fifth line of the generalized Hooke's Law (2.134) then follows:

$$\tau_{23} = C_{14}\varepsilon_{11} + C_{24}\varepsilon_{22} + C_{34}\varepsilon_{33} + C_{44}\gamma_{23} + C_{45}\gamma_{13} + C_{46}\gamma_{12},$$
$$\tau_{13} = C_{15}\varepsilon_{11} + C_{25}\varepsilon_{22} + C_{35}\varepsilon_{33} + C_{45}\gamma_{23} + C_{55}\gamma_{13} + C_{56}\gamma_{12}. \quad (2.136)$$

Using the mirrored reference system $\bar{x}_1, \bar{x}_2, \bar{x}_3$ on the other hand yields

$$\bar{\tau}_{23} = -C_{14}\varepsilon_{11} - C_{24}\varepsilon_{22} - C_{34}\varepsilon_{33} + C_{44}\gamma_{23} + C_{45}\gamma_{13} - C_{46}\gamma_{12},$$
$$\bar{\tau}_{13} = -C_{15}\varepsilon_{11} - C_{25}\varepsilon_{22} - C_{35}\varepsilon_{33} + C_{45}\gamma_{23} + C_{55}\gamma_{13} - C_{56}\gamma_{12}. \quad (2.137)$$

However, a constitutive law has to be invariant to a change of the reference system as well as to a mirroring as performed above so that the differences between (2.136) and (2.137) can only be eliminated if the following stiffnesses disappear:

$$C_{14} = C_{24} = C_{34} = C_{46} = C_{15} = C_{25} = C_{35} = C_{56} = 0. \quad (2.138)$$

The generalized Hooke's law for a monoclinic material is therefore:

$$\begin{pmatrix} \sigma_{11} \\ \sigma_{22} \\ \sigma_{33} \\ \tau_{23} \\ \tau_{13} \\ \tau_{12} \end{pmatrix} = \begin{bmatrix} C_{11} & C_{12} & C_{13} & 0 & 0 & C_{16} \\ C_{12} & C_{22} & C_{23} & 0 & 0 & C_{26} \\ C_{13} & C_{23} & C_{33} & 0 & 0 & C_{36} \\ 0 & 0 & 0 & C_{44} & C_{45} & 0 \\ 0 & 0 & 0 & C_{45} & C_{55} & 0 \\ C_{16} & C_{26} & C_{36} & 0 & 0 & C_{66} \end{bmatrix} \begin{pmatrix} \varepsilon_{11} \\ \varepsilon_{22} \\ \varepsilon_{33} \\ \gamma_{23} \\ \gamma_{13} \\ \gamma_{12} \end{pmatrix}. \quad (2.139)$$

The inverted form reads:

$$
\begin{pmatrix} \varepsilon_{11} \\ \varepsilon_{22} \\ \varepsilon_{33} \\ \gamma_{23} \\ \gamma_{13} \\ \gamma_{12} \end{pmatrix} = \begin{bmatrix} S_{11} & S_{12} & S_{13} & 0 & 0 & S_{16} \\ S_{12} & S_{22} & S_{23} & 0 & 0 & S_{26} \\ S_{13} & S_{23} & S_{33} & 0 & 0 & S_{36} \\ 0 & 0 & 0 & S_{44} & S_{45} & 0 \\ 0 & 0 & 0 & S_{45} & S_{55} & 0 \\ S_{16} & S_{26} & S_{36} & 0 & 0 & S_{66} \end{bmatrix} \begin{pmatrix} \sigma_{11} \\ \sigma_{22} \\ \sigma_{33} \\ \tau_{23} \\ \tau_{13} \\ \tau_{12} \end{pmatrix} . \tag{2.140}
$$

Examples of monoclinic material behavior are found with fibrous materials such as wood or unidirectional fiber-reinforced plastics whose fiber direction does not coincide with one of the reference axes. Monoclinic material behavior is obviously described by 13 material constants. It can be seen that with monoclinic material some coupling effects remain. Thus, in addition to the coupling of the normal stresses σ_{11}, σ_{22} and σ_{33} with the three strains ε_{11}, ε_{22} and ε_{33} the connection with the shear strain γ_{12} remains, which, however, can be explained quite clearly by looking at a mono-clinic material using the example of a unidirectional fiber-reinforced plastic under a plane tensile load σ_{11}. The deformation process shows not only the immediately obvious displacements in x_1- and x_2-direction and the associated strains ε_{11} and ε_{22} but also a change in the cuboid shape into the shape of a parallelepiped, which indicates an additional shear strain γ_{12}. Its occurrence can be explained immediately by the different stiffnesses parallel to the fibers and in the transverse direction. We refer to this effect as the so-called shear coupling. In addition, the constitutive law (2.139) and (2.140) shows a coupling between the two shear stresses τ_{23} and τ_{13} with both shear strains γ_{23} and γ_{13} which is another special feature of monoclinic material behavior.

2.6.3   Orthogonal Anisotropy/Orthotropy

An important case of anisotropy is the so-called orthogonal anisotropy, or orthotropy (see Fig. 2.14). Orthotropy assumes that the considered material exhibits mirror symmetries with respect to the planes $x_1 = 0$, $x_2 = 0$, $x_3 = 0$, where from this point on the axes x_1, x_2, x_3 are the principal material axes. It can be shown quite simply that for this special case of anisotropy further stiffnesses in the generalized Hooke's law have to become zero as follows:

$$
C_{14} = C_{24} = C_{34} = C_{46} = C_{15} = C_{25} = C_{35} = C_{56} = C_{16} = C_{26} = C_{36} = C_{45} = 0. \tag{2.141}
$$

The generalized Hooke's law then takes on the following form:

Fig. 2.14 Orthotropic material

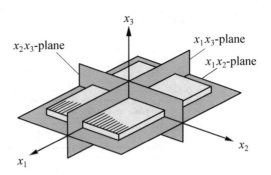

$$
\begin{pmatrix} \sigma_{11} \\ \sigma_{22} \\ \sigma_{33} \\ \tau_{23} \\ \tau_{13} \\ \tau_{12} \end{pmatrix}
=
\begin{bmatrix}
C_{11} & C_{12} & C_{13} & 0 & 0 & 0 \\
C_{12} & C_{22} & C_{23} & 0 & 0 & 0 \\
C_{13} & C_{23} & C_{33} & 0 & 0 & 0 \\
0 & 0 & 0 & C_{44} & 0 & 0 \\
0 & 0 & 0 & 0 & C_{55} & 0 \\
0 & 0 & 0 & 0 & 0 & C_{66}
\end{bmatrix}
\begin{pmatrix} \varepsilon_{11} \\ \varepsilon_{22} \\ \varepsilon_{33} \\ \gamma_{23} \\ \gamma_{13} \\ \gamma_{12} \end{pmatrix},
\tag{2.142}
$$

and in its inverted form:

$$
\begin{pmatrix} \varepsilon_{11} \\ \varepsilon_{22} \\ \varepsilon_{33} \\ \gamma_{23} \\ \gamma_{13} \\ \gamma_{12} \end{pmatrix}
=
\begin{bmatrix}
S_{11} & S_{12} & S_{13} & 0 & 0 & 0 \\
S_{12} & S_{22} & S_{23} & 0 & 0 & 0 \\
S_{13} & S_{23} & S_{33} & 0 & 0 & 0 \\
0 & 0 & 0 & S_{44} & 0 & 0 \\
0 & 0 & 0 & 0 & S_{55} & 0 \\
0 & 0 & 0 & 0 & 0 & S_{66}
\end{bmatrix}
\begin{pmatrix} \sigma_{11} \\ \sigma_{22} \\ \sigma_{33} \\ \tau_{23} \\ \tau_{13} \\ \tau_{12} \end{pmatrix}.
\tag{2.143}
$$

The complete description of orthotropic material behavior requires the knowledge of nine independent material parameters. Important examples for orthotropic materials are fiber-reinforced plastics, e.g. with unidirectional fiber reinforcement, but also the natural material wood or reinforced concrete. A prerequisite for this is that the fiber or reinforcement direction is oriented in the direction of one of the reference axes. The assignment of the material matrices in (2.142) and (2.143) is considerably simplified compared to the cases of complete anisotropy and monotropy. The previously discussed coupling effects do not occur in the case of orthotropic material behavior. However, it is important to note that the remaining entries of the matrices in (2.142) and (2.143) are independent of each other and thus generally different. Therefore, orthotropy still has a direction-dependent behavior, even if coupling effects do not occur here.

For orthotropy the following relations apply between the stiffnesses C_{ij} and the compliances S_{ij}:

$$C_{11} = \frac{S_{23}^2 - S_{22}S_{33}}{S_{11}S_{23}^2 - S_{11}S_{22}S_{33} - 2S_{12}S_{13}S_{23} + S_{22}S_{13}^2 + S_{33}S_{12}^2},$$

$$C_{22} = \frac{S_{13}^2 - S_{11}S_{33}}{S_{11}S_{23}^2 - S_{11}S_{22}S_{33} - 2S_{12}S_{13}S_{23} + S_{22}S_{13}^2 + S_{33}S_{12}^2},$$

$$C_{33} = \frac{S_{12}^2 - S_{11}S_{22}}{S_{11}S_{23}^2 - S_{11}S_{22}S_{33} - 2S_{12}S_{13}S_{23} + S_{22}S_{13}^2 + S_{33}S_{12}^2},$$

$$C_{12} = \frac{S_{12}S_{33} - S_{13}S_{23}}{S_{11}S_{23}^2 - S_{11}S_{22}S_{33} - 2S_{12}S_{13}S_{23} + S_{22}S_{13}^2 + S_{33}S_{12}^2},$$

$$C_{13} = \frac{S_{13}S_{22} - S_{12}S_{23}}{S_{11}S_{23}^2 - S_{11}S_{22}S_{33} - 2S_{12}S_{13}S_{23} + S_{22}S_{13}^2 + S_{33}S_{12}^2},$$

$$C_{23} = \frac{S_{11}S_{23} - S_{12}S_{13}}{S_{11}S_{23}^2 - S_{11}S_{22}S_{33} - 2S_{12}S_{13}S_{23} + S_{22}S_{13}^2 + S_{33}S_{12}^2},$$

$$C_{44} = \frac{1}{S_{44}}, \quad C_{55} = \frac{1}{S_{55}}, \quad C_{66} = \frac{1}{S_{66}}. \tag{2.144}$$

2.6.4 Transverse Isotropy

The so-called transverse isotropy is present when the material under consideration has a preferred direction (here the x_1−axis) and isotropy is present with respect to a plane, e.g. with respect to the x_2x_3−plane. Such a case is shown in Fig. 2.15. Transverse isotropy is thus expressed by the fact that the material behavior is invariant to a rotation of the reference system around the principal axis x_1, which is equivalent to the fact that the considered material behaves identically in any direction perpendicular to the x_1− axis which explains the term transverse isotropy. Every plane that includes the x_1−axis is thus a plane of symmetry. This case can be applied, assuming an even distribution of the fibers, e.g. for many unidirectionally reinforced plastics if x_1 indicates the fiber direction. The assignment of the stiffness matrix $\underline{\underline{C}}$ and the

Fig. 2.15 Transverse isotropy

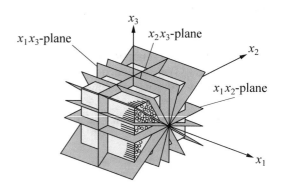

compliance matrix $\underline{\underline{S}}$ is identical to the case of orthotropy, but the following identities can additionally be derived for the stiffnesses C_{ij}:

$$C_{13} = C_{12}, \quad C_{33} = C_{22}, \quad C_{66} = C_{55}, \quad C_{44} = \frac{1}{2}\left(C_{22} - C_{23}\right). \tag{2.145}$$

Hooke's generalized law then reads:

$$
\begin{pmatrix} \sigma_{11} \\ \sigma_{22} \\ \sigma_{33} \\ \tau_{23} \\ \tau_{13} \\ \tau_{12} \end{pmatrix}
=
\begin{bmatrix}
C_{11} & C_{12} & C_{12} & 0 & 0 & 0 \\
C_{12} & C_{22} & C_{23} & 0 & 0 & 0 \\
C_{12} & C_{23} & C_{22} & 0 & 0 & 0 \\
0 & 0 & 0 & \frac{1}{2}(C_{22}-C_{23}) & 0 & 0 \\
0 & 0 & 0 & 0 & C_{55} & 0 \\
0 & 0 & 0 & 0 & 0 & C_{55}
\end{bmatrix}
\begin{pmatrix} \varepsilon_{11} \\ \varepsilon_{22} \\ \varepsilon_{33} \\ \gamma_{23} \\ \gamma_{13} \\ \gamma_{12} \end{pmatrix},
\tag{2.146}
$$

and in its inverted form:

$$
\begin{pmatrix} \varepsilon_{11} \\ \varepsilon_{22} \\ \varepsilon_{33} \\ \gamma_{23} \\ \gamma_{13} \\ \gamma_{12} \end{pmatrix}
=
\begin{bmatrix}
S_{11} & S_{12} & S_{12} & 0 & 0 & 0 \\
S_{12} & S_{22} & S_{23} & 0 & 0 & 0 \\
S_{12} & S_{23} & S_{22} & 0 & 0 & 0 \\
0 & 0 & 0 & 2\,(S_{22}-S_{23}) & 0 & 0 \\
0 & 0 & 0 & 0 & S_{55} & 0 \\
0 & 0 & 0 & 0 & 0 & S_{55}
\end{bmatrix}
\begin{pmatrix} \sigma_{11} \\ \sigma_{22} \\ \sigma_{33} \\ \tau_{23} \\ \tau_{13} \\ \tau_{12} \end{pmatrix}.
\tag{2.147}
$$

Hence, with transverse isotropy only five independent material constants remain.

2.6.5 Isotropy

Finally, the case of the isotropy which represents the highest level of material symmetry is to be discussed. In the case of isotropy the material behaves identically in every direction, any axis is also a principal axis, and any plane is a plane of symmetry. The generalized Hooke's law then takes the following form:

$$
\begin{pmatrix} \sigma_{11} \\ \sigma_{22} \\ \sigma_{33} \\ \tau_{23} \\ \tau_{13} \\ \tau_{12} \end{pmatrix}
=
\begin{bmatrix}
C_{11} & C_{12} & C_{12} & 0 & 0 & 0 \\
C_{12} & C_{11} & C_{12} & 0 & 0 & 0 \\
C_{12} & C_{12} & C_{11} & 0 & 0 & 0 \\
0 & 0 & 0 & \frac{1}{2}(C_{11}-C_{12}) & 0 & 0 \\
0 & 0 & 0 & 0 & \frac{1}{2}(C_{11}-C_{12}) & 0 \\
0 & 0 & 0 & 0 & 0 & \frac{1}{2}(C_{11}-C_{12})
\end{bmatrix}
\begin{pmatrix} \varepsilon_{11} \\ \varepsilon_{22} \\ \varepsilon_{33} \\ \gamma_{23} \\ \gamma_{13} \\ \gamma_{12} \end{pmatrix}.
\tag{2.148}
$$

The inverted form reads:

$$\begin{pmatrix} \varepsilon_{11} \\ \varepsilon_{22} \\ \varepsilon_{33} \\ \gamma_{23} \\ \gamma_{13} \\ \gamma_{12} \end{pmatrix} = \begin{bmatrix} S_{11} & S_{12} & S_{12} & 0 & 0 & 0 \\ S_{12} & S_{11} & S_{12} & 0 & 0 & 0 \\ S_{12} & S_{12} & S_{11} & 0 & 0 & 0 \\ 0 & 0 & 0 & 2(S_{11}-S_{12}) & 0 & 0 \\ 0 & 0 & 0 & 0 & 2(S_{11}-S_{12}) & 0 \\ 0 & 0 & 0 & 0 & 0 & 2(S_{11}-S_{12}) \end{bmatrix} \begin{pmatrix} \sigma_{11} \\ \sigma_{22} \\ \sigma_{33} \\ \tau_{23} \\ \tau_{13} \\ \tau_{12} \end{pmatrix}. \quad (2.149)$$

The description of isotropic material behavior then requires only two independent material constants. Materials with isotropic behavior as relevant for lightweight engineering are many different metals and plastics.

2.6.6 Engineering Constants

Besides the use of the stiffnesses C_{ij} and the compliances S_{ij} to describe the behavior of a linear elastic (anisotropic) material, the description by the so-called engineering constants is very common. In the three-dimensional case when considering orthotropy there are 12 engineering constants in total:

- Three generalized moduli of elasticity E_{11}, E_{22}, E_{33},
- Six generalized Poisson's ratios $\nu_{12}, \nu_{13}, \nu_{23}, \nu_{21}, \nu_{31}, \nu_{32}$,
- three generalized shear moduli G_{23}, G_{13}, G_{12}.

The generalized Hooke's Law in its inverted form (2.143) can be represented as follows when using the engineering constants:

$$\begin{pmatrix} \varepsilon_{11} \\ \varepsilon_{22} \\ \varepsilon_{33} \\ \gamma_{23} \\ \gamma_{13} \\ \gamma_{12} \end{pmatrix} = \begin{bmatrix} \dfrac{1}{E_{11}} & -\dfrac{\nu_{21}}{E_{22}} & -\dfrac{\nu_{31}}{E_{33}} & 0 & 0 & 0 \\ -\dfrac{\nu_{12}}{E_{11}} & \dfrac{1}{E_{22}} & -\dfrac{\nu_{32}}{E_{33}} & 0 & 0 & 0 \\ -\dfrac{\nu_{13}}{E_{11}} & -\dfrac{\nu_{23}}{E_{22}} & \dfrac{1}{E_{33}} & 0 & 0 & 0 \\ 0 & 0 & 0 & \dfrac{1}{G_{23}} & 0 & 0 \\ 0 & 0 & 0 & 0 & \dfrac{1}{G_{13}} & 0 \\ 0 & 0 & 0 & 0 & 0 & \dfrac{1}{G_{12}} \end{bmatrix} \begin{pmatrix} \sigma_{11} \\ \sigma_{22} \\ \sigma_{33} \\ \tau_{23} \\ \tau_{13} \\ \tau_{12} \end{pmatrix}. \quad (2.150)$$

The comparison of (2.143) with (2.150) leads to the following relations between the engineering constants and the compliances S_{ij}:

$$S_{11} = \frac{1}{E_{11}}, \quad S_{22} = \frac{1}{E_{22}}, \quad S_{33} = \frac{1}{E_{33}},$$

$$S_{44} = \frac{1}{G_{23}}, \quad S_{55} = \frac{1}{G_{13}}, \quad S_{66} = \frac{1}{G_{12}},$$

$$S_{12} = -\frac{\nu_{12}}{E_{11}} = -\frac{\nu_{21}}{E_{22}}, \quad S_{23} = -\frac{\nu_{23}}{E_{22}} = -\frac{\nu_{32}}{E_{33}}, \quad S_{13} = -\frac{\nu_{31}}{E_{33}} = -\frac{\nu_{13}}{E_{11}}. \quad (2.151)$$

Since the compliance matrix $\underline{\underline{S}}$ must be symmetrical also in the representation in the engineering constants, the following identities can be determined:

$$\frac{\nu_{12}}{E_{11}} = \frac{\nu_{21}}{E_{22}}, \quad \frac{\nu_{23}}{E_{22}} = \frac{\nu_{32}}{E_{33}}, \quad \frac{\nu_{31}}{E_{33}} = \frac{\nu_{13}}{E_{11}}. \tag{2.152}$$

It turns out that only nine of the twelve engineering constants are independent of each other.

The stiffnesses C_{ij} can be expressed by the engineering constants as follows:

$$C_{11} = \frac{(1 - \nu_{23}\nu_{32})E_{11}}{1 - \nu_{12}\nu_{21} - \nu_{23}\nu_{32} - \nu_{31}\nu_{13} - 2\nu_{21}\nu_{13}\nu_{32}},$$

$$C_{22} = \frac{(1 - \nu_{31}\nu_{13})E_{11}}{1 - \nu_{12}\nu_{21} - \nu_{23}\nu_{32} - \nu_{31}\nu_{13} - 2\nu_{21}\nu_{13}\nu_{32}},$$

$$C_{33} = \frac{(1 - \nu_{21}\nu_{12})E_{33}}{1 - \nu_{12}\nu_{21} - \nu_{23}\nu_{32} - \nu_{31}\nu_{13} - 2\nu_{21}\nu_{13}\nu_{32}},$$

$$C_{44} = G_{23}, \quad C_{55} = G_{13}, \quad C_{66} = G_{12},$$

$$C_{12} = \frac{(\nu_{12} + \nu_{32}\nu_{13})E_{22}}{1 - \nu_{12}\nu_{21} - \nu_{23}\nu_{32} - \nu_{31}\nu_{13} - 2\nu_{21}\nu_{13}\nu_{32}}$$

$$= \frac{(\nu_{21} + \nu_{31}\nu_{23})E_{11}}{1 - \nu_{12}\nu_{21} - \nu_{23}\nu_{32} - \nu_{31}\nu_{13} - 2\nu_{21}\nu_{13}\nu_{32}},$$

$$C_{13} = \frac{(\nu_{13} + \nu_{12}\nu_{23})E_{33}}{1 - \nu_{12}\nu_{21} - \nu_{23}\nu_{32} - \nu_{31}\nu_{13} - 2\nu_{21}\nu_{13}\nu_{32}}$$

$$= \frac{(\nu_{31} + \nu_{21}\nu_{32})E_{11}}{1 - \nu_{12}\nu_{21} - \nu_{23}\nu_{32} - \nu_{31}\nu_{13} - 2\nu_{21}\nu_{13}\nu_{32}},$$

$$C_{23} = \frac{(\nu_{23} + \nu_{21}\nu_{13})E_{33}}{1 - \nu_{12}\nu_{21} - \nu_{23}\nu_{32} - \nu_{31}\nu_{13} - 2\nu_{21}\nu_{13}\nu_{32}}$$

$$= \frac{(\nu_{32} + \nu_{12}\nu_{31})E_{11}}{1 - \nu_{12}\nu_{21} - \nu_{23}\nu_{32} - \nu_{31}\nu_{13} - 2\nu_{21}\nu_{13}\nu_{32}}, \tag{2.153}$$

as can be shown directly by inverting (2.150).

In the case of isotropy only three engineering constants remain. These are the modulus of elasticity E, the shear modulus G and the Poisson's number ν where certain dependencies can be identified because only two of these three engineering constants are independent of each other. The elastic compliances S_{ij} can be described as follows:

$$S_{11} = S_{22} = S_{33} = \frac{1}{E},$$

$$S_{12} = S_{13} = S_{23} = -\frac{\nu}{E},$$

$$S_{44} = S_{55} = S_{66} = \frac{1}{G}, \tag{2.154}$$

and for the elastic stiffnesses C_{ij} we have:

$$C_{11} = C_{22} = C_{33} = \frac{(1-\nu)\,E}{(1+\nu)\,(1-2\nu)},$$

$$C_{12} = C_{13} = C_{23} = \frac{\nu E}{(1+\nu)\,(1-2\nu)},$$

$$C_{44} = C_{55} = C_{66} = G. \tag{2.155}$$

The following relation exists between the isotropic engineering constants:

$$G = \frac{E}{2(1+\nu)}, \tag{2.156}$$

so that also when using the engineering constants only two independent material constants remain. The material law (2.150) then takes on the following form:

$$
\begin{pmatrix} \varepsilon_{11} \\ \varepsilon_{22} \\ \varepsilon_{33} \\ \gamma_{23} \\ \gamma_{13} \\ \gamma_{12} \end{pmatrix}
=
\begin{bmatrix}
\dfrac{1}{E} & -\dfrac{\nu}{E} & -\dfrac{\nu}{E} & 0 & 0 & 0 \\[4pt]
-\dfrac{\nu}{E} & \dfrac{1}{E} & -\dfrac{\nu}{E} & 0 & 0 & 0 \\[4pt]
-\dfrac{\nu}{E} & -\dfrac{\nu}{E} & \dfrac{1}{E} & 0 & 0 & 0 \\[4pt]
0 & 0 & 0 & \dfrac{1}{G} & 0 & 0 \\[4pt]
0 & 0 & 0 & 0 & \dfrac{1}{G} & 0 \\[4pt]
0 & 0 & 0 & 0 & 0 & \dfrac{1}{G}
\end{bmatrix}
\begin{pmatrix} \sigma_{11} \\ \sigma_{22} \\ \sigma_{33} \\ \tau_{23} \\ \tau_{13} \\ \tau_{12} \end{pmatrix}.
\tag{2.157}
$$

2.6.7 Value Ranges for the Material Parameters

Certain value ranges can be specified for the essential types of the material symmetries discussed here, both for the stiffnesses C_{ij} and the corresponding engineering constants which we will discuss here in more detail (see also Altenbach et al. (1996)). Starting point of the considerations is the specific strain energy U_0 according to (2.119) which we can write down in a vector-matrix notation for the case of orthotropy as:

$$
U_0 = \frac{1}{2}
\begin{pmatrix} \varepsilon_{11} \\ \varepsilon_{22} \\ \varepsilon_{33} \\ \gamma_{23} \\ \gamma_{13} \\ \gamma_{12} \end{pmatrix}^T
\begin{bmatrix}
C_{11} & C_{12} & C_{13} & 0 & 0 & 0 \\
C_{12} & C_{22} & C_{23} & 0 & 0 & 0 \\
C_{13} & C_{23} & C_{33} & 0 & 0 & 0 \\
0 & 0 & 0 & C_{44} & 0 & 0 \\
0 & 0 & 0 & 0 & C_{55} & 0 \\
0 & 0 & 0 & 0 & 0 & C_{66}
\end{bmatrix}
\begin{pmatrix} \varepsilon_{11} \\ \varepsilon_{22} \\ \varepsilon_{33} \\ \gamma_{23} \\ \gamma_{13} \\ \gamma_{12} \end{pmatrix}.
\tag{2.158}
$$

The specific strain energy U_0 is a positive definite function. Thus, it only assumes the value zero if all strain components become zero. In all other cases U_0 always has a positive value.

We now perform a thought experiment and set all strain components to zero except the normal strain ε_{11}. Then we get for U_0:

$$U_0 = \frac{1}{2} C_{11} \varepsilon_{11}^2. \tag{2.159}$$

Since U_0 must always be greater than zero, regardless of the present value for ε_{11} the stiffness C_{11} must also be greater than zero: $C_{11} > 0$. If this thought experiment is repeated successively for all other strain components we find that this statement must apply to all stiffnesses C_{ii} (with $i = 1, 2, ..., 6$), i.e:

$$C_{ii} > 0, \quad \text{with} \quad i = 1, 2, ..., 6. \tag{2.160}$$

As a result, the stiffness matrix $\underline{\underline{C}}$ is always non-singular, has a positive determinant and can be inverted. The matrix inverse to $\underline{\underline{C}}$, i.e. the compliance matrix $\underline{\underline{S}}$, is then a symmetric matrix and positively definite.

Quite analogously, the thought experiments $\varepsilon_{11} \neq 0, \varepsilon_{22} \neq 0, \varepsilon_{33} = 0$ and $\varepsilon_{11} = 0, \varepsilon_{22} \neq 0, \varepsilon_{33} \neq 0$ and $\varepsilon_{11} \neq 0, \varepsilon_{22} = 0, \varepsilon_{33} \neq 0$ and $\varepsilon_{11} \neq 0, \varepsilon_{22} \neq 0, \varepsilon_{33} \neq 0$ lead to the conclusions that the submatrices

$$\begin{bmatrix} C_{11} & C_{12} \\ C_{12} & C_{22} \end{bmatrix}, \begin{bmatrix} C_{22} & C_{23} \\ C_{23} & C_{33} \end{bmatrix}, \begin{bmatrix} C_{11} & C_{13} \\ C_{13} & C_{33} \end{bmatrix}, \begin{bmatrix} C_{11} & C_{12} & C_{13} \\ C_{12} & C_{22} & C_{23} \\ C_{13} & C_{23} & C_{33} \end{bmatrix} \tag{2.161}$$

each have positive diagonal terms and are not singular. They can be inverted, have a determinant greater than zero and their inverts are symmetric and not singular. Hence:

$$\begin{vmatrix} C_{11} & C_{12} \\ C_{12} & C_{22} \end{vmatrix} = C_{11} C_{22} - C_{12}^2 > 0,$$

$$\begin{vmatrix} C_{22} & C_{23} \\ C_{23} & C_{33} \end{vmatrix} = C_{22} C_{33} - C_{23}^2 > 0,$$

$$\begin{vmatrix} C_{11} & C_{13} \\ C_{13} & C_{33} \end{vmatrix} = C_{11} C_{33} - C_{13}^2 > 0,$$

$$\begin{vmatrix} C_{11} & C_{12} & C_{13} \\ C_{12} & C_{22} & C_{23} \\ C_{13} & C_{23} & C_{33} \end{vmatrix} = C_{11} \begin{vmatrix} C_{22} & C_{23} \\ C_{23} & C_{33} \end{vmatrix} + C_{22} \begin{vmatrix} C_{11} & C_{13} \\ C_{13} & C_{33} \end{vmatrix} + C_{33} \begin{vmatrix} C_{11} & C_{12} \\ C_{12} & C_{22} \end{vmatrix} > 0. \tag{2.162}$$

In the case of transverse isotropy (see Eq. (2.146)) we obtain:

$$C_{11} > 0, \quad C_{22} > 0, \quad C_{55} > 0, \quad C_{22} - C_{23} > 0,$$

$$\begin{vmatrix} C_{11} & C_{12} \\ C_{12} & C_{22} \end{vmatrix} = C_{11}C_{22} - C_{12}^2 > 0,$$

$$\begin{vmatrix} C_{22} & C_{23} \\ C_{23} & C_{22} \end{vmatrix} = C_{22}^2 - C_{23}^2 > 0,$$

$$\begin{vmatrix} C_{11} & C_{12} & C_{12} \\ C_{12} & C_{22} & C_{23} \\ C_{12} & C_{23} & C_{22} \end{vmatrix} = C_{11} \begin{vmatrix} C_{22} & C_{23} \\ C_{23} & C_{22} \end{vmatrix} + 2C_{22} \begin{vmatrix} C_{11} & C_{12} \\ C_{12} & C_{22} \end{vmatrix} > 0. \qquad (2.163)$$

For isotropy we have:

$$C_{11} > 0, \quad C_{11} - C_{12} > 0,$$

$$\begin{vmatrix} C_{11} & C_{12} \\ C_{12} & C_{11} \end{vmatrix} = C_{11}^2 - C_{12}^2 > 0,$$

$$\begin{vmatrix} C_{11} & C_{12} & C_{12} \\ C_{12} & C_{11} & C_{12} \\ C_{12} & C_{12} & C_{11} \end{vmatrix} = 3C_{11} \begin{vmatrix} C_{11} & C_{12} \\ C_{12} & C_{11} \end{vmatrix} > 0. \qquad (2.164)$$

Quite analogous correlations can also be derived for the engineering constants. In the case of orthotropy, both the three generalized elastic moduli E_{11}, E_{22}, E_{33} and the three shear moduli G_{12}, G_{13}, G_{23} are always positive:

$$E_{11} > 0, \quad E_{22} > 0, \quad E_{33} > 0, \quad G_{12} > 0, \quad G_{13} > 0, \quad G_{23} > 0. \quad (2.165)$$

From the representation (2.150) for the generalized Hooke's Law the following can be derived:

$$\nu_{12}^2 < \frac{E_{11}}{E_{22}}, \quad \nu_{21}^2 < \frac{E_{22}}{E_{11}},$$

$$\nu_{23}^2 < \frac{E_{22}}{E_{33}}, \quad \nu_{32}^2 < \frac{E_{33}}{E_{22}},$$

$$\nu_{13}^2 < \frac{E_{11}}{E_{33}}, \quad \nu_{31}^2 < \frac{E_{33}}{E_{11}}. \qquad (2.166)$$

In addition we have:

$$1 - \nu_{12}\nu_{21} - \nu_{23}\nu_{32} - \nu_{13}\nu_{31} - 2\nu_{12}\nu_{13}\nu_{32} > 0. \qquad (2.167)$$

For the case of isotropy with $E > 0$, $G = \frac{E}{2(1+\nu)} > 0$:

$$\begin{vmatrix} \dfrac{1}{E} & -\dfrac{\nu}{E} \\ -\dfrac{\nu}{E} & \dfrac{1}{E} \end{vmatrix} = \frac{1}{E^2}\left(1 - \nu^2\right) > 0. \tag{2.168}$$

This leads to the following conclusion:

$$\nu^2 < 1. \tag{2.169}$$

From

$$\begin{vmatrix} \dfrac{1}{E} & -\dfrac{\nu}{E} & -\dfrac{\nu}{E} \\ -\dfrac{\nu}{E} & \dfrac{1}{E} & -\dfrac{\nu}{E} \\ -\dfrac{\nu}{E} & -\dfrac{\nu}{E} & \dfrac{1}{E} \end{vmatrix} = \frac{1}{E^3}\left(1 - 2\nu\right)\left(1 + \nu^2\right) > 0 \tag{2.170}$$

we obtain

$$\nu < \frac{1}{2}. \tag{2.171}$$

This gives us the following permissible interval for the Poisson's ratio:

$$-1 < \nu < \frac{1}{2}. \tag{2.172}$$

2.6.8 Alternative Representation of Isotropic Materials

In some cases, the use of alternative elasticity quantities for isotropic materials can be advantageous which we will discuss briefly below (see also Becker and Gross (2002)). Isotropic materials behave the same in all directions. This means that the stiffness or elasticity tensor C_{ijkl} must be an isotropic tensor. We want to write C_{ijkl} in the following form:

$$C_{ijkl} = \lambda \delta_{ij}\delta_{kl} + \mu \left(\delta_{ik}\delta_{jl} + \delta_{il}\delta_{jk}\right) + \kappa \left(\delta_{ik}\delta_{jl} - \delta_{il}\delta_{jk}\right). \tag{2.173}$$

Herein δ_{ij} is the so-called Kronecker symbol with the properties $\delta_{ij} = 0$ if $i \neq j$ and $\delta_{ij} = 1$ if $i = j$. The quantities λ, μ and κ are constants which will be discussed in the following.

It can easily be shown that due to the symmetry of the stiffness tensor the third component in (2.173) is omitted so the following expression remains:

$$C_{ijkl} = \lambda \delta_{ij} \delta_{kl} + \mu \left(\delta_{ik} \delta_{jl} + \delta_{il} \delta_{jk} \right). \tag{2.174}$$

Hooke's Law in the form $\sigma_{ij} = C_{ijkl} \varepsilon_{kl}$ then assumes the following form:

$$\sigma_{ij} = \lambda \delta_{ij} \delta_{kl} \varepsilon_{kl} + \mu \left(\delta_{ik} \delta_{jl} + \delta_{il} \delta_{jk} \right) \varepsilon_{kl}, \tag{2.175}$$

respectively

$$\sigma_{ij} = \lambda \varepsilon_{kk} \delta_{ij} + 2\mu \varepsilon_{ij}. \tag{2.176}$$

In fully expanded form:

$$\begin{aligned}
\sigma_{11} &= \lambda \left(\varepsilon_{11} + \varepsilon_{22} + \varepsilon_{33} \right) + 2\mu \varepsilon_{11}, \\
\sigma_{22} &= \lambda \left(\varepsilon_{11} + \varepsilon_{22} + \varepsilon_{33} \right) + 2\mu \varepsilon_{22}, \\
\sigma_{33} &= \lambda \left(\varepsilon_{11} + \varepsilon_{22} + \varepsilon_{33} \right) + 2\mu \varepsilon_{33}, \\
\sigma_{12} &= 2\mu \varepsilon_{12}, \\
\sigma_{13} &= 2\mu \varepsilon_{13}, \\
\sigma_{23} &= 2\mu \varepsilon_{23}.
\end{aligned} \tag{2.177}$$

The two elastic constants remaining here are referred to as the so-called Lamé[11] constants. We now decompose the strain state into a pure volumetric part and a part that describes a pure shape change:

$$\varepsilon_{ij} = \frac{1}{3} \varepsilon_{kk} \delta_{ij} + e_{ij}. \tag{2.178}$$

Herein, the first term is the spherical tensor of the strain state, and the second term describes the deviator. Inserting into Hooke's law (2.176) then results in

$$\sigma_{ij} = \left(\lambda + \frac{2}{3} \mu \right) \varepsilon_{kk} \delta_{ij} + 2\mu e_{ij}. \tag{2.179}$$

Thus the first term on the right side of this equation can be understood as the spherical tensor with $\frac{1}{3} \sigma_{kk} \delta_{ij}$. The second term again is the deviational part and can be interpreted as s_{ij}. We now introduce the new elasticity quantity

$$K = \lambda + \frac{2}{3} \mu \tag{2.180}$$

---

[11] Gabriel Lamé, 1795–1870, French physicist.

which is commonly addressed as the compression modulus and obtain:

$$\sigma_{kk} = 3K\varepsilon_{kk}, \quad s_{ij} = 2\mu e_{ij}. \tag{2.181}$$

These terms can be understood as follows. The first term in (2.181) describes the material behavior in the case of a pure volume change. The second term, on the other hand, characterizes the material behavior in the case of a pure shape change.

The relations (2.181) can also be inverted so that the strains can be expressed by the stresses. It follows:

$$
\begin{aligned}
\varepsilon_{ij} &= \frac{1}{3}\varepsilon_{kk}\delta_{ij} + e_{ij} \\
&= \frac{1}{9K}\sigma_{kk}\delta_{ij} + \frac{1}{2\mu}s_{ij} \\
&= \frac{1}{3(3\lambda + 2\mu)}\sigma_{kk}\delta_{ij} + \frac{1}{2\mu}\left(\sigma_{ij} - \frac{1}{3}\sigma_{kk}\delta_{ij}\right) \\
&= -\frac{\lambda}{2\mu(3\lambda + 2\mu)}\sigma_{kk}\delta_{ij} + \frac{1}{2\mu}\sigma_{ij}.
\end{aligned}
\tag{2.182}
$$

We want to consider the special case of uniaxial stress a little closer from which we can extract relations between the Lamé constants λ and μ and the engineering constants E and ν. Assume that the uniaxial tensile stress state with the non-vanishing stress σ_{11} is present. Then ε_{11} and ε_{22} result as follows:

$$
\begin{aligned}
\varepsilon_{11} &= -\frac{\lambda}{2\mu(3\lambda + 2\mu)}\sigma_{11} + \frac{1}{2\mu}\sigma_{11} = \frac{\lambda + \mu}{\mu(3\lambda + 2\mu)}\sigma_{11}, \\
\varepsilon_{22} &= -\frac{\lambda}{2\mu(3\lambda + 2\mu)}\sigma_{11}.
\end{aligned}
\tag{2.183}
$$

On the other hand for the pure tensile stress state when using the engineering constants E and ν we have:

$$\varepsilon_{11} = \frac{1}{E}\sigma_{11}, \quad \varepsilon_{22} = -\frac{\nu}{E}\sigma_{11}. \tag{2.184}$$

The comparison between (2.183) and (2.184) then results in the desired connection between the Lamé constants λ and μ and the engineering constants E and ν:

$$E = \frac{\mu(3\lambda + 2\mu)}{\lambda + \mu}, \quad \nu = \frac{\lambda}{2(\lambda + \mu)}. \tag{2.185}$$

We now consider a pure shear stress state with the non-vanishing shear stress σ_{12}:

$$\varepsilon_{12} = \frac{1}{2\mu}\sigma_{12}, \tag{2.186}$$

or when using the shear modulus G:

$$\varepsilon_{12} = \frac{1}{2G}\sigma_{12}. \tag{2.187}$$

Comparison between (2.186) and (2.187) yields:

$$G = \mu. \tag{2.188}$$

The inverted relations to (2.185) and (2.188) read:

$$\lambda = \frac{\nu E}{(1+\nu)(1-2\nu)}, \quad \mu = G = \frac{E}{2(1+\nu)}. \tag{2.189}$$

We can also write the compression modulus K according to (2.180) as:

$$K = \lambda + \frac{2}{3}\mu = \frac{E}{3(1-2\nu)}. \tag{2.190}$$

Hooke's law is obtained when using the engineering constants as

$$\varepsilon_{ij} = -\frac{\nu}{E}\sigma_{kk}\delta_{ij} + \frac{1+\nu}{E}\sigma_{ij}, \tag{2.191}$$

respectively in expanded form:

$$\begin{aligned}
E\varepsilon_{11} &= \sigma_{11} - \nu(\sigma_{22} + \sigma_{33}), \\
E\varepsilon_{22} &= \sigma_{22} - \nu(\sigma_{11} + \sigma_{33}), \\
E\varepsilon_{33} &= \sigma_{33} - \nu(\sigma_{11} + \sigma_{22}), \\
\gamma_{12} &= \frac{2(1+\nu)}{E}\sigma_{12} = \frac{1}{G}\sigma_{12}, \\
\gamma_{13} &= \frac{2(1+\nu)}{E}\sigma_{13} = \frac{1}{G}\sigma_{13}, \\
\gamma_{23} &= \frac{2(1+\nu)}{E}\sigma_{23} = \frac{1}{G}\sigma_{23}.
\end{aligned} \tag{2.192}$$

At this point it is useful to consider again the specific strain energy $U_0 = \frac{1}{2}\sigma_{ij}\varepsilon_{ij}$:

$$\begin{aligned}
U_0 &= \frac{1}{2}\left(\frac{1}{3}\sigma_{kk}\delta_{ij} + s_{ij}\right)\left(\frac{1}{3}\varepsilon_{kk}\delta_{ij} + e_{ij}\right) \\
&= \frac{1}{6}\sigma_{kk}\varepsilon_{ll} + \frac{1}{2}s_{ij}e_{ij} \\
&= \frac{1}{2}K\varepsilon_{kk}^2 + Ge_{ij}e_{ij}.
\end{aligned} \tag{2.193}$$

This result can be interpreted as follows. The first term of the last line in (2.193) can be interpreted as the so-called volume change energy U_V, i.e:

$$U_V = \frac{1}{2} K \varepsilon_{kk}^2 = \frac{1}{18K} \sigma_{kk}^2.$$

(2.194)

The second term in (2.193) can be interpreted as the so-called shape change energy U_S, that is:

$$U_S = G e_{ij} e_{ij} = \frac{1}{4G} s_{ij} s_{ij}.$$

(2.195)

Since the specific strain energy is positively definite, this must also apply to both U_V and U_S. This means that $U_V > 0$ applies if $\varepsilon_{kk} \neq 0$. Additionally $U_S > 0$ is valid if all $e_{ij} \neq 0$. This can only be true if the following conditions are met:

$$K = \frac{E}{3(1 - 2\nu)} > 0, \quad G = \frac{E}{2(1 + \nu)} > 0.$$

(2.196)

This results in the restrictions of the value ranges for the modulus of elasticity E and the Poisson's ratio ν already determined in the previous section:

$$E > 0, \quad -1 \leq \nu \leq \frac{1}{2}.$$

(2.197)

2.7 Transformation Rules

So far we have assumed that all state variables - stresses, strains and displacements as well as the material constants S_{ij} and C_{ij} - are related to the material principal axes x_1, x_2, x_3 (so-called on-axis-system). We will now consider the question of what changes the above-mentioned quantities undergo when the coordinate system x_1, x_2, x_3 is transformed into another orthogonal coordinate system $\bar{x}_1$, $\bar{x}_2$, $\bar{x}_3$ (so-called off-axis system) by a pure rotation, whereby the origin of both systems is identical and the rotation is performed around the fixed x_3−axis, so that $x_3 = \bar{x}_3$. In addition, we want to assume that the right angles between the reference axes x_2 and x_3 are preserved in the transformed state so that this is a pure rotation with the angle θ (see Fig. 2.16). We will now pursue the question of how the elastic properties of an anisotropic material change if we perform a coordinate transformation as described and search for the compliances and stiffnesses S_{ij} and C_{ij} with respect to the off-axis reference system $\bar{x}_1$, $\bar{x}_2$, $\bar{x}_3$. The transformation of the stresses and the strains from the coordinate system x_1, x_2, x_3 into the new coordinate system $\bar{x}_1$, $\bar{x}_2$, $\bar{x}_3$ is carried out as follows (see also Eq. (2.20) and (2.84)):

$$\underline{\bar{\sigma}} = \underline{\underline{T}}_\sigma \underline{\sigma}, \quad \underline{\bar{\varepsilon}} = \underline{\underline{T}}_\varepsilon \underline{\varepsilon},$$

(2.198)

Fig. 2.16 Rotation of the coordinate system

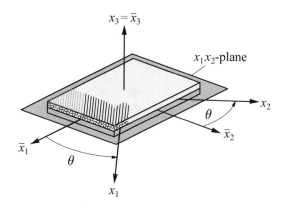

where

$$\underline{\underline{T}}_\sigma^{-1} = \underline{\underline{T}}_\varepsilon^T, \quad \underline{\underline{T}}_\varepsilon^{-1} = \underline{\underline{T}}_\sigma^T. \tag{2.199}$$

If we use the generalized Hooke's Law $\underline{\sigma} = \underline{\underline{C}}\varepsilon$ in the first equation in (2.198), we get:

$$\bar{\underline{\sigma}} = \underline{\underline{T}}_\sigma \underline{\underline{C}}\varepsilon. \tag{2.200}$$

Inserting the relation $\underline{\varepsilon} = \underline{\underline{T}}_\varepsilon^{-1}\bar{\varepsilon}$, which results from the inversion of the second equation in (2.198) yields:

$$\bar{\underline{\sigma}} = \underline{\underline{T}}_\sigma \underline{\underline{C}}\underline{\underline{T}}_\varepsilon^{-1}\bar{\varepsilon}, \tag{2.201}$$

or when using (2.199):

$$\bar{\underline{\sigma}} = \underline{\underline{T}}_\sigma \underline{\underline{C}}\underline{\underline{T}}_\sigma^T\bar{\varepsilon}. \tag{2.202}$$

With (2.202) there is thus a connection between the stresses and strains in the off-axis system. Starting from the second expression in (2.198) one can proceed in an identical manner and obtain the expression inverse to (2.202) as follows:

$$\bar{\underline{\varepsilon}} = \underline{\underline{T}}_\varepsilon \underline{\underline{S}}\underline{\underline{T}}_\varepsilon^T\bar{\underline{\sigma}}. \tag{2.203}$$

The expressions $\underline{\underline{T}}_\sigma \underline{\underline{C}}\underline{\underline{T}}_\sigma^T$ and $\underline{\underline{T}}_\varepsilon \underline{\underline{S}}\underline{\underline{T}}_\varepsilon^T$ that occur in Eqs. (2.202) and (2.203) constitute the transformed stiffness matrix $\bar{\underline{\underline{C}}}$ and the transformed compliance matrix $\bar{\underline{\underline{S}}}$:

$$\bar{\underline{\underline{S}}} = \underline{\underline{T}}_\varepsilon \underline{\underline{S}}\underline{\underline{T}}_\varepsilon^T,$$
$$\bar{\underline{\underline{C}}} = \underline{\underline{T}}_\sigma \underline{\underline{C}}\underline{\underline{T}}_\sigma^T. \tag{2.204}$$

With the transformation matrices $\underline{\underline{T}}_\sigma$ and $\underline{\underline{T}}_\varepsilon$ according to (2.20) and (2.84) the following transformation rules result after elementary matrix operations for the stiffnesses C_{ij} and the compliances S_{ij} in the case of orthotropic material. The following transformation rules result for the compliances S_{ij}:

$$
\begin{aligned}
\bar{S}_{11} &= S_{11} \cos^4 \theta + S_{22} \sin^4 \theta + 2S_{12} \cos^2 \theta \sin^2 \theta + S_{66} \cos^2 \theta \sin^2 \theta, \\
\bar{S}_{22} &= S_{11} \sin^4 \theta + 2S_{12} \cos^2 \theta \sin^2 \theta + S_{22} \cos^4 \theta + S_{66} \cos^2 \theta \sin^2 \theta, \\
\bar{S}_{12} &= (S_{11} + S_{22}) \cos^2 \theta \sin^2 \theta + S_{12} \left(\cos^4 \theta + \sin^4 \theta \right) - S_{66} \cos^2 \theta \sin^2 \theta, \\
\bar{S}_{66} &= 4(S_{11} + S_{22}) \cos^2 \theta \sin^2 \theta - 8S_{12} \cos^2 \theta \sin^2 \theta + S_{66} \left(\cos^2 \theta - \sin^2 \theta \right)^2, \\
\bar{S}_{16} &= 2S_{11} \cos^3 \theta \sin \theta + 2S_{12} \left(\cos \theta \sin^3 \theta - \cos^3 \theta \sin \theta \right) \\
&\quad - 2S_{22} \cos \theta \sin^3 \theta + S_{66} \left(\cos \theta \sin^3 \theta - \cos^3 \theta \sin \theta \right), \\
\bar{S}_{26} &= 2S_{11} \cos \theta \sin^3 \theta + 2S_{12} \left(\cos^3 \theta \sin \theta - \cos \theta \sin^3 \theta \right) \\
&\quad - 2S_{22} \cos^3 \theta \sin \theta + S_{66} \left(\cos^3 \theta \sin \theta - \cos \theta \sin^3 \theta \right), \\
\bar{S}_{13} &= S_{13} \cos^2 \theta + S_{23} \sin^2 \theta, \\
\bar{S}_{23} &= S_{13} \sin^2 \theta + S_{23} \cos^2 \theta, \\
\bar{S}_{33} &= S_{33}, \\
\bar{S}_{36} &= 2S_{13} \cos \theta \sin \theta - 2S_{23} \cos \theta \sin \theta, \\
\bar{S}_{44} &= S_{44} \cos^2 \theta + S_{55} \sin^2 \theta, \\
\bar{S}_{45} &= S_{55} \cos \theta \sin \theta - S_{44} \cos \theta \sin \theta, \\
\bar{S}_{55} &= S_{44} \sin^2 \theta + S_{55} \cos^2 \theta.
\end{aligned}
\tag{2.205}
$$

All other compliances $\bar{S}_{ij}$ not listed here become zero with this type of transformation for orthotropic material. For the generalized Hooke's Law $\bar{\varepsilon} = \underline{\underline{S}} \bar{\sigma}$ in the off-axis system we have:

$$
\begin{pmatrix} \bar{\varepsilon}_{11} \\ \bar{\varepsilon}_{22} \\ \bar{\varepsilon}_{33} \\ \bar{\gamma}_{23} \\ \bar{\gamma}_{13} \\ \bar{\gamma}_{12} \end{pmatrix} = \begin{bmatrix} \bar{S}_{11} & \bar{S}_{12} & \bar{S}_{13} & 0 & 0 & \bar{S}_{16} \\ \bar{S}_{12} & \bar{S}_{22} & \bar{S}_{23} & 0 & 0 & \bar{S}_{26} \\ \bar{S}_{13} & \bar{S}_{23} & \bar{S}_{33} & 0 & 0 & \bar{S}_{36} \\ 0 & 0 & 0 & \bar{S}_{44} & \bar{S}_{45} & 0 \\ 0 & 0 & 0 & \bar{S}_{45} & \bar{S}_{55} & 0 \\ \bar{S}_{16} & \bar{S}_{26} & \bar{S}_{36} & 0 & 0 & \bar{S}_{66} \end{bmatrix} \begin{pmatrix} \bar{\sigma}_{11} \\ \bar{\sigma}_{22} \\ \bar{\sigma}_{33} \\ \bar{\sigma}_{23} \\ \bar{\sigma}_{13} \\ \bar{\sigma}_{12} \end{pmatrix} .
\tag{2.206}
$$

It can be seen that the material law for orthotropic material changes through this coordinate transformation into a representation corresponding to a monoclinic material with only one single plane of symmetry at $x_3 = 0$.

In the same way the transformed stiffnesses $\bar{C}_{ij}$ are obtained as follows

$$\bar{C}_{11} = C_{11} \cos^4 \theta + C_{22} \sin^4 \theta + 2C_{12} \cos^2 \theta \sin^2 \theta + 4C_{66} \cos^2 \theta \sin^2 \theta,$$
$$\bar{C}_{22} = C_{11} \sin^4 \theta + C_{22} \cos^4 \theta + 2C_{12} \cos^2 \theta \sin^2 \theta + 4C_{66} \cos^2 \theta \sin^2 \theta,$$
$$\bar{C}_{12} = (C_{11} + C_{22}) \cos^2 \theta \sin^2 \theta + C_{12} \left(\cos^4 \theta + \sin^4 \theta\right) - 4C_{66} \cos^2 \theta \sin^2 \theta,$$
$$\bar{C}_{66} = (C_{11} + C_{22}) \cos^2 \theta \sin^2 \theta - 2C_{12} \cos^2 \theta \sin^2 \theta + C_{66} \left(\cos^2 \theta - \sin^2 \theta\right)^2,$$
$$\bar{C}_{16} = C_{11} \cos^3 \theta \sin \theta + C_{12} \left(\cos \theta \sin^3 \theta - \cos^3 \theta \sin \theta\right)$$
$$\quad - C_{22} \cos \theta \sin^3 \theta + 2C_{66} \left(\cos \theta \sin^3 \theta - \cos^3 \theta \sin \theta\right),$$
$$\bar{C}_{26} = C_{11} \cos \theta \sin^3 \theta + C_{12} \left(\cos^3 \theta \sin \theta - \cos \theta \sin^3 \theta\right)$$
$$\quad - C_{22} \cos^3 \theta \sin \theta + 2C_{66} \left(\cos^3 \theta \sin \theta - \cos \theta \sin^3 \theta\right),$$
$$\bar{C}_{13} = C_{13} \cos^2 \theta + C_{23} \sin^2 \theta,$$
$$\bar{C}_{23} = C_{13} \sin^2 \theta + C_{23} \cos^2 \theta,$$
$$\bar{C}_{33} = C_{33},$$
$$\bar{C}_{36} = C_{13} \cos \theta \sin \theta - C_{23} \cos \theta \sin \theta,$$
$$\bar{C}_{44} = C_{44} \cos^2 \theta + C_{55} \sin^2 \theta,$$
$$\bar{C}_{45} = C_{55} \cos \theta \sin \theta - C_{44} \cos \theta \sin \theta,$$
$$\bar{C}_{55} = C_{44} \sin^2 \theta + C_{55} \cos^2 \theta. \tag{2.207}$$

All other transformed stiffnesses $\bar{C}_{ij}$ disappear for orthotropic material with this type of transformation. The transformed material law $\bar{\underline{\sigma}} = \underline{\underline{\bar{C}}}\,\bar{\underline{\varepsilon}}$ in the off-axis system results as:

$$\begin{pmatrix} \bar{\sigma}_{11} \\ \bar{\sigma}_{22} \\ \bar{\sigma}_{33} \\ \bar{\tau}_{23} \\ \bar{\tau}_{13} \\ \bar{\tau}_{12} \end{pmatrix} = \begin{bmatrix} \bar{C}_{11} & \bar{C}_{12} & \bar{C}_{13} & 0 & 0 & \bar{C}_{16} \\ \bar{C}_{12} & \bar{C}_{22} & \bar{C}_{23} & 0 & 0 & \bar{C}_{26} \\ \bar{C}_{13} & \bar{C}_{23} & \bar{C}_{33} & 0 & 0 & \bar{C}_{36} \\ 0 & 0 & 0 & \bar{C}_{44} & \bar{C}_{45} & 0 \\ 0 & 0 & 0 & \bar{C}_{45} & \bar{C}_{55} & 0 \\ \bar{C}_{16} & \bar{C}_{26} & \bar{C}_{36} & 0 & 0 & \bar{C}_{66} \end{bmatrix} \begin{pmatrix} \bar{\varepsilon}_{11} \\ \bar{\varepsilon}_{22} \\ \bar{\varepsilon}_{33} \\ \bar{\gamma}_{23} \\ \bar{\gamma}_{13} \\ \bar{\gamma}_{12} \end{pmatrix}. \tag{2.208}$$

At this point it is important to note that the 13 transformed stiffnesses $\bar{C}_{ij}$ and compliances $\bar{S}_{ij}$ in the off-axis system naturally depend on the nine independent quantities C_{ij} and S_{ij} in the on-axis system. Thus, even when using an off-axis system there are only nine independent variables.

Analogous transformation rules can be specified for the engineering constants. In the case of a pure rotation of the coordinate system around the fixed x_3-axis we have:

$$\bar{E}_{11} = \frac{1}{\dfrac{\cos^4\theta}{E_{11}} + \dfrac{\sin^4\theta}{E_{22}} - 2\dfrac{\nu_{12}}{E_{11}}\cos^2\theta\sin^2\theta + \dfrac{\cos^2\theta\sin^2\theta}{G_{12}}},$$

$$\bar{E}_{22} = \frac{1}{\dfrac{\sin^4\theta}{E_{11}} + \dfrac{\cos^4\theta}{E_{22}} - 2\dfrac{\nu_{12}}{E_{11}}\cos^2\theta\sin^2\theta + \dfrac{\cos^2\theta\sin^2\theta}{G_{12}}},$$

$$\bar{E}_{33} = E_{33}, \quad \bar{G}_{23} = \frac{1}{\dfrac{\cos^2\theta}{G_{23}} + \dfrac{\sin^2\theta}{G_{13}}}, \quad \bar{G}_{13} = \frac{1}{\dfrac{\sin^2\theta}{G_{23}} + \dfrac{\cos^2\theta}{G_{13}}},$$

$$\bar{G}_{12} = \frac{1}{4\left(\dfrac{1}{E_{11}} + \dfrac{1}{E_{22}}\right)\cos^2\theta\sin^2\theta + 8\dfrac{\nu_{12}}{E_{11}}\cos^2\theta\sin^2\theta + \dfrac{\left(\cos^2\theta - \sin^2\theta\right)^2}{G_{12}}},$$

$$\bar{\nu}_{12} = -\frac{\left(\dfrac{1}{E_{11}} + \dfrac{1}{E_{22}}\right)\cos^2\theta\sin^2\theta - \dfrac{\nu_{12}}{E_{11}}\left(\cos^4\theta + \sin^4\theta\right) - \dfrac{\cos^2\theta\sin^2\theta}{G_{12}}}{\dfrac{\cos^4\theta}{E_{11}} + \dfrac{\sin^4\theta}{E_{22}} - 2\dfrac{\nu_{12}}{E_{11}}\cos^2\theta\sin^2\theta + \dfrac{\cos^2\theta\sin^2\theta}{G_{12}}},$$

$$\bar{\nu}_{21} = -\frac{\left(\dfrac{1}{E_{11}} + \dfrac{1}{E_{22}}\right)\cos^2\theta\sin^2\theta - \dfrac{\nu_{12}}{E_{11}}\left(\cos^4\theta + \sin^4\theta\right) - \dfrac{\cos^2\theta\sin^2\theta}{G_{12}}}{\dfrac{\sin^4\theta}{E_{11}} + \dfrac{\cos^4\theta}{E_{22}} - 2\dfrac{\nu_{12}}{E_{11}}\cos^2\theta\sin^2\theta + \dfrac{\cos^2\theta\sin^2\theta}{G_{12}}},$$

$$\bar{\nu}_{13} = \frac{\dfrac{\nu_{31}}{E_{33}}\cos^2\theta + \dfrac{\nu_{23}}{E_{22}}\sin^2\theta}{\dfrac{\cos^4\theta}{E_{11}} + \dfrac{\sin^4\theta}{E_{22}} - 2\dfrac{\nu_{12}}{E_{11}}\cos^2\theta\sin^2\theta + \dfrac{\cos^2\theta\sin^2\theta}{G_{12}}},$$

$$\bar{\nu}_{23} = \frac{\dfrac{\nu_{31}}{E_{33}}\sin^2\theta + \dfrac{\nu_{23}}{E_{22}}\cos^2\theta}{\dfrac{\sin^4\theta}{E_{11}} + \dfrac{\cos^4\theta}{E_{22}} - 2\dfrac{\nu_{12}}{E_{11}}\cos^2\theta\sin^2\theta + \dfrac{\cos^2\theta\sin^2\theta}{G_{12}}},$$

$$\bar{\nu}_{31} = E_{33}\left(\dfrac{\nu_{31}}{E_{33}}\cos^2\theta + \dfrac{\nu_{23}}{E_{22}}\sin^2\theta\right),$$

$$\bar{\nu}_{32} = E_{33}\left(\dfrac{\nu_{13}}{E_{11}}\sin^2\theta + \dfrac{\nu_{23}}{E_{22}}\cos^2\theta\right).$$

$$(2.209)$$

2.8   Hygrothermal Problems

. Besides purely mechanical stresses, hygrothermal influences can also play a decisive
role in lightweight engineering. First, we consider a body which undergoes a change
ΔT of its initial temperature T_0 and assumes the current temperature T. It is obvious
that this results in the thermal strains ε_{ij}^T. In many technically relevant cases it can be
assumed that there is a linear relationship between the temperature change ΔT and
the temperature strains that occur. The temperature strains ε_{ij}^T are thus assumed to be
proportional to the temperature change $\Delta T = T - T_0$. The corresponding material
constants which establish the proportionality between ε_{ij}^T and ΔT are the so-called
coefficients of thermal expansion α_{ij}^T. Hence:

$$\varepsilon_{ij}^T = \alpha_{ij}^T \Delta T. \tag{2.210}$$

Since we have not yet made any restrictions with respect to possible material symmetries, the thermal expansion coefficients α_{ij}^T can all be different, i.e. they describe
an anisotropic thermal expansion behavior. In addition, a temperature change ΔT
can also cause shear strains in an anisotropic body which of course is not known
from isotropic materials.

If not only thermal strains ε_{ij}^T occur but also mechanical strains arise, then the
total strains ε_{ij} result from the sum of these two parts:

$$\varepsilon_{ij} = S_{ijkl}\sigma_{kl} + \alpha_{ij}^T \Delta T, \tag{2.211}$$

and in a vector-matrix notation:

$$\begin{pmatrix} \varepsilon_{11} \\ \varepsilon_{22} \\ \varepsilon_{33} \\ \gamma_{23} \\ \gamma_{13} \\ \gamma_{12} \end{pmatrix} = \begin{bmatrix} S_{11} & S_{12} & S_{13} & S_{14} & S_{15} & S_{16} \\ S_{21} & S_{22} & S_{23} & S_{24} & S_{25} & S_{26} \\ S_{31} & S_{32} & S_{33} & S_{34} & S_{35} & S_{36} \\ S_{41} & S_{42} & S_{43} & S_{44} & S_{45} & S_{46} \\ S_{51} & S_{52} & S_{53} & S_{54} & S_{55} & S_{56} \\ S_{61} & S_{62} & S_{63} & S_{64} & S_{65} & S_{66} \end{bmatrix} \begin{pmatrix} \sigma_{11} \\ \sigma_{22} \\ \sigma_{33} \\ \sigma_{23} \\ \sigma_{13} \\ \sigma_{12} \end{pmatrix} + \Delta T \begin{pmatrix} \alpha_{11}^T \\ \alpha_{22}^T \\ \alpha_{33}^T \\ \alpha_{23}^T \\ \alpha_{13}^T \\ \alpha_{12}^T \end{pmatrix}. \tag{2.212}$$

The inverse relation reads:

$$\begin{pmatrix} \sigma_{11} \\ \sigma_{22} \\ \sigma_{33} \\ \sigma_{23} \\ \sigma_{13} \\ \sigma_{12} \end{pmatrix} = \begin{bmatrix} C_{11} & C_{12} & C_{13} & C_{14} & C_{15} & C_{16} \\ C_{21} & C_{22} & C_{23} & C_{24} & C_{25} & C_{26} \\ C_{31} & C_{32} & C_{33} & C_{34} & C_{35} & C_{36} \\ C_{41} & C_{42} & C_{43} & C_{44} & C_{45} & C_{46} \\ C_{51} & C_{52} & C_{53} & C_{54} & C_{55} & C_{56} \\ C_{61} & C_{62} & C_{63} & C_{64} & C_{65} & C_{66} \end{bmatrix} \begin{pmatrix} \varepsilon_{11} - \alpha_{11}^T \Delta T \\ \varepsilon_{22} - \alpha_{22}^T \Delta T \\ \varepsilon_{33} - \alpha_{33}^T \Delta T \\ \gamma_{23} - \alpha_{23}^T \Delta T \\ \gamma_{13} - \alpha_{13}^T \Delta T \\ \gamma_{12} - \alpha_{12}^T \Delta T \end{pmatrix}. \tag{2.213}$$

Apparently only the difference between the total strains and the thermal strains
causes stresses, this difference being exactly equal to the mechanical strains.

In the case of the orthotropy it can be seen that the thermal expansion coefficients α_{23}^T, α_{13}^T, α_{12}^T become zero. Thus:

$$
\begin{pmatrix} \varepsilon_{11} \\ \varepsilon_{22} \\ \varepsilon_{33} \\ \gamma_{23} \\ \gamma_{13} \\ \gamma_{12} \end{pmatrix} = \begin{bmatrix} S_{11} & S_{12} & S_{13} & 0 & 0 & 0 \\ S_{21} & S_{22} & S_{23} & 0 & 0 & 0 \\ S_{31} & S_{32} & S_{33} & 0 & 0 & 0 \\ 0 & 0 & 0 & S_{44} & 0 & 0 \\ 0 & 0 & 0 & 0 & S_{55} & 0 \\ 0 & 0 & 0 & 0 & 0 & S_{66} \end{bmatrix} \begin{pmatrix} \sigma_{11} \\ \sigma_{22} \\ \sigma_{33} \\ \sigma_{23} \\ \sigma_{13} \\ \sigma_{12} \end{pmatrix} + \Delta T \begin{pmatrix} \alpha_{11}^T \\ \alpha_{22}^T \\ \alpha_{33}^T \\ 0 \\ 0 \\ 0 \end{pmatrix}. \tag{2.214}
$$

It should again be pointed out that in case of complete anisotropy in general also all coefficients of thermal expansion α_{ij}^T occur. However, the complexity of the thermomechanical behavior is considerably reduced when considering an isotropic material. In this case all α_{ij}^T for $i \neq j$ are zero. Furthermore, the remaining coefficients of thermal expansion are identical and $\alpha_{11}^T = \alpha_{22}^T = \alpha_{33}^T = \alpha^T$ applies. For the total strains we then have:

$$
\begin{pmatrix} \varepsilon_{11} \\ \varepsilon_{22} \\ \varepsilon_{33} \\ \gamma_{23} \\ \gamma_{13} \\ \gamma_{12} \end{pmatrix} = \begin{bmatrix} \dfrac{1}{E} & -\dfrac{\nu}{E} & -\dfrac{\nu}{E} & 0 & 0 & 0 \\ -\dfrac{\nu}{E} & \dfrac{1}{E} & -\dfrac{\nu}{E} & 0 & 0 & 0 \\ -\dfrac{\nu}{E} & -\dfrac{\nu}{E} & \dfrac{1}{E} & 0 & 0 & 0 \\ 0 & 0 & 0 & \dfrac{1}{G} & 0 & 0 \\ 0 & 0 & 0 & 0 & \dfrac{1}{G} & 0 \\ 0 & 0 & 0 & 0 & 0 & \dfrac{1}{G} \end{bmatrix} \begin{pmatrix} \sigma_{11} \\ \sigma_{22} \\ \sigma_{33} \\ \sigma_{23} \\ \sigma_{13} \\ \sigma_{12} \end{pmatrix} + \Delta T \begin{pmatrix} \alpha^T \\ \alpha^T \\ \alpha^T \\ 0 \\ 0 \\ 0 \end{pmatrix}. \tag{2.215}
$$

Apparently, the temperature strains occurring in an isotropic structure result in a pure strain state, whereby the strains are the same in all directions and no change in shape is associated with a temperature change.

Hygroscopic stresses, e.g. in the case of fiber composite laminates, pose a problem very similar to thermal stresses. Many typical matrix materials of modern fiber composites absorb moisture from or release moisture to the ambient air. If the ambient moisture content is not equal to the stress-free state of the lightweight material, so-called moisture strains ε_{ij}^H occur, where the letter 'H' stands for hygroscopy. If a linear relationship is assumed, the following applies:

$$
\varepsilon_{ij}^H = \beta_{ij} \Delta c. \tag{2.216}
$$

Herein, the quantities β_{ij} are the so-called moisture expansion coefficients which are specified in the unit $\frac{1}{\%}$. The quantity Δc is the moisture concentration which is defined as follows:

$$
\Delta c = \frac{\text{Mass of absorbed fluid}}{\text{Mass of dry volume}}. \tag{2.217}
$$

The moisture concentration Δc is usually given in [%]. Note that anisotropic materials generally exhibit an anisotropic hygroscopic behavior.

2.9 Cylindrical Coordinates

In many cases it may be advantageous to employ cylindrical coordinates r, φ, z instead of cartesian coordinates x, y, z (Fig. 2.17, left). To derive the local equilibrium conditions we consider the free body diagram of Fig. 2.17, right, in which the stress components $\sigma_{rr}, \sigma_{\varphi\varphi}, \sigma_{zz}, \tau_{r\varphi} = \tau_{\varphi r}, \tau_{z\varphi} = \tau_{\varphi z}, \tau_{rz} = \tau_{zr}$ are depicted. Volume forces f_r, f_φ, f_z are not shown for reasons of clarity. We consider the equilibrium of forces in radial direction r and obtain

$$
\left(\sigma_{rr} + \frac{\partial \sigma_{rr}}{\partial r}dr\right)(r + dr)\,d\varphi dz - \sigma_{rr}r d\varphi dz
$$
$$
+ \left(\tau_{rz} + \frac{\partial \tau_{rz}}{\partial z}dz\right)r dr d\varphi - \tau_{rz}r dr d\varphi
$$
$$
- \left(\sigma_{\varphi\varphi} + \frac{\partial \sigma_{\varphi\varphi}}{\partial \varphi}d\varphi\right)dr dz \sin\left(\frac{d\varphi}{2}\right) - \sigma_{\varphi\varphi}dr dz \sin\left(\frac{d\varphi}{2}\right)
$$
$$
+ \left(\tau_{r\varphi} + \frac{\partial \tau_{r\varphi}}{\partial \varphi}d\varphi\right)dr dz \cos\left(\frac{d\varphi}{2}\right) - \tau_{r\varphi}dr dz \cos\left(\frac{d\varphi}{2}\right) + f_r r dr d\varphi dz = 0.
$$

$$
\tag{2.218}
$$

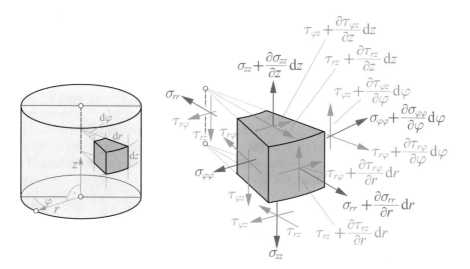

Fig. 2.17 Cylindrical coordinate system (left), free body diagram of an infinitesimal element (right)

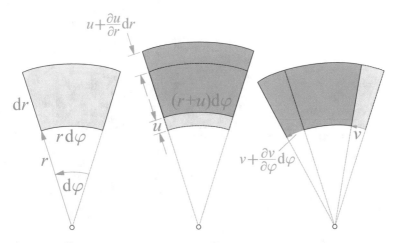

Fig. 2.18 Top view of an infinitesimal element, determination of the strains ε_{rr} and $\varepsilon_{\varphi\varphi}$

Neglecting higher order terms and considering $\sin\left(\frac{d\varphi}{2}\right) \simeq \frac{d\varphi}{2}$ and $\cos\left(\frac{d\varphi}{2}\right) \simeq 1$ eventually yields:

$$\frac{\partial\sigma_{rr}}{\partial r} + \frac{1}{r}\frac{\partial\tau_{r\varphi}}{\partial\varphi} + \frac{\partial\tau_{rz}}{\partial z} + \frac{\sigma_{rr} - \sigma_{\varphi\varphi}}{r} + f_r = 0. \tag{2.219}$$

Analogously the two remaining equilibrium conditions can be formed:

$$\frac{\partial\tau_{r\varphi}}{\partial r} + \frac{1}{r}\frac{\sigma_{\varphi\varphi}}{\partial\varphi} + \frac{\partial\tau_{\varphi z}}{\partial z} + 2\frac{\tau_{r\varphi}}{r} + f_\varphi = 0,$$

$$\frac{\partial\tau_{rz}}{\partial r} + \frac{1}{r}\frac{\partial\tau_{\varphi z}}{\partial\varphi} + \frac{\partial\sigma_{zz}}{\partial z} + \frac{\tau_{rz}}{r} + f_z = 0. \tag{2.220}$$

The strain components ε_{rr}, $\varepsilon_{\varphi\varphi}$, ε_{zz}, $\gamma_{r\varphi}$, γ_{rz}, $\gamma_{\varphi z}$ can be expressed by means of the kinematic equations by the displacements u, v, w with respect to r, φ, z. To derive the corresponding equations, we consider the infinitesimal sectional element in top view, as shown in Figs. 2.18 and 2.19. The two deformation states shown in Fig. 2.18 lead to the two strain components ε_{rr} and $\varepsilon_{\varphi\varphi}$. The radial strain ε_{rr} can be directly determined from Fig. 2.18, middle, as:

$$\varepsilon_{rr} = \frac{\left(u + \frac{\partial u}{\partial r}dr\right) - u}{dr} = \frac{\partial u}{\partial r}. \tag{2.221}$$

The strain ε_{rr} is thus determined exclusively by the displacement u, the displacement v does not enter the expression (2.221).

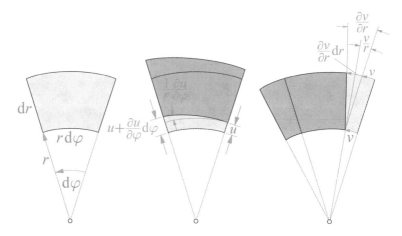

Fig. 2.19 Top view of an infinitesimal element, determination of the shear strain $\gamma_{r\varphi}$

Similarly, from Fig. 2.18, middle, we can determine the proportion of the tangential strain $\varepsilon_{\varphi\varphi}$ due to the displacement u:

$$\varepsilon_{\varphi\varphi} = \frac{(r+u)\,\mathrm{d}\varphi - r\mathrm{d}\varphi}{r\mathrm{d}\varphi} = \frac{u}{r}. \tag{2.222}$$

In addition, according to Fig. 2.18, right, another part due to the tangential displacement v as follows results:

$$\varepsilon_{\varphi\varphi} = \frac{\left(v + \dfrac{\partial v}{\partial \varphi}\mathrm{d}\varphi\right) - v}{r\mathrm{d}\varphi} = \frac{1}{r}\frac{\partial v}{\partial \varphi}. \tag{2.223}$$

In total we have for $\varepsilon_{\varphi\varphi}$:

$$\varepsilon_{\varphi\varphi} = \frac{1}{r}\frac{\partial v}{\partial \varphi} + \frac{u}{r}. \tag{2.224}$$

The determination of the shear strain $\gamma_{r\varphi}$ follows from Fig. 2.19. We have:

$$\gamma_{r\varphi} = \frac{\partial v}{\partial r} + \frac{1}{r}\frac{\partial u}{\partial \varphi} - \frac{v}{r}. \tag{2.225}$$

The derivation of the remaining strain components will not be presented at this point. In all, the following kinematic equations in cylindrical coordinates result:

$$\varepsilon_{rr} = \frac{\partial u}{\partial r}, \quad \varepsilon_{\varphi\varphi} = \frac{1}{r}\frac{\partial v}{\partial \varphi} + \frac{u}{r}, \quad \varepsilon_{zz} = \frac{\partial w}{\partial z},$$

$$\gamma_{r\varphi} = \frac{\partial v}{\partial r} + \frac{1}{r}\frac{\partial u}{\partial \varphi} - \frac{v}{r}, \quad \gamma_{rz} = \frac{\partial w}{\partial r} + \frac{\partial u}{\partial z}, \quad \gamma_{\varphi z} = \frac{\partial v}{\partial z} + \frac{1}{r}\frac{\partial w}{\partial \varphi}. \quad (2.226)$$

In an analogous manner the generalized Hooke's Law can be represented in cylindrical coordinates as follows, whereby we limit ourselves here to the representation of cylindrical orthotropy:

$$\begin{pmatrix} \sigma_{rr} \\ \sigma_{\varphi\varphi} \\ \sigma_{zz} \\ \tau_{\varphi z} \\ \tau_{rz} \\ \tau_{r\varphi} \end{pmatrix} = \begin{bmatrix} C_{11} & C_{12} & C_{13} & 0 & 0 & 0 \\ C_{12} & C_{22} & C_{23} & 0 & 0 & 0 \\ C_{13} & C_{23} & C_{33} & 0 & 0 & 0 \\ 0 & 0 & 0 & C_{44} & 0 & 0 \\ 0 & 0 & 0 & 0 & C_{55} & 0 \\ 0 & 0 & 0 & 0 & 0 & C_{66} \end{bmatrix} \begin{pmatrix} \varepsilon_{rr} \\ \varepsilon_{\varphi\varphi} \\ \varepsilon_{zz} \\ \gamma_{\varphi z} \\ \gamma_{rz} \\ \gamma_{r\varphi} \end{pmatrix}. \quad (2.227)$$

In inverted form:

$$\begin{pmatrix} \varepsilon_{rr} \\ \varepsilon_{\varphi\varphi} \\ \varepsilon_{zz} \\ \gamma_{\varphi z} \\ \gamma_{rz} \\ \gamma_{r\varphi} \end{pmatrix} = \begin{bmatrix} S_{11} & S_{12} & S_{13} & 0 & 0 & 0 \\ S_{12} & S_{22} & S_{23} & 0 & 0 & 0 \\ S_{13} & S_{23} & S_{33} & 0 & 0 & 0 \\ 0 & 0 & 0 & S_{44} & 0 & 0 \\ 0 & 0 & 0 & 0 & S_{55} & 0 \\ 0 & 0 & 0 & 0 & 0 & S_{66} \end{bmatrix} \begin{pmatrix} \sigma_{rr} \\ \sigma_{\varphi\varphi} \\ \sigma_{zz} \\ \tau_{\varphi z} \\ \tau_{rz} \\ \tau_{r\varphi} \end{pmatrix}. \quad (2.228)$$

Using engineering constants this results in:

$$\begin{pmatrix} \varepsilon_{rr} \\ \varepsilon_{\varphi\varphi} \\ \varepsilon_{zz} \\ \gamma_{\varphi z} \\ \gamma_{rz} \\ \gamma_{r\varphi} \end{pmatrix} = \begin{bmatrix} \dfrac{1}{E_{rr}} & -\dfrac{\nu_{\varphi r}}{E_{\varphi\varphi}} & -\dfrac{\nu_{zr}}{E_{zz}} & 0 & 0 & 0 \\ -\dfrac{\nu_{r\varphi}}{E_{rr}} & \dfrac{1}{E_{\varphi\varphi}} & -\dfrac{\nu_{z\varphi}}{E_{zz}} & 0 & 0 & 0 \\ -\dfrac{\nu_{rz}}{E_{rr}} & -\dfrac{\nu_{\varphi z}}{E_{\varphi\varphi}} & \dfrac{1}{E_{zz}} & 0 & 0 & 0 \\ 0 & 0 & 0 & \dfrac{1}{G_{\varphi z}} & 0 & 0 \\ 0 & 0 & 0 & 0 & \dfrac{1}{G_{rz}} & 0 \\ 0 & 0 & 0 & 0 & 0 & \dfrac{1}{G_{r\varphi}} \end{bmatrix} \begin{pmatrix} \sigma_{rr} \\ \sigma_{\varphi\varphi} \\ \sigma_{zz} \\ \tau_{\varphi z} \\ \tau_{rz} \\ \tau_{r\varphi} \end{pmatrix}. \quad (2.229)$$

References

Altenbach H, Altenbach J, Rikards R (1996) Einführung in die Mechanik der Laminat- und Sandwichtragwerke. Deutscher Verlag der Grundstoffindustrie, Stuttgart, Germany

Ambartsumyan SA (1970) Theory of anisotropic plates. Technomic Publishing Co. Inc, Stamford, USA

Ashton JE, Whitney JM (1970) Theory of laminated plates. Technomic Publishing Co. Inc, Stamford, USA

Becker W, Gross D (2002) Mechanik elastischer Körper und Strukturen, Springer. Berlin et al., Germany

Eschenauer H, Schnell W (1986) Elastizitätstheorie I, Second edition, Bibliographisches Institut. Mannheim et al, Germany

Frick A, Klamser H (1990) Untersuchung der Verschiebungs- und Spannungsverteilungen an unbelasteten Rändern multidirektionaler Laminate. VDI Verlag, Düsseldorf, Germany

Göldner H, Altenbach J, Eschke K, Garz KF, Sähn S (1979) Lehrbuch Höhere Festigkeitslehre, vol 1. Physik Verlag, Weinheim, Germany

Göldner H, Altenbach J, Eschke K, Garz KF, Sähn S (1985) Lehrbuch Höhere Festigkeitslehre, vol 2. Physik Verlag, Weinheim, Germany

Gross D, Hauger W, Schnell W, Wriggers P (1995) Technische Mechanik 4: Hydromechanik, Elemente der Höheren Mechanik, Numerische Methoden, Second edition, Springer. Berlin et al, Germany

Hahn HG (1985) Elastizitätstheorie. BG Teubner Verlag, Stuttgart, Germany

Jones RM (1975) Mechanics of composite materials. Scripta Book Co., Washington, USA

Mittelstedt C, Becker W (2016) Strukturmechanik ebener Laminate. Technische Universität Darmstadt, Germany, Studienbereich Mechanik

Reddy JN (2004) Mechanics of laminated composite plates and shells, 2nd edn. CRC Press Boca Raton et al, USA

Tsai SW, Hahn HT (1980) Introduction to composite materials. Technomic Publishing Co., Inc., Lancaster et al., UK

Chapter 3
Plane Problems

3.1 Introduction

The relations discussed in the previous chapter apply to arbitrary problems of elasticity. However, the practical implementation shows that finding a closed-form solution for problems in the framework of a three-dimensional stress and strain analysis for elastic and anisotropic material behavior is only possible in a few special cases. In most cases, engineers are thus confronted with the task of making simplifying assumptions for given problems and simplifying a lightweight engineering problem to an extent that accordingly simplified analysis methods can be employed. In many cases, especially with the typical thin-walled structures in lightweight engineering, it is a very common approach to reduce a three-dimensional problem to two dimensions. In this way, not only is the number of state variables (displacements, strains, stresses) to be determined decreased, but also the complexity of the equations to be treated is significantly reduced which often makes it possible to solve problems which are not possible to solve when considering a three-dimensional problem. At the same time, the necessary computational effort is reduced, which is of considerable importance especially in the practical work of an engineer in lightweight engineering. Besides the general terms of plane strain and plane stress states, this concerns especially the idealizing concepts of the so-called surface structures (i.e. disks, plates and shells). We will deal with these concepts in detail in the further course of this chapter, but it should already be noted that the mathematical treatment of surface structures and the associated constructive design possibilities are not part of this book.

C. Mittelstedt, *Structural Mechanics in Lightweight Engineering*,
https://doi.org/10.1007/978-3-030-75193-7_3

3.2   Surface Structures

3.2.1   Plane Surface Structures: Disks and Plates

A disk (see Fig. 3.1, left) is a thin plane structure where the thickness h is much smaller than its characteristic dimension l in its plane. A disk is loaded exclusively by forces in its plane that are assumed to be constant over the thickness h as well as by volume forces in its plane. A plate (see Fig. 3.1, right) on the other hand is a thin planar load bearing structure which is loaded by loads perpendicular to its middle plane. In addition, individual forces in the z-direction and individual moments about the x- and y-axis may occur. In general, disks and plates are identical with respect to their geometry, they differ only in the type of applied load.

The thickness (i.e. the dimension in the z-direction) of the plane structures considered here is denoted by h and is significantly smaller than the characteristic inplane dimensions. The plane structure is halved at each point by the so-called middle plane. The type of load indicates whether it is a disk problem or whether the structure must be treated as a plate. Analogous to the elementary beam and rod theory, disk and plate problems can be considered separately if the problem involves small deformations (geometrically linear problem), even if inplane and out-of-plane loads occur simultaneously.

If the structure under consideration is a disk that is exclusively subjected to loads independent of z in the xy–plane, then only the stress components σ_{xx}, σ_{yy} and τ_{xy} occur (Fig. 3.2). Whether in addition a normal stress σ_{zz} occurs depends on whether it is a disk in the so-called plane stress state or in the so-called plane strain state. We will concretize these terms at a later point. Independent of this, the stresses σ_{xx}, σ_{yy} and τ_{xy} in a disk are constant over the thickness h and thus independent of z, so that $\sigma_{xx} = \sigma_{xx}(x, y)$, $\sigma_{yy} = \sigma_{yy}(x, y)$ and $\tau_{xy} = \tau_{xy}(x, y)$. Thus, the stresses in a disk can also be represented by the resulting force fluxes N_{xx}^0, N_{yy}^0, N_{xy}^0, which can be determined by integrating the stresses σ_{xx}, σ_{yy} and τ_{xy} with respect to z:

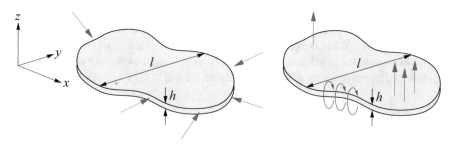

Fig. 3.1 Disk (left) and plate (right)

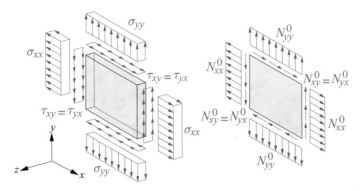

Fig. 3.2 Disk element with the associated stress components (left); reduction to the disk middle plane, force fluxes (right)

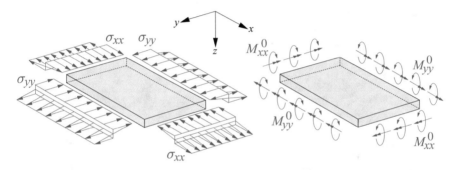

Fig. 3.3 Normal stresses (left) and resultant bending moment fluxes (right) in a plate

$$N_{xx}^0 = \int_{-\frac{h}{2}}^{+\frac{h}{2}} \sigma_{xx} dz, \quad N_{yy}^0 = \int_{-\frac{h}{2}}^{+\frac{h}{2}} \sigma_{yy} dz, \quad N_{xy}^0 = \int_{-\frac{h}{2}}^{+\frac{h}{2}} \tau_{xy} dz. \tag{3.1}$$

The force fluxes N_{xx}^0, N_{yy}^0, N_{xy}^0 are thus distributed forces in the unit of a force per length unit. These force fluxes occur exactly in the middle plane of the disk. They are assigned with the superscript index 0 to indicate that they are related to the center plane of the disk.

In a plate, however, shear stresses τ_{yz} and τ_{xz} in the thickness direction z must also occur in addition to the plane stress components σ_{xx}, σ_{yy} and τ_{xy} for reasons of equilibrium. The state of stress that occurs in a plate and the corresponding force and moment fluxes are shown in Figs. 3.3, 3.4 and 3.5.

Since a plate, unlike a disk, is subjected to loads that cause the plate to bend, the plane stress components σ_{xx}, σ_{yy} and τ_{xy} will have linearly varying distributions through the plate thickness. The shear stresses τ_{yz} and τ_{xz}, on the other hand, will show a parabolic distribution through the thickness of the plate. Consequently, in addition to M_{xx}^0 and M_{yy}^0 (the so-called bending moment fluxes) and M_{xy}^0 (the so-called twisting moment flux), the transverse shear force fluxes Q_x and Q_y will occur

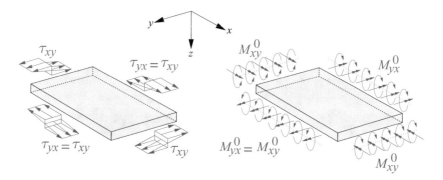

Fig. 3.4 Plane shear stresses (left) and resultant twisting moment fluxes (right) in a plate

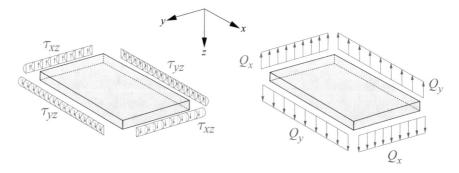

Fig. 3.5 Transverse shear stresses (left) and resultant transverse shear force fluxes (right) in a plate

as well where in the latter case the superscripted index 0 is usually omitted. The force and moment fluxes can be determined from the stress components by integration over the thickness as follows:

$$M_{xx}^0 = \int_{-\frac{h}{2}}^{+\frac{h}{2}} \sigma_{xx} z \, \mathrm{d}z, \quad M_{yy}^0 = \int_{-\frac{h}{2}}^{+\frac{h}{2}} \sigma_{yy} z \, \mathrm{d}z, \quad M_{xy}^0 = \int_{-\frac{h}{2}}^{+\frac{h}{2}} \tau_{xy} z \, \mathrm{d}z,$$

$$Q_x = \int_{-\frac{h}{2}}^{+\frac{h}{2}} \tau_{xz} \, \mathrm{d}z, \quad Q_y = \int_{-\frac{h}{2}}^{+\frac{h}{2}} \tau_{yz} \, \mathrm{d}z. \tag{3.2}$$

It follows that the transverse shear force fluxes have the unit of a force per length unit, whereas the moment fluxes $M_{xx}^0, M_{yy}^0, M_{xy}^0$ have the unit of a moment per length unit, e.g. Nm/m = N. The superscript index 0 again indicates that these are quantities related to the middle plane of the considered plate structure.

3.2.2 Curved Surface Structures: Shells

The surface structures discussed so far (disks and plates) are plane structures. A surface structure, however, does not necessarily have to be plane in order to be classified as such. In many technical applications, especially in lightweight construction and design, thin-walled surface structures occur in a curved form. Such thin shell structures have a number of very advantageous properties so that they are used very frequently, especially in lightweight engineering. Chapter 1 contains some examples of shell structures from modern aircraft construction. The most important difference of the shell compared to disks and plates is that in the case of a shell it is often not possible to distinguish between a pure membrane and a pure bending effect and in general these load-bearing effects occur simultaneously as a result of the curvature of the structure regardless of the type of load. As a consequence, the load bearing behaviour of a shell is generally much more complex than the behaviour of disks and plates. The structural analysis of shell structures is not dealt with in this book.

Finally, it should be noted that in engineering practice the use of the terms flat plate or curved shell is very common. However, the previous definitions should have made it clear that these designations contain redundancies. A plate is by definition always flat, just as a shell is by its very definition always curved.

3.3 Plane State of Strain

For the motivation of the plane strain state we first consider a linear-elastic and isotropic structure where the two displacements u and v occur in the x−direction and y-direction, but the third displacement component w in the z-direction does not arise. In addition, we will consider here the case that the two displacements u and v are not dependent on z, so that

$$u = u(x, y), \quad v = v(x, y), \quad w = 0. \tag{3.3}$$

If such a displacement field exists, then all strain components containing the index z must disappear:

$$\gamma_{xz} = \gamma_{yz} = \varepsilon_{zz} = 0. \tag{3.4}$$

If these conditions are fulfilled, then we speak of the so called plane strain state with respect to the xy−plane. The only remaining strains in this case are the two normal strains ε_{xx}, ε_{yy} and the shear strain γ_{xy}. These remaining strain components are then exclusively functions of the plane coordinates x and y and result from the kinematic relations (2.70) assuming geometric linerity as:

$$\varepsilon_{xx}(x, y) = \frac{\partial u}{\partial x}, \quad \varepsilon_{yy}(x, y) = \frac{\partial v}{\partial y}, \quad \gamma_{xy}(x, y) = \frac{\partial u}{\partial y} + \frac{\partial v}{\partial x}. \tag{3.5}$$

Fig. 3.6 Elastic body
between two rigid blocks

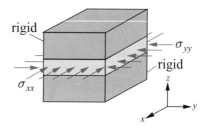

Plane strain states have significant technical importance and occur e.g. in structures whose form and load are not functions of the z-direction and where the displacement w is prevented. A simple generic example is shown in Fig. 3.6 with an elastic body under the stresses σ_{xx} and σ_{yy} where the displacement w is prevented by two rigid blocks.

Hooke's Law (2.148) for isotropic material immediately leads to the conclusion that in a plane strain state the two shear stresses τ_{yz} and τ_{xz} which act in z-direction must become zero. On the other hand, the normal stress σ_{zz} in general does not become zero due to the strain and displacement constraints in z-direction. From the third line in (2.157):

$$\varepsilon_{zz} = -\frac{\nu}{E}\sigma_{xx} - \frac{\nu}{E}\sigma_{yy} + \frac{1}{E}\sigma_{zz} = 0. \tag{3.6}$$

Once the stresses σ_{xx} and σ_{yy} are known, we can determine the normal stress σ_{zz} which arises in the plane strain state in addition to the stress components in the xy-plane as:

$$\sigma_{zz} = \nu\left(\sigma_{xx} + \sigma_{yy}\right). \tag{3.7}$$

With the help of this expression the normal stress σ_{zz} can be eliminated from (2.157), and the following form of the material law remains in the plane strain state:

$$\varepsilon_{xx} = \frac{1 - \nu^2}{E}\left(\sigma_{xx} - \frac{\nu}{1 - \nu}\sigma_{yy}\right),$$

$$\varepsilon_{yy} = \frac{1 - \nu^2}{E}\left(\sigma_{yy} - \frac{\nu}{1 - \nu}\sigma_{xx}\right),$$

$$\gamma_{xy} = \frac{2(1 + \nu)}{E}\tau_{xy}. \tag{3.8}$$

At this point it is useful to define a replacement modulus of elasticity $\overline{E}$ and a replacement Poisson's ratio $\overline{\nu}$ as follows:

$$\overline{E} = \frac{E}{1 - \nu^2}, \quad \overline{\nu} = \frac{\nu}{1 - \nu}. \tag{3.9}$$

From (3.8) we get:

$$\varepsilon_{xx} = \frac{1}{E}\left(\sigma_{xx} - \bar{\nu}\sigma_{yy}\right),$$

$$\varepsilon_{yy} = \frac{1}{E}\left(\sigma_{yy} - \bar{\nu}\sigma_{xx}\right),$$

$$\gamma_{xy} = \frac{2\left(1 + \bar{\nu}\right)}{\bar{E}}\tau_{xy}. \tag{3.10}$$

As a consequence, in the plane state of strain the strains as well as the stresses are independent of the thickness coordinate z. If we assume that there are no volume forces in the z-direction, then we can show that the third equilibrium condition in (2.45) is automatically fulfilled. The two remaining conditions in (2.45) are then reduced to:

$$\frac{\partial \sigma_{xx}}{\partial x} + \frac{\partial \tau_{xy}}{\partial y} + f_x = 0,$$

$$\frac{\partial \tau_{xy}}{\partial x} + \frac{\partial \sigma_{yy}}{\partial y} + f_y = 0. \tag{3.11}$$

From the compatibility conditions (2.87) only the following condition remains:

$$\frac{\partial^2 \varepsilon_{xx}}{\partial y^2} + \frac{\partial^2 \varepsilon_{yy}}{\partial x^2} - \frac{\partial^2 \gamma_{xy}}{\partial x \partial y} = 0. \tag{3.12}$$

A given problem in a plane state of strain is then completely described by the equations (3.5), (3.8), (3.11) and (3.12) as well as given boundary conditions. It is also worth mentioning here that no approximations or simplifications beyond the assumption of the independence of all state variables of z have been introduced so far. The equations (3.5), (3.8), (3.11) and (3.12) thus represent exact equations within the framework of the assumed linear elasticity theory of small deformations for isotropic structures.

3.4 Plane State of Stress

A further technically very important simplifying assumption in the framework of lightweight engineering is the assumption of a plane state of stress. Although we will show that the resulting equations are no longer exact in the sense of the theory of elasticity, experience has shown that in many cases this special case of analysis of thin-walled structures provides very useful results which justifies a detailed discussion.

A plane state of stress is found in good approximation e.g. in disks (see Fig. 3.1), i.e. mainly in very thin plane surface structures that are loaded exclusively by forces in their plane. If this is the case, then it can be assumed that the stresses σ_{zz}, τ_{xz} and τ_{yz} (i.e. all stress components with an index z) not only disappear at the free surfaces at $z = -\frac{h}{2}$ and $z = \frac{h}{2}$ but, due to the small thickness h compared to the inplane dimensions, also disappear over the entire thickness:

$$\sigma_{zz} = \tau_{xz} = \tau_{yz} = 0. \tag{3.13}$$

The plane stress state of a disk is thus characterized by the fact that only the inplane stress components σ_{xx}, σ_{yy} and τ_{xy} remain which are also independent of the thickness coordinate z:

$$\sigma_{xx} = \sigma_{xx}(x, y), \quad \sigma_{yy} = \sigma_{yy}(x, y), \quad \tau_{xy} = \tau_{xy}(x, y). \tag{3.14}$$

The generalized Hooke's Law (2.157) then assumes the following form in a plane state of stress:

$$\varepsilon_{xx} = \frac{1}{E}\left(\sigma_{xx} - \nu\sigma_{yy}\right),$$

$$\varepsilon_{yy} = \frac{1}{E}\left(\sigma_{yy} - \nu\sigma_{xx}\right),$$

$$\varepsilon_{zz} = -\frac{\nu}{E}\left(\sigma_{xx} + \sigma_{yy}\right),$$

$$\gamma_{xy} = \frac{2(1+\nu)}{E}\tau_{xy},$$

$$\gamma_{xz} = \gamma_{yz} = 0. \tag{3.15}$$

The comparison with the constitutive equations (3.10) for the plane strain state shows that both formula sets are completely identical, except for the quantities $\overline{E}$ and $\overline{\nu}$ which appear in (3.10). Furthermore, it can be seen that in the present case, quite in contrast to the plane strain state, the transverse normal strain ε_{zz} in the thickness direction z can currently also occur.

The constitutive law transformed according to the stress components reads for the plane stress state:

$$\sigma_{xx} = \frac{E}{1-\nu^2}\left(\varepsilon_{xx} + \nu\varepsilon_{yy}\right),$$

$$\sigma_{yy} = \frac{E}{1-\nu^2}\left(\varepsilon_{yy} + \nu\varepsilon_{xx}\right),$$

$$\tau_{xy} = G\gamma_{xy}. \tag{3.16}$$

The kinematic equations currently read:

$$\varepsilon_{xx} = \frac{\partial u}{\partial x}, \quad \varepsilon_{yy} = \frac{\partial v}{\partial y}, \quad \gamma_{xy} = \frac{\partial u}{\partial y} + \frac{\partial v}{\partial x}, \tag{3.17}$$

and the equilibrium conditions reduce to:

$$\frac{\partial \sigma_{xx}}{\partial x} + \frac{\partial \tau_{xy}}{\partial y} + f_x = 0,$$

$$\frac{\partial \tau_{xy}}{\partial x} + \frac{\partial \sigma_{yy}}{\partial y} + f_y = 0. \tag{3.18}$$

The only remaining compatibility condition reads:

$$\frac{\partial^2 \varepsilon_{xx}}{\partial y^2} + \frac{\partial^2 \varepsilon_{yy}}{\partial x^2} - \frac{\partial^2 \gamma_{xy}}{\partial x \partial y} = 0. \tag{3.19}$$

In order to describe the plane state of stress completely, it is also necessary to consider given boundary conditions in addition to the basic equations listed here.

The comparison of the present set of equations with the one concerning the plane state of strain shows that the equations are completely identical except for the elastic constants E and ν or $\overline{E}$ and $\overline{\nu}$, respectively. If, for example, there is a solution for a given boundary value problem within the plane strain state, then the corresponding solution for the plane stress state can be determined simply by exchanging the material parameters $\overline{E}$ and $\overline{\nu}$ for E and ν.

3.5 Stress Transformation

3.5.1 Introduction

We will now turn to the question which change of the stress state results if instead of considering a section parallel to the global reference axes x, y, z we examine an arbitrary section through the considered body point and thus perform a coordinate transformation as already discussed in Chap. 2. At this point we want to use the terms ξ, η for the two planar orthogonal axes (Fig. 3.7) rotated by the angle θ about the

Fig. 3.7 Coordinate transformation

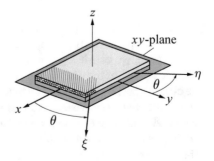

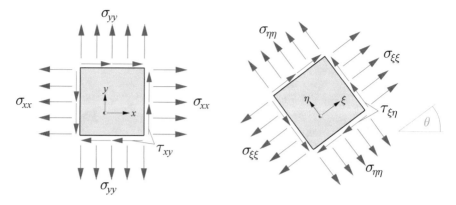

Fig. 3.8 Stress transformation: Initial state (left), rotation by an angle θ (right)

Fig. 3.9 Infinitesimal disk element

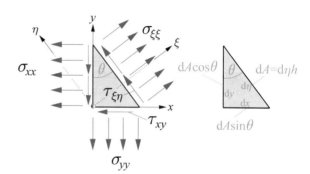

fixed z-axis to indicate that this transformation is performed for an arbitrary angle and does not necessarily have to be a distinct angle such as the principal material axes (as indicated in Fig. 3.7) of an anisotropic material. Using Eq. (2.20) we can determine the following relationship:

$$\begin{pmatrix} \sigma_{\xi\xi} \\ \sigma_{\eta\eta} \\ \tau_{\xi\eta} \end{pmatrix} = \begin{bmatrix} \cos^2 \theta & \sin^2 \theta & 2\cos\theta\sin\theta \\ \sin^2 \theta & \cos^2 \theta & -2\cos\theta\sin\theta \\ -\cos\theta\sin\theta & \cos\theta\sin\theta & \cos^2\theta - \sin^2\theta \end{bmatrix} \begin{pmatrix} \sigma_{xx} \\ \sigma_{yy} \\ \tau_{xy} \end{pmatrix}. \qquad (3.20)$$

The corresponding stress states are shown in Fig. 3.8. However, this result also allows for a particularly illustrative interpretation as will be explained below. For this purpose, we consider an infinitesimally small triangular sectional element that we have cut out of the disk in question (Fig. 3.9). Let the element be separated from the disk in question in such a way that we have two surfaces parallel to the reference axes x and y and one surface parallel to the rotated axis η. The stresses released on the individual surfaces act as shown. The plane penetrated by the $\xi-$ axis is called dA, it is calculated as d$A =$ dηh, if h is the thickness of the disk under consideration. The two surfaces oriented parallel to x and y with the lengths dx and dy then result as $dA \sin \theta$ and $dA \cos \theta$, respectively. If the equilibrium of forces is formulated in the direction of the reference axis ξ, then:

$$\sigma_{\xi\xi} dA - \sigma_{xx} dA \cos \theta \cos \theta - \sigma_{yy} dA \sin \theta \sin \theta - \tau_{xy} dA \cos \theta \sin \theta - \tau_{xy} dA \sin \theta \cos \theta = 0, \tag{3.21}$$

where we have already made use of the equality of the shear stresses $\tau_{xy} = \tau_{yx}$. After elementary transformations we obtain for $\sigma_{\xi\xi}$:

$$\sigma_{\xi\xi} = \sigma_{xx} \cos^2 \theta + \sigma_{yy} \sin^2 \theta + 2\tau_{xy} \sin \theta \cos \theta. \tag{3.22}$$

In an analogous manner equilibrium of forces with respect to the η-axis yields:

$$\tau_{\xi\eta} = -\sigma_{xx} \sin \theta \cos \theta + \sigma_{yy} \sin \theta \cos \theta + \tau_{xy} \left(\cos^2 \theta - \sin^2 \theta \right). \tag{3.23}$$

At another suitable free-body diagram where the η−axis penetrates the intersection dA (not shown here), a formulation for the normal stress $\sigma_{\eta\eta}$ can be determined as follows:

$$\sigma_{\eta\eta} = \sigma_{xx} \sin^2 \theta + \sigma_{yy} \cos^2 \theta - 2\tau_{xy} \sin \theta \cos \theta. \tag{3.24}$$

A common representation of the Eqs. (3.22), (3.23) and (3.24) can be achieved by using the relationships

$$\cos^2 \theta = \frac{1}{2} \left(1 + \cos 2\theta \right), \quad \sin^2 \theta = \frac{1}{2} \left(1 - \cos 2\theta \right),$$
$$2 \sin \theta \cos \theta = \sin 2\theta, \quad \cos^2 \theta - \sin^2 \theta = \cos 2\theta, \tag{3.25}$$

which eventually leads to:

$$\sigma_{\xi\xi} = \frac{1}{2} \left(\sigma_{xx} + \sigma_{yy} \right) + \frac{1}{2} \left(\sigma_{xx} - \sigma_{yy} \right) \cos 2\theta + \tau_{xy} \sin 2\theta,$$
$$\sigma_{\eta\eta} = \frac{1}{2} \left(\sigma_{xx} + \sigma_{yy} \right) - \frac{1}{2} \left(\sigma_{xx} - \sigma_{yy} \right) \cos 2\theta - \tau_{xy} \sin 2\theta,$$
$$\tau_{\xi\eta} = -\frac{1}{2} \left(\sigma_{xx} - \sigma_{yy} \right) \sin 2\theta + \tau_{xy} \cos 2\theta. \tag{3.26}$$

These are the transformation equations for the stress components in a state of plane stress. They can be used to determine the stresses $\sigma_{\xi\xi}$, $\sigma_{\eta\eta}$ and $\tau_{\xi\eta}$ in any reference system ξ, η from given stresses σ_{xx}, σ_{yy} and τ_{xy} in the reference system x, y.

The invariants of the stress state, already introduced in Chap. 2 with (2.27), are also valid in case of a plane stress state:

$$I_1 = \sigma_{xx} + \sigma_{yy}, \quad I_2 = \tau_{xy}^2 - \sigma_{xx}\sigma_{yy}, \quad I_3 = \sigma_{xx}\sigma_{yy} - \tau_{xy}^2, \tag{3.27}$$

where the two invariants I_2 and I_3 are identical except for their signs.

A special case of the plane stress state is given if only the two normal stresses σ_{xx} and σ_{yy} are present, but no shear stresses, and the two normal stresses are furthermore identical: $\sigma_{xx} = \sigma_{yy}$. In this case, the transformation equations (3.26) can be used to show that the normal stresses are identical in any given section and, moreover, always assume the same value. This stress state is also called hydrostatic stress state, due to the fact that the stress state in a point in a resting fluid is also the same in all directions.

3.5.2   Principal Stresses

After we have derived transformation equations for the plane stress state for arbitrary section directions in a point of a solid, we now want to consider the question under which direction the stresses assume extreme values and how large these extreme values are. First, we consider the two normal stresses and solve the following two extreme value problems:

$$\frac{d\sigma_{\xi\xi}}{d\theta} = 0, \quad \frac{d\sigma_{\eta\eta}}{d\theta} = 0. \tag{3.28}$$

Both equations lead to the same result for the angle θ_p under which the extreme normal stresses occur:

$$\tan 2\theta_p = \frac{2\tau_{xy}}{\sigma_{xx} - \sigma_{yy}}. \tag{3.29}$$

The direction marked by the angle θ_p is also referred to as the so-called principal direction, and the corresponding axes ξ and η as the principal axes. Since the tan-function is a periodic function that is periodic with π, (3.29) always results in two angles θ_p and $\theta_p + \frac{\pi}{2}$ under which the normal stresses assume extreme values. These two directions are perpendicular to each other and, as we will show later, entail equal results. If we insert the result (3.29) into the transformation equations (3.26), the following equations for the determination of the extremal normal stresses result. For the sake of clarity, we will refer to the extremal stresses as σ_1 and σ_2:

$$\sigma_{1,2} = \frac{\sigma_{xx} + \sigma_{yy}}{2} \pm \sqrt{\left(\frac{\sigma_{xx} - \sigma_{yy}}{2}\right)^2 + \tau_{xy}^2}. \tag{3.30}$$

The two extremal normal stresses σ_1 and σ_2, one of which represents the maximum value and the other the minimum value, are the so-called principal stresses in accordance with the directions under which they occur. They are numbered so that the greater principal stress is σ_1, i.e. $\sigma_1 > \sigma_2$.

Furthermore, it is shown that if θ_p or $\theta_p + \frac{\pi}{2}$ is inserted into the third transformation equation in (3.26), a vanishing shear stress $\tau_{\xi\eta}$ is obtained. Obviously, in such directions in which the normal stresses assume their extreme values the shear stresses always disappear.

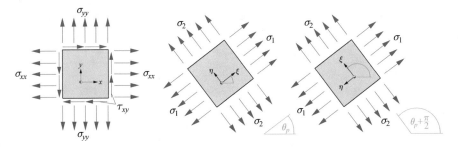

Fig. 3.10 Equality of the principal directions θ_p and $\theta_p + \frac{\pi}{2}$

The equality of the two principal directions θ_p and $\theta_p + \frac{\pi}{2}$ is briefly illustrated in Fig. 3.10. It can be seen that the principal stresses occurring in the two principal directions are always identical, regardless of whether the principal axis system is considered under θ_p or under $\theta_p + \frac{\pi}{2}$. Only the axes ξ, η swap their positions which, however, has no influence on the resulting stress state.

Besides the question in which direction the two extremal normal stresses arise it is also of fundamental interest to determine in which direction the extreme shear stresses occur and which values they assume. For this we consider the following extreme value problem:

$$\frac{d\tau_{\xi\eta}}{d\theta} = 0, \tag{3.31}$$

which, after elementary transformations, leads to the following equation for the associated angle θ'_p:

$$\tan 2\theta'_p = -\frac{\sigma_{xx} - \sigma_{yy}}{2\tau_{xy}}. \tag{3.32}$$

It turns out that when determining the principal direction for the extremal shear stresses, just like when considering the principal stresses, one always encounters two possible and equal directions, namely the angle θ'_p and the direction $\theta'_p + \frac{\pi}{2}$. In addition, the comparison between (3.29) and (3.32) shows that there is obviously a connection as follows:

$$\tan 2\theta'_p = -\frac{1}{\tan 2\theta_p}. \tag{3.33}$$

so that the two directions $2\theta'_p$ and $2\theta_p$ are oriented perpendicularly to each other. As a result, this means that the direction θ'_p of the extremal shear stresses is rotated by 45° compared to the direction θ_p of the principal normal stresses. The associated extremal shear stress which we want to denote as the principal shear stress $\tau_{\max}$ results from inserting θ'_p according to (3.32) into the transformation equations (3.26). It follows:

$$\tau_{\max} = \pm\sqrt{\left(\frac{\sigma_{xx} - \sigma_{yy}}{2}\right)^2 + \tau_{xy}^2}. \tag{3.34}$$

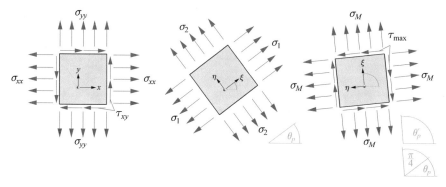

Fig. 3.11 Initial stress state (left), principal normal stresses (middle), extremal shear stresses (right)

Analogous to the spatial case (cf. Eq. (2.30)) it can also be represented with the help of the principal stresses σ_1 and σ_2 as:

$$\tau_{\max} = \pm\frac{1}{2}\left(\sigma_1 - \sigma_2\right). \tag{3.35}$$

From the transformation equations (3.26), after inserting θ'_p for the normal stresses we achieve the result that both normal stresses are identical under this direction, and the resulting value which we want to denote as σ_M can be calculated as follows:

$$\sigma_M = \frac{1}{2}\left(\sigma_{xx} + \sigma_{yy}\right). \tag{3.36}$$

Due to the invariance of the sum of the two normal stresses this can also be represented as follows:

$$\sigma_M = \frac{1}{2}\left(\sigma_1 + \sigma_2\right). \tag{3.37}$$

Thus, in sections in which the extremal shear stresses occur, the normal stresses generally do not disappear.

The different stress states discussed so far are shown in Fig. 3.11.

3.5.3 Mohr's Circle

The transformation rules discussed so far as well as the knowledge gained therefrom with regard to the various stress states allow for a particularly clear interpretation in addition to the mathematical treatment, namely with the help of Mohr's circle (see also Fig. 2.6). We obtain the describing circle equation starting from the transformation equations (3.26) by squaring the first and third equations and adding them up, thus eliminating the angle θ. We obtain:

$$\left[\sigma_{\xi\xi} - \frac{1}{2}\left(\sigma_{xx} + \sigma_{yy}\right)\right]^2 + \tau_{\xi\eta}^2 = \left(\frac{\sigma_{xx} - \sigma_{yy}}{2}\right)^2 + \tau_{xy}^2. \tag{3.38}$$

If the stresses σ_{xx}, σ_{yy} and τ_{xy} are given in the initial state, then the right side of Eq. (3.38) represents a constant quantity that we abbreviate with r^2. We obtain:

$$\left[\sigma_{\xi\xi} - \frac{1}{2}\left(\sigma_{xx} + \sigma_{yy}\right)\right]^2 + \tau_{\xi\eta}^2 = r^2, \tag{3.39}$$

where

$$r^2 = \left(\frac{\sigma_{xx} - \sigma_{yy}}{2}\right)^2 + \tau_{xy}^2. \tag{3.40}$$

Using (3.36) yields:

$$\left(\sigma_{\xi\xi} - \sigma_M\right)^2 + \tau_{\xi\eta}^2 = r^2. \tag{3.41}$$

We would also obtain the same result if we had used the second and third equations of the transformation equations (3.26). We can therefore omit the indices ξ and η for the moment and have:

$$(\sigma - \sigma_M)^2 + \tau^2 = r^2. \tag{3.42}$$

Apparently this equation represents a circle equation in the $\sigma\tau$-plane. The corresponding circle is the so-called Mohr's stress circle. It has the radius r and the center point $(\sigma_M, 0)$.

From the relations derived so far, Mohr's stress circle can be constructed if the stresses in any initial state with the values σ_{xx}, σ_{yy} and τ_{xy} are known. For this purpose a normal stress axis σ and a shear stress axis τ perpendicular to it are introduced. First, the two normal stresses σ_{xx}, σ_{yy} on the σ−axis are drawn. The shear stresses are then plotted over the two points determined in this way, where τ_{xy} is plotted over σ_{xx} with the correct sign and at σ_{yy} with the opposite sign. With that, the two points (σ_{xx}, τ_{xy}) and $(\sigma_{yy}, -\tau_{xy})$ are fixed and can be connected with a straight line. The resulting straight line intersects the σ−axis at σ_M. This point of intersection is then, by definition, the center point of Mohr's stress circle, and the length of the straight line between the two points on the circle is $2r$. With a known center point and the two determined points (σ_{xx}, τ_{xy}), $(\sigma_{yy}, -\tau_{xy})$, Mohr's stress circle can then be drawn. It is shown in Fig. 3.12. The state of stress under any section of the considered disk is represented by a point on the stress circle. Once the stress circle is given as shown in Fig. 3.12, the stress state can be determined graphically for any other direction. It is obvious that the principal normal stresses σ_1 and σ_2 as well as the principal shear stresses τ_{max} can be determined directly from Mohr's circle (Figs. 3.13 and 3.14). It should be noted that due to the involved equations the double angle θ must always be applied, and in the case of a positive angle always in a clockwise manner. In the same way, any other stress state under the angle θ can be determined from Mohr's stress circle (Fig. 3.15).

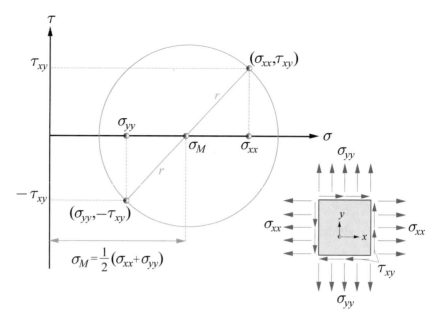

Fig. 3.12 Mohr's circle (left) and associated stress state (right)

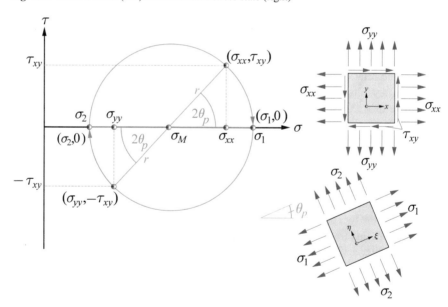

Fig. 3.13 Determination of σ_1 and σ_2 using Mohr's stress circle

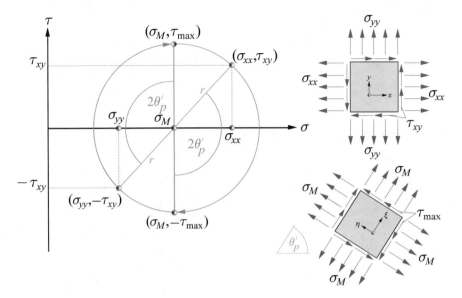

Fig. 3.14 Determination of τ_{max} using Mohr's stress circle

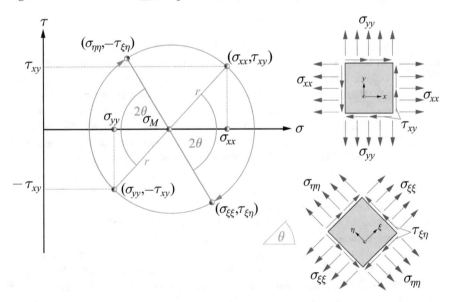

Fig. 3.15 Determination of the stress state under an arbitrary angle θ using Mohr's stress circle

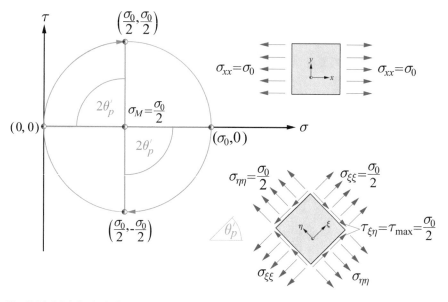

Fig. 3.16 Mohr's circle for a disk under uniaxial tensile load $\sigma_{xx} = \sigma_0$

Some interesting special cases of the plane state of stress in a disk which can be explained in a particularly illustrative way using Mohr's stress circle are briefly discussed below. The first case is a disk that is under uniaxial tension σ_{xx} in the x–direction, where σ_{xx} has the value $\sigma_{xx} = \sigma_0$. The two stresses σ_{yy} and τ_{xy} are zero. The resulting stress circle is shown in Fig. 3.16. In this case the circle touches the τ–axis on its left side, and the given initial stress state is already the principal stress state so that $\theta_p = 0°$. Here $\sigma_{xx} = \sigma_0$ already represents the maximum principal stress σ_1 while $\sigma_{yy} = 0$ is the second and here the minimum principal stress σ_2. The center of the circle is then located at $(\sigma_M, 0) = \left(\frac{\sigma_0}{2}, 0\right)$. Consequently, based on Mohr's stress circle the angle θ'_p takes on the value 45°. The maximum shear stress $\tau_{\max}$ then takes on the value $\tau_{\max} = \frac{\sigma_0}{2}$, as do the two normal stresses $\sigma_{\xi\xi}$ and $\sigma_{\eta\eta}$. The corresponding stress state is shown in Fig. 3.16, bottom right.

Another technically extremely important special case is when the shear stress τ_{xy} is given with the value τ_0 and the normal stresses σ_{xx} and σ_{yy} are zero. This case is also referred to as a pure shear stress state. The corresponding Mohr's stress circle is given in Fig. 3.17. In this case the center point of Mohr's circle lies exactly on the origin of the $\sigma\tau$ reference system, and the circle is intersected exactly in the middle by both axes. It is immediately evident that the given pure shear stress state also represents the principal shear stress state, so that $\theta'_p = 0°$ applies and the given shear stress τ_0 is also the maximum shear stress $\tau_{\max}$. Accordingly, the angle θ_p with the value 45° is present, and the resulting principal normal stress state is also indicated in Fig. 3.17. The two principal stresses then take on the values $\sigma_{\xi\xi} = \sigma_1 = \tau_0$ and $\sigma_{\eta\eta} = \sigma_2 = -\tau_0$. In this state which is shown in Fig. 3.17, bottom right, the shear

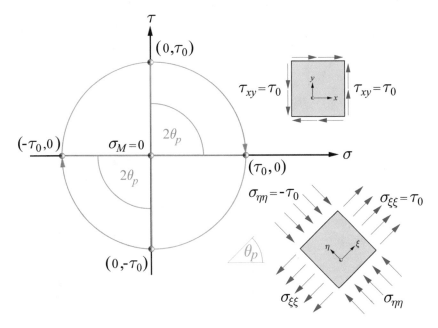

Fig. 3.17 Mohr's stress circle for a disk under pure shear $\tau_{xy} = \tau_0$

stress τ_{xy} disappears. Apparently, a pure shear stress state is equivalent to a pure normal stress state that is rotated by 45° compared to the reference system and in which one normal stress occurs as a tensile stress with the value τ_0, while the other normal stress is a compressive stress of the same amount.

As a last special case we want to consider the so-called hydrostatic stress state, in which the two normal stresses σ_{xx} and σ_{yy} are given with the value σ_0, but no shear stress occurs. The corresponding Mohr's stress circle is given in Fig. 3.18. Obviously Mohr's circle is reduced to a point on the σ−axis which can be found at $\sigma_M = \sigma_0$. A transformation of the stress state thus provides the same stress state for any angle θ, so that in this case every angle results in the principal directions with the principal normal stresses. The two given normal stresses σ_{xx} and σ_{yy} are thus simultaneously the principal stresses σ_1 and σ_2 in any axis system, both of which have the same value $\sigma_1 = \sigma_2 = \sigma_0$.

3.6   Strain Transformation

In addition to considering the transformation behavior of the stresses σ_{xx}, σ_{yy} and τ_{xy} there is an inherent interest in clarifying the corresponding transformation rules for the strain components ε_{xx}, ε_{yy} and γ_{xy}, which for the present plane case are summarized in the infinitesimal strain tensor $\underline{\underline{\varepsilon}}$:

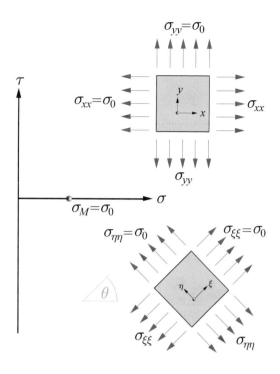

$$\underline{\underline{\varepsilon}} = \begin{bmatrix} \varepsilon_{xx} & \frac{1}{2}\gamma_{xy} \\ \frac{1}{2}\gamma_{xy} & \varepsilon_{yy} \end{bmatrix}. \tag{3.43}$$

Similar to the transformation properties of Cauchy's stress tensor $\underline{\underline{\sigma}}$, analogous transformation rules also apply to the infinitesimal strain tensor $\underline{\underline{\varepsilon}}$ so that in a coordinate system ξ, η rotated by the angle θ with respect to the x, y reference system the following strain components result:

$$\varepsilon_{\xi\xi} = \frac{1}{2}\left(\varepsilon_{xx} + \varepsilon_{yy}\right) + \frac{1}{2}\left(\varepsilon_{xx} - \varepsilon_{yy}\right)\cos 2\theta + \frac{1}{2}\gamma_{xy}\sin 2\theta,$$

$$\varepsilon_{\eta\eta} = \frac{1}{2}\left(\varepsilon_{xx} + \varepsilon_{yy}\right) - \frac{1}{2}\left(\varepsilon_{xx} - \varepsilon_{yy}\right)\cos 2\theta - \frac{1}{2}\gamma_{xy}\sin 2\theta,$$

$$\frac{1}{2}\gamma_{\xi\eta} = -\frac{1}{2}\left(\varepsilon_{xx} - \varepsilon_{yy}\right)\sin 2\theta + \frac{1}{2}\gamma_{xy}\cos 2\theta. \tag{3.44}$$

Exactly as in the case of the plane stress state, a clearly defined angle θ_p can also be determined for the state of strain under which extremal normal strains (the so-called principal strains) ε_1 and ε_2 occur. The angle θ_p follows as:

Fig. 3.19 Orthotropic
material

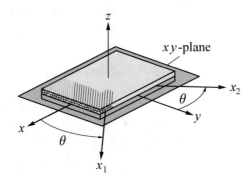

$$\tan 2\theta_p = \frac{\gamma_{xy}}{\varepsilon_{xx} - \varepsilon_{yy}}. \tag{3.45}$$

The principal normal strains result as:

$$\varepsilon_{1,2} = \frac{\varepsilon_{xx} + \varepsilon_{yy}}{2} \pm \sqrt{\left(\frac{\varepsilon_{xx} - \varepsilon_{yy}}{2}\right)^2 + \left(\frac{1}{2}\gamma_{xy}\right)^2}. \tag{3.46}$$

In exactly the same way Mohr's circle can be created for the state of strain which, however, is not discussed at this point.

3.7 Formulation for Orthotropic Materials

Many materials in lightweight engineering such as fiber composite materials are orthotropic or have transversely isotropic properties. In this section we are assuming orthotropic material properties, from which all relationships for transverse isotropy can be deducted as a special case. In the following the transformed axes ξ and η are the principal axes of the considered orthotropic material (on-axis system), so that in this section we will refer to the reference axes x_1 and x_2 instead of ξ and η. The reference axes x, y, z are an arbitrary reference system (off-axis system), where x and y are perpendicular to each other and the axes x and x_1 or y and x_2 each rotated against each other by the angle θ about the fixed $z-$ axis (Fig. 3.19).

3.7.1 Plane State of Stress

Hooke's generalized law (2.143) can be given for an orthotropic disk in the principal axis system as follows:

$$
\begin{pmatrix} \varepsilon_{11} \\ \varepsilon_{22} \\ \gamma_{12} \end{pmatrix} = \begin{bmatrix} S_{11} & S_{12} & 0 \\ S_{12} & S_{22} & 0 \\ 0 & 0 & S_{66} \end{bmatrix} \begin{pmatrix} \sigma_{11} \\ \sigma_{22} \\ \tau_{12} \end{pmatrix}. \tag{3.47}
$$

When using engineering constants this reads:

$$
\begin{pmatrix} \varepsilon_{11} \\ \varepsilon_{22} \\ \gamma_{12} \end{pmatrix} = \begin{bmatrix} \dfrac{1}{E_{11}} & -\dfrac{\nu_{21}}{E_{22}} & 0 \\ -\dfrac{\nu_{12}}{E_{11}} & \dfrac{1}{E_{22}} & 0 \\ 0 & 0 & \dfrac{1}{G_{12}} \end{bmatrix} \begin{pmatrix} \sigma_{11} \\ \sigma_{22} \\ \tau_{12} \end{pmatrix}. \tag{3.48}
$$

In a plane state of stress a normal strain ε_{zz} can occur which can be deduced from (2.143) by imposing $\sigma_{zz} = 0$:

$$
\varepsilon_{zz} = S_{13}\sigma_{11} + S_{23}\sigma_{22}. \tag{3.49}
$$

The inverse relation to (3.48) reads:

$$
\begin{pmatrix} \sigma_{11} \\ \sigma_{22} \\ \tau_{12} \end{pmatrix} = \begin{bmatrix} C_{11} & C_{12} & C_{13} & 0 \\ C_{12} & C_{22} & C_{23} & 0 \\ 0 & 0 & 0 & C_{66} \end{bmatrix} \begin{pmatrix} \varepsilon_{11} \\ \varepsilon_{22} \\ \varepsilon_{zz} \\ \gamma_{12} \end{pmatrix}. \tag{3.50}
$$

Since the normal stress σ_{zz} has to disappear in the plane stress state, the strain ε_{zz} can be eliminated from (3.50). For this purpose, the generalized Hooke's law in the form (2.142) is considered, and from the third line of (2.142) we get with $\sigma_{zz} = 0$:

$$
0 = C_{13}\varepsilon_{11} + C_{23}\varepsilon_{22} + C_{33}\varepsilon_{zz}. \tag{3.51}
$$

We can solve this expression for ε_{zz} and obtain:

$$
\varepsilon_{zz} = -\frac{C_{13}}{C_{33}}\varepsilon_{11} - \frac{C_{23}}{C_{33}}\varepsilon_{22}. \tag{3.52}
$$

Inserting into the first two equations of (3.50) yields:

$$
\sigma_{11} = \left(C_{11} - \frac{C_{13}^2}{C_{33}} \right)\varepsilon_{11} + \left(C_{12} - \frac{C_{13}C_{23}}{C_{33}} \right)\varepsilon_{22},
$$
$$
\sigma_{22} = \left(C_{12} - \frac{C_{13}C_{23}}{C_{33}} \right)\varepsilon_{11} + \left(C_{22} - \frac{C_{23}^2}{C_{33}} \right)\varepsilon_{22}. \tag{3.53}
$$

The third equation in (3.50) remains unchanged.

It is convenient to introduce the so-called reduced stiffnesses $Q_{11}, Q_{22}, Q_{12}, Q_{66}$ at this point so that (3.50) takes the following form:

$$\begin{pmatrix} \sigma_{11} \\ \sigma_{22} \\ \tau_{12} \end{pmatrix} = \begin{bmatrix} Q_{11} & Q_{12} & 0 \\ Q_{12} & Q_{22} & 0 \\ 0 & 0 & Q_{66} \end{bmatrix} \begin{pmatrix} \varepsilon_{11} \\ \varepsilon_{22} \\ \gamma_{12} \end{pmatrix}. \tag{3.54}$$

In a symbolic notation we can write (3.54) as:

$$\underline{\sigma} = \underline{\underline{Q}}\,\underline{\varepsilon}. \tag{3.55}$$

The reduced stiffnesses $Q_{11}, Q_{22}, Q_{12}, Q_{66}$ are defined as follows:

$$Q_{11} = C_{11} - \frac{C_{13}^2}{C_{33}}, \quad Q_{12} = C_{12} - \frac{C_{13}C_{23}}{C_{33}}, \quad Q_{22} = C_{22} - \frac{C_{23}^2}{C_{33}}, \quad Q_{66} = C_{66}. \tag{3.56}$$

Formulated in the engineering constants we have:

$$Q_{11} = \frac{E_{11}}{1 - \nu_{12}\nu_{21}}, \quad Q_{22} = \frac{E_{22}}{1 - \nu_{12}\nu_{21}}, \quad Q_{12} = \frac{\nu_{12}E_{22}}{1 - \nu_{12}\nu_{21}}, \quad Q_{66} = G_{12}. \tag{3.57}$$

The engineering constants $E_{11}, E_{22}, G_{12}, \nu_{12}$ and ν_{21} are assumed to be associated with the material principal axes.

The relations (3.47) or (3.48) and (3.54) refer to the principal material axes of the orthotropic material under consideration. Of course, an orthotropic material such as e.g. a fiber-reinforced plastic can be arranged in any orientation with respect to the reference axes x, y, so that it is important to consider such a case, namely the material oriented under the angle θ in the xy−plane in a plane state of stress. The transformation rules with regard to the stresses and the strains can be taken directly from the previous explanations and are as follows:

$$\begin{pmatrix} \sigma_{xx} \\ \sigma_{yy} \\ \tau_{xy} \end{pmatrix} = \begin{bmatrix} \cos^2\theta & \sin^2\theta & -2\cos\theta\sin\theta \\ \sin^2\theta & \cos^2\theta & 2\cos\theta\sin\theta \\ \cos\theta\sin\theta & -\cos\theta\sin\theta & \cos^2\theta - \sin^2\theta \end{bmatrix} \begin{pmatrix} \sigma_{11} \\ \sigma_{22} \\ \tau_{12} \end{pmatrix}$$

$$= \underline{\underline{T}} \begin{pmatrix} \sigma_{11} \\ \sigma_{22} \\ \tau_{12} \end{pmatrix}, \tag{3.58}$$

and

$$\begin{pmatrix} \varepsilon_{11} \\ \varepsilon_{22} \\ \gamma_{12} \end{pmatrix} = \begin{bmatrix} \cos^2\theta & \sin^2\theta & \cos\theta\sin\theta \\ \sin^2\theta & \cos^2\theta & -\cos\theta\sin\theta \\ -2\cos\theta\sin\theta & 2\cos\theta\sin\theta & \cos^2\theta - \sin^2\theta \end{bmatrix} \begin{pmatrix} \bar{\varepsilon}_{11} \\ \bar{\varepsilon}_{22} \\ \bar{\gamma}_{12} \end{pmatrix}$$

$$= \underline{\underline{T}}^T \begin{pmatrix} \varepsilon_{xx} \\ \varepsilon_{yy} \\ \gamma_{xy} \end{pmatrix}. \tag{3.59}$$

Note that the angle θ relates to the transformation of the principal material axes x_1 and x_2 with respect to the reference axes x and y. From (3.58) and (3.59) the following relation between the stresses and the strains with respect to the reference axes x and y can be derived:

$$\begin{pmatrix} \sigma_{xx} \\ \sigma_{yy} \\ \tau_{xy} \end{pmatrix} = \underline{\underline{T}}\,\underline{\underline{Q}}\,\underline{\underline{T}}^T \begin{pmatrix} \varepsilon_{xx} \\ \varepsilon_{yy} \\ \gamma_{xy} \end{pmatrix} = \underline{\underline{\bar{Q}}} \begin{pmatrix} \varepsilon_{xx} \\ \varepsilon_{yy} \\ \gamma_{xy} \end{pmatrix}. \tag{3.60}$$

Therein, $\underline{\underline{\bar{Q}}} = \underline{\underline{T}}\,\underline{\underline{Q}}\,\underline{\underline{T}}^T$ is the matrix of the transformed reduced stiffnesses. It is generally fully occupied, and we obtain:

$$\begin{pmatrix} \sigma_{xx} \\ \sigma_{yy} \\ \tau_{xy} \end{pmatrix} = \begin{bmatrix} \bar{Q}_{11} & \bar{Q}_{12} & \bar{Q}_{16} \\ \bar{Q}_{12} & \bar{Q}_{22} & \bar{Q}_{26} \\ \bar{Q}_{16} & \bar{Q}_{26} & \bar{Q}_{66} \end{bmatrix} \begin{pmatrix} \varepsilon_{xx} \\ \varepsilon_{yy} \\ \gamma_{xy} \end{pmatrix}. \tag{3.61}$$

The quantities $\bar{Q}_{11}, \bar{Q}_{22}, \bar{Q}_{12}, \bar{Q}_{66}, \bar{Q}_{16}, \bar{Q}_{26}$ are the so-called transformed reduced stiffnesses. Evaluation of the matrix product $\underline{\underline{T}}\,\underline{\underline{Q}}\,\underline{\underline{T}}^T$ results in the following transformation rules for the transformed reduced stiffnesses:

$$\begin{aligned}
\bar{Q}_{11} &= Q_{11}\cos^4\theta + 2(Q_{12} + 2Q_{66})\cos^2\theta\sin^2\theta + Q_{22}\sin^4\theta, \\
\bar{Q}_{22} &= Q_{11}\sin^4\theta + 2(Q_{12} + 2Q_{66})\cos^2\theta\sin^2\theta + Q_{22}\cos^4\theta, \\
\bar{Q}_{12} &= (Q_{11} + Q_{22} - 4Q_{66})\cos^2\theta\sin^2\theta + Q_{12}\left(\cos^4\theta + \sin^4\theta\right), \\
\bar{Q}_{66} &= (Q_{11} + Q_{22} - 2Q_{12} - 2Q_{66})\cos^2\theta\sin^2\theta + Q_{66}\left(\cos^4\theta + \sin^4\theta\right), \\
\bar{Q}_{16} &= (Q_{11} - Q_{12} - 2Q_{66})\cos^3\theta\sin\theta + (Q_{12} - Q_{22} + 2Q_{66})\cos\theta\sin^3\theta, \\
\bar{Q}_{26} &= (Q_{11} - Q_{12} - 2Q_{66})\cos\theta\sin^3\theta + (Q_{12} - Q_{22} + 2Q_{66})\cos^3\theta\sin\theta.
\end{aligned} \tag{3.62}$$

A graphic representation of the transformed reduced stiffnesses $\bar{Q}_{11}, \bar{Q}_{22}, \bar{Q}_{12}, \bar{Q}_{66}, \bar{Q}_{16}, \bar{Q}_{26}$ depending on the angle θ for $0° \leq \theta \leq 90°$ is shown in Fig. 3.20. The on-axis material data of a typical carbon fiber reinforced plastic are given with $E_{11} = 138000\text{MPa}$, $E_{22} = 8960\text{MPa}$, $G_{12} = 7100\text{MPa}$ and $\nu_{12} = 0.30$. Here x_1 is the principal material axis with the dominant properties which in this case is the fiber direction. The results contained in Fig. 3.20 show a behavior that is typical for an orthotropic material. The reduced stiffnesses $\bar{Q}_{11}$ and $\bar{Q}_{22}$ show their maximum and minimum values at $\theta = 0°$. This is immediately clear if one keeps in mind that the fiber direction in the given example is the x_1-direction which is identical to the x-axis at $\theta = 0°$. The situation is reversed when the angle $\theta = 90°$ is considered. Then the fibers point in y-direction, and accordingly $\bar{Q}_{11}$ and $\bar{Q}_{22}$ assume their minimum and maximum values.

A further interesting feature is the reduced stiffness $\bar{Q}_{66}$ which is a measure of the shear stiffness of an orthotropic disk. The maximum value of this stiffness is reached at $\theta = 45°$. Hence, if a lightweight design requires achieving the highest possible shear rigidity when using an orthotropic material, then the alignment of the orthotropic material under the angle $\theta = 45°$ is particularly useful.

Lastly, the transformed reduced stiffnesses $\bar{Q}_{16}$ and $\bar{Q}_{26}$ which couple normal stresses with shear strains and shear stresses with normal strains require discussion. These quantities become zero under at the angles $\theta = 0°$ and $\theta = 90°$ but assume nonzero values for all other considered angles. It turns out that even when considering disk structures, a certain coupling effect, namely the so-called shear coupling, occurs in which, for example, a pure tensile load causes a shear strain γ_{xy} in addition to the two normal strains ε_{xx} and ε_{yy} (Fig. 3.21). Such coupling effects are typical especially for fiber-reinforced plastics and must always be appropriately taken into account in a structural analysis.

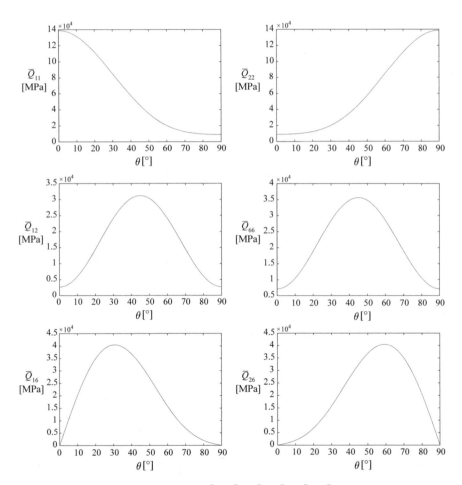

Fig. 3.20 Transformed reduced stiffnesses $\bar{Q}_{11}, \bar{Q}_{22}, \bar{Q}_{12}, \bar{Q}_{66}, \bar{Q}_{16}, \bar{Q}_{26}$ as functions of the angle θ for $0° \leq \theta \leq 90°$

Fig. 3.21 Shear coupling at the example of a fiber reinforced plastic

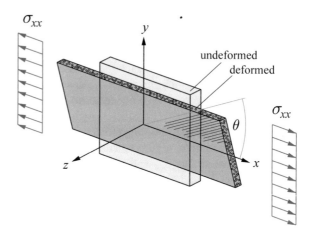

3.7.2 Plane State of Strain

In the case of an orthotropic disk in a plane state of strain the displacement w and the strain ε_{zz} in the direction of the thickness are prohibited. The corresponding relations (3.3), (3.4), (3.5), (3.11) and (3.12) also apply to an orthotropic disk. It can be easily shown that the generalized Hooke's law can be written as follows:

$$
\begin{pmatrix} \sigma_{11} \\ \sigma_{22} \\ \tau_{12} \end{pmatrix} = \begin{bmatrix} C_{11} & C_{12} & 0 \\ C_{12} & C_{22} & 0 \\ 0 & 0 & C_{66} \end{bmatrix} \begin{pmatrix} \varepsilon_{11} \\ \varepsilon_{22} \\ \gamma_{12} \end{pmatrix} ,
\tag{3.63}
$$

and in inverted form:

$$
\begin{pmatrix} \varepsilon_{11} \\ \varepsilon_{22} \\ \gamma_{12} \end{pmatrix} = \begin{bmatrix} R_{11} & R_{12} & 0 \\ R_{12} & R_{22} & 0 \\ 0 & 0 & R_{66} \end{bmatrix} \begin{pmatrix} \sigma_{11} \\ \sigma_{22} \\ \tau_{12} \end{pmatrix} .
\tag{3.64}
$$

Therein the quantities R_{ij} $(i, j = 1, 2, 6)$ are the so-called reduced compliances. They can be determined from the compliances S_{ij} as follows:

$$
R_{ij} = S_{ij} - \frac{S_{i3} S_{j3}}{S_{33}}.
\tag{3.65}
$$

The normal stress σ_{zz} in the thickness direction of the considered disk that generally also occurs in a plane state of strain can be determined as:

$$
\sigma_{zz} = -\frac{1}{S_{33}} \left(S_{13} \sigma_{11} + S_{23} \sigma_{22} \right).
\tag{3.66}
$$

Fig. 3.22 Polar coordinates r, φ

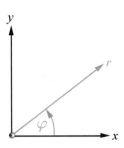

The transformation rules concerning the reduced compliances R_{ij} can be determined in much the same way as for the reduced stiffnesses Q_{ij}. We will not go into deeper details at this point.

3.8 Formulation for Polar Coordinates

In many cases, plane problems can be described advantageously by using polar coordinates r, φ (Fig. 3.22). The relation between x, y and r, φ can be given as:

$$x = r \cos \varphi, \quad y = r \sin \varphi, \quad r = \sqrt{x^2 + y^2}, \quad \varphi = \arctan \left(\frac{y}{x} \right). \qquad (3.67)$$

Let us consider such a problem initially in the plane stress state. Then the equations already derived in Chapter 2 for the spatial case in cylindrical coordinates can be reused directly by neglecting all quantities that concern the thickness direction z. Hence, from the equilibrium conditions (2.219) and (2.220) we have:

$$\frac{\partial \sigma_{rr}}{\partial r} + \frac{1}{r} \frac{\partial \tau_{r\varphi}}{\partial \varphi} + \frac{\sigma_{rr} - \sigma_{\varphi\varphi}}{r} + f_r = 0,$$

$$\frac{\partial \tau_{r\varphi}}{\partial r} + \frac{1}{r} \frac{\partial \sigma_{\varphi\varphi}}{\partial \varphi} + 2 \frac{\tau_{r\varphi}}{r} + f_\varphi = 0. \qquad (3.68)$$

The kinematic equations (2.226) reduce to:

$$\varepsilon_{rr} = \frac{\partial u}{\partial r}, \quad \varepsilon_{\varphi\varphi} = \frac{1}{r} \frac{\partial v}{\partial \varphi} + \frac{u}{r}, \quad \gamma_{r\varphi} = \frac{\partial v}{\partial r} + \frac{1}{r} \frac{\partial u}{\partial \varphi} - \frac{v}{r}, \qquad (3.69)$$

and the only remaining compatibility equation reads:

$$\frac{\partial^2 \varepsilon_{\varphi\varphi}}{\partial r^2} + \frac{1}{r^2} \frac{\partial^2 \varepsilon_{rr}}{\partial \varphi^2} + \frac{2}{r} \frac{\varepsilon_{\varphi\varphi}}{\partial r} - \frac{1}{r} \frac{\partial \varepsilon_{rr}}{\partial r} = \frac{1}{r} \frac{\partial^2 \gamma_{r\varphi}}{\partial r \partial \varphi} + \frac{1}{r^2} \frac{\partial \gamma_{r\varphi}}{\partial \varphi}. \qquad (3.70)$$

Hooke's law is reduced to the following form in the case of the plane stress state:

$$\begin{pmatrix} \sigma_{rr} \\ \sigma_{\varphi\varphi} \\ \tau_{r\varphi} \end{pmatrix} = \begin{bmatrix} Q_{11} & Q_{12} & 0 \\ Q_{12} & Q_{22} & 0 \\ 0 & 0 & Q_{66} \end{bmatrix} \begin{pmatrix} \varepsilon_{rr} \\ \varepsilon_{\varphi\varphi} \\ \gamma_{r\varphi} \end{pmatrix}, \tag{3.71}$$

and in its inverted form:

$$\begin{pmatrix} \varepsilon_{rr} \\ \varepsilon_{\varphi\varphi} \\ \gamma_{r\varphi} \end{pmatrix} = \begin{bmatrix} S_{11} & S_{12} & 0 \\ S_{12} & S_{22} & 0 \\ 0 & 0 & S_{66} \end{bmatrix} \begin{pmatrix} \sigma_{rr} \\ \sigma_{\varphi\varphi} \\ \tau_{r\varphi} \end{pmatrix}. \tag{3.72}$$

If the case of the plane state of strain is considered, then the reduced compliances R_{ij} $(i, j = 1, 2, 6)$ according to (3.65) must be applied. The normal stress σ_{zz} follows analogously to (3.66).

In the case of an isotropic problem, the material law in the plane stress state is as follows:

$$\begin{pmatrix} \varepsilon_{rr} \\ \varepsilon_{\varphi\varphi} \\ \gamma_{r\varphi} \end{pmatrix} = \begin{bmatrix} \dfrac{1}{E} & -\dfrac{\nu}{E} & 0 \\ -\dfrac{\nu}{E} & \dfrac{1}{E} & 0 \\ 0 & 0 & \dfrac{1}{G} \end{bmatrix} \begin{pmatrix} \sigma_{rr} \\ \sigma_{\varphi\varphi} \\ \tau_{r\varphi} \end{pmatrix}. \tag{3.73}$$

In a plane state of strain, the equivalent stiffness values $\bar{E}$ and $\bar{\nu}$ are to be applied.

If the stresses σ_{xx}, σ_{yy}, τ_{xy} are given in Cartesian coordinates, then the corresponding components σ_{rr}, $\sigma_{\varphi\varphi}$, $\tau_{r\varphi}$ in polar coordinates can be determined using the transformation relations (3.26):

$$\sigma_{rr} = \frac{1}{2}\left(\sigma_{xx} + \sigma_{yy}\right) + \frac{1}{2}\left(\sigma_{xx} - \sigma_{yy}\right)\cos 2\theta + \tau_{xy}\sin 2\theta,$$

$$\sigma_{\varphi\varphi} = \frac{1}{2}\left(\sigma_{xx} + \sigma_{yy}\right) - \frac{1}{2}\left(\sigma_{xx} - \sigma_{yy}\right)\cos 2\theta - \tau_{xy}\sin 2\theta,$$

$$\tau_{r\varphi} = -\frac{1}{2}\left(\sigma_{xx} - \sigma_{yy}\right)\sin 2\theta + \tau_{xy}\cos 2\theta. \tag{3.74}$$

The sum of the two normal stresses is always invariant, i.e. $\sigma_{xx} + \sigma_{yy} = \sigma_{rr} + \sigma_{\varphi\varphi}$ applies.

Bibliography

Altenbach H, Altenbach J, Rikards R (1996) Einführung in die Mechanik der Laminat- und Sand-wichtragwerke. Deutscher Verlag der Grundstoffindustrie, Stuttgart, Germany

Ambartsumyan SA (1970) Theory of anisotropic plates. Technomic Publishing Co. Inc, Stamford, USA

Ashton JE, Whitney JM (1970) Theory of laminated plates. Technomic Publishing Co. Inc, Stamford, USA

Gross D, Hauger W, Schnell W, Wriggers P (1995) Technische Mechanik 4: Hydromechanik, Elemente der Höheren Mechanik, Numerische Methoden, 2nd edn. Springer Berlin et al, Germany

Jones RM (1975) Mechanics of composite materials. Scripta Book Co., Washington, USA

Lekhnitskii SG (1968) Anisotropic plates, Gordon and Breach. New York et al, USA

Mittelstedt C, Becker W (2016) Strukturmechanik ebener Laminate. Technische Universität Darmstadt, Germany, Studienbereich Mechanik

Reddy JN (2004) Mechanics of laminated composite plates and shells, 2nd edn. CRC Press Boca Raton et al, USA

Tsai SW, Hahn HT (1980) Introduction to composite materials. Technomic Publishing Co., Inc., Lancaster et al., UK

Chapter 4
Strength Criteria for Isotropic Materials

4.1 Introduction

One of the most important tasks of engineers in lightweight engineering practice is to be able to predict the strength of a structure, usually using means of analysis based on the stress state σ_{ij} that occurs in a lightweight structure. To predict the strength of a structure (i.e. making a statement about whether a given stress state is admissible or not, and what type of failure can be expected), the so-called failure criteria are employed to which this chapter is dedicated. Failure criteria are used to evaluate a given stress state with regard to its criticality. Many of the failure criteria that we will discuss below are of an older date but have proven themselves in practical use and are therefore still broadly employed. We will restrict our considerations in this chapter exclusively to the static analysis of lightweight structures.

First of all, the concept of so-called theoretical strength is considered. We assume a solid body that consists of a theoretically perfect material, i.e. free of defects and perfectly homogeneous. In the case of such a perfect material, it is assumed that its basic building blocks such as atoms, ions or molecules are perfectly regularly arranged in a periodic way. A failure of such a material then manifests itself in a loosening of the bonds between the basic building blocks of the material. The binding force F between two neighboring basic building blocks can be described approximately as:

$$F = -\frac{a}{r^m} + \frac{b}{r^n},\tag{4.1}$$

wherein $m < n$, and the constants a and b as well as m and n are to be chosen adequately. We consider the state of equilibrium between two adjacent building blocks, characterized by the distance $r = d_0$ (Fig. 4.1). If we now assume small deflections from this equilibrium position, then we can assume a linear approximation of the force $F(r)$. In the macroscopic sense, this corresponds to the application of

© The Author(s), under exclusive license to Springer Nature Switzerland AG 2021
C. Mittelstedt, *Structural Mechanics in Lightweight Engineering*,
https://doi.org/10.1007/978-3-030-75193-7_4

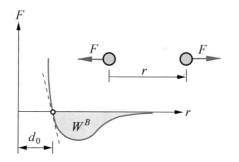

Fig. 4.1 Force-distance behaviour

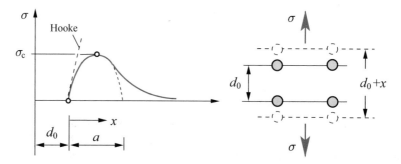

Fig. 4.2 Idealized derivation of the theoretical strength

Hooke's law to the mapping of the binding force F. If we now assume a perfectly regular lattice structure of the material under consideration, then the mechanical stress σ in the tensile range $x = r - d_0 > 0$ can be approximated by

$$\sigma \approx \sigma_c \sin\left(\frac{\pi x}{a}\right). \tag{4.2}$$

Here σ_c is the so-called cohesive strength or theoretical strength (Fig. 4.2).

If we assume small strains $\varepsilon = \frac{x}{d_0}$ and if we also assume that $d_0 \approx a$, then we have:

$$\sigma \approx \sigma_c \frac{\pi x}{a} = \sigma_c \pi \varepsilon, \tag{4.3}$$

so that

$$\sigma_c \pi = E \tag{4.4}$$

or

$$\sigma_c \approx \frac{E}{\pi}. \tag{4.5}$$

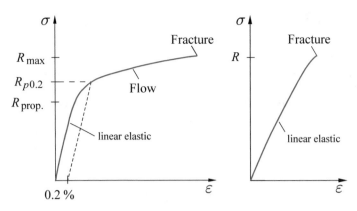

Fig. 4.3 Ductile and brittle material behavior

It can be concluded that the strength of a material and its elastic stiffness are closely related, so that the theoretical strength is in the order of magnitude of about one third of the modulus of elasticity E. If we consider the metallic material steel, with a modulus of elasticity of $E = 210,000$ MPa, a theoretical strength of about $\sigma_c \approx 70,000$ MPa can be concluded. Experimental results, however, show that this value overestimates the actual strength of polycrystalline steels by far. The reason for the massive deviations from the real strength is the inhomogeneous microstructure of the material as well as the occurrence of defects on the microstructural level, e.g. of microcracks, micro-holes, pores, crystalline grains, defects in the crystal lattice, and the like.

With regard to the strength assessment of a structure, some attention must also be paid to the stress-strain behavior and the resulting failure behavior. Basically, materials are divided into ductile and brittle behavior (see Fig. 4.3). Ductile material behavior is characterized by the fact that when a certain stress, the so-called proportionality stress R_{prop}, is exceeded, permanent deformations occur. This process is called plastic flow. The actual fracture, usually referred to as ductile fracture, only occurs with comparatively large inelastic deformations. From the point of view of micromechanics, plastic flow can be explained by the formation of pores and cavities and their coalescence. In lightweight engineering, this process is treated purely phenomenologically, i.e. by comparing an occurring stress state with a permissible stress or a limit stress. Here the so-called technical yield strength $R_{p\,0.2}$ is determined. This is the stress at which a plastic strain of 0.2% remains after the structure is relieved of stress.

Brittle material behavior, on the other hand, is usually associated with the fact that e.g. in a uniaxial tensile test no particularly noticeable deformation occurs until failure (usually referred to as brittle failure or fracture), and the sample then suddenly fails when the tensile strength of the material is reached. Typically, the deformations that occur before failure are reversible and not associated with plastic deformations.

It can already be stated at this point that both ductile and brittle material behavior as well as the resulting failure patterns depend on the stress state. Accordingly, a uniaxial tensile test will show a different failure pattern than will be the case with a four-point bending test. If there is a largely hydrostatic tensile stress state, then even ductile materials show a comparatively brittle behavior. Conversely, brittle materials often show a comparatively ductile behavior in a compressive hydrostatic stress state.

The engineering statement as to whether a given structure tends to fail under a given state of stress or strain is made with the help of so-called strength or failure criteria. Strength or failure criteria take the following very general form. If the state of stress is used to evaluate a given situation, then:

$$F(\sigma_{ij}) = 1. \tag{4.6}$$

Herein, F is the so-called failure function the concrete mathematical form of which depends on the employed failure criterion. In some cases it can also be useful to use a formulation of the kind

$$G = F - 1 = 0. \tag{4.7}$$

In general, this formulation of a failure criterion leads to a corresponding so-called failure (hyper-)surface in the six-dimensional space of the stresses or in the three-dimensional space of the principal stresses.

If the state of strain is employed for the estimation of the failure behavior then the following form of a failure criterion can be employed:

$$f(\varepsilon_{ij}) = 1. \tag{4.8}$$

Over time, a number of failure criteria have been developed which have proven themselves in various areas of engineering application and in a large number of cases are able to predict failure with good accuracy. A selection is presented in the following.

4.2 Principal Stress Hypothesis

Let the behavior of a material be given by its tensile strength R^t (t = tension) and its compressive strength R^c (c = compression). Within the framework of the so-called principal stress hypothesis according to Rankine,[1] Lamé[2] and Navier[3] it is now postulated that failure will occur if either the largest positive principal stress reaches the value R^t or when the smallest principal stress reaches the value $-|R^c|$. The

---

[1] William John Macquorn Rankine, 1820–1872, Scottish physicist.

[2] Gabriel Lamé, 1795–1870, French physicist.

[3] Claude Louis Marie Henri Navier, 1785–1836, French mathematician and physicist.

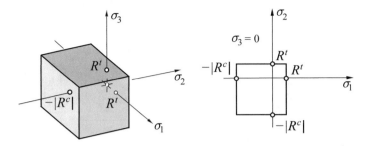

Fig. 4.4 Principal stress hypothesis

corresponding failure surface in the principal stress space then forms a cube (Fig. 4.4). If a plane stress state is considered, then the cube surface results in a square in the $\sigma_1 - \sigma_2$ plane.

The principal stress hypothesis most closely describes a brittle material failure. A major disadvantage, however, is that any interaction between the principal stresses is not taken into account. However, a comparison with experimental findings shows that this does not agree with reality for many materials, so that the principal stress hypothesis is only applicable to a very limited extent.

4.3 Principal Strain Hypothesis

The principal strain hypothesis according to Saint-Venant[4] and Bach[5] represents an improvement of the principal stress hypothesis insofar as a certain interaction of the principal stresses via transverse contraction is taken into account. The principal strain hypothesis assumes failure if the largest principal strain assumes a limit strain value e^t. With the corresponding tensile strength $R^t = E\,e^t$ the following failure conditions result:

$$\sigma_1 - \nu(\sigma_2 + \sigma_3) = R^t, \quad \sigma_2 - \nu(\sigma_3 + \sigma_1) = R^t, \quad \sigma_3 - \nu(\sigma_1 + \sigma_2) = R^t. \quad (4.9)$$

The correspondingly composed failure surface represents a three-sided pyramid around the hydrostatic axis ($\sigma_1 = \sigma_2 = \sigma_3$) with the vertex

$$\sigma_1 = \sigma_2 = \sigma_3 = \frac{R^t}{1 - 2\nu}. \quad (4.10)$$

---

[4]Adhémar Jean Claude Barré de Saint-Venant, 1797–1886, French scientist.
[5]Carl Julius von Bach, 1847–1931, German engineer.

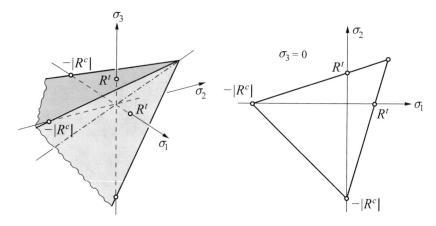

Fig. 4.5 Principal strain hypothesis

In the case of uniaxial compression failure would occur at $R^c = \dfrac{R^t}{\nu}$ (Fig. 4.5). However, here too there is a discrepancy to experimental findings for most materials, so that the principal strain hypothesis also shows only a very limited applicability in engineering practice.

4.4 Beltrami Strain Energy Hypothesis

The strain energy hypothesis according to Beltrami[6] is a hypothesis that was developed for the prediction of the onset of plastic flowing of metallic materials. Here, failure is synonymous with the onset of flowing, and the onset of flowing is assumed if the linear-elastic strain energy density U_0 reaches a critical value U_c, i.e. $U_0 = U_c$. The critical value U_c is related to the uniaxial equivalent stress R_p as follows:

$$U_c = \frac{R_p^2}{2E}. \tag{4.11}$$

The strain energy density can be given as the sum of the volume change energy density U_V and the deformation energy density U_S, i.e. $U_0 = U_V + U_S$, where:

$$U_V = \frac{1}{2} \frac{E}{3(1-2\nu)} \varepsilon_{kk}^2, \quad U_S = \frac{E}{2(1+\nu)} \varepsilon'_{ij}\varepsilon'_{ij}. \tag{4.12}$$

The deformation energy density can be expressed in the principal stresses using the principal strains so that the following representation results:

---

[6]Eugenio Beltrami, 1835–1900, Italian mathematician.

$$U_0 = \frac{1}{2}\frac{E}{3(1-2\nu)}(\varepsilon_1+\varepsilon_2+\varepsilon_3)^2 + \frac{E}{2(1+\nu)}\left[\left(\frac{2\varepsilon_1-\varepsilon_2-\varepsilon_3}{3}\right)^2\right.$$

$$\left.+\left(\frac{2\varepsilon_2-\varepsilon_1-\varepsilon_3}{3}\right)^2+\left(\frac{2\varepsilon_3-\varepsilon_1-\varepsilon_2}{3}\right)^2\right]=\frac{1}{18K}\sigma_{kk}^2+\frac{1}{4\mu}\sigma'_{ij}\sigma'_{ij}$$

$$=\frac{3(1-2\nu)}{18E}(\sigma_1+\sigma_2+\sigma_3)^2+\frac{2(1+\nu)}{4E}\left[\left(\frac{2\sigma_1-\sigma_2-\sigma_3}{3}\right)^2\right.$$

$$\left.+\left(\frac{2\sigma_2-\sigma_1-\sigma_3}{3}\right)^2+\left(\frac{2\sigma_3-\sigma_1-\sigma_2}{3}\right)^2\right]$$

$$=\frac{1-2\nu}{6E}(\sigma_1+\sigma_2+\sigma_3)^2+\frac{1+\nu}{18E}\left[6\sigma_1^2+6\sigma_2^2+6\sigma_3^2-6\sigma_1\sigma_2-6\sigma_2\sigma_3-6\sigma_1\sigma_3\right]$$

$$=\frac{1-2\nu}{6E}(\sigma_1+\sigma_2+\sigma_3)^2+\frac{1+\nu}{6E}\left[(\sigma_1-\sigma_2)^2+(\sigma_2-\sigma_3)^2+(\sigma_3-\sigma_1)^2\right]$$

$$=\frac{1}{2E}\left[\frac{1-2\nu}{3}I_1^2+2(1+\nu)J_2\right]. \tag{4.13}$$

The failure criterion then reads:

$$(1+\nu)\left[(\sigma_1-\sigma_2)^2+(\sigma_2-\sigma_3)^2+(\sigma_3-\sigma_1)^2\right]+(1-2\nu)(\sigma_1+\sigma_2+\sigma_3)^2=3R_p^2. \tag{4.14}$$

The corresponding failure surface is an ellipsoid of revolution around the hydro-static axis, the vertices are at the points

$$\sigma_1=\sigma_2=\sigma_3=\pm\frac{R_p}{\sqrt{3(1-2\nu)}}. \tag{4.15}$$

As a consequence of this failure criterion, failure occurs under the same hydrostatic tension and hydrostatic compression, which, however, often contradicts experimental findings.

4.5 Von Mises Strain Energy Hypothesis

The strain energy hypothesis according to von Mises[7] (often also referred to as von Mises flow condition)   results from Beltrami's hypothesis if the volume change energy density U_V is omitted. Accordingly, it is postulated that plastic flow occurs when the deformation energy density U_S reaches a critical value U_C. From the uniaxial stress state, the uniaxial equivalent yield stress R_p (von Mises equivalent stress) can be deduced as follows:

---

[7]Richard Edler von Mises, 1883–1953, Austrian mathematician.

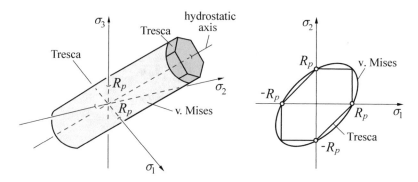

Fig. 4.6 Von Mises strain energy hypothesis

$$U_C = \frac{1+\nu}{3E} R_p^2 = \frac{1+\nu}{E} J_2. \tag{4.16}$$

The second invariant of the stress deviator can be expressed using the principal stresses of the given stress state. Then the following form of the failure criterion results:

$$(\sigma_1 - \sigma_2)^2 + (\sigma_2 - \sigma_3)^2 + (\sigma_3 - \sigma_1)^2 = 2R_p^2, \tag{4.17}$$

or:

$$R_p^2 = \sqrt{\frac{3}{2} \sigma_{ij}' \sigma_{ij}'} = \sqrt{3J_2}. \tag{4.18}$$

Geometrically, this can be interpreted as a cylinder surface in the space of the principal stresses, whose central axis coincides with the hydrostatic axis $\sigma_1 = \sigma_2 = \sigma_3$ (Fig. 4.6). In the special case of a plane stress state with $\sigma_3 = 0$, the strain energy hypothesis according to von Mises changes into the following form:

$$\sigma_1^2 + \sigma_2^2 - \sigma_1\sigma_2 = R_p^2. \tag{4.19}$$

The associated failure surface results in a flow curve in the form of an ellipse.

In the case of metallic materials, the strain energy hypothesis according to von Mises mostly shows good agreement with experimental findings and is therefore widely used in engineering practice.

4.6 Tresca Yield Criterion

A yield criterion very similar to the strain energy hypothesis is the yield criterion according to Tresca.[8] It assumes plastic flow if the maximum shear stress of a given stress state assumes a critical value. Using the uniaxial equivalent stress R_p this leads to:

$$\frac{\sigma_1 - \sigma_3}{2} = \pm\frac{R_p}{2}, \qquad \frac{\sigma_2 - \sigma_1}{2} = \pm\frac{R_p}{2}, \qquad \frac{\sigma_3 - \sigma_2}{2} = \pm\frac{R_p}{2}. \qquad (4.20)$$

The associated yield area in the principal stress space is a hexagonal cylinder composed of sub-planes, the central axis of which is the hydrostatic axis. It turns out that this hexagonal cylinder is exactly inscribed in the von Mises cylinder (Fig. 4.6). As a result, there are usually no great differences between the results according to the Tresca yield criterion and those of the strain energy hypothesis so that Tresca's yield criterion is comparatively popular, e.g. in the area of sheet metal forming.

4.7 Coulomb-Mohr Hypothesis

A criterion suitable for geological and granular materials is the so-called Coulomb-Mohr hypothesis.[9] It is based on failure in the form of slipping processes on internal surfaces in the material. Both the normal stress σ and the shear stress τ acting on the surface play decisive roles, the former especially when it is a compressive stress. Similar to Coulomb's law of friction sliding along a surface is assumed if:

$$|\tau| = -\sigma \tan \rho. \qquad (4.21)$$

Herein ρ is the angle of friction. In order to prevent that sliding is not already postulated for arbitrarily small shear stresses in the case of $\sigma = 0$, but only when a finite shear stress is present and so that the surfaces under consideration can also bear a certain tensile stress, the following modification of the slip condition is appropriate:

$$|\tau| = -\sigma \tan \rho + c. \qquad (4.22)$$

The quantity c represents the cohesion . This form of the slip condition is also known as the Coulomb-Mohr hypothesis. Geometrically, it can be expressed in the σ-τ-plane as being composed of two straight lines that enclose the permissible Mohr stress circles. The Coulomb-Mohr hypothesis leads to the following condition:

---

[8] Henri Édouard Tresca, 1814–1885, French engineer.

[9] Charles Augustin de Coulomb, 1736–1806, French physicist, and Christian Otto Mohr, 1835-1918, German engineer.

$$\frac{|\sigma_1 - \sigma_3|}{2} = \left[\frac{c}{\tan \rho} - \frac{\sigma_1 + \sigma_3}{2} \right] \sin \rho. \tag{4.23}$$

There is a connection with the uniaxial tensile strength R^t if the uniaxial tensile stress state $\sigma_1 = R^t$, $\sigma_3 = 0$ is considered:

$$R^t = \frac{2c \cos \rho}{1 + \sin \rho}. \tag{4.24}$$

The relation with the uniaxial compressive strength R^c follows from considering the uniaxial compressive stress state $\sigma_1 = 0$, $\sigma_3 = -R^c$:

$$R^c = \frac{2c \cos \rho}{1 - \sin \rho}. \tag{4.25}$$

If the tensile strength and compressive strength are known, the constants c and ρ can be determined from these two relations. A representation of the Coulomb-Mohr hypothesis is shown in Fig. 4.7. The failure surface forms a six-sided pyramid around the hydrostatic axis in the three-dimensional principal stress space, where the principal stresses are not necessarily ordered according to size. The apex of the pyramid is found at $\sigma_1 = \sigma_2 = \sigma_3 = \frac{c}{\tan \rho}$. If, on the other hand, one regards σ_1 as

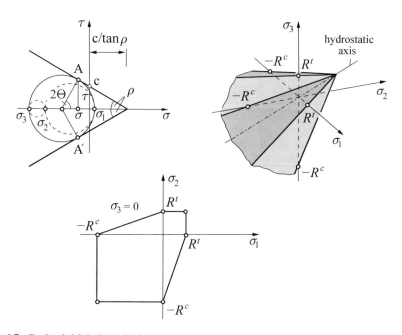

Fig. 4.7 Coulomb-Mohr hypothesis

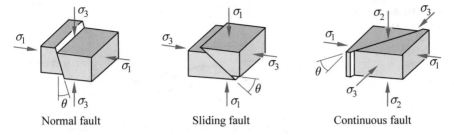

Fig. 4.8 Different types of faults in geology

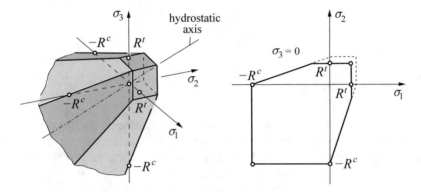

Fig. 4.9 Tension cutoffs

the largest and σ_3 as the smallest principal stress, then the direction of the normal to the failure slip plane results in the σ_1-σ_3-plane in the direction $\theta_{1,2} = \pm (45° - \rho/2)$.

The Coulomb-Mohr hypothesis makes it possible to explain different types of faults in geology (Fig. 4.8), whereby it is assumed that all principal stresses are compressive stresses with $|\sigma_3| \geq |\sigma_2| \geq |\sigma_1|$. Experience shows, however, that the Coulomb-Mohr hypothesis only leads to useful results in the compression range for many materials, whereas other failure mechanisms become relevant in the tensile range. This can be taken into account by modifying the failure surface, namely by introducing normal stress sections (so-called tension cutoffs) in the failure criterion (Fig. 4.9). From the failure hypothesis according to Coulomb-Mohr one arrives at the general Mohr's failure hypothesis, if instead of the Coulomb friction a more general approach of the kind

$$|\tau| = h(\sigma) \tag{4.26}$$

is applied. Therein, the function $h(\sigma)$ requires experimental investigation. Geometrically, Mohr's failure hypothesis can be interpreted as an envelope of the permissible Mohr's circles in the $\sigma - \tau$-plane (Fig. 4.10).

Fig. 4.10 General Mohr's
failure hypothesis

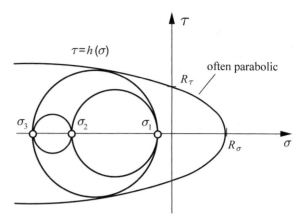

4.8   Drucker-Prager Hypothesis

The Drucker-Prager hypothesis[10] represents a modification of the Coulomb-Mohr
hypothesis. Therein, the normal stress belonging to a certain cutting direction is
replaced by the first invariant of the stress tensor. In addition, the von Mises equivalent stress is used instead of the shear stress in the form of the root of the second
invariant of the stress deviator. The Drucker-Prager failure criterion then reads:

$$\alpha I_1 + \sqrt{J_2} - k = 0. \tag{4.27}$$

Here I_1 is the first invariant of the stress tensor, and J_2 is the second invariant of
the stress deviator. The quantities α and k are material parameters.

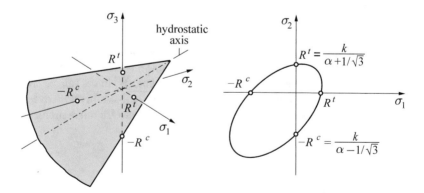

Fig. 4.11 Drucker-Prager hypothesis

[10]Daniel Charles Drucker, 1918–2001, American engineer; William Prager (born as Willy Prager),
1903–1980, German-US-American mathematician.

The Drucker-Prager hypothesis represents a failure surface in the principal stress space in the form of a circular cone around the hydrostatic axis with the apex $\sigma_1 = \sigma_2 = \sigma_3 = \frac{k}{3\alpha}$ (Fig. 4.11). For the special case of the plane stress state with $\sigma_3 = 0$ it is simplified to an ellipse in the σ_1-σ_2-plane. For the special case $\alpha = 0$ it assumes the form of the yield condition according to von Mises .

4.9 Cuntze's Failure Mode Concept

Some general and essential requirements must be placed on failure criteria. On the one hand, for a simple application in engineering practice it should always be ensured that the criterion is easy to formulate and can be handled with ease in engineering practice. In addition, it should always be guaranteed that a failure criterion is robust in terms of numerical feasibility. Of course, it must always be required that a failure criterion has a solid physical background and only requires standard strength values.

Most failure criteria ($F >=< 1$) are based on so-called global failure conditions ($F = 1$). Typically, several failure modes are connected with one another in the framework of such failure conditions. As a consequence, if a change is made in one failure mode area, failure in another mode is also affected which has no physical justification. For this reason, Cuntze[11] follows a strict failure mode-related formulation. Each failure mode is described here for itself, so that in the end just as many conditions have to be established as there are failure modes. While the usual form of a failure condition can be given in the form $F\left(\underline{\sigma}, \underline{R}\right) = 1$ and thus represents a single global condition, Cuntze uses a set of failure conditions of the form

$$F\left(\underline{\sigma}, R^{\text{mode}}\right) = 1. \tag{4.28}$$

Here $\underline{\sigma}$ includes the set of all stress tensor components, $\underline{R}$ represents the strength values of the considered material combined in a vector. The Cuntze failure mode concept only assigns the associated strength to each mode. The number of strength values to be provided depends on the type of material symmetry, i.e. whether it is an isotropic, transversely isotropic or orthotropic material. The number of independent elasticity constants then corresponds to the number of independent strength values. If the material is brittle, the type of failure is fracture.

The failure mode concept according to Cuntze (short: FMC) postulates that the number of independent failure modes corresponds to the number of independent strength values and also to the number of independent elastic constants. When formulating the mode failure conditions, it is often useful to use invariants associated with material symmetry. The selection of the invariant(s) to be applied for the failure mode under consideration depends on the physical mechanism, which is illustrated below using the example of a dense isotropic brittle material. Such materials typically fail in uniaxial tension due to fracture or normal fracture (Fig. 4.12, left), accompa-

---

[11]Ralf Cuntze, German engineer.

Fig. 4.12 Normal fracture under uniaxial tensile load (left), shear fracture under uniaxial compressive load (right)

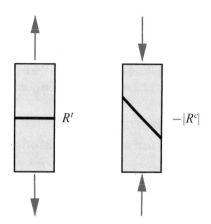

nied by small deformations and a fairly smooth fracture surface. On the other hand, it is also characteristic for such materials that in the case of uniaxial compression (Fig. 4.12, right) shear fracture occurs. Such a fracture behavior is described by two failure modes as indicated, as well as by two independent strength values and two independent individual stiffness values $((E, \nu)$ or $(E, G))$.

Another example is an isotropic, porous and brittle material. Examples of such a material are foams or fiber-reinforced ceramics. Typically, such materials will fail under uniaxial tension due to normal fracture. In this case, it can be assumed that such a fracture will be accompanied by very little deformation, whereas a rather rough fracture surface will result. In contrast to the previous example, there will be no shear fracture under uniaxial compression but rather a failure in the form of crushing, so that in the end an accumulation of fragments remains. Here, too, there are again two failure modes, two independent strength values and two independent stiffnesses.

Consider a brittle or isotropic material with two failure mode conditions (= fracture conditions):

$$\text{Normal fracture:} \qquad \frac{\frac{1}{3}I_1 + \sqrt{\frac{4}{3}J_2}}{R^t} = 1, \qquad (4.29)$$

$$\text{Shear fracture:} \qquad a_\tau^d \frac{3J_2}{R^{c2}} + \left(a_\tau^d - 1\right)\frac{I_1}{R^c} = 1. \qquad (4.30)$$

The strength parameter a_τ^d must be determined either from multi-axial compression tests or from uniaxial compression tests with measurement of the fracture angle.

The individual terms occurring in the above conditions represent the so-called stress efforts of the material in the individual modes. For the current example e.g. for the normal fraction we have:

$$Eff^{\text{mode}} = Eff^{\text{NB}} = \frac{\frac{1}{3}I_1 + \sqrt{\frac{4}{3}J_2}}{R^t}. \qquad (4.31)$$

In order to obtain a closed, multidimensional failure surface the mode failure conditions are superimposed with the help of an interaction equation. This results in a closed-form representation but without the disadvantage of the global formulation. This is done by introducing a serial failure model in which the efforts of all mode efforts activated by the current stress state are summarized as a failure equation:

$$Eff^m = \left(Eff^{\text{mode 1}}\right)^m + \left(Eff^{\text{mode 2}}\right)^m + \ldots = 1 = 100\%. \qquad (4.32)$$

The interaction coefficient m can be applied with values between 2.5 and 3. The above formulation leads to failure surfaces that are rotationally symmetrical about the hydrostatic axis.

The Cuntze failure mode concept plays a rather subordinate role in the treatment of isotropic materials, but is of great importance in the analysis of fiber-reinforced plastics.

Bibliography

Anderson TL (1995) Fracture mechanics—fundamentals and applications, 2nd edn. CRC Press, Boca Raton, USA

Gross D, Seelig T (2007) Bruchmechanik - mit einer Einführung in die Mikromechanik, 4th edn. Springer, Berlin, Heidelberg

Kanninen MF, Popelar CH (1985) Advanced fracture mechanics. Oxford University Press, USA

Lemaitre J (1992) A course on damage mechanics. Springer, Berlin, Heidelberg

Mittelstedt C, Becker W (2016) Strukturmechanik ebener Laminate. Technische Universität Darmstadt, Darmstadt, Studienbereich Mechanik

Part II
Thin-Walled Beam Structures

Chapter 5
Beams Under Normal Forces and Bending Moments

5.1 Introduction

In this chapter we consider beam structures in the framework of linear elasticity. We limit ourselves at this point to those beam structures in which only normal forces, bending moments and transverse shear forces occur (Fig. 5.1, left). Torsion as well as the related internal forces, stresses and deformations are discussed at a later stage in Chaps. 7 and 8. The beam under consideration is loaded by the two line loads q_y and q_z, wherein the indices indicate the direction of the line loads. Both line loads can be arbitrary functions of the longitudinal coordinate x. In addition, there is an arbitrary number of point loads/single forces F_y and F_z which may apply at any point x of the beam length. Single bending moments can also occur but are not shown in Fig. 5.1 for the sake of clarity. A torsional load in the form of single torques/torsional moments M_T and distributed torques/torsional moments m_T is excluded for the time being.

We will restrict the considerations in this chapter exclusively to very thin beam cross-sections which we can represent in very good approximation by their respective mean lines/skeleton lines (Fig. 5.2). The mean lines/skeleton lines halve the cross-section thickness at each point.

The external loads result in internal stresses which in turn are represented by integral quantities, namely the resultant forces and moments. The internal forces (Fig. 5.1, left) are the normal force N, the two bending moments M_y and M_z as well as the two transverse shear forces V_y and V_z. The associated displacements and rotations are the three displacements u, v and w in $x-$, $y-$ and $z-$direction, and the two rotations φ_y and φ_z (the so-called bending angles), as shown in Fig. 5.1, right. The rotation ϑ of the cross-section about the $x-$axis as relevant when torsion is considered is excluded here.

It should be pointed out at this point that the heading of this chapter contains a certain inconsistency. Generally, a one-dimensional straight structural element under axial load where normal forces N occur without any bending effects is commonly addressed as a rod. However, if the applied load induces (potentially biaxial) bending

C. Mittelstedt, *Structural Mechanics in Lightweight Engineering*,
https://doi.org/10.1007/978-3-030-75193-7_5

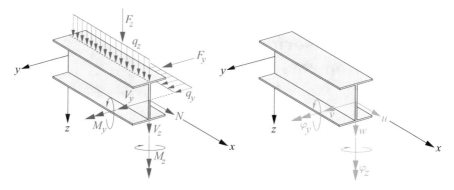

Fig. 5.1 Loads, internal forces and moments of a beam (left), associated displacements and rotations (right)

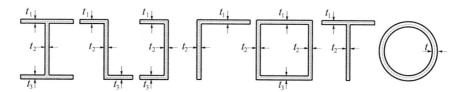

Fig. 5.2 Thin-walled beams and their skeleton lines (dashed lines)

and the associated internal moments M_y and M_z as well as forces V_y and V_z, then such one-dimensional structural elements are commonly addressed as beams. In the sense of a uniform treatment of both classes of structural elements in this chapter, we want to speak of beams under normal forces and bending moments.

The internal forces N, V_y, V_z and moments M_y, M_z are related to the stress components (normal stress σ_{xx} and shear stress τ) as indicated in Fig. 5.3. The loads q_z and F_z cause the transverse shear force V_z and the bending moment M_y. The loads q_y and F_y, however, give rise to M_z and V_y. In addition a normal force N may also occur. The Fig. 5.3, top right, shows that the normal force N causes the normal stress σ_{xx} which, under certain conditions to be discussed, is constantly distributed over the cross-section of the beam. In addition, the bending moment M_y causes the normal stress σ_{xx}, which extends linearly over the cross section (Fig. 5.3, middle left). The transverse shear force V_z is the cause of the occurrence of the shear stress τ (Fig. 5.3, middle right) which shows a linear distribution in the flanges and a parabolic distribution in the web of the beam. The direction of action of τ is indicated by arrows. Since the line of action of τ is always parallel to the skeleton line of the cross-section this stress component is not indexed here. Further, the bending moment M_z may also occur, which corresponds to the normal stress σ_{xx} which also runs linear across the cross-section as indicated in Fig. 5.3, bottom left. In addition the transverse shear force V_y occurs which is the cause for the shear stress τ (Fig. 5.3, bottom right). It is then the main task of engineers to determine these stresses and to prove their maximum values against failure of the material. In addition, the deformations of the beam structure under consideration must be

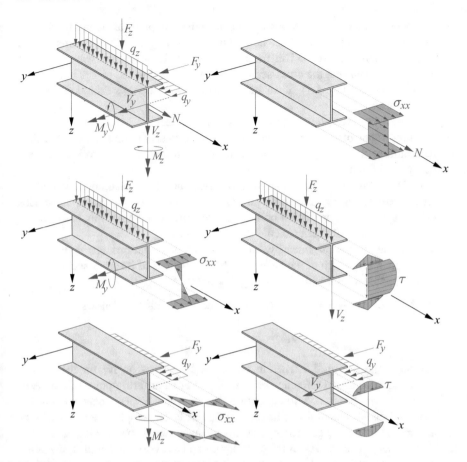

Fig. 5.3 Forces and moments of a beam and the associated stress components

verified against permissible maximum values. It must also be verified that the beam exhibits a sufficient resistance against a global or local buckling failure. In this chapter we dedicate ourselves to the determination of the normal stresses σ_{xx} as well as the development of the necessary analytical methods as well as the calculation of the deformations of rod and beam structures.

5.2 Basic Equations for an Arbitrary Reference System

We assume the beam theory according to Euler and Bernoulli[1] to be valid and make the following assumptions:

---

[1]Leonhard Euler, 1707–1783, and Jakob Bernoulli, 1654–1705, both Swiss mathematicians and physicists.

- We assume that the cross-sections of the beam remain plane also in the deformed state. Further, it is assumed that a normal to the beam's center line remains normal also in the deformed state. This means that cross-sections do not warp when the beam is bent and that the cross-sections are orthogonal to the center line/neutral axis of the beam (Fig. 5.4). Shear strains in the beams are thus excluded from the considerations.
- We assume linear elasticity so that Hooke's law is valid.
- The dimensions of the cross-section are assumed to be significantly smaller than the length of the beam, so that an idealization within the framework of a beam theory is justified.
- In all that follows we assume that the cross-sectional shape is completely retained during the beam deformations. We also assume that the cross-sectional shape does not change at load application points. In a specific application, this has to be be ensured by means of adequate constructive measures.
- We assume geometric linearity, meaning that deformations are significantly smaller than the length of the beam and the dimensions of its cross-section.
- In this chapter we only consider straight beam structures.

Hooke's law can be written for the current beam situation as follows:

$$\sigma_{xx} = E\varepsilon_{xx}, \quad \tau = G\gamma. \tag{5.1}$$

Starting point is Fig. 5.5 wherein we employ the notation $N_{\bar{x}}$ for the normal force N. We introduce a reference system $\bar{x}$, $\bar{y}$, $\bar{z}$ at an arbitrary point of the beam as indicated. Our aim is now to determine the transverse displacement $\bar{w}_P$ and the longitudinal displacement $\bar{u}_P$ of an arbitrary point P. The displacement $\bar{v}_P$ will be considered at a later stage. The displacement quantities to be considered are the displacements $\bar{u}$, $\bar{v}$, $\bar{w}$ as well as the rotations $\varphi_{\bar{y}}$ and $\varphi_{\bar{z}}$. We consider the kinematics of the beam according to Fig. 5.6 in which a beam element is given in the initial undeformed state as well as in the deformed state. Let h be the height of the cross-section. The beam is loaded by a normal force and a bending load in such a way that the element considered in Fig. 5.6 is subjected to both a transverse displacement $\bar{w}$ and a longitudinal displacement $\bar{u}$. In the deformed state, there is also an inclination $\bar{w}'$. The corresponding displacments of the point P are denoted as $\bar{w}_P$ and $\bar{u}_P$. Due to the rotation $-\varphi_{\bar{y}}$, the origin of the reference system will undergo a longitudinal displacement $\bar{u}$ which differs from the displacement $\bar{u}_P$ of point P. It is now the aim to determine the difference between these two displacement values which we denote as $\Delta\bar{u}$. In order to calculate $\Delta\bar{u}$ we consider the detail shown in Fig. 5.6, right. We have:

$$\sin(-\varphi_{\bar{y}}) = \frac{\Delta\bar{u}}{\bar{z}_P}. \tag{5.2}$$

Since we assume small displacements and thus also small rotations $\varphi_{\bar{y}}$ and $\varphi_{\bar{z}}$, we can employ the approximation $\sin(-\varphi_{\bar{y}}) \simeq -\varphi_{\bar{y}}$ so that:

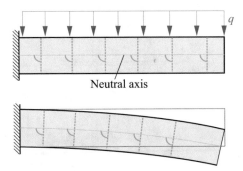

Fig. 5.4 Kinematic assumptions of the Euler-Bernoulli beam theory: cross-sections do not warp and remain normal to the beam's center line

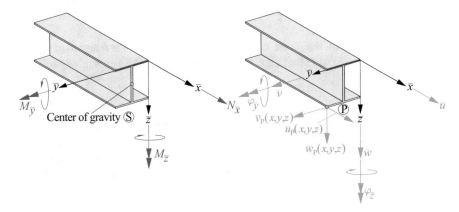

Fig. 5.5 Forces, moments and kinematic quantities with respect to an arbitrary reference system $\bar{x}, \bar{y}, \bar{z}$; displacements $\bar{u}_P$, $\bar{v}_P$ and $\bar{w}_P$ of an arbitrary point P

$$\Delta\bar{u} = -\bar{z}_P \varphi_{\bar{y}}. \tag{5.3}$$

Since we also assume that no warping of the cross-section due to the bending deformation occurs and that the normal hypothesis also applies, we equate the inclination $\bar{w}'$ and the angle of rotation $-\varphi_{\bar{y}}$, so that:

$$\Delta\bar{u} = \bar{z}_P \bar{w}'. \tag{5.4}$$

The longitudinal displacement $\bar{u}_P$ of point P can thus be written as:

$$\bar{u}_P = \bar{u} - \Delta\bar{u} = \bar{u} - \bar{z}_P \bar{w}'. \tag{5.5}$$

This connection applies to the case of the so-called uniaxial bending, i.e. for the case that only deformations $\bar{u}$ and $\bar{w}$ occur in the $\bar{x}\bar{z}$−plane. We will specify the conditions under which this applies at a later point.

Fig. 5.6 Kinematics of the
Euler-Bernoulli beam theory

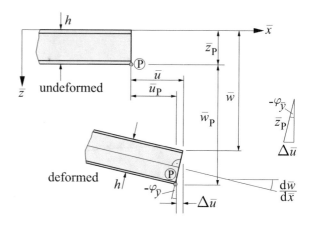

If, on the other hand, the given load leads to the displacements $\bar{u}$ and $\bar{w}$ as well as to the displacement $\bar{v}$ (so-called biaxial bending), then Eq. (5.5) expands as follows:

$$\bar{u}_P = \bar{u} - \bar{z}_P \bar{w}' - \bar{y}_p \bar{v}', \tag{5.6}$$

wherein $\bar{y}_p$ is the distance of point P from the origin of the reference system $\bar{x}, \bar{y}, \bar{z}$.

Finally, a statement about the relation between the displacement $\bar{w}$ of the coordinate origin and the displacement $\bar{w}_P$ of the point P has to be made. Since we assume that the cross-sectional dimensions do not change due to the deformation of the beam, we can equate $\bar{w}$ and $\bar{w}_P$ as a good approximation:

$$\bar{w}_P = \bar{w}. \tag{5.7}$$

Analogously this holds for the displacements in the $\bar{y}$−direction:

$$\bar{v}_P = \bar{v}. \tag{5.8}$$

Now that we have clarified the displacement field of the beam we can look at the strain field. Since we are assuming that the beam is slender with a significantly large length compared to the cross-sectional dimensions and given that we have also excluded any shear strains, the only occurring strain component is the normal strain $\varepsilon_{\bar{x}\bar{x}P}$. It is calculated using the infinitesimal strain tensor (Eq. (2.70)) as follows:

$$\varepsilon_{\bar{x}\bar{x}P} = \frac{\mathrm{d}\bar{u}_P}{\mathrm{d}\bar{x}} = \bar{u}'_P = \bar{u}' - \bar{z}_P \bar{w}'' - \bar{y}_p \bar{v}''. \tag{5.9}$$

Since the relations derived so far apply to any point on the cross-section, we want to drop the index P in all that follows. With the help of Hooke's law (5.1) we obtain the normal stress $\sigma_{\bar{x}\bar{x}}$ as:

$$\sigma_{\bar{x}\bar{x}} = E\varepsilon_{\bar{x}\bar{x}} = E\left(\bar{u}' - \bar{z}\bar{w}'' - \bar{y}\bar{v}''\right). \tag{5.10}$$

The normal stress $\sigma_{\bar{x}\bar{x}}$ is related to the normal force and the bending moments of the beam as follows:

$$N_{\bar{x}} = \int_A \sigma_{\bar{x}\bar{x}} dA, \quad M_{\bar{y}} = \int_A \sigma_{\bar{x}\bar{x}} \bar{z} dA, \quad M_{\bar{z}} = -\int_A \sigma_{\bar{x}\bar{x}} \bar{y} dA. \tag{5.11}$$

Insertion of (5.10) then yields:

$$N_{\bar{x}} = E \int_A \left(\bar{u}' - \bar{z}\bar{w}'' - \bar{y}\bar{v}''\right) dA,$$
$$= E\bar{u}' \int_A dA - E\bar{w}'' \int_A \bar{z} dA - E\bar{v}'' \int_A \bar{y} dA,$$
$$M_{\bar{y}} = E \int_A \left(\bar{u}' - \bar{z}\bar{w}'' - \bar{y}\bar{v}''\right) \bar{z} dA,$$
$$= E\bar{u}' \int_A \bar{z} dA - E\bar{w}'' \int_A \bar{z}^2 dA - E\bar{v}'' \int_A \bar{y}\bar{z} dA,$$
$$M_{\bar{z}} = -E \int_A \left(\bar{u}' - \bar{z}\bar{w}'' - \bar{y}\bar{v}''\right) \bar{y} dA,$$
$$= -E\bar{u}' \int_A \bar{y} dA + E\bar{w}'' \int_A \bar{y}\bar{z} dA + E\bar{v}'' \int_A \bar{y}^2 dA. \tag{5.12}$$

At this point we introduce typical and frequently recurring area integrals as follows. The cross-sectional area A can be calculated as:

$$A = \int_A dA. \tag{5.13}$$

The area integrals of first order which also appear in (5.12) can be interpreted as the so-called static moments $S_{\bar{y}}$ and $S_{\bar{z}}$:

$$S_{\bar{z}} = \int_A \bar{y} dA, \quad S_{\bar{y}} = \int_A \bar{z} dA. \tag{5.14}$$

They result in the unit of a length to the power of three.

The second-order area integrals occurring in (5.12) are the so-called moments of inertia $I_{\bar{y}\bar{y}}$, $I_{\bar{z}\bar{z}}$ as well as the so-called deviation moment $I_{\bar{y}\bar{z}}$:

$$I_{\bar{y}\bar{y}} = \int_A \bar{z}^2 dA, \quad I_{\bar{z}\bar{z}} = \int_A \bar{y}^2 dA, \quad I_{\bar{y}\bar{z}} = \int_A \bar{y}\bar{z} dA. \tag{5.15}$$

While the two moments of inertia $I_{\bar{y}\bar{y}}$ and $I_{\bar{z}\bar{z}}$ always assume positive values, the deviation moment $I_{\bar{y}\bar{z}}$ may have both positive and negative values or in special cases (which will be discussed later) may even become zero. All area integrals of the second order have the unit of a length to the fourth power.

With the definitions (5.13), (5.14) and (5.15) we can rewrite the equation system (5.12) as follows:

$$N_{\bar{x}} = EA\bar{u}' - ES_{\bar{y}}\bar{w}'' - ES_{\bar{z}}\bar{v}'',$$
$$M_{\bar{y}} = ES_{\bar{y}}\bar{u}' - EI_{\bar{y}\bar{y}}\bar{w}'' - EI_{\bar{y}\bar{z}}\bar{v}'',$$
$$M_{\bar{z}} = -ES_{\bar{z}}\bar{u}' + EI_{\bar{y}\bar{z}}\bar{w}'' + EI_{\bar{z}\bar{z}}\bar{v}''. \tag{5.16}$$

In a vector-matrix notation we have:

$$\begin{pmatrix} N_{\bar{x}} \\ M_{\bar{y}} \\ -M_{\bar{z}} \end{pmatrix} = E \begin{bmatrix} A & S_{\bar{y}} & S_{\bar{z}} \\ S_{\bar{y}} & I_{\bar{y}\bar{y}} & I_{\bar{y}\bar{z}} \\ S_{\bar{z}} & I_{\bar{y}\bar{z}} & I_{\bar{z}\bar{z}} \end{bmatrix} \begin{pmatrix} \bar{u}' \\ -\bar{w}'' \\ -\bar{v}'' \end{pmatrix}. \tag{5.17}$$

If an arbitrary reference coordinate system $\bar{x}$, $\bar{y}$, $\bar{z}$ is used, the normal force $N_{\bar{x}}$ and the bending moments $M_{\bar{y}}$, $M_{\bar{z}}$ are coupled with all occurring displacement components $\bar{u}$, $\bar{v}$, $\bar{w}$, which makes the determination of these quantities quite complicated. Solving for the displacement components and inserting in (5.10) then enables the determination of the normal stress $\sigma_{\bar{x}\bar{x}}$. The resulting expression, however, is not given at this point since it is not useful for practical application. The set of equations (5.16) or (5.17) is also called the constitutive law of the beam, here with respect to the reference system $\bar{x}$, $\bar{y}$, $\bar{z}$.

The determination of the area integrals according to (5.13), (5.14) and (5.15) is illustrated at the example of the cross-section as given in Fig. 5.7, left. The I-cross-section with the constant wall thickness t consists of the two flanges (width $2b$) and the web (height h), for which local reference axes s_i $(i = 1, 2, ..., 5)$ are assigned which are oriented along the skeleton line of the cross-section.

We first determine the cross-sectional area A according to (5.13). In the area integral appearing there, the expression dA can be replaced by $t\,ds$, so that:

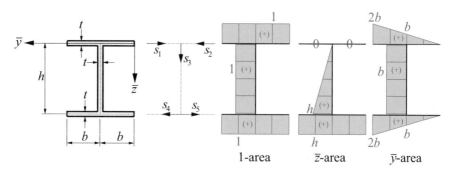

Fig. 5.7 On the calculation of the area integrals at the example of an I-cross-section

$$A = \int_A dA = \sum_{i=1}^{5} \int_0^{l_i} t_i ds_i. \tag{5.18}$$

Herein, t_i is the wall thickness of the segment i. The size l_i represents the length of the segment under consideration. Since the cross-section consists of individual segments, we have split the integral in (5.18) into a sum of integrals. Using this specific example, we get:

$$A = \int_0^b t ds_1 + \int_0^b t ds_2 + \int_0^h t ds_3 + \int_0^b t ds_4 + \int_0^b t ds_5. \tag{5.19}$$

The integrals appearing here are elementary and very easy to solve. Nevertheless, we want to introduce a formalism at this point that will make the analysis much easier, especially in all further calculations. We define the so-called 1-area of the cross-section, as shown in Fig. 5.7, middle, and define a constant auxiliary value with the amount 1 at every point of the cross-section. The integrals in (5.19) can then, after factoring out the wall thickness t, be written as:

$$A = t \left(\int_0^b 1 \cdot 1 \cdot ds_1 + \int_0^b 1 \cdot 1 \cdot ds_2 + \int_0^h 1 \cdot 1 \cdot ds_3 + \int_0^b 1 \cdot 1 \cdot ds_4 + \int_0^b 1 \cdot 1 \cdot ds_5 \right). \tag{5.20}$$

This means that in each segment we multiply a rectangular area with the height 1 by another rectangular area with the height 1 and form the integral over this product. Colloquially, one also says that two rectangles are superimposed. The corresponding value of the integral can be read off in Fig. 5.8 using a so-called integral table. The value of the integral $\int_0^l f(s)g(s)ds$ with the two functions $f(s)$ and $g(s)$ results in FGl for constant functions $f(s), g(s)$ where $F = G = 1$. The integral thus assumes the value l. For the current analysis this means:

$$A = t (b + b + h + b + b) = t (4b + h). \tag{5.21}$$

The area integrals with which we have to deal in the case of beam structures usually consist of certain very often recurring standard cases, for which the results of the calculation of e.g. in (5.13), (5.14) and (5.15) can be given in a tabular form as shown in Fig. 5.8. We want to demonstrate this using the example of two linear functions $f(x)$ and $g(x)$ which are defined within the limits of $s = 0$ and $s = l$ with the boundary values $f(s = 0) = 0$, $f(s = l) = F$, $g(s = 0) = 0$, $g(s = l) = G$:

$$f(s) = \frac{F}{l}s, \quad g(s) = \frac{G}{l}s. \tag{5.22}$$

The integral

$$\int_0^l f(s)g(s)ds \tag{5.23}$$

$f(s)$ \\ $g(s)$	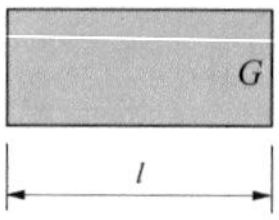	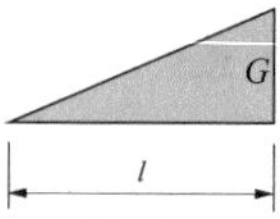	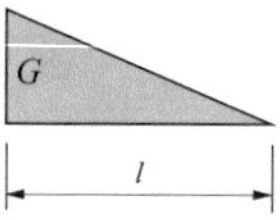	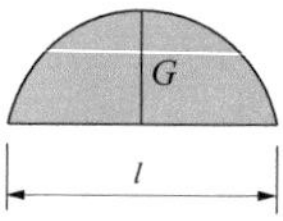
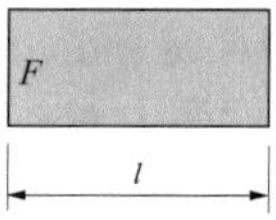 F	FGl	$\frac{1}{2}FGl$	$\frac{1}{2}FGl$	$\frac{2}{3}FGl$
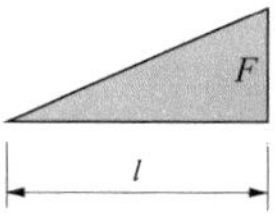 F	$\frac{1}{2}FGl$	$\frac{1}{3}FGl$	$\frac{1}{6}FGl$	$\frac{1}{3}FGl$
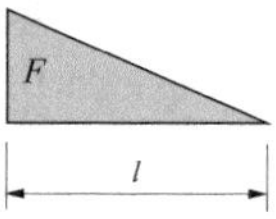 F	$\frac{1}{2}FGl$	$\frac{1}{6}FGl$	$\frac{1}{3}FGl$	$\frac{1}{3}FGl$
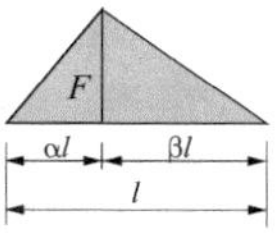 F, αl, βl	$\frac{1}{2}FGl$	$\frac{1}{6}FGl(1+\alpha)$	$\frac{1}{6}FGl(1+\beta)$	$\frac{1}{3}FGl(1+\alpha\beta)$
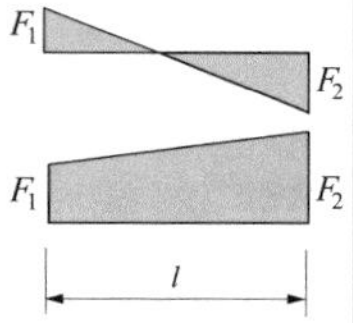 F_1, F_2	$\frac{l}{2}(F_1+F_2)G$	$\frac{l}{6}(F_1+2F_2)G$	$\frac{l}{6}(2F_1+F_2)G$	$\frac{l}{3}(F_1+F_2)G$
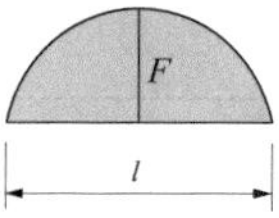 F	$\frac{2}{3}FGl$	$\frac{1}{3}FGl$	$\frac{1}{3}FGl$	$\frac{8}{15}FGl$
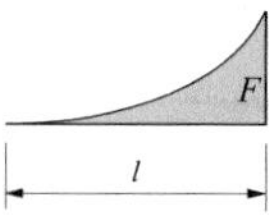 F	$\frac{1}{3}FGl$	$\frac{1}{4}FGl$	$\frac{1}{12}FGl$	$\frac{1}{5}FGl$
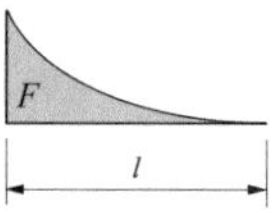 F	$\frac{1}{3}FGl$	$\frac{1}{12}FGl$	$\frac{1}{4}FGl$	$\frac{1}{5}FGl$
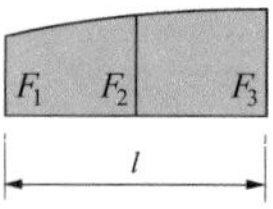 F_1, F_2, F_3	$\frac{Gl}{6}(F_1+4F_2+F_3)$	$\frac{Gl}{6}(2F_2+F_3)$	$\frac{Gl}{6}(F_1+2F_2)$	$\frac{Gl}{15}(F_1+8F_2+F_3)$

(rows 6–9 labelled: Quadratic parabolas)

Fig. 5.8 Integral table $\int_0^l f(s)g(s)\,\mathrm{d}s$ for a selection of typical cases

then assumes the value $\frac{1}{3}FGl$. Figure 5.8 contains some standard cases as relevant for our purposes.

Next we consider the calculation of the two static moments $S_{\bar{y}}$ and $S_{\bar{z}}$. Here, too, we introduce the auxiliary value 1 as follows:

$$
\begin{aligned}
S_{\bar{z}} &= \int_A 1 \cdot \bar{y} dA \\
&= t \left(\int_0^b 1 \cdot \bar{y} ds_1 + \int_0^b 1 \cdot \bar{y} ds_2 + \int_0^h 1 \cdot \bar{y} ds_3 + \int_0^b 1 \cdot \bar{y} ds_4 + \int_0^b 1 \cdot \bar{y} ds_5 \right), \\
S_{\bar{y}} &= \int_A 1 \cdot \bar{z} dA \\
&= t \left(\int_0^b 1 \cdot \bar{z} ds_1 + \int_0^b 1 \cdot \bar{z} ds_2 + \int_0^h 1 \cdot \bar{z} ds_3 + \int_0^b 1 \cdot \bar{z} ds_4 + \int_0^b 1 \cdot \bar{z} ds_5 \right). \quad (5.24)
\end{aligned}
$$

Carrying out the necessary integrations by using the integral table in Fig. 5.8 in connection with the $\bar{y}-$ and $\bar{z}-$areas as shown in Fig. 5.7 then results in:

$$
S_{\bar{z}} = bt \left(4b + h \right), \quad S_{\bar{y}} = th \left(\frac{h}{2} + 2b \right). \quad (5.25)
$$

Finally we consider the moments of inertia $I_{\bar{y}\bar{y}}$, $I_{\bar{z}\bar{z}}$ and the deviation moment $I_{\bar{y}\bar{z}}$. A multiplication by the auxiliary value 1 is not necessary in this case, as we can make use of the fact that in the second-order area integrals two functions $\bar{y}$ or $\bar{z}$ are multiplied with each other the integration of which can be treated by using the integral table of Fig. 5.8 with the $\bar{y}-$ and $\bar{z}-$areas shown in Fig. 5.7:

$$
\begin{aligned}
I_{\bar{y}\bar{y}} &= \int_A \bar{z}\bar{z} dA = th^2 \left(\frac{h}{3} + 2b \right), \\
I_{\bar{z}\bar{z}} &= \int_A \bar{y}\bar{y} dA = tb^2 \left(\frac{16}{3}b + h \right), \\
I_{\bar{y}\bar{z}} &= \int_A \bar{y}\bar{z} dA = tbh \left(\frac{h}{2} + 2b \right). \quad (5.26)
\end{aligned}
$$

The constitutive law (5.17) can be represented for the present case as:

$$
\begin{pmatrix} N_{\bar{x}} \\ M_{\bar{y}} \\ -M_{\bar{z}} \end{pmatrix} = E \begin{bmatrix} t(4b+h) & th\left(\frac{h}{2}+2b\right) & bt(4b+h) \\ th\left(\frac{h}{2}+2b\right) & th^2\left(\frac{h}{3}+2b\right) & tbh\left(\frac{h}{2}+2b\right) \\ bt(4b+h) & tbh\left(\frac{h}{2}+2b\right) & tb^2\left(\frac{16}{3}b+h\right) \end{bmatrix} \begin{pmatrix} \bar{u}' \\ -\bar{w}'' \\ -\bar{v}'' \end{pmatrix}. \quad (5.27)
$$

We also consider the Z-cross-section shown in Fig. 5.9 with regard to the determination of the area integrals. We obtain the following values:

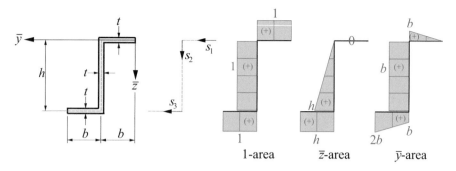

Fig. 5.9 On the calculation of the area integrals at the example of a Z-cross-section

$$A = \int_A 1 \cdot 1 dA = t\,(2b + h)\,,$$

$$S_{\bar{y}} = \int_A 1 \cdot \bar{z} dA = th\left(\frac{h}{2} + b\right),$$

$$S_{\bar{z}} = \int_A 1 \cdot \bar{y} dA = tb\,(2b + h)\,,$$

$$I_{\bar{y}\bar{y}} = \int_A \bar{z}\bar{z} dA = th^2\left(\frac{h}{3} + b\right),$$

$$I_{\bar{z}\bar{z}} = \int_A \bar{y}\bar{y} dA = tb^2\left(\frac{8b}{3} + h\right),$$

$$I_{\bar{y}\bar{z}} = \int_A \bar{y}\bar{z} dA = tbh\left(\frac{h}{2} + \frac{3b}{2}\right). \tag{5.28}$$

A remark should be made regarding the accuracy of the analysis procedure shown here. For illustration we consider the Z-cross-section as given in Fig. 5.10. The type of calculation introduced above, i.e. the strict orientation with respect to the skeleton line of the cross-section, means that the cross-section is computationally broken down into individual segments which always have overlaps or unconsidered areas at their intersection points. Using the example of the Z-cross-section as given in Fig. 5.10, these are the two intersection points of the flanges with the web. In each case a small sub-area occurs which is considered twice in the calculation, but also a small sub-area which is not considered at all. This is an inevitable error in the calculation simply due to geometric reasons. However, experience shows that the error committed in this way can be tolerated, provided that (as assumed here) we are dealing with very thin-walled cross-sections, so that we want to assume that our results are sufficiently accurate.

Fig. 5.10 Accuracy of the
area integrals of thin-walled
beams

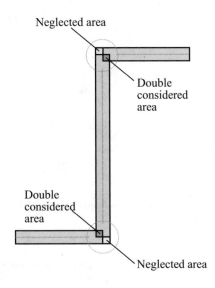

5.3 First Cross-Sectional Normalization: Center of Gravity S

Using the equations (5.16) or (5.17) we were able to show that with an arbitrary choice of the reference system it is possible to establish the constitutive law for the beam, which, however, is quite complicated due to the complete coupling of the normal force and the bending moments with all three displacements. A significant simplification of the relations is achieved if the reference system is moved to a defined point in which the two static moments disappear. We will show below that this is the center of gravity S of the cross-section[2] under consideration, provided that we are dealing with a cross-section consisting of a homogeneous material.

We move the reference system into the center of gravity S of the cross section (Fig. 5.11) which is located at $(\bar{y} = \bar{y}_S, \bar{z} = \bar{z}_S)$. In the following we want to denote the reference system related to the center of gravity S with $\hat{x}, \hat{y}, \hat{z}$. Similarly, we denote the normal force and bending moments as $N_{\hat{x}}, M_{\hat{y}}, M_{\hat{z}}$ and the displacement components as $\hat{u}, \hat{v}, \hat{w}$. The following relations exist between the reference axes $\bar{x}, \bar{y}, \bar{z}$ and $\hat{x}, \hat{y}, \hat{z}$:

$$\hat{x} = \bar{x}, \quad \hat{y} = \bar{y} - \bar{y}_S, \quad \hat{z} = \bar{z} - \bar{z}_S. \tag{5.29}$$

Transforming the axis system into the center of gravity corresponds to a pure translation of the reference system in the $\bar{y}\bar{z}-$ plane. The longitudinal axes $\bar{x}$ and $\hat{x}$ are identical for this kind of axis transformation.

---

[2]From the German expression 'Schwerpunkt' for the center of gravity.

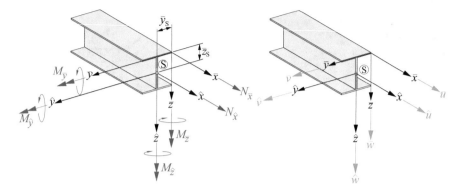

Fig. 5.11 First cross-sectional normalization: reference system in the center of gravity S with the reference axes $\hat{x}$, $\hat{y}$, $\hat{z}$

A prerequisite for transforming the reference system into the center of gravity S is the disappearance of the two static moments $S_{\hat{z}}$ and $S_{\hat{y}}$:

$$S_{\hat{z}} = \int_A \hat{y}\,\mathrm{d}A = \int_A \bar{y}\,\mathrm{d}A - \bar{y}_S \int_A \mathrm{d}A = 0,$$

$$S_{\hat{y}} = \int_A \hat{z}\,\mathrm{d}A = \int_A \bar{z}\,\mathrm{d}A - \bar{z}_S \int_A \mathrm{d}A = 0. \tag{5.30}$$

This can be solved directly for $\bar{y}_S$, $\bar{z}_S$, and we get:

$$\bar{y}_S = \frac{\displaystyle\int_A \bar{y}\,\mathrm{d}A}{\displaystyle\int_A \mathrm{d}A} = \frac{S_{\bar{z}}}{A}, \quad \bar{z}_S = \frac{\displaystyle\int_A \bar{z}\,\mathrm{d}A}{\displaystyle\int_A \mathrm{d}A} = \frac{S_{\bar{y}}}{A}. \tag{5.31}$$

The main advantage of using the reference system $\hat{x}$, $\hat{y}$, $\hat{z}$ in the center of gravity is that the equations (5.16) are considerably simplified by the disappearance of the static moments in which now the normal force and the bending effect are decoupled.

We calculate the moments of inertia $I_{\hat{y}\hat{y}}$, $I_{\hat{z}\hat{z}}$ and the deviation moment $I_{\hat{y}\hat{z}}$. For $I_{\hat{y}\hat{y}}$ we obtain:

$$I_{\hat{y}\hat{y}} = \int_A \hat{z}^2\,\mathrm{d}A = \int_A (\bar{z} - \bar{z}_S)^2\,\mathrm{d}A$$

$$= \int_A \bar{z}^2\,\mathrm{d}A - 2\bar{z}_S \int_A \bar{z}\,\mathrm{d}A + \bar{z}_S^2 \int_A \mathrm{d}A = I_{\bar{y}\bar{y}} - 2\bar{z}_S S_{\bar{y}} + \bar{z}_S^2 A. \tag{5.32}$$

Using $S_{\bar{y}} = \bar{z}_s A$ yields:

$$I_{\hat{y}\hat{y}} = I_{\bar{y}\bar{y}} - \bar{z}_S^2 A. \tag{5.33}$$

Analogous expressions can also be derived for $I_{\bar{z}\bar{z}}$ and $I_{\hat{y}\hat{z}}$, and we get the so-called Steiner theorem[3] or parallel axis theorem:

$$I_{\hat{y}\hat{y}} = I_{\bar{y}\bar{y}} - \bar{z}_S^2 A, \quad I_{\hat{z}\hat{z}} = I_{\bar{z}\bar{z}} - \bar{y}_S^2 A, \quad I_{\hat{y}\hat{z}} = I_{\bar{y}\bar{z}} - \bar{y}_S \bar{z}_S A. \tag{5.34}$$

Steiner's theorem describes the change in the moments of inertia when translating any Cartesian reference system into the center of gravity S of the cross section. It should be noted that the signs of those components in (5.34) that contain the cross-sectional area A are reversed when the coordinate system is shifted from the center of gravity S to any other point. It is also important to note that the moments of inertia with respect to the center of gravity are minimal.

We now demonstrate the application of Steiner's theorem at the example of the two cross-sections already discussed (see Figs. 5.7 and 5.9). For the I-cross section of Fig. 5.7 the calculation of the center of gravity with the help of the integral table of Fig. 5.8 results in:

$$\bar{y}_S = \frac{\displaystyle\int_A 1 \cdot \bar{y} dA}{\displaystyle\int_A dA} = \frac{t\left(\frac{1}{2} \cdot 2b \cdot 1 \cdot 2b \cdot 2 + 1 \cdot h \cdot b\right)}{t\left(1 \cdot 1 \cdot 2b \cdot 2 + 1 \cdot 1 \cdot h\right)} = b,$$

$$\bar{z}_S = \frac{\displaystyle\int_A 1 \cdot \bar{z} dA}{\displaystyle\int_A dA} = \frac{t\left(\frac{1}{2} \cdot h \cdot 1 \cdot h + 1 \cdot 2b \cdot 1 \cdot h\right)}{t\left(1 \cdot 1 \cdot 2b \cdot 2 + 1 \cdot 1 \cdot h\right)} = \frac{h}{2}. \tag{5.35}$$

From this, the moments of inertia and the deviation moment can be determined with the aid of Steiner's theorem as follows:

$$I_{\hat{y}\hat{y}} = I_{\bar{y}\bar{y}} - \bar{z}_S^2 A = \frac{1}{12} th^3 + tbh^2,$$

$$I_{\hat{z}\hat{z}} = I_{\bar{z}\bar{z}} - \bar{y}_S^2 A = \frac{4}{3} tb^3,$$

$$I_{\hat{y}\hat{z}} = I_{\bar{y}\bar{z}} - \bar{y}_S \bar{z}_S A = 0. \tag{5.36}$$

An interesting observation is that in the case of the present doubly symmetrical I-cross section, the deviation moment $I_{\hat{y}\hat{z}}$ vanishes. This will be addressed again at a later point.

We further demonstrate the calculation of the moments of inertia $I_{\hat{y}\hat{y}}$, $I_{\hat{z}\hat{z}}$ and the deviation moment $I_{\hat{y}\hat{z}}$ at the example of the Z cross-section as given in Fig. 5.9. The application of the integral table of Fig. 5.8 results in the following coordinates of the center of gravity S:

---

[3] Jakob Steiner (1796–1863), Swiss mathematician.

$$\bar{y}_S = \frac{\displaystyle\int_A 1 \cdot \bar{y} \, dA}{\displaystyle\int_A dA} = b, \quad \bar{z}_S = \frac{\displaystyle\int_A 1 \cdot \bar{z} \, dA}{\displaystyle\int_A dA} = \frac{h}{2}. \tag{5.37}$$

Steiner's theorem then gives the moments of inertia and the deviation moment as follows:

$$I_{\hat{y}\hat{y}} = I_{\bar{y}\bar{y}} - \bar{z}_S^2 A = \frac{1}{12} t h^3 + \frac{1}{2} t b h^2,$$

$$I_{\hat{z}\hat{z}} = I_{\bar{z}\bar{z}} - \bar{y}_S^2 A = \frac{2}{3} t b^3,$$

$$I_{\hat{y}\hat{z}} = I_{\bar{y}\bar{z}} - \bar{y}_S \bar{z}_S A = \frac{1}{2} t b^2 h. \tag{5.38}$$

For this cross-section the deviation moment $I_{\hat{y}\hat{z}}$ obviously does not vanish.

In some cases the center of gravity S does not have to be explicitly calculated, but is already known at the beginning of the analysis (for the two cross-sections treated here so far, the position of the center of gravity can be deduced from engineering intuition). In such cases the above illustrated methodology to determine the necessary area integrals still applies. We first consider the I cross-section again as shown in Fig. 5.12. We carry out the calculation of all area integrals again as follows:

$$A = \int_A 1 \cdot 1 dA = t \left(1 \cdot 1 \cdot 1 \cdot 2b \cdot 2 + 1 \cdot 1 \cdot 1 \cdot h \right) = t \left(4b + h \right),$$

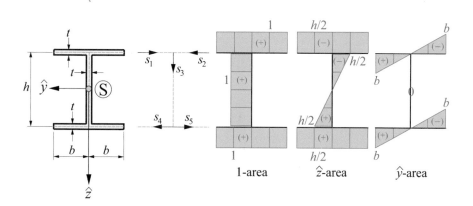

Fig. 5.12 On the calculation of the area integrals at the example of an I-cross-section with respect to the center of gravity S

$$S_{\hat{y}} = \int_A 1 \cdot \hat{z} dA$$

$$= t \left(1 \cdot 1 \cdot \left(-\frac{h}{2} \right) \cdot 2b + 1 \cdot 1 \cdot \frac{h}{2} \cdot 2b + \frac{1}{2} \cdot 1 \cdot \frac{h}{2} \cdot \left(-\frac{h}{2} \right) + \frac{1}{2} \cdot 1 \cdot \frac{h}{2} \cdot \frac{h}{2} \right)$$

$$= 0,$$

$$S_{\hat{z}} = \int_A 1 \cdot \hat{y} dA = t \left(\frac{1}{2} \cdot b \cdot 1b \cdot 2 + \frac{1}{2} \cdot b \cdot 1 \, (-b) \cdot 2 \right) = 0,$$

$$I_{\hat{y}\hat{y}} = \int_A \hat{z}\hat{z} dA = t \left(1 \cdot 2b \cdot \left(-\frac{h}{2} \right)^2 + 1 \cdot 2b \cdot \left(\frac{h}{2} \right)^2 + \frac{1}{3} \cdot \frac{h}{2} \left(-\frac{h}{2} \right)^2 + \frac{1}{3} \cdot \left(\frac{h}{2} \right)^3 \right)$$

$$= tbh^2 + \frac{th^3}{12},$$

$$I_{\hat{z}\hat{z}} = \int_A \hat{y}\hat{y} dA = t \left(\frac{1}{3} \cdot b \cdot b^2 \cdot 2 + \frac{1}{3} \cdot b \cdot (-b)^2 \cdot 2 \right) = \frac{4}{3} tb^3,$$

$$I_{\hat{y}\hat{z}} = \int_A \hat{y}\hat{z} dA$$

$$= t \left(\frac{1}{3} \cdot b \cdot \left(-\frac{h}{2} \right) \cdot b + \frac{1}{3} \cdot b \cdot \left(-\frac{h}{2} \right)(-b) + \frac{1}{3} \cdot b \cdot \frac{h}{2} \cdot b + \frac{1}{3} \cdot b \cdot \frac{h}{2} \cdot (-b) \right) = 0. \quad (5.39)$$

The analysis is further illustrated here at the example of the Z-cross-section for the moments of inertia as well as the deviation moment (Fig. 5.13).

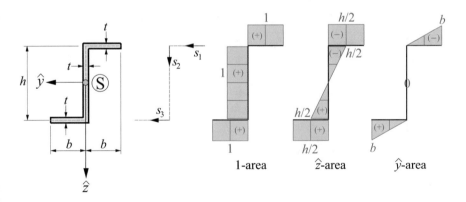

Fig. 5.13 On the calculation of the area integrals at the example of a Z-cross-section with respect to the center of gravity S

$$I_{\hat{y}\hat{y}} = \int_A \hat{z}\hat{z}\,\mathrm{d}A = t\left(1 \cdot b \cdot \left(-\frac{h}{2}\right)^2 + \frac{1}{3}\cdot\frac{h}{2}\cdot\left(-\frac{h}{2}\right)^2 + \frac{1}{3}\cdot\left(\frac{h}{2}\right)^3 + 1\cdot b \cdot \left(\frac{h}{2}\right)^2\right)$$

$$= \frac{1}{2}tbh^2 + \frac{1}{12}th^3,$$

$$I_{\hat{z}\hat{z}} = \int_A \hat{y}\hat{y}\,\mathrm{d}A = t\left(\frac{1}{3}\cdot b \cdot (-b)^2 + \frac{1}{3}\cdot b^3\right) = \frac{2}{3}tb^3,$$

$$I_{\hat{y}\hat{z}} = \int_A \hat{y}\hat{z}\,\mathrm{d}A = t\left(\frac{1}{2}\cdot b \cdot \left(-\frac{h}{2}\right)\cdot(-b) + \frac{1}{2}\cdot b \cdot \frac{h}{2}\cdot b\right) = \frac{1}{2}tb^2h. \tag{5.40}$$

Finally, the normal force and the bending moments are to be related to the center of gravity S. Using Fig. 5.11 we can easily determine the following relations:

$$N_{\hat{x}} = N_{\bar{x}}, \quad M_{\hat{y}} = M_{\bar{y}} - N_{\bar{x}}\bar{z}_S, \quad M_{\hat{z}} = M_{\bar{z}} + N_{\bar{x}}\bar{y}_S. \tag{5.41}$$

Thus the constitutive relations (5.16) change into the following form:

$$N_{\hat{x}} = EA\hat{u}',$$
$$M_{\hat{y}} = -EI_{\hat{y}\hat{y}}\hat{w}'' - EI_{\hat{y}\hat{z}}\hat{v}'',$$
$$-M_{\hat{z}} = -EI_{\hat{y}\hat{z}}\hat{w}'' - EI_{\hat{z}\hat{z}}\hat{v}'', \tag{5.42}$$

or

$$\begin{pmatrix} N_{\hat{x}} \\ M_{\hat{y}} \\ -M_{\hat{z}} \end{pmatrix} = E \begin{bmatrix} A & 0 & 0 \\ 0 & I_{\hat{y}\hat{y}} & I_{\hat{y}\hat{z}} \\ 0 & I_{\hat{y}\hat{z}} & I_{\hat{z}\hat{z}} \end{bmatrix} \begin{pmatrix} \hat{u}' \\ -\hat{w}'' \\ -\hat{v}'' \end{pmatrix}. \tag{5.43}$$

This can be rearranged for the derivatives of the displacement components as follows:

$$\hat{u}' = \frac{N_{\hat{x}}}{EA},$$
$$\hat{v}'' = \frac{1}{E}\frac{I_{\hat{y}\hat{z}}M_{\hat{y}} + I_{\hat{y}\hat{y}}M_{\hat{z}}}{I_{\hat{y}\hat{y}}I_{\hat{z}\hat{z}} - I_{\hat{y}\hat{z}}^2},$$
$$\hat{w}'' = -\frac{1}{E}\frac{I_{\hat{z}\hat{z}}M_{\hat{y}} + I_{\hat{y}\hat{z}}M_{\hat{z}}}{I_{\hat{y}\hat{y}}I_{\hat{z}\hat{z}} - I_{\hat{y}\hat{z}}^2}, \tag{5.44}$$

or

$$\begin{pmatrix} \hat{u}' \\ -\hat{w}'' \\ -\hat{v}'' \end{pmatrix} = \frac{1}{E} \begin{bmatrix} \dfrac{1}{A} & 0 & 0 \\ 0 & \dfrac{I_{\hat{z}\hat{z}}}{\hat{I}} & -\dfrac{I_{\hat{y}\hat{z}}}{\hat{I}} \\ 0 & -\dfrac{I_{\hat{y}\hat{z}}}{\hat{I}} & \dfrac{I_{\hat{y}\hat{y}}}{\hat{I}} \end{bmatrix} \begin{pmatrix} N_{\hat{x}} \\ M_{\hat{y}} \\ -M_{\hat{z}} \end{pmatrix}, \tag{5.45}$$

with $\hat{I} = I_{\hat{y}\hat{y}}I_{\hat{z}\hat{z}} - I_{\hat{y}\hat{z}}^2$.

Apparently, after the first cross-sectional normalization, the bending and stretching of the beam are decoupled from each other which significantly simplifies the analysis. In a practical application one will therefore strive to relate all calculations to a reference system that is located in the center of gravity S of the cross-section. However, it is obvious that the two bending actions, expressed by the two displacements $\hat{v}$ and $\hat{w}$ as well as the two bending moments $M_{\hat{y}}$ and $M_{\hat{z}}$, still are coupled to each other. In the following section we will examine how this (typically undesirable) coupling can also be eliminated.

Insertion of (5.44) into Hooke's law

$$\sigma_{\hat{x}\hat{x}} = E\left(\hat{u}' - \hat{z}\hat{w}'' - \hat{y}\hat{v}''\right) \tag{5.46}$$

gives the following expression for the normal stress $\sigma_{\hat{x}\hat{x}}$ from which the stresses at any point $\hat{y}$ and $\hat{z}$ of the cross-section can be calculated if $N_{\hat{x}}$, $M_{\hat{y}}$, $M_{\hat{z}}$ and the necessary area integrals are given:

$$\sigma_{\hat{x}\hat{x}} = \frac{N_{\hat{x}}}{A} + \frac{I_{\hat{z}\hat{z}}M_{\hat{y}} + I_{\hat{y}\hat{z}}M_{\hat{z}}}{I_{\hat{y}\hat{y}}I_{\hat{z}\hat{z}} - I_{\hat{y}\hat{z}}^2}\hat{z} - \frac{I_{\hat{y}\hat{z}}M_{\hat{y}} + I_{\hat{y}\hat{y}}M_{\hat{z}}}{I_{\hat{y}\hat{y}}I_{\hat{z}\hat{z}} - I_{\hat{y}\hat{z}}^2}\hat{y}. \tag{5.47}$$

An important quantity when determining the normal stress $\sigma_{\hat{x}\hat{x}}$ is the so-called zero stress line. Eq. (5.47) formally describes a plane in the two-dimensional $\hat{y}\hat{z}$−space to describe the distribution of $\sigma_{\hat{x}\hat{x}}$ over the cross-section of the beam. Setting (5.47) to zero then allows one of the two variables $\hat{y}$ and $\hat{z}$ to be expressed as a function of the other. Formally, this describes a straight line on which the normal stress $\sigma_{\hat{x}\hat{x}}$ attains zero values.

5.4   Second Cross-Sectional Normalization: Principal Axes

The first cross-sectional normalization, namely the transformation of an arbitrary reference system into the center of gravity of the cross-section by translation of the reference system, decouples the beam stretching from beam bending. However, the two bending effects, characterized by the two bending moments $M_{\hat{y}}$ and $M_{\hat{z}}$, are still coupled to each other by the deviation moment $I_{\hat{y}\hat{z}}$. Another significant simplification of the relations is obtained when a particular reference system, namely the so-called principal axis system, is determined and all quantities are related to this system. The aim is to find the reference system in which the deviation moment $I_{\hat{y}\hat{z}}$ becomes zero. In this case, as one can easily deduce from (5.42) or (5.43), the two bending effects will also decouple from each other. This particular reference system with its origin in the center of gravity S, the so-called principal axes, is rotated by the angle φ_0 about the $\hat{x}$−axis compared to the reference system $\hat{x}$, $\hat{y}$, $\hat{z}$. This coordinate rotation by the angle φ_0 is also referred to as the so-called second cross-sectional normalization.

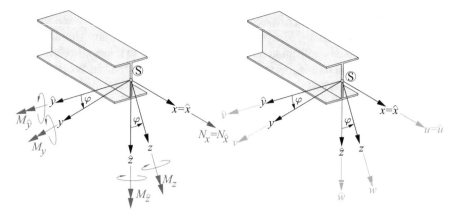

Fig. 5.14 Rotation of the reference system $\hat{x}, \hat{y}, \hat{z}$ by the angle φ into the reference system x, y, z

Fig. 5.15 Relation between $\hat{y}, \hat{z}$ and y, z

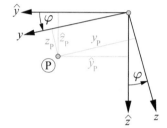

Under the angle φ_0, the two moments of inertia I_{yy} and I_{zz} become extremal as we will show later.

We begin our considerations with a rotation of the coordinate system $\hat{x}, \hat{y}, \hat{z}$ by an arbitrary angle φ into the new reference system x, y, z, as shown in Fig. 5.14. The transformation of the coordinates can be derived from Fig. 5.15 by considering a point P with the coordinates $\hat{y}_P, \hat{z}_P$ respectively y_P, z_P:

$$y = \hat{y}\cos\varphi + \hat{z}\sin\varphi, \quad z = -\hat{y}\sin\varphi + \hat{z}\cos\varphi. \tag{5.48}$$

In vector-matrix notation we have:

$$\begin{pmatrix} y \\ z \end{pmatrix} = \begin{bmatrix} \cos\varphi & \sin\varphi \\ -\sin\varphi & \cos\varphi \end{bmatrix} \begin{pmatrix} \hat{y} \\ \hat{z} \end{pmatrix}. \tag{5.49}$$

The matrix occurring therein is also referred to as the so-called transformation matrix.

Since the coordinate system is rotated about the $\hat{x}$−axis, the axes x and $\hat{x}$ are identical:

$$x = \hat{x}. \tag{5.50}$$

We now consider the deviation moment I_{yz} in the rotated coordinate system and get:

$$
\begin{aligned}
I_{yz} &= \int_A yz \, dA = \int_A (\hat{y} \cos \varphi + \hat{z} \sin \varphi)(-\hat{y} \sin \varphi + \hat{z} \cos \varphi) \, dA \\
&= -\sin \varphi \cos \varphi \int_A \hat{y}^2 \, dA + \cos^2 \varphi \int_A \hat{y}\hat{z} \, dA - \sin^2 \varphi \int_A \hat{y}\hat{z} \, dA + \sin \varphi \cos \varphi \int_A \hat{z}^2 \, dA \\
&= -I_{\hat{z}\hat{z}} \sin \varphi \cos \varphi + I_{\hat{y}\hat{z}} \left(\cos^2 \varphi - \sin^2 \varphi \right) + I_{\hat{y}\hat{y}} \sin \varphi \cos \varphi \\
&= I_{\hat{y}\hat{z}} \left(\cos^2 \varphi - \sin^2 \varphi \right) + \left(I_{\hat{y}\hat{y}} - I_{\hat{z}\hat{z}} \right) \sin \varphi \cos \varphi.
\end{aligned}
\tag{5.51}
$$

Using $\cos^2 \varphi = \frac{1}{2}(1 + \cos 2\varphi)$, $\sin^2 \varphi = \frac{1}{2}(1 - \cos 2\varphi)$ and $2 \sin \varphi \cos \varphi = \sin 2\varphi$ leads to:

$$
I_{yz} = \frac{1}{2} \left(I_{\hat{y}\hat{y}} - I_{\hat{z}\hat{z}} \right) \sin 2\varphi + I_{\hat{y}\hat{z}} \cos 2\varphi.
\tag{5.52}
$$

We now require the disappearance of the deviation moment I_{yz} in order to determine the principal axis angle φ_0. If we set (5.52) to zero we obtain:

$$
\tan 2\varphi_0 = \frac{2 I_{\hat{y}\hat{z}}}{I_{\hat{z}\hat{z}} - I_{\hat{y}\hat{y}}}.
\tag{5.53}
$$

Thus, an equation for the principal axis angle φ_0 is found at which the deviation moment disappears and, analogous to the determination of the principal stresses in the plane stress state (see Chap. 3), the moments of inertia become extremal. Correlations analogous to (5.52) can also be found for the two moments of inertia I_{yy} and I_{zz}. In summary, we get the following classical transformation equations for an arbitrary angle φ:

$$
\begin{aligned}
I_{yy} &= \frac{1}{2} \left(I_{\hat{y}\hat{y}} + I_{\hat{z}\hat{z}} \right) + \frac{1}{2} \left(I_{\hat{y}\hat{y}} - I_{\hat{z}\hat{z}} \right) \cos 2\varphi - I_{\hat{y}\hat{z}} \sin 2\varphi, \\
I_{zz} &= \frac{1}{2} \left(I_{\hat{y}\hat{y}} + I_{\hat{z}\hat{z}} \right) - \frac{1}{2} \left(I_{\hat{y}\hat{y}} - I_{\hat{z}\hat{z}} \right) \cos 2\varphi + I_{\hat{y}\hat{z}} \sin 2\varphi, \\
I_{yz} &= \frac{1}{2} \left(I_{\hat{y}\hat{y}} - I_{\hat{z}\hat{z}} \right) \sin 2\varphi + I_{\hat{y}\hat{z}} \cos 2\varphi.
\end{aligned}
\tag{5.54}
$$

Since the tan −function in (5.53) is periodic, there is also another angle $\varphi_0 + \frac{\pi}{2}$ in addition to the angle φ_0 under which the deviation moment I_{yz} disappears and the moments of inertia I_{yy} and I_{zz} become extremal (so-called principal moments of inertia). Since these two principal axis systems are orthogonal to each other the consideration of the angle $\varphi = \varphi_0 + \frac{\pi}{2}$ does not lead to any further conclusions.

The values of the principal moments of inertia are obtained from the consideration of (5.53) and (5.54), if one also includes the identities

$$\cos 2\varphi_0 = \frac{1}{\sqrt{1 + \tan^2 2\varphi_0}} = \frac{I_{\hat{z}\hat{z}} - I_{\hat{y}\hat{y}}}{\sqrt{\left(I_{\hat{z}\hat{z}} - I_{\hat{y}\hat{y}}\right)^2 + 4I_{\hat{y}\hat{z}}^2}},$$

$$\sin 2\varphi_0 = \frac{\tan 2\varphi_0}{\sqrt{1 + \tan^2 2\varphi_0}} = \frac{2I_{\hat{y}\hat{z}}}{\sqrt{\left(I_{\hat{z}\hat{z}} - I_{\hat{y}\hat{y}}\right)^2 + 4I_{\hat{y}\hat{z}}^2}}. \tag{5.55}$$

This leads to the following principal moments of inertia:

$$I_{1,2} = \frac{I_{\hat{y}\hat{y}} + I_{\hat{z}\hat{z}}}{2} \pm \sqrt{\left(\frac{I_{\hat{z}\hat{z}} - I_{\hat{y}\hat{y}}}{2}\right)^2 + I_{\hat{y}\hat{z}}^2}. \tag{5.56}$$

The nomenclature is such that I_1 denotes the maximum moment of inertia, whereas I_2 represents the minimum value.

It should be noted that the comparison with the transformation equations as outlined in Chap. 3 shows that the structure of the equations (5.54) corresponds to that of the stress transformation relations. Thus, analogous to the Mohr's stress circle, one can also create an inertia circle that follows from Mohr's stress circle by replacing the two normal stresses σ_{xx} and σ_{yy} by the two moments of inertia $I_{\hat{y}\hat{y}}$ and $I_{\hat{z}\hat{z}}$ and the shear stress τ_{xy} by the deviation moment $I_{\hat{y}\hat{z}}$. A representation of the inertia circle is given in Fig. 5.16.

It can be shown that the Eq. (5.53) for the principal axis angle φ_0 can also be obtained from the requirement that the two moments of inertia I_{yy} and I_{zz} assume extremal values, i.e.:

$$\frac{\mathrm{d}I_{yy}}{\mathrm{d}\varphi} = 0, \quad \frac{\mathrm{d}I_{zz}}{\mathrm{d}\varphi} = 0. \tag{5.57}$$

Some general rules (Fig. 5.17) can be set up on the basis of symmetry properties of beam cross-sections:

Fig. 5.16 Inertia circle

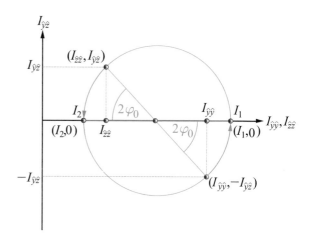

- If a cross-section has an axis of symmetry, then one of the two principal axes y, z runs parallel to this axis of symmetry. This means that the principal axis angle φ_0 always becomes zero, provided that there is at least one axis of symmetry for the cross section under consideration and one of the two reference axes runs parallel to this axis of symmetry. In this case, however, the center of gravity S needs to be determined and cannot usually be deduced directly from engineering intuition. Exemplary cross-sections can be found in Fig. 5.17, upper row.
- If a cross-section has two axes of symmetry, then these axes automatically represent the two principal axes. In such cases the deviation moment $I_{\hat{y}\hat{z}}$ becomes zero so that the principal axis angle φ_0 must necessarily be zero. With such cross-sections, the second cross-sectional normalization can be omitted, and the first cross-sectional normalization is also no longer necessary since the center of gravity S of the cross-section is always located at the intersection of the two axes of symmetry. Examples of cross-sections with two symmetry axes can be found in Fig. 5.17, middle.

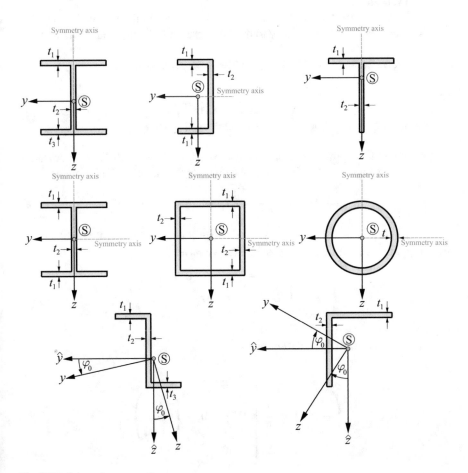

Fig. 5.17 Selected cross-sections

- If the considered cross-section has no symmetry properties, then both the first and the second cross-section normalization must be carried out, i.e. the center of gravity must be determined and the principal axis angle φ_0 must also be calculated. Two examples of such cross-sections are given in 5.17, lower row.

It can also be shown that the sum of the two moments of inertia I_{yy} and I_{zz} or $I_{\hat{y}\hat{y}}$ and $I_{\hat{z}\hat{z}}$, regardless of the choice of the reference system, always assumes the same value and thus represents an invariant:

$$I_{yy} + I_{zz} = I_{\hat{y}\hat{y}} + I_{\hat{z}\hat{z}} = I_p. \tag{5.58}$$

The abbreviation I_p denotes the so-called polar moment of inertia, i.e. the sum of both moments of inertia. The polar moment of inertia is therefore an invariant quantity. One can also show that the expression $\left[\frac{1}{2}\left(I_{yy} - I_{zz}\right)\right]^2 + I_{xy}^2$ represents a further invariant.

We also transform the bending moments $M_{\hat{y}}$ and $M_{\hat{z}}$ to the reference system x, y, z. This can be done analogously to the transformation (5.48), and we achieve:

$$M_y = M_{\hat{y}} \cos \varphi + M_{\hat{z}} \sin \varphi, \quad M_z = -M_{\hat{y}} \sin \varphi + M_{\hat{z}} \cos \varphi, \tag{5.59}$$

or:

$$\begin{pmatrix} M_y \\ M_z \end{pmatrix} = \begin{bmatrix} \cos \varphi & \sin \varphi \\ -\sin \varphi & \cos \varphi \end{bmatrix} \begin{pmatrix} M_{\hat{y}} \\ M_{\hat{z}} \end{pmatrix}. \tag{5.60}$$

As shown, if the principal axis system x, y, z is considered there is no deviation moment at the angle φ_0. The constitutive equations then assume the following form:

$$N = EAu', \quad M_y = -EI_{yy}w'', \quad M_z = EI_{zz}v'', \tag{5.61}$$

where from this point on we drop the index x for the normal force N for the sake of simplicity. In a vector matrix representation we have:

$$\begin{pmatrix} N \\ M_y \\ -M_z \end{pmatrix} = E \begin{bmatrix} A & 0 & 0 \\ 0 & I_{yy} & 0 \\ 0 & 0 & I_{zz} \end{bmatrix} \begin{pmatrix} u' \\ -w'' \\ -v'' \end{pmatrix}. \tag{5.62}$$

In its inverted form this relation reads:

$$\begin{pmatrix} u' \\ -w'' \\ -v'' \end{pmatrix} = \frac{1}{E} \begin{bmatrix} \dfrac{1}{A} & 0 & 0 \\ 0 & \dfrac{1}{I_{yy}} & 0 \\ 0 & 0 & \dfrac{1}{I_{zz}} \end{bmatrix} \begin{pmatrix} N \\ M_y \\ -M_z \end{pmatrix}. \tag{5.63}$$

Obviously, employing the principal axis system is advantageous because it simplifies the constitutive relationships significantly—normal force and bending moments are uncoupled from each other, and the two bending effects can also be treated separately. However, it should be pointed out again that this separate consideration of the individual effects is only possible in the framework of geometric linearity as assumed here (i.e. the assumption of small deformations).

From Hooke's law

$$\sigma_{xx} = E\left(u' - zw'' - yv''\right) \qquad (5.64)$$

the following expression for the determination of the normal stress σ_{xx} at every point y, z of the cross-section can be derived:

$$\sigma_{xx} = \frac{N}{A} + \frac{M_y}{I_{yy}}z - \frac{M_z}{I_{zz}}y. \qquad (5.65)$$

The comparison with (5.47) shows a significant simplification. The coordinates y and z can then be determined according to the $y-$ and $z-$areas in Fig. 5.18 which result from the coordinate transformation (5.48). The zero stress line can again be determined by setting the resulting expression of the normal stress σ_{xx} to zero.

Fig. 5.18 $\hat{y}-$ and $\hat{z}-$area; $y-$ and $z-$area of a Z-cross-section

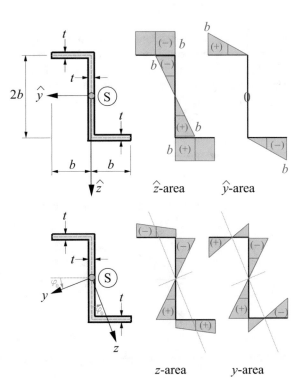

If the beam is only subjected to bending about one of the two principal axes y or z, then the maximum normal stresses $\sigma_{xx,\text{max}}$ relevant for the stress analysis result as:

$$|\sigma|_{xx,\text{max}} = \frac{N}{A} + \frac{M_y}{W_y}, \tag{5.66}$$

or

$$|\sigma|_{xx,\text{max}} = \frac{N}{A} - \frac{M_z}{W_z}. \tag{5.67}$$

Here the quantities W_y and W_z are the so-called resistance moments of the given cross-section which result as follows:

$$W_y = \frac{I_{yy}}{|z|_{\text{max}}}, \quad W_z = \frac{I_{zz}}{|y|_{\text{max}}}, \tag{5.68}$$

where z_{max} and y_{max} are the coordinates of the point that is the furthest away from the center of gravity S.

5.5 Selected Basic Cases

In many cases, cross-sections as relevant in lightweight engineering are composed of elementary segments which can also be treated separately and then combined to form the total cross-section. We also want to discuss this very classic procedure and provide it as an alternative to the analysis method shown above using integral tables. First, however, we want to look at some very elementary basic cross-sections.

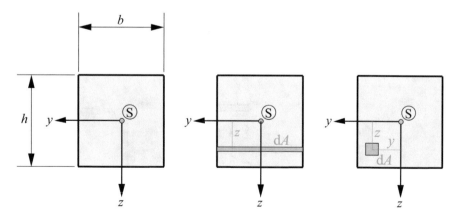

Fig. 5.19 Determination of the moments of inertia at the example of a rectangular cross-section

Let us consider the rectangular cross-section (width b, height h), as shown in Fig. 5.19, left. We can denote the coordinate axes as x, y, z since the axes shown here of course constitute the principal axis system. We start by determining the moment of inertia I_{yy}, defined as

$$I_{yy} = \int_A z^2 dA. \tag{5.69}$$

The area element dA can currently be expressed as $dA = bdz$ (Fig. 5.19, middle), so that:

$$I_{yy} = \int_{-\frac{h}{2}}^{\frac{h}{2}} z^2 b dz = \left.\frac{bz^3}{3}\right|_{-\frac{h}{2}}^{\frac{h}{2}} = \frac{bh^3}{12}. \tag{5.70}$$

In the same way the moment of inertia I_{zz} can be determined:

$$I_{zz} = \frac{hb^3}{12}. \tag{5.71}$$

The determination of the deviation moment

$$I_{yz} = \int_A yz dA \tag{5.72}$$

can be performed using the area element $dA = dydz$ (Fig. 5.19, right) and leads to:

$$I_{yz} = \int_{-\frac{h}{2}}^{+\frac{h}{2}} \int_{-\frac{b}{2}}^{+\frac{b}{2}} yz dy dz = 0. \tag{5.73}$$

From these elementary formulas, further equations for cross-sections as they are typical in lightweight engineering such as the box cross-section given in Fig. 5.20 can be derived. The box cross-section has the width b, the height h, and the constant wall thickness t. We use $\bar{b} = b - 2t$ and $\bar{h} = h - 2t$ as auxiliary variables. The box

Fig. 5.20 Determination of the moments of inertia at the example of a rectangular box cross-section

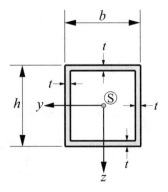

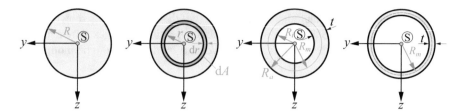

Fig. 5.21 Determination of the moments of inertia for several forms of circular cross-sections

cross-section can be mathematically interpreted in such a way that the moments of inertia of the inscribed cross-section with the width $\bar{b}$ and the height $\bar{h}$ are subtracted from the moments of inertia of the full cross-section (width b, height h):

$$I_{yy} = \frac{bh^3}{12} - \frac{\bar{b}\bar{h}^3}{12} = \frac{1}{12}\left(bh^3 - \bar{b}\bar{h}^3\right). \tag{5.74}$$

Analogously we have for I_{zz}:

$$I_{zz} = \frac{hb^3}{12} - \frac{\bar{h}\bar{b}^3}{12} = \frac{1}{12}\left(hb^3 - \bar{h}\bar{b}^3\right). \tag{5.75}$$

The deviation moment I_{yz} becomes zero in the current coordinate system.

As a further elementary case we want to consider the circular cross-section (radius R) as shown in Fig. 5.21. We determine the polar moment of inertia I_p as

$$I_p = \int_A r^2 \mathrm{d}A. \tag{5.76}$$

We interpret the area element $\mathrm{d}A$ as a circular ring with the thickness $\mathrm{d}r$, so that $\mathrm{d}A = 2\pi r\mathrm{d}r$. Then:

$$I_p = 2\pi \int_0^R r^3 \mathrm{d}r = \frac{\pi R^4}{2}. \tag{5.77}$$

Since the polar moment of inertia represents the sum of I_{yy} and I_{zz} and the circular cross-section is doubly symmetrical (which leads to $I_{yy} = I_{zz}$), the following applies:

$$I_{yy} = I_{zz} = \frac{\pi R^4}{4}. \tag{5.78}$$

The deviation moment I_{yz} is zero due to the symmetry properties of the cross-section.

The moments of inertia for a circular ring cross-section (wall thickness t) with the inner radius R_i and the outer radius R_a can be derived accordingly. The procedure is analogous to the situation shown in Fig. 5.20 for a box cross-section. We obtain:

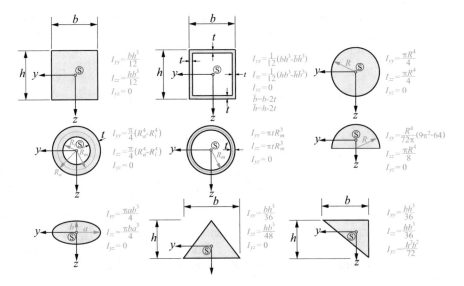

Fig. 5.22 Moments of inertia of selected cross-sections

$$I_{yy} = I_{zz} = \frac{\pi R_a^4}{4} - \frac{\pi R_i^4}{4} = \frac{\pi}{4}\left(R_a^4 - R_i^4\right). \tag{5.79}$$

As an auxiliary variable we also define the mean radius R_m as

$$R_m = \frac{1}{2}\left(R_a + R_i\right). \tag{5.80}$$

Then $R_a^4 - R_i^4 = 4t R_m^3 \left(1 + \frac{t^2}{4R_m^2}\right)$ holds. If we now consider a very thin-walled circular ring cross-section with $t \ll R_m$, then we have:

$$I_{yy} = I_{zz} = \pi t R_m^3. \tag{5.81}$$

Further elementary cases can be dealt with in a similar way which, however, is not presented here in detail. Figure 5.22 gives an overview of some selected cross-sectional shapes.

5.6 Analysis of Arbitrarily Segmented Cross-Sections

We now want to consider how Steiner's theorem can be employed to determine the moments of inertia for arbitrarily segmented thin-walled cross-sections such as the examples given in Fig. 5.2. We have already used the fact that we can calculate the area integrals of cross-sections which consist of several sub-segments from the sum of the area integrals of the individual segments. For a cross-section with n individual segments, the following generally applies:

$$I_{\hat{y}\hat{y}} = \int_A \hat{z}^2 \mathrm{d}A = \sum_{i=1}^{i=n} \int_{A_i} \hat{z}^2 \mathrm{d}A_i = \sum_{i=1}^{i=n} I_{\hat{y}\hat{y},i},$$

$$I_{\hat{z}\hat{z}} = \int_A \hat{y}^2 \mathrm{d}A = \sum_{i=1}^{i=n} \int_{A_i} \hat{y}^2 \mathrm{d}A_i = \sum_{i=1}^{i=n} I_{\hat{z}\hat{z},i},$$

$$I_{\hat{y}\hat{z}} = \int_A \hat{y}\hat{z}\,\mathrm{d}A = \sum_{i=1}^{i=n} \int_{A_i} \hat{y}\hat{z}\,\mathrm{d}A_i = \sum_{i=1}^{i=n} I_{\hat{y}\hat{z},i}. \tag{5.82}$$

Herein it is assumed that the partial moments of inertia $I_{\hat{y}\hat{y},i}$, $I_{\hat{z}\hat{z},i}$, $I_{\hat{y}\hat{z},i}$ are related to an axis system $\hat{y}$, $\hat{z}$ located in the center of gravity. In what follows we will deal with the question how these partial moments of inertia can be calculated in a suitable manner.

We consider again the axis translation of Fig. 5.11 where we now shift the coordinate system $\hat{y}$, $\hat{z}$ away from the center of gravity by the distance $\bar{y}_S$ and $\bar{z}_S$ into an arbitrary point outside of the center of gravity S. With

$$\bar{y} = \hat{y} + \bar{y}_S, \quad \bar{z} = \hat{z} + \bar{z}_S \tag{5.83}$$

the following alternative representation of Steiner's theorem results:

$$I_{\bar{y}\bar{y}} = I_{\hat{y}\hat{y}} + \bar{z}_S^2 A,$$

$$I_{\bar{z}\bar{z}} = I_{\hat{z}\hat{z}} + \bar{y}_S^2 A,$$

$$I_{\bar{y}\bar{z}} = I_{\hat{y}\hat{z}} + \bar{y}_S \bar{z}_S A. \tag{5.84}$$

This representation of Steiner's theorem expresses that the area integrals $I_{\hat{y}\hat{y}}$, $I_{\hat{z}\hat{z}}$, $I_{\hat{y}\hat{z}}$ change by the amount $\bar{z}_S^2 A$, $\bar{y}_S^2 A$, $\bar{y}_S \bar{z}_S A$ with a translation of the reference system out of the center of gravity and change into the expressions $I_{\bar{y}\bar{y}}$, $I_{\bar{z}\bar{z}}$, $I_{\bar{y}\bar{z}}$. Obviously the values $I_{\hat{y}\hat{y}}$, $I_{\hat{z}\hat{z}}$, $I_{\hat{y}\hat{z}}$ are minimal in the center of gravity compared to any other point of the cross-section.

This formulation of Steiner's theorem has a significant practical value insofar as it is based on the area integrals with respect to the center of gravity S of the cross-section. In practical applications, one will often first calculate the position of the center of gravity S and then determine the area integrals or, in the case of segmented cross-sections (Fig. 5.23), calculate them by adding up the contribu-

Fig. 5.23 Segmented
cross-section

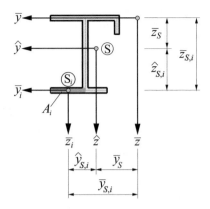

tions of the individual cross-sectional segments. With a cross-section consisting of n
elementary segments, the cross-sectional area A can be determined as the sum of the
partial areas A_i as follows:

$$A = \int_A dA = \sum_{i=1}^{i=n} A_i. \tag{5.85}$$

The static moments with respect to the arbitrarily chosen reference system $\bar{y}$, $\bar{z}$
can be written as:

$$S_{\bar{y}} = \int_A \bar{z} dA = \sum_{i=1}^{i=n} A_i \bar{z}_{S,i}, \quad S_{\bar{z}} = \int_A \bar{y} dA = \sum_{i=1}^{i=n} A_i \bar{y}_{S,i}, \tag{5.86}$$

where $\bar{y}_{S,i}$ and $\bar{z}_{S,i}$ are the centroid coordinates of segment i. The position of the
center of gravity is then as follows:

$$\bar{y}_S = \frac{\int_A \bar{y} dA}{\int_A dA} = \frac{\sum_{i=1}^{i=n} A_i \bar{y}_{S,i}}{\sum_{i=1}^{i=n} A_i}, \quad \bar{z}_S = \frac{\int_A \bar{z} dA}{\int_A dA} = \frac{\sum_{i=1}^{i=n} A_i \bar{z}_{S,i}}{\sum_{i=1}^{i=n} A_i}. \tag{5.87}$$

The area integrals $I_{\hat{y}\hat{y}}$, $I_{\hat{z}\hat{z}}$, $I_{\hat{y}\hat{z}}$ with respect to the center of gravity S then result as
the sum of all moments of inertia $I_{\bar{y}\bar{y},i}$, $I_{\bar{z}\bar{z},i}$, $I_{\bar{y}\bar{z},i}$ as well as the associated coordinates
$\hat{y}_{S,i}$, $\hat{z}_{S,i}$ multiplied with the cross-sectional areas A_i:

Fig. 5.24 Application of Steiner's theorem at the example of an I-section

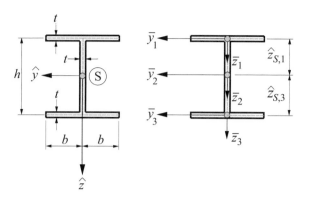

$$I_{\hat{y}\hat{y}} = \sum_{i=1}^{i=n} I_{\bar{y}\bar{y},i} + \sum_{i=1}^{i=n} \hat{z}_{S,i}^2 A_i,$$

$$I_{\hat{z}\hat{z}} = \sum_{i=1}^{i=n} I_{\bar{z}\bar{z},i} + \sum_{i=1}^{i=n} \hat{y}_{S,i}^2 A_i,$$

$$I_{\hat{y}\hat{z}} = \sum_{i=1}^{i=n} I_{\bar{y}\bar{z},i} + \sum_{i=1}^{i=n} \hat{y}_{S,i}\hat{z}_{S,i} A_i. \tag{5.88}$$

Here $I_{\bar{y}\bar{y},i}$, $I_{\bar{z}\bar{z},i}$, $I_{\bar{y}\bar{z},i}$ are the moments of inertia and the deviation moment of segment i with respect to its local axes $\bar{y}_i$, $\bar{z}_i$. The quantities $\hat{y}_{S,i}$ and $\hat{z}_{S,i}$ are the distances of the center of gravity S_i of sub-segment i, measured from the overall center of gravity of the cross-section. After determining the moments of inertia and the deviation moment as outlined, the second cross-sectional normalization can then take place as already shown.

We want to illustrate the analysis procedure using the two cross-sectional shapes already discussed and first consider the I cross-section shown in Fig. 5.24, left. We divide the cross-section into the segments 1 (upper flange), 2 (web), and 3 (lower flange) and assign the local reference systems $\bar{y}_1$, $\bar{z}_1$, $\bar{y}_2$, $\bar{z}_2$ and $\bar{y}_3$, $\bar{z}_3$, which are located in the centroids of the individual segments, as shown in Fig. 5.24, right. The centers of gravity of the individual segments have the coordinates $\hat{y}_{S,1} = 0$, $\hat{z}_{S,1} = -\frac{h}{2}$, $\hat{y}_{S,2} = 0$, $\hat{z}_{S,2} = 0$ and $\hat{y}_{S,3} = 0$, $\hat{z}_{S,3} = \frac{h}{2}$ with regard to the center of gravity S (see Fig. 5.24, right). The moments of inertia of the individual segments are:

$$I_{\bar{y}\bar{y},1} = I_{\bar{y}\bar{y},3} = \frac{2bt^3}{12} \simeq 0, \quad I_{\bar{y}\bar{y},2} = \frac{th^3}{12},$$

$$I_{\bar{z}\bar{z},1} = I_{\bar{z}\bar{z},3} = \frac{t(2b)^3}{12} = \frac{2tb^3}{3}, \quad I_{\bar{z}\bar{z},2} = \frac{ht^3}{12} \simeq 0,$$

$$I_{\bar{y}\bar{z},1} = I_{\bar{y}\bar{z},2} = I_{\bar{y}\bar{z},3} = 0. \tag{5.89}$$

Fig. 5.25 Application of Steiner's theorem at the example of a Z-section

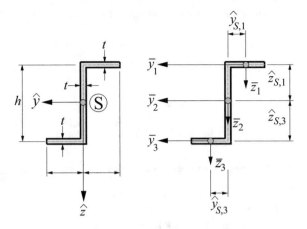

The cross-sectional areas A_i result in:

$$A_1 = A_3 = 2tb, \quad A_2 = th. \tag{5.90}$$

Steiner's theorem (5.88) then yields:

$$
\begin{aligned}
I_{\hat{y}\hat{y}} &= \frac{th^3}{12} + \left(-\frac{h}{2}\right)^2 2tb + \left(\frac{h}{2}\right)^2 2tb = \frac{th^3}{12} + tbh^2, \\
I_{\hat{z}\hat{z}} &= \frac{2tb^3}{3} + \frac{2tb^3}{3} = \frac{4tb^3}{3}, \\
I_{\hat{y}\hat{z}} &= 0.
\end{aligned}
\tag{5.91}
$$

It turns out that these results are identical with the previously determined values.

As a further example we consider again the Z-cross-section as shown in Fig. 5.25, left. We divide this cross-section into three segments, as shown in Fig. 5.25, right. The centers of gravity of the segments can be found at the positions $\hat{y}_{S,1} = -\frac{b}{2}$, $\hat{z}_{S,1} = -\frac{h}{2}$, $\hat{y}_{S,2} = 0$, $\hat{z}_{S,2} = 0$ and $\hat{y}_{S,3} = \frac{b}{2}$, $\hat{z}_{S,3} = \frac{h}{2}$, measured from the center of gravity S. The moments of inertia of the segments result as:

$$
\begin{aligned}
I_{\bar{y}\bar{y},1} = I_{\bar{y}\bar{y},3} &= \frac{bt^3}{12} \simeq 0, \quad I_{\bar{y}\bar{y},2} = \frac{th^3}{12}, \\
I_{\bar{z}\bar{z},1} = I_{\bar{z}\bar{z},3} &= \frac{tb^3}{12}, \quad I_{\bar{z}\bar{z},2} = \frac{ht^3}{12} \simeq 0, \\
I_{\bar{y}\bar{z},1} = I_{\bar{y}\bar{z},2} &= I_{\bar{y}\bar{z},3} = 0.
\end{aligned}
\tag{5.92}
$$

The cross-sectional areas A_i can be determined as follows:

$$A_1 = A_3 = tb, \quad A_2 = th. \tag{5.93}$$

Steiner's theorem then results in:

$$I_{\hat{y}\hat{y}} = \frac{th^3}{12} + \left(-\frac{h}{2}\right)^2 tb + \left(\frac{h}{2}\right)^2 tb = \frac{th^3}{12} + \frac{tbh^2}{2},$$

$$I_{\hat{z}\hat{z}} = \frac{tb^3}{12} + \frac{tb^3}{12} + \left(-\frac{b}{2}\right)^2 tb + \left(\frac{b}{2}\right)^2 tb = \frac{2tb^3}{3},$$

$$I_{\hat{y}\hat{z}} = \left(-\frac{b}{2}\right)\left(-\frac{h}{2}\right) tb + \frac{b}{2}\frac{h}{2}tb = \frac{tb^2h}{2}. \tag{5.94}$$

These results also agree with the previously determined values.

5.7 Calculation of Beam Deflections

In addition to calculating the normal stress σ_{xx} in beam structures under normal force and bending, it is also of interest for the construction and design of such structures to determine their deflections under a given load. If we assume that the chosen reference system is the principal axis system x, y, z, then the three displacements u, v, w may occur. Here we devote ourselves to the determination of the two beam deflections v and w. We start the considerations be restricting ourselves to the case that there is only one bending moment M_y so that only the deflection w occurs. The results can then easily be transferred to the determination of the deflection v.

First we consider the equilibrium conditions for a beam with the variable bending stiffness EI_{yy} subjected to the line load q_z (Fig. 5.26, left). Since we want to derive the equilibrium conditions in a general form, no boundary conditions are specified at this point. The reference system x, y, z is assumed to be the principal axis system. We consider the infinitesimal beam element of Fig. 5.26, right, with the length dx and establish the equilibrium conditions. It should be noted that in addition to the bending moment M_y, the transverse shear force V_z also occurs. The influence of the transverse shear force and the resulting shear stresses and deformations will be discussed in detail in later chapters. The beam equilibrium conditions result in:

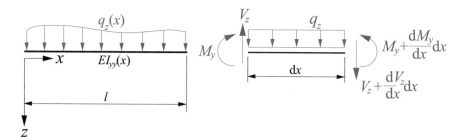

Fig. 5.26 Equilibrium for an infinitesimal beam element

Fig. 5.27 Typical boundary
conditions for a beam

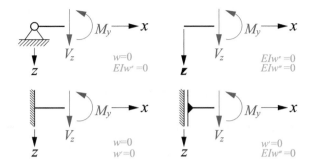

$$\frac{dV_z}{dx} = V_z' = -q_z, \quad \frac{dM_y}{dx} = M_y' = V_z. \tag{5.95}$$

From the constitutive law (5.61) it follows with $M_y' = V_z$ and $V_z' = -q_z$:

$$M_y = -EI_{yy}w'', \quad V_z = -\left(EI_{yy}w''\right)', \quad \left(EI_{yy}w''\right)'' = q_z. \tag{5.96}$$

If the beam under consideration has constant properties in the longitudinal direction x, i.e. it has a constant bending stiffness EI_{yy}, then these equations are simplified as follows:

$$M_y = -EI_{yy}w'', \quad V_z = -EI_{yy}w''', \quad EI_{yy}w'''' = q_z. \tag{5.97}$$

The equilibrium conditions with regard to the bending moment M_z, the transverse shear force V_y and the line load q_y are analogous:

$$\frac{dV_y}{dx} = V_y' = -q_y, \quad \frac{dM_z}{dx} = M_z' = -V_y. \tag{5.98}$$

With the constitutive law (5.61) we achieve:

$$M_z = EI_{zz}v'', \quad V_y = -\left(EI_{zz}v''\right)', \quad \left(EI_{zz}v''\right)'' = q_y. \tag{5.99}$$

In the case of a prismatic beam with a constant bending stiffness EI_{zz} this yields:

$$M_z = EI_{zz}v'', \quad V_y = -EI_{zz}v''', \quad EI_{zz}v'''' = q_y. \tag{5.100}$$

In the following we want to consider the case of bending in the $xz-$plane more closely. The equations (5.97) show that the determination of the beam deflections or the so-called bending line $w(x)$ can be done by integrating these equations. There are several possible starting points. If, for instance, the bending moment line $M_y(x)$ has already been determined, then the bending line $w(x)$ can be determined from twofold integration of $M_y(x)$. In the same way, one can also employ the transverse shear force line $V_z(x)$ or start with the external load $q_z(x)$. Let us consider the last case:

$$EI_{yy}w'''' = q_z,$$

$$EI_{yy}w''' = -V_z = \int q_z dx + C_1,$$

$$EI_{yy}w'' = -M_y = \int \int q_z dx dx + C_1 x + C_2,$$

$$EI_{yy}w' = \int \int \int q_z dx dx dx + \frac{1}{2}C_1 x^2 + C_2 x + C_3,$$

$$EI_{yy}w = \int \int \int \int q_z dx dx dx dx + \frac{1}{6}C_1 x^3 + \frac{1}{2}C_2 x^2 + C_3 x + C_4. \quad (5.101)$$

The bending line $w(x)$ can thus be obtained by fourfold integration of the applied load q_z. The integration constants C_1, C_2, C_3, C_4 are to be determined from the given boundary conditions. Some typical boundary conditions are shown in Fig. 5.27. The given boundary conditions can be formulated as follows:

- If one end of the beam is simply supported, both the deflection w and the bending moment $M_y = -EIw''$ disappear (Fig. 5.27, top left).
- At a free, unloaded end of the beam without any support, both the transverse shear force $V_z = -EIw'''$ and the bending moment $M_y = -EIw''$ become zero (Fig. 5.27, top right).
- If, on the other hand, the beam end is fully clamped, neither a deflection w nor an inclination w' of the bending line are possible (Fig. 5.27, bottom left).
- In the case of a guided support (Fig. 5.27, bottom right), no inclination of the bending line w' and no transverse shear force $V_z = -EIw'''$ can occur.

The calculation is briefly illustrated at the example of the cantilever beam given in Fig. 5.28. A bar of length l and constant bending stiffness EI_{yy} is given which is loaded at the free end by the force F. The bending line of the beam is obtained by fourfold integration of the beam differential equation (5.97):

$$EI_{yy}w'''' = q_z(x) = 0,$$

$$EI_{yy}w''' = -V_z(x) = C_1,$$

$$EI_{yy}w'' = -M_y(x) = C_1 x + C_2,$$

$$EI_{yy}w' = \frac{1}{2}C_1 x^2 + C_2 x + C_3,$$

$$EI_{yy}w = \frac{1}{6}C_1 x^3 + \frac{1}{2}C_2 x^2 + C_3 x + C_4. \quad (5.102)$$

The integration constants $C_1, ..., C_4$ result from the boundary conditions which are given as follows:

$$w(x = 0) = 0, \quad w'(x = 0) = 0, \quad V_z(x = l) = F, \quad M_y(x = l) = 0. \quad (5.103)$$

Evaluation yields the following integration constants $C_1, ..., C_4$:

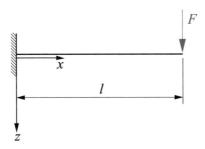

Fig. 5.28 Exemplary cantilever beam

$$C_1 = -F, \quad C_2 = Fl, \quad C_3 = 0, \quad C_4 = 0. \tag{5.104}$$

The bending line $w(x)$ then results as:

$$w(x) = \frac{1}{EI_{yy}} \left(-\frac{F}{6}x^3 + \frac{F}{2}lx^2 \right) = \frac{Fl^3}{6EI_{yy}} \left[3\left(\frac{x}{l}\right)^2 - \left(\frac{x}{l}\right)^3 \right]. \tag{5.105}$$

The maximum deflection $w_{\max}$ occurs at the end of the beam at $x = l$:

$$w_{\max} = \frac{Fl^3}{3EI_{yy}}. \tag{5.106}$$

The integration prescribed in (5.101) assumes that all parameters involved - M_y, V_z, q_z, w, w' - are continuous over the length of the beam. In the case of a multi-span bar, i.e. the situation of a bar that e.g. is supported by intermediate supports or which is interrupted by joints or loaded by point forces or moments, then the beam is to be mathematically divided into areas within which the above-mentioned quantities are distributed continuously. An exemplary multi-span bar is shown in Fig. 5.29. The integration prescribed in (5.101) must then be carried out separately in each sub-area. Using the example of Fig. 5.29 there are four sub-areas with a total of 16 integration constants to be determined. In the case of multi-span beams, in addition to the boundary conditions (see Fig. 5.27), the so-called transition conditions must be observed at the transitions between the sub-areas. Figure 5.30 shows some exemplary transition conditions that can typically occur for multi-span beams. We consider the transition between the sub-areas i and $i + 1$ which are loaded by the line loads q_i and q_{i+1} and which exhibit the bending stiffnesses EI_i and EI_{i+1}, respectively. Each transition point allows the unambiguous definition of four transition conditions.

For illustration purposes we want to consider the example given in Fig. 5.29 and formulate the boundary and transition conditions for this multi-span beam. For this purpose we introduce the local axes x_i ($i = 1, 2, 3, 4$) as shown in Fig. 5.31. For simplicity, let the bending stiffness EI_{yy} of the beam be constant over its entire length. The boundary and transition conditions for this multi-span beam are as follows. The following boundary conditions apply at the left support at $x_1 = 0$:

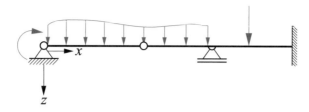

Fig. 5.29 Exemplary multi-span beam

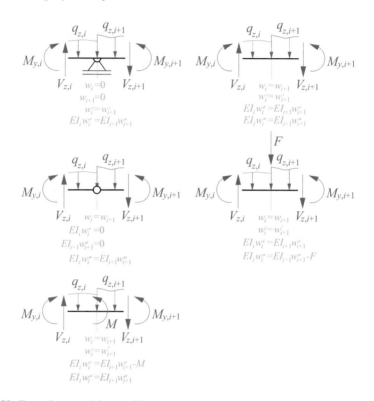

Fig. 5.30 Exemplary transition conditions

Fig. 5.31 Multi-span beam with local reference axes

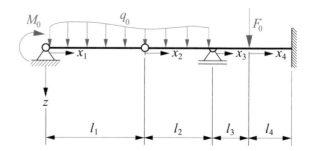

$$w_1(x_1 = 0) = 0,$$
$$M_1(x_1 = 0) = M_0, \tag{5.107}$$

where M_1 is the bending moment in section 1 and the second boundary condition can also be represented as:

$$w_1''(x_1 = 0) = -\frac{M_0}{EI_{yy}}. \tag{5.108}$$

At the hinge location $x_1 = l_1$ or $x_2 = 0$, the transition conditions are as follows:

$$w_1(x_1 = l_1) = w_2(x_2 = 0),$$
$$V_1(x_1 = l_1) = V_2(x_2 = 0),$$
$$M_1(x_1 = l_1) = 0,$$
$$M_2(x_2 = 0) = 0. \tag{5.109}$$

The moment and shear force conditions can also be specified as:

$$w_1'''(x_1 = l_1) = w_2'''(x_2 = 0),$$
$$w_1''(x_1 = l_1) = 0,$$
$$w_2''(x_2 = 0) = 0. \tag{5.110}$$

At the supported point at $x_2 = l_2$ or $x_3 = 0$, the following transition conditions can be formulated:

$$w_2(x_2 = l_2) = 0,$$
$$w_3(x_3 = 0) = 0,$$
$$w_2'(x_2 = l_2) = w_3'(x_3 = 0),$$
$$M_2(x_2 = l_2) = M_3(x_3 = 0). \tag{5.111}$$

The moment condition can also be written in the following form:

$$w_2''(x_2 = l_2) = w_3''(x_3 = 0). \tag{5.112}$$

At the force introduction point at $x_3 = l_3$ or $x_4 = 0$ the following conditions apply:

$$w_3(x_3 = l_3) = w_4(x_4 = 0),$$
$$w_3'(x_3 = l_3) = w_4'(x_4 = 0),$$
$$V_3(x_3 = l_3) - V_4(x_4 = 0) = F_0,$$
$$M_3(x_3 = l_3) = M_4(x_4 = 0). \tag{5.113}$$

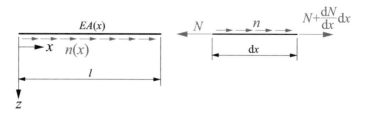

Fig. 5.32 Equilibrium for an infinitesimal rod element

The shear force condition and the moment condition occurring here can also be represented as:

$$w_3'''(x_3 = l_3) - w_4'''(x_4 = 0) = -\frac{F_0}{EI_{yy}},$$

$$w_3''(x_3 = l_3) = w_4''(x_4 = 0). \qquad (5.114)$$

At the clamping point $x_4 = l_4$ the following boundary conditions apply:

$$w_4(x_4 = l_4) = 0,$$

$$w_4'(x_4 = l_4) = 0. \qquad (5.115)$$

This provides 16 conditions for determining the 16 integration constants.

5.8   Rod Structures

If a beam is loaded in such a way that only a normal force N is generated, we also speak of a so-called rod or bar. This is always the case when axial loads in the form of single forces as well as line loads act centrally, i.e. their line of action coincides with the centroid line of the rod. The only stress component occurring in a rod is the normal stress σ_{xx} which is constantly distributed over the cross-sectional area (Fig. 5.3, top right) where we assume that the axis system constitutes the principal axes of the cross-section. The normal stress σ_{xx} can be calculated according to (5.65) as

$$\sigma_{xx} = \frac{N}{A}, \qquad (5.116)$$

i.e. as the quotient of normal force N and cross-sectional area A.

The constitutive law for the rod results directly from (5.61) as:

$$N = EAu'. \qquad (5.117)$$

Fig. 5.33 Determination of the normal force N and the longitudinal displacement u at the example of a clamped straight rod under distributed load n and a single force F

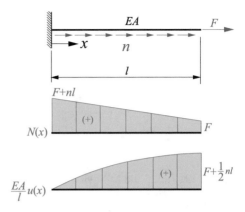

The infinitesimal equilibrium is briefly considered here for an infinitesimal rod element with the length dx (Fig. 5.32). A member of length l is given which can exhibit a variable stiffness $EA(x)$ and which is subjected to the variable line load $n(x)$. The equilibrium of forces at the infinitesimal rod element in the direction of the rod gives:

$$N - ndx - N - \frac{dN}{dx}dx = 0, \tag{5.118}$$

which can be rearranged to

$$N' = -n. \tag{5.119}$$

Hence, the first derivative of the normal force results in the negative applied axial line load n. With the constitutive law (5.117) it follows:

$$(EAu')' = -n. \tag{5.120}$$

If the applied load $n(x)$ is given, then the longitudinal displacement $u(x)$ of the rod results from twofold integration:

$$(EAu')' = -n,$$
$$EAu' = N = -\int ndx + C_1,$$
$$EAu = -\int \int ndxdx + C_1x + C_2. \tag{5.121}$$

The two constants of integration are then to be determined from the given boundary conditions.

The procedure is briefly explained using the example of Fig. 5.33. Consider a straight rod of length l with the constant extensional stiffness EA. The rod is loaded by the constant line load n as well as by a single force F at its free end. The integration

Fig. 5.34 Typical boundary
conditions for rod structures

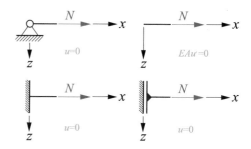

rule (5.121) takes the following form:

$$(EAu')' = -n,$$

$$EAu' = N = -\int n\mathrm{d}x + C_1 = -nx + C_1,$$

$$EAu = -\int\int n\mathrm{d}x\mathrm{d}x + C_1 x + C_2 = -\frac{1}{2}nx^2 + C_1 x + C_2. \quad (5.122)$$

The boundary conditions to be applied here are:

$$u(x = 0) = 0, \quad N(x = l) = F. \quad (5.123)$$

The integration constants C_1 and C_2 result as:

$$C_1 = F + nl, \quad C_2 = 0. \quad (5.124)$$

The normal force distribution $N(x)$ and the rod displacement $u(x)$ can thus be
specified as:

$$N(x) = F + n(l - x),$$

$$u(x) = \frac{1}{EA}\left[-\frac{1}{2}nx^2 + (F + nl)x\right]. \quad (5.125)$$

They are also shown in Fig. 5.33. With the normal force diagram N determined
in this way, the normal stress σ_{xx} acting in the cross section can be calculated using
(5.116) at every point x of the rod.

Figure 5.34 contains some typical cases of boundary conditions for rod structures.
At each end of a rod a boundary condition can always be specified, either for u or
for N.

If a rod structure is considered in which the state variables are not continuous and
therefore the integration rule (5.121) cannot be carried out over the entire length of
the rod, then corresponding transition conditions must be formulated in addition to
the boundary conditions. A simple example for this case is shown in Fig. 5.35. A

Fig. 5.35 Clamped rod loaded by a single force F

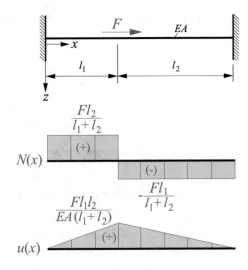

Fig. 5.36 Exemplary transition conditions for rod structures

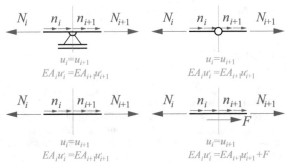

straight rod with constant axial stiffness EA and the lengths l_1 and l_2 is given. The rod is loaded by a single axial force F at the point $x = l_1$.

Due to the discontinuous distributions of $N(x)$ and $u(x)$ we divide the bar into the two parts enclosed in the intervals $0 \leq x \leq l_1$ and $l_1 \leq x \leq l_1 + l_2$. In the interval $0 \leq x \leq l_1$ the following results:

$$
\begin{aligned}
EAu_1'' &= 0, \\
EAu_1' &= C_1, \\
EAu_1 &= C_1 x + C_2.
\end{aligned}
\tag{5.126}
$$

For $l_1 \leq x \leq l_1 + l_2$ we have:

$$
\begin{aligned}
EAu_2'' &= 0, \\
EAu_2' &= C_3, \\
EAu_2 &= C_3 x + C_4.
\end{aligned}
\tag{5.127}
$$

The following boundary and transition conditions apply for the determination of the constants $C_1, \ldots, C_4$:

$$u_1(x = 0) = 0,$$
$$u_1(x = l_1) = u_2(x = l_1),$$
$$E A u_1'(x = l_1) = E A u_2'(x = l_1) + F,$$
$$u_2(x = l_1 + l_2) = 0. \tag{5.128}$$

From this the constants $C_1, \ldots, C_4$ can be determined as:

$$C_1 = \frac{F l_2}{l_1 + l_2}, \quad C_2 = 0, \quad C_3 = F\left(\frac{l_2}{l_1 + l_2} - 1\right), \quad C_4 = F l_1. \tag{5.129}$$

The diagrams for $N(x)$ and $u(x)$ are also shown in Fig. 5.35.

Figure 5.36 contains a representation of typical transition conditions for rod structures.

Bibliography

Bauchau OA, Craig JI (2009) Structural analysis: with applications to aerospace structures. Springer Dordrecht et al, The Netherlands

Czerwenka G, Schnell W (1970) Einführung in die Rechenmethoden des Leichtbaus, Band 1, Bibliographisches Institut, Mannheim et al., Germany

Gross D, Hauger W, Schröder J (2014) Technische Mechanik 2: Elastostatik, Twelfth. Springer, Berlin et al

Hanswille G (1995) Vorlesungen über Stahl- und Verbundbau. Bergische Universität Wuppertal, Germany, Department of Civil Engineering

Kindmann R, Frickel J (2017) Elastische und plastische Querschnittstragfähigkeit, Online edition 2017, https://www.kindmann.de, last retrieved March 2020

Linke M, Nast E (2015) Festigkeitslehre für den Leichtbau. Springer Berlin et al, Germany

Megson THG (1999) Aircraft structures for engineering students, 3rd edn. Arnold, London, UK

Petersen C (1997) Stahlbau 3. Auflage, Vieweg Verlag, Braunschweig et al., Germany

Roik KH (1978) Vorlesungen über Stahlbau, Ernst & Sohn. Berlin et al, Germany

Wiedemann J (2007) Leichtbau 1: Elemente. Springer Berlin et al, Germany

Wlassow WS (1964) Dünnwandige elastische Stäbe, vol I. VEB Verlag für Bauwesen, Berlin, Germany

Wlassow WS (1965) Dünnwandige elastische Stäbe, vol II. VEB Verlag für Bauwesen, Berlin, Germany

Chapter 6
Beams Under Transverse Shear Forces

6.1 Introduction

In the previous Chap. 5 we have dealt with straight beams under normal force and bending moments and mainly dealt with the determination of the normal stress σ_{xx} as a result of the normal force N and the two bending moments M_y. However, the consideration of the equilibrium conditions (see also Fig. 5.26) has shown that transverse shear forces V_y and V_z are required in order to maintain equilibrium. Fig. 5.3 shows that transverse shear forces are related to the shear stresses τ. This Chap. is devoted to the determination of shear stresses in thin-walled beam cross-sections. However, a problem that arises within the framework of the Euler-Bernoulli beam theory is the assumption that cross-sections remain warping free even in the deformed state and that the normal hypothesis is assumed to be valid (Fig. 5.4). These assumptions lead to vanishing shear strains γ. However, the shear stress τ is related to the shear strain γ via Hooke's law

$$\tau = G\gamma \tag{6.1}$$

which leads to vanishing shear stresses τ if the shear strain γ is set to zero. This is a contradiction of the Euler-Bernoulli beam theory which is inevitable. Consequently, there is no constitutive relationship available for determining the shear stress τ so that statements about the shear stress τ must be obtained from equilibrium considerations.

In the following we consider the determination of the shear stresses due to transverse shear forces in thin-walled cross-sections, and we want to assume that the external loads all act in the shear center axis M. The exact consequences and relevance of this definition will be discussed later, and we will consider the determination of the shear center M in detail in this chapter. It is sufficient to state at this point that the external loads are applied in such a way that they do not cause torsion of the beam. Furthermore, we want to assume that the reference axes x, y, z are the principal axes of the cross-section under consideration and that external loads act in the principal

C. Mittelstedt, *Structural Mechanics in Lightweight Engineering*,
https://doi.org/10.1007/978-3-030-75193-7_6

axis directions y and z. As in the previous Chap. 5 we consider straight thin-walled beams, and we assume that the shear stresses τ are constant across the thickness of the individual cross-sectional segments and are oriented in the tangential direction.

In this chapter we assume linear elasticity and small deformations. The material under consideration is isotropic and the cross-section consists of a homogeneous material, so that the material behavior is identical at every point of the cross-section.

6.2 Shear Stresses in Open Cross-Sections

6.2.1 Basic Equations

For the derivation of the basic equations we consider the situation given in Fig. 6.1. Consider a straight beam as shown which for the moment is only loaded by the line load $q_z(x)$ whereby the transverse shear force $V_z(x)$ and the bending moment $M_y(x)$ result which are functions of the longitudinal beam axis x. We now cut out an infinitesimal element of the length dx and consider the free body diagram as shown in Fig. 6.1, middle. Due to the given load, the bending moment M_y will cause the normal stress σ_{xx} which is distributed linearly over z. Since the stress σ_{xx}, in addition to its variable distribution over the cross-section, is also a function of the longitudinal coordinate x, at the positive cutting edge the infinitesimal increase $\frac{d\sigma_{xx}}{dx}dx$ is applied. The same applies to the transverse shear force V_z and the bending moment M_y. We now want to determine the distribution of the shear stress τ over the cross-section for a given normal stress σ_{xx}. For this purpose, we introduce the circumferential

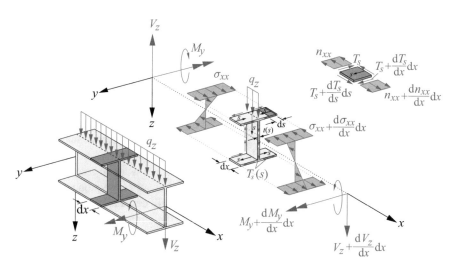

Fig. 6.1 Determination of the shear flow $T_s(s)$

coordinate s as indicated in Fig. 6.1. Consequently, in all that follows we denote the shear stress τ as τ_s since it always runs tangential to the walls of the cross-section and is thus a function of s: $\tau_s = \tau_s(s)$. The wall thickness t of the cross-section may also be a function of s, i.e. $t = t(s)$.

It is advisable not to work with the shear stress τ_s for the time being, but rather to introduce the so-called shear flow $T_s(s)$. The shear flow represents the resultant of the shear stress τ_s over the cross-section thickness t. Since we assume that the shear stress τ_s is constant over the wall thickness of the thin-walled cross-sections considered here we may write:

$$T_s(s) = \tau_s(s)t(s). \tag{6.2}$$

Accordingly, $T_s(s)$ is given in the unit of a force per length unit.

We now consider the free body diagram as given in Fig. 6.1, right, in which an infinitesimal element of width ds is shown that we have separated from the element in Fig. 6.1, middle. In addition to the shear flow $T_s(s)$ we also have the resulting force flow n_{xx} which is the resultant of the normal stress σ_{xx} and can be determined as:

$$n_{xx}(s) = \sigma_{xx}t(s). \tag{6.3}$$

If we formulate the equilibrium of forces in the $x-$direction at the element shown in Fig. 6.1, then we get:

$$\left(n_{xx} + \frac{dn_{xx}}{dx}dx\right)ds - n_{xx}ds + \left(T_s + \frac{dT_s}{ds}ds\right)dx - T_s dx = 0. \tag{6.4}$$

This can be solved as follows:

$$\frac{dn_{xx}}{dx}dxds + \frac{dT_s}{ds}dsdx = 0. \tag{6.5}$$

Dividing this expression by dx and performing an integration over s from an arbitrary starting point $s = s_A$ to a point s yields:

$$\int_{s_A}^{s} \frac{dT_s}{ds}ds = T_s(s) - T_s(s = s_A) = -\int_{s_A}^{s} \frac{dn_{xx}}{dx}ds. \tag{6.6}$$

Assuming biaxial bending from this point on (i.e. both bending moments M_y and M_z occur) and expressing the force flow n_{xx} using (6.3) with

$$\sigma_{xx} = \frac{N}{A} + \frac{M_y}{I_{yy}}z - \frac{M_z}{I_{zz}}y \tag{6.7}$$

(cf. Eq. (5.65)) yields:

$$n_{xx} = \left(\frac{N}{A} + \frac{M_y}{I_{yy}} z - \frac{M_z}{I_{zz}} y \right) t, \tag{6.8}$$

and

$$\frac{dn_{xx}}{dx} = n'_{xx} = \left(\frac{N'}{A} + \frac{M'_y}{I_{yy}} z - \frac{M'_z}{I_{zz}} y \right) t. \tag{6.9}$$

We want to assume here and in all further explanations that the normal force N in the beam is constant over x, so that $N' = 0$. Let us also use the relations (5.95) and (5.98), i.e.

$$\frac{dM_y}{dx} = M'_y = V_z, \quad \frac{dM_z}{dx} = M'_z = -V_y, \tag{6.10}$$

then from (6.9) we get:

$$n'_{xx} = \left(\frac{V_z}{I_{yy}} z + \frac{V_y}{I_{zz}} y \right) t. \tag{6.11}$$

From (6.6):

$$
\begin{aligned}
T_s(s) - T_s(s = s_A) &= - \int_{s_A}^{s} \left(\frac{V_z}{I_{yy}} z + \frac{V_y}{I_{zz}} y \right) t \, ds \\
&= - \int_{s_A}^{s} \frac{V_z}{I_{yy}} t z \, ds - \int_{s_A}^{s} \frac{V_y}{I_{zz}} t y \, ds \\
&= - \frac{V_z}{I_{yy}} \int_{s_A}^{s} t z \, ds - \frac{V_y}{I_{zz}} \int_{s_A}^{s} t y \, ds.
\end{aligned} \tag{6.12}
$$

The integrals appearing here can be understood as the static moments $S_y(s)$ and $S_z(s)$ with respect to a point s of the element under consideration, so that:

$$T_s(s) - T_s(s = s_A) = - \frac{V_z}{I_{yy}} S_y(s) - \frac{V_y}{I_{zz}} S_z(s). \tag{6.13}$$

This is an expression for the determination of the shear flow $T(s)$, depending on the transverse shear forces V_y and V_z, the two moments of inertia I_{yy} and I_{zz}, as well as the two static moments $S_y(s)$ and $S_z(s)$ of the cross-section between the point s and the starting point $s = s_A$ of the integration path. In addition, the boundary value $T_s(s = s_A)$ appears here which has yet to be determined. If the shear flow $T_s(s)$ is known at every point s, then the shear stress $\tau_s(s)$ can be calculated by dividing by the wall thickness:

$$\tau_s(s) = \frac{T_s(s)}{t(s)}. \tag{6.14}$$

In this section we exclusively devote ourselves to the determination of the shear stress $\tau_s(s)$ or the shear flow $T_s(s)$ at open cross-sections, for example at I-, C -, Z-, T- or L-cross-sections. Closed cross-sections (such as box or circular ring cross-

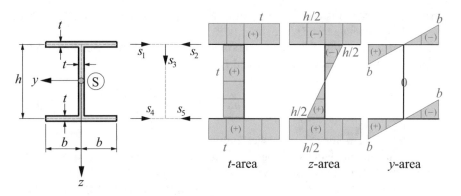

Fig. 6.2 Double symmetric I-beam with the local axes s_i ($i = 1, 2, ..., 5$)

sections) are dealt with at a later point. In the case of an open cross-section, the integration prescribed in (6.13) which comes into play in the two static moments S_y and S_z, will begin at a free end, so that—if, as given here, no tangential boundary forces are to be considered—the boundary value $T_s(s = s_A)$ may be set to zero. We thus have:

$$T_s(s) = -\frac{V_z}{I_{yy}} S_y - \frac{V_y}{I_{zz}} S_z. \tag{6.15}$$

In the following we want to consider the practical execution of the calculation of the shear stress $\tau_s(s)$ and consider the I-beam in Fig. 6.2. Since we assume a thin-walled cross-section, we again use the skeleton line for the analysis. The cross-section has the web height h and the flange width $2b$, the wall thickness is assumed to be constant with the value t. The principal axes y and z have their origin in the center of gravity S of the cross-section as shown.

To carry out the analysis, we introduce the local circumferential reference axes s_i, where i is the number of the cross-sectional segment under consideration. Hence, in order to cover the entire cross-section five local circumferential axes s_i ($i = 1, 2, ..., 5$) must be introduced. The starting points and directions of these local reference axes are arbitrary and at the discretion of the user, but should always be chosen from the point of view of expediency.

We introduce the t area, i.e. the graphic representation of the wall thickness at each point of the cross-section. Furthermore, we determine the $z-$area and the $y-$area, as shown in Fig. 6.2. The integrations necessary for the determination of the static moments

$$S_y(s) = \int_{s_A}^{s} tz\,\mathrm{d}s, \quad S_z(s) = \int_{s_A}^{s} ty\,\mathrm{d}s \tag{6.16}$$

can then be carried out advantageously with the help of the integral table in Fig. 5.8, whereby the integration usually begins at free ends and is then performed segment by segment until the entire cross-section has been covered. It should be noted that when the integration is carried out against the directions of the specified reference axes s_i, a sign change takes place for τ_s or T_s.

For the I-beam in Fig. 6.2 the moments of inertia can be taken directly from (5.91), whereby we use the designations I_{yy} and I_{zz}:

$$I_{yy} = \frac{th^3}{12} + tbh^2, \quad I_{zz} = \frac{4tb^3}{3}. \tag{6.17}$$

The deviation moment I_{yz} vanishes in this case.

We begin the calculations for segment 1, described by the local axis s_1, where $0 \le s_1 \le b$. The static moment $S_y(s_1)$ takes on the value $S_y(s_1 = 0) = 0$ at the point $s_1 = 0$. If one considers the position $s_1 = b$, then the integration rule (6.16) in connection with the integral table of Fig. 5.8 gives:

$$S_y(s_1 = b) = \int_0^b tz\,ds_1 = 1 \cdot t \cdot \left(-\frac{h}{2}\right) \cdot b = -\frac{tbh}{2}, \tag{6.18}$$

where two rectangles were superimposed. The static moment $S_y(s_1)$ runs linearly between the positions $s_1 = 0$ and $s_1 = b$ as can be easily verified. We get identical values for the static moment $S_y(s_2)$ in segment 2, i.e. $S_y(s_2 = 0) = 0$ and $S_y(s_2 = b) = -\frac{tbh}{2}$.

The starting point of the integration in segment 3 is at the intersection point between the web and the two upper flanges. Therefore, the two values of the static moments $S_y(s_1 = b)$ and $S_y(s_2 = b)$ add up at $s_3 = 0$, i.e .:

$$S_y(s_3 = 0) = S_y(s_1 = b) + S_y(s_2 = b) = -tbh. \tag{6.19}$$

The value of $S_y(s_3)$ at the height of the center of gravity of the cross-section, i.e. at $s_3 = \frac{h}{2}$, then amounts to:

$$S_y\left(s_3 = \frac{h}{2}\right) = -tbh + \int_0^{\frac{h}{2}} t(s_3)z\,ds_3 = -tbh + \frac{1}{2} \cdot t \cdot \left(-\frac{h}{2}\right) \cdot \frac{h}{2} = -th\left(b + \frac{h}{8}\right). \tag{6.20}$$

The static moment $S_y(s_3)$ at the lower end of segment 3, i.e. at $s_3 = h$, results as:

$$S_y(s_3 = h) = -tbh + \int_0^h t(s_3)z\,ds_3 = -tbh + \frac{1}{2} \cdot t \cdot \left(-\frac{h}{2}\right) \cdot \frac{h}{2} + \frac{1}{2} \cdot t \cdot \frac{h}{2} \cdot \frac{h}{2} = -tbh. \tag{6.21}$$

Since a linear function is integrated when determining $S_y(s_3)$, a parabolic distribution of $S_y(s_3)$ over the web height h results. It can be seen that the static moment S_y assumes a maximum value at the the the center of gravity S of the cross-section.

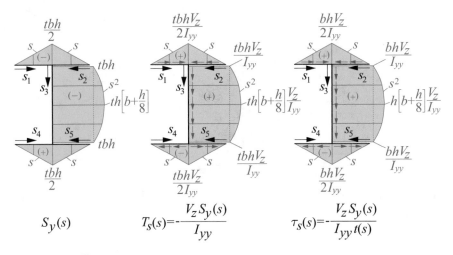

$$S_y(s) \qquad T_s(s) = -\frac{V_z S_y(s)}{I_{yy}} \qquad \tau_s(s) = -\frac{V_z S_y(s)}{I_{yy}\, t(s)}$$

Fig. 6.3 Static moment $S_y(s)$, shear flow $T_s(s)$ and shear stress $\tau_s(s)$ of an I-beam under transverse shear force V_z

Lastly we consider areas 4 and 5, i.e. the two lower flanges. The static moment $S_y(s_4)$ at the point $s_4 = 0$ is zero, i.e. $S_y(s_4 = 0) = 0$. The value of $S_y(s_4)$ at $s_4 = b$, on the other hand, results in $S_y(s_4 = b) = \frac{tbh}{2}$. Similarly, we get the values $S_y(s_5 = 0) = 0$ and $S_y(s_5 = b) = \frac{tbh}{2}$ in segment 5.

A graphic representation of the distribution of the static moment $S_y(s)$ can be found in Fig. 6.3, left. The resultant shear flow $T_s(s)$ and shear stress distributions $\tau_s(s)$ are shown in Fig. 6.3, middle and right.

We can derive the following general rules from the considerations so far:

- At the free profile ends, the shear flow is identically zero, provided that no tangential shear loads are introduced.
- The distribution of T_s or τ_s is directly coupled with the distribution of the $y-$ and the $z-$line, whereby we want to refer to z here for explanation. If $z = 0$, then T_s and τ_s are constant values. If, on the other hand, there is a constant, non-zero value for z, then the functional course of T_s and τ_s is linear over the local reference axis s. In the case of a linearly variable $z-$line, T_s and τ_s result as parabolic over s.
- It can be shown that T_s due to a transverse force V_z takes on extreme values exactly where the $y-$axis intersects the skeleton line of the cross-section. This applies analogously to T_s as a result of V_y, here the extreme value occurs where the $z-$axis intersects the skeleton line.
- If the cross-section under consideration exhibits an axis of symmetry, then the course of T_s and τ_s is also symmetrical due to the transverse shear force acting in the axis of symmetry. In addition, T_s and τ_s have zero values as a result of V_z at the intersections of the $z-$axis with the skeleton line, provided that z is a symmetry axis. Analogous conclusions can be drawn for the shear flow T_s as a result of V_y.

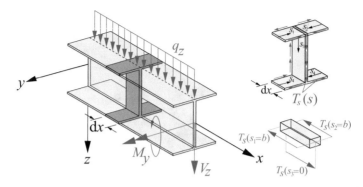

Fig. 6.4 Verification of the results at the intersection point between flanges and web at the example of an I-beam

- A simple verification can be carried out by using the fact that the resultant of the shear stress τ_s in all segments must equal the acting transverse shear force as will be shown later with an example.
- Another verification can be carried out at nodes, i.e. at those points where several cross-sectional segments intersect. In such nodes the sum of all resulting shear forces must be zero. An example is the I-beam in Fig. 6.2. We consider Fig. 6.4 in which the section element of Fig. 6.1 with the length dx is depicted. We now cut out another element exactly at the intersection point of the two flanges and the web and consider the equilibrium of the resultants of the shear flows acting at this point. The sum of forces in $x-$direction then results in $(T_s(s_1 = b) + T_s(s_2 = b) - T_s(s_3 = 0))\,dx = 0$. Using the results in Fig. 6.3 we can easily verify that this requirement is met for this example.

Some qualitative shear flow distributions $T_s(s)$ for selected cross-sections under the transverse shear force V_z are given in Fig. 6.5.

If a cross-section is given in which the reference system is not the principal axis system, the procedure shown in Fig. 6.6 can be employed. We consider a Z-cross section, the two reference systems $\hat{y}$, $\hat{z}$ (centroid axes according to the first cross-sectional normalization) and y, z (principal axes under the angle φ_0 according to the second cross-sectional normalization) are also shown. The cross-section is loaded by the transverse shear force $V_{\hat{z}}$. To determine the resulting shear flow, the shear force $V_{\hat{z}}$ is first divided along the principal axes y, z as follows:

$$V_y = V_{\hat{z}} \sin \varphi_0, \quad V_z = V_{\hat{z}} \cos \varphi_0. \tag{6.22}$$

Using the transformation rule (cf. Eq. (5.48))

$$y = \hat{y} \cos \varphi_0 + \hat{z} \sin \varphi_0, \quad z = -\hat{y} \sin \varphi_0 + \hat{z} \cos \varphi_0 \tag{6.23}$$

employing the $y-$ and the $z-$area as shown in Fig. 6.6, the static moments S_y, S_z and the two moments of inertia I_{yy}, I_{zz} can be determined. From this Eq. (6.15) can be used to calculate the shear flow distribution $T_s(s)$.

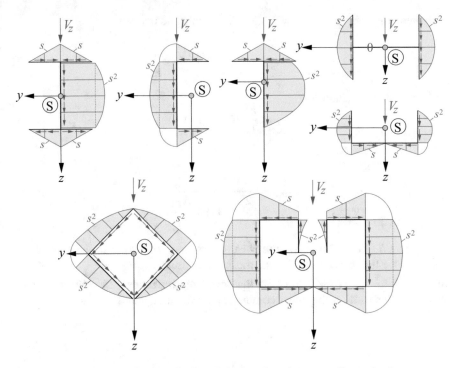

Fig. 6.5 Qualitative shear flow distributions $T_s(s)$ for selected cross-sections under the transverse shear force V_z

Fig. 6.6 Determination of the shear flow $T_s(s)$ for a Z-cross-section under the transverse shear force $V_{\hat{z}}$

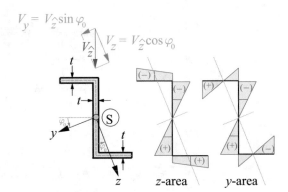

6.2.2   Simplified Analysis of an I-Cross-Section

In this section we want to briefly discuss a simplification common in lightweight engineering when determining the shear stress in double symmetrical I-cross-sections under the shear load V_z using the example in Fig. 6.3. The maximum shear stress $\tau_{\max}$ for this cross-section can be calculated with the flange area $A_{Fl} = 2bt$ and the web area $A_W = ht$ as well as with the maximum static moment

$$S_{y,\max} = -\frac{h}{2}\left(A_{Fl} + \frac{1}{4}A_W\right) \tag{6.24}$$

and the moment of inertia

$$I_{yy} = \frac{h^2}{12}\left(6A_{Fl} + A_W\right) \tag{6.25}$$

as:

$$\tau_{\max} = \frac{3V_z}{2A_W}\frac{1 + 4\dfrac{A_{Fl}}{A_W}}{1 + 6\dfrac{A_{Fl}}{A_W}}. \tag{6.26}$$

Figure 6.7 shows the maximum shear stress $\tau_{\max}$ (occurring in the centroid of the web, see Fig. 6.3, divided by the mean shear stress $\tau_m = \frac{V_z}{ht} = \frac{V_z}{A_{St}}$, plotted as a function of the ratio b/h. It turns out that for increasing ratios $\frac{b}{h}$ the values $\tau_{\max}$ and τ_m align. This justifies the rather common approximation that the mean shear stress

$$\tau_m = \frac{V_z}{ht} = \frac{V_z}{A_W} \tag{6.27}$$

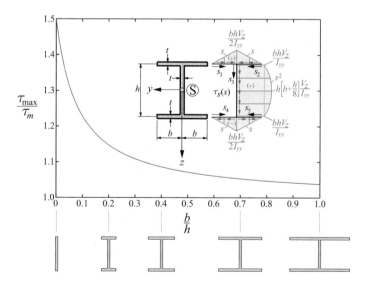

Fig. 6.7 Relative maximum shear stress of a double symmetric I-cross section, plotted against the ratio b/h

instead of the maximum shear stress according to Fig. 6.3 is employed for justification. This means that the shear stress $\tau_s(s)$ which is distributed parabolically over the web height is approximated by a value τ_m that is constant over the web height.

6.2.3 Unit Shear Flow

At this point we want to define the so-called unit shear flow. It can be seen from Eq. (6.15) that the distribution of $T_s(s)$ is identical to the distribution of the two static moments $S_y(s)$ and $S_z(s)$, and the terms $\frac{V_z}{I_{yy}}$ and $\frac{V_y}{I_{zz}}$ represent constant multipliers. We therefore introduce the two so-called unit shear flows $T_z(s) = -S_y(s)$ and $T_y(s) = -S_z(s)$ which will prove to be useful when determining the so-called shear center M at a later point in this chapter. With $\frac{V_z}{I_{yy}} = 1$ and $\frac{V_y}{I_{zz}} = 1$ they are defined as

$$T_z(s) = -S_y(s) = -\int_A z\,dA, \quad T_y(s) = -S_z(s) = -\int_A y\,dA. \tag{6.28}$$

6.3 Shear Stresses in Closed Cross-Sections

6.3.1 Single-Cell Cross-Sections

In this section, we want to address the question how the shear flow $T_s(s)$ and the shear stresses $\tau_s(s)$ due to transverse shear forces in closed cross-sections can be determined. To motivate the procedure we consider the closed box cross-section in Fig. 6.8, left, which is subjected to a transverse shear force V_z. The resulting shear flow $T_s(s)$ is indicated by arrows. We start the discussion with Eq. (6.13), which we show here again in a slightly rearranged form:

$$T_s(s) = -\frac{V_z}{I_{yy}}S_y(s) - \frac{V_y}{I_{zz}}S_z(s) + T_s(s = s_A). \tag{6.29}$$

The problem that arises with a closed cross-section is that, in contrast to an open cross-section, we cannot set the term $T_s(s = s_A)$ a priori to zero due to lack of free ends where $T_s(s = s_A) = 0$ would apply. We therefore have to work with compatibility conditions which we want to discuss in the following. We are initially assuming an exclusive loading by the transverse shear force V_z, but the following discussion can be extended quite easily to the additional shear force V_y. The following applies for the shear flow $T_s(s)$:

$$T_s(s) = -\frac{V_z}{I_{yy}}S_y(s) + T_s(s = s_A). \tag{6.30}$$

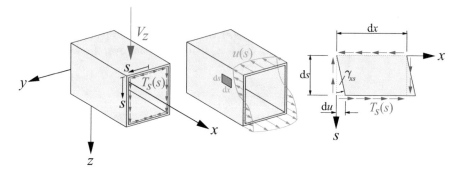

Fig. 6.8 Determination of the shear flow $T_s(s)$ in a symmetric closed cross-section under transverse shear force V_z

We consider the cross-sectional warping according to Fig. 6.8, middle, expressed by the longitudinal displacement $u(s)$ which is a function of the circumferential coordinate s. Using the infinitesimal element in Fig. 6.8, right, we can derive the following expression for the resulting shear strain γ_{xs}:

$$\tan \gamma_{xs} = \frac{du}{ds}. \tag{6.31}$$

Assuming small angles with $\tan(\gamma_{xs}) \simeq \gamma_{xs}$ we have for du:

$$du = \gamma_{xs}(s)ds. \tag{6.32}$$

With Hooke's law

$$\tau_s(s) = \frac{T_s(s)}{t(s)} = G\gamma_{xs}(s) \tag{6.33}$$

Equation (6.32) yields:

$$du = \frac{T_s(s)}{Gt(s)}ds, \tag{6.34}$$

and after integration:

$$u(s) = \int_s \frac{T_s(s)}{Gt(s)}ds. \tag{6.35}$$

Insertion of (6.30) leads to:

$$u(s) = -\int_s \frac{V_z S_y(s)}{GI_{yy}t(s)}ds + \int_s \frac{T_s(s = s_A)}{Gt(s)}ds. \tag{6.36}$$

We can now use the fact that the integration of the longitudinal displacement $u(s)$ over the entire cross-sectional circumference must be zero, since no discontinuities in the displacement $u(s)$ can occur in a closed cross-section. We express this by a ring integral as follows:

$$\oint u(s)ds = -\frac{V_z}{GI_{yy}} \oint \frac{S_y(s)}{t(s)}ds + \frac{T_s(s = s_A)}{G} \oint_s \frac{ds}{t(s)} = 0.$$ (6.37)

We can rearrange this expression immediately for the value $T_s(s = s_A)$ and get:

$$T_s(s = s_A) = \frac{\dfrac{V_z}{I_{yy}} \oint \dfrac{S_y(s)}{t(s)}ds}{\oint \dfrac{ds}{t(s)}},$$ (6.38)

whereby (6.30) takes on the following form:

$$T_s(s) = -\frac{V_z}{I_{yy}} \left[S_y(s) - \frac{\oint \dfrac{S_y(s)}{t(s)}ds}{\oint \dfrac{ds}{t(s)}} \right].$$ (6.39)

However, the shown procedure reaches its limits when dealing with multi-cell closed cross-sections. We therefore want to discuss another possibility of analysis, namely in the form of a statically indeterminate calculation. We look again at the already discussed closed, thin-walled box cross-section and make an imaginary cut along the longitudinal direction at an arbitrary point (Fig. 6.9, top). The shear flow $T_{s0}(s)$ in the open cross-section as a result of the shear force V_z can be easily determined. However, due to the cross-sectional warping $u(s)$, a displacement discontinuity Δu_{10} will occur at the point of the longitudinal section which of course cannot occur on the actually closed cross-section. We can calculate this displacement discontinuity Δu_{10} by integrating over the entire cross-sectional circumference as follows:

$$\Delta u_{10} = \oint \frac{T_{s0}(s)}{Gt(s)}ds = -\frac{V_z}{GI_{yy}} \oint \frac{S_y(s)}{t(s)}ds.$$ (6.40)

In order to compensate for this incompatibility, we apply the circumferential constant shear flow T_{s1} to the cross-section, which we initially consider as unit shear flow $T_{s1} = 1$ and which will also lead to a displacement discontinuity Δu_{11}. This discontinuity can be calculated with $T_{s1} = 1$ as:

$$\Delta u_{11} = \oint \frac{T_{s1}}{Gt(s)}ds = \frac{1}{G} \oint \frac{ds}{t(s)}.$$ (6.41)

The shear flow T_{s1} is applied here as a unit shear flow, but its final amount is still unknown at this point. We can now easily determine T_{s1} by requiring that the sum

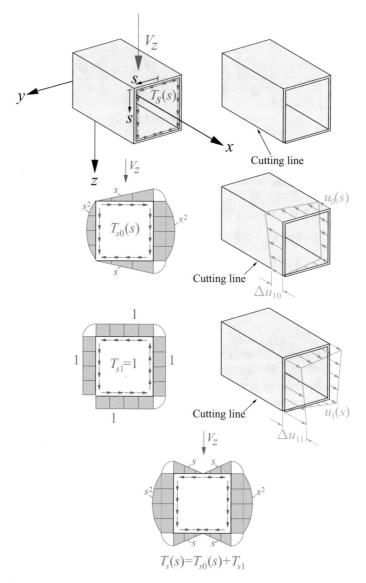

Fig. 6.9 Shear flow determination $T_s(s)$ in a symmetric closed box cross-section under the transverse shear force V_z

of Δu_{10} and Δu_{11} becomes zero, where we multiply the displacement discontinuity Δu_{11} with the (as yet unknown) shear flow T_{s1}:

$$\Delta u_{10} + T_{s1}\Delta u_{11} = 0. \tag{6.42}$$

Since the two displacement values Δu_{10} and Δu_{11} are known with (6.40) and (6.41) we can solve this equation immediately for T_{s1}:

$$T_{s1} = -\frac{\Delta u_{10}}{\Delta u_{11}} = \frac{V_z}{I_{yy}} \frac{\oint \dfrac{S_y(s)}{t(s)} ds}{\oint \dfrac{ds}{t(s)}}. \tag{6.43}$$

The shear flow T_s in the closed cross-section then represents the sum of $T_{s0}(s)$ and T_{s1}:

$$T_s(s) = T_{s0}(s) + T_{s1} = -\frac{V_z}{I_{yy}} \left[S_y(s) - \frac{\oint \dfrac{S_y(s)}{t(s)} ds}{\oint \dfrac{ds}{t(s)}} \right]. \tag{6.44}$$

This expression is in agreement with (6.39). The shear flow T_{s1} is therefore identical to the initially unknown boundary value $T_s(s = s_A)$.

We illustrate the analysis approach at the example of the square box cross-section of constant wall thickness t as given in Fig. 6.10 and employ the statically indeterminate calculation for this cross-section. The moment of inertia I_{yy} can be written as:

$$I_{yy} = \int_A z^2 dA = t \left[b \cdot \left(-\frac{b}{2} \right) \cdot \left(-\frac{b}{2} \right) + b \cdot \frac{b}{2} \cdot \frac{b}{2} \right.$$
$$\left. + 2 \cdot \frac{1}{3} \cdot \frac{b}{2} \cdot \left(-\frac{b}{2} \right) \cdot \left(-\frac{b}{2} \right) + 2 \cdot \frac{1}{3} \cdot \frac{b}{2} \cdot \frac{b}{2} \cdot \frac{b}{2} \right] = \frac{2tb^3}{3}. \tag{6.45}$$

The distribution of the static moment $S_y(s)$ for the case of a cut in the middle of the upper flange together with the local circumferential coordinates s_i ($i = 1, 2, ..., 5$) is shown in Fig. 6.11. We now calculate the displacement discontinuity Δu_{10} due to the shear flow $T_{s0}(s)$ according to (6.40):

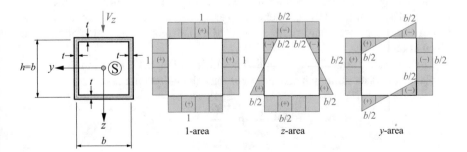

Fig. 6.10 Symmetric closed square box cross-section subjected to the transverse shear force V_z

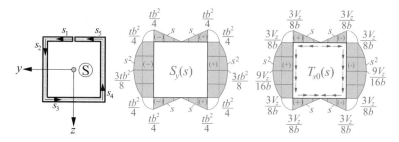

Fig. 6.11 Open square box cross-section under the transverse shear force V_z with the local circumferential axes s_i ($i = 1, 2, ..., 5$) as well as static moment $S_y(s)$ and shear flow $T_{s0}(s)$

$$\Delta u_{10} = \oint \frac{T_{s0}(s)}{Gt(s)} \mathrm{d}s = -\frac{V_z}{GI_{yy}} \oint \frac{S_y(s)}{t(s)} \mathrm{d}s = 0. \tag{6.46}$$

This result can be explained directly with the symmetry of the distribution of $S_y(s)$. The discontinuity in the longitudinal displacement due to the unit shear flow T_{s1} then results from (6.41) as:

$$\Delta u_{11} = \oint \frac{T_{s1}}{Gt(s)} \mathrm{d}s = \frac{1}{G} \oint \frac{\mathrm{d}s}{t(s)} = \frac{4b}{Gt}. \tag{6.47}$$

The compatibility condition (6.42) results in the unknown shear flow T_{s1} as $T_{s1} = 0$. Obviously, the shear flow $T_{s0}(s)$, which we determined on the open cross-section, is also the final shear flow $T_s(s)$ that is the objective of the analysis. The present special case with $T_{s1} = 0$ is due to the fact that we made the cut on the symmetry axis of the cross section in which the acting transverse shear force V_z also acts. Therefore the static moment $S_y(s)$ and thus also the shear flow $T_{s0}(s)$ is symmetrical with respect to the relevant axis of symmetry, whereby the integral $\oint \frac{S_y(s)}{t(s)} \mathrm{d}s$ in (6.40) becomes zero and thus the displacement discontinuity Δu_{10} disappears. We will show later that this fact can be used to advantage in the analysis of closed cross-sections with symmetry properties.

We repeat the above example and open the cross-section to carry out the statically indeterminate calculation as shown in Fig. 6.12, top left, and as already indicated in Fig. 6.9. The resulting static moment $S_y(s)$ and the shear flow $T_{s0}(s)$ are also shown in Fig. 6.12, top center and right. A detailed explanation of the calculation is not given at this point.

The unit shear flow T_{s1} to be applied is shown in Fig. 6.12, bottom left. The two displacement discontinuities necessary for the calculation then result as:

$$\Delta u_{10} = -\frac{V_z}{GI_{yy}} \oint \frac{S_y(s)}{t(s)} \mathrm{d}s = -\frac{3V_z}{2Gt},$$

$$\Delta u_{11} = \frac{1}{G} \oint \frac{\mathrm{d}s}{t(s)} = \frac{4b}{Gt}. \tag{6.48}$$

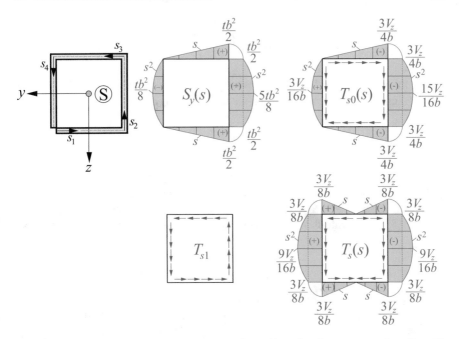

Fig. 6.12 Symmetric closed square box cross-section subjected to the transverse shear force V_z, alternative statically indeterminate analysis

From the compatibility condition $\Delta u_{10} + T_{s1} \Delta u_{11} = 0$ the statically indeterminate shear flow

$$T_{s1} = \frac{3V_z}{8b} \tag{6.49}$$

can be determined, and the final shear flow follows as $T_s(s) = T_{s0}(s) + T_{s1}$ as given in Fig. 6.12, bottom right. Apparently this result is in agreement with the result as given in Fig. 6.11 for T_{s0}.

6.3.2 Multi-Cell Cross-Sections

If a multi-cell closed cross-section under the transverse force V_z is given, then the procedure described above can be generalized in this regard. We discuss the closed cross-section consisting of two cells as given in Fig. 6.13, top left. We open this cross-section by cutting open each cell, as indicated in Fig. 6.13, top right. This creates an entirely open cross-section at which the shear flow $T_{s0}(s)$ can be calculated. Due to the opening of the cross-section, however, displacement discontinuities are possible which can be calculated as follows. First of all, there is the displacement discontinuity

Δu_{10} in cell 1:

$$\Delta u_{10} = \oint \frac{T_{s0}(s)}{Gt(s)}\,\mathrm{d}s. \tag{6.50}$$

The integration over the circumference prescribed here refers exclusively to the circumference of cell 1. Analogously, we have for Δu_{20}:

$$\Delta u_{20} = \oint \frac{T_{s0}(s)}{Gt(s)}\,\mathrm{d}s. \tag{6.51}$$

This is to be integrated over the circumference of the second cell.

As before, a unit shear flow $T_{s1} = 1$ is applied in cell 1, as shown in Fig. 6.14, top. As before, this results in a displacement discontinuity Δu_{11}. The same is performed on cell 2 (Fig. 6.14, middle). The two displacement discontinuities are calculated as:

$$\Delta u_{11} = \oint \frac{1}{Gt(s)}\,\mathrm{d}s, \quad \Delta u_{22} = \oint \frac{1}{Gt(s)}\,\mathrm{d}s, \tag{6.52}$$

where the integrations are to be carried out exclusively on cell 1 and cell 2, respectively. The indexing of the displacement discontinuities indicate both the location (first index) and the cause (second index) for the discontinuity. Accordingly, e.g. Δu_{11} is the displacement discontinuity in cell 1 as a result of the unit shear flow T_{s1}. The values Δu_{ii} ($i = 1, 2, ..., n$ for n cells) are always positive.

At the same time, however, it must be noted that due to the two unit shear flows T_{s1} and T_{s2} not only discontinuities in the longitudinal displacements occur within

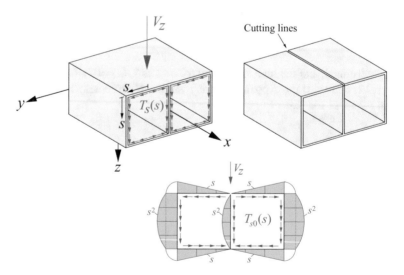

Fig. 6.13 Determination of the shear flow $T_{s0}(s)$ for a box cross-section consisting of two cells under the transverse shear force V_z

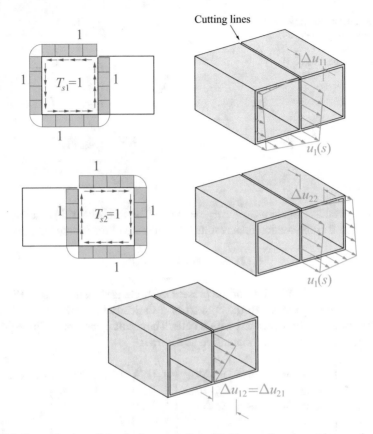

Fig. 6.14 Determination of the displacement discontinuities on the symmetric open box cross-section consisting of two cells

the two cells, but also an incompatibility between the two cells will usually arise which we want to denote as $\Delta u_{12} = \Delta u_{21}$:

$$\Delta u_{12} = \Delta u_{21} = -\oint \frac{ds}{Gt(s)}. \tag{6.53}$$

Herein the integration is to be carried out exclusively via the cell walls common to both cells. The values Δu_{ij} ($i, j = 1, 2, ..., n$ for n cells, $i \neq j$) are always assigned with a negative sign.

Analogous to (6.42), two compatibility conditions can now be formulated from which the two shear flows T_{s1} and T_{s2} can be determined:

$$\Delta u_{10} + T_{s1}\Delta u_{11} + T_{s2}\Delta u_{12} = 0,$$
$$\Delta u_{20} + T_{s1}\Delta u_{12} + T_{s2}\Delta u_{22} = 0. \tag{6.54}$$

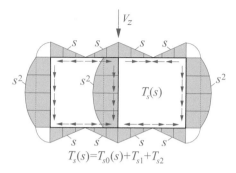

Fig. 6.15 Shear flow $T_s(s)$
for the two-cell symmetric
closed cross-section under
the transverse shear force V_z

Therein we have made use of $\Delta u_{12} = \Delta u_{21}$. The final shear flow acting on the two-cell closed cross-section then results from the sum of the individual shear flows:

$$T_s(s) = T_{s0}(s) + T_{s1} + T_{s2}. \qquad (6.55)$$

The qualitative distribution for the present example is shown in Fig. 6.15.

The procedure described in the context of a statically indeterminate calculation can be extended to cross-sections with n cells. The resulting compatibility equations can be given in a vector-matrix notation as follows:

$$\begin{bmatrix} \Delta u_{11} & \Delta u_{12} & \Delta u_{13} & \cdots & \Delta u_{1n} \\ \Delta u_{12} & \Delta u_{22} & \Delta u_{23} & \cdots & \Delta u_{2n} \\ \Delta u_{13} & \Delta u_{23} & \Delta u_{33} & \cdots & \Delta u_{3n} \\ \vdots & \vdots & \vdots & \ddots & \vdots \\ \Delta u_{1n} & \Delta u_{2n} & \Delta u_{3n} & \cdots & \Delta u_{nn} \end{bmatrix} \begin{pmatrix} T_{s1} \\ T_{s2} \\ T_{s3} \\ \vdots \\ T_{sn} \end{pmatrix} = - \begin{pmatrix} \Delta u_{10} \\ \Delta u_{20} \\ \Delta u_{30} \\ \vdots \\ \Delta u_{n0} \end{pmatrix}, \qquad (6.56)$$

or in a symbolic form:

$$\underline{\underline{\Delta T}} = -\underline{\Delta}_0. \qquad (6.57)$$

The matrix $\underline{\underline{\Delta}}$ is always symmetrical, and the final shear flow $T_s(s)$ results analogously to (6.55) from a summation as follows:

$$T_s(s) = T_{s0}(s) + T_{s1} + T_{s2} + \ldots + T_{sn}. \qquad (6.58)$$

We illustrate the procedure using the three-cell cross-section given in Fig. 6.16. Since this is a thin-walled cross-section, we will strictly relate all further explanations to the skeleton line.

To determine the shear flow $T_s(s)$ or the shear stress $\tau_s(s)$, we carry out a statically indeterminate calculation and first make the cross-section statically determinate by performing three (arbitrary) cuts as shown in Fig. 6.17, top. The cuts are placed in

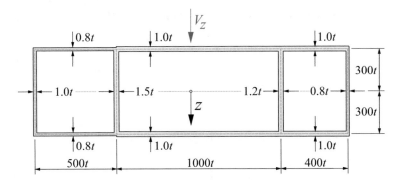

Fig. 6.16 Three-cell cross-section under transverse shear force V_z

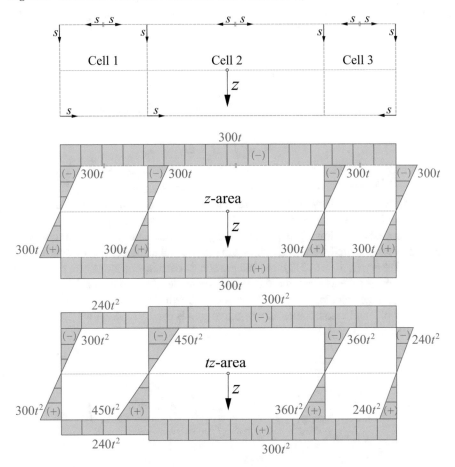

Fig. 6.17 Skeleton line and local axes (top), z−area (middle), tz−area (bottom)

each cell in the center of the upper limiting skeleton line. The circumferential axes
s employed here are also indicated in Fig. 6.17.

To calculate the moment of inertia

$$I_{yy} = \int_A z^2 dA = \int_s z^2 t ds \qquad (6.59)$$

we first determine the $z-$area. It is shown in Fig. 6.17, middle. To analyze the moment
of inertia, we formulate the integration in (6.59) as follows:

$$I_{yy} = \int_s z \cdot (tz) ds, \qquad (6.60)$$

so that we superimpose the $z-$area with the $tz-$area as the product of the
$z-$coordinate and the wall thickness t. The $tz-$area is shown in Fig. 6.17, bottom.
Using the integral table in Fig. 5.8, the calculation yields:

$$I_{yy} = 4,05 \cdot 10^8 t^4. \qquad (6.61)$$

The static moment S_y then results at the opened cross-section

$$S_y = \int_{s_A}^s zt(s) ds \qquad (6.62)$$

as shown in Fig. 6.18, top. The resultant shear flow is shown in Fig. 6.18, bottom.
To determine the shear flow at the closed cross-section, a circumferential unit shear
flow $T_{si} = 1$ ($i = 1, 2, 3$) is assigned to each cell as shown in Fig. 6.19.

To describe the following steps in the analysis we also introduce the circumferential axes $\bar{s}$ as indicated. When calculating the individual displacement discontinuities
Δu_{i0} and Δu_{ij} ($i, j = 1, 2, 3$), the integration is carried out along the $\bar{s}-$axes. If
the directions of the $s-$axis and the $\bar{s}-$axis coincide, the calculation is carried out
according to the calculation equations already shown. In cell sections in which the
directions of s and $\bar{s}$ do not match, a negative sign is used. For Δu_{10} we get (see also
Fig. 6.19):

$$\begin{aligned}
\Delta u_{10} &= -\frac{V_z}{GI_{yy}} \oint \frac{S_y(s)}{t(s)} ds \\
&= -\frac{V_z}{GI_{yy}} \left[\frac{1}{t_{a-1}} \int_{\bar{s}_a}^{\bar{s}_1} S_y(s) d\bar{s} + \frac{1}{t_{1-2}} \int_{\bar{s}_1}^{\bar{s}_2} S_y(s) d\bar{s} + \frac{1}{t_{2-3}} \int_{\bar{s}_2}^{\bar{s}_3} S_y(s) d\bar{s} \right. \\
&\quad \left. - \frac{1}{t_{3-4}} \int_{\bar{s}_3}^{\bar{s}_4} S_y(s) d\bar{s} - \frac{1}{t_{4-a}} \int_{\bar{s}_4}^{\bar{s}_a} S_y(s) d\bar{s} \right].
\end{aligned} \qquad (6.63)$$

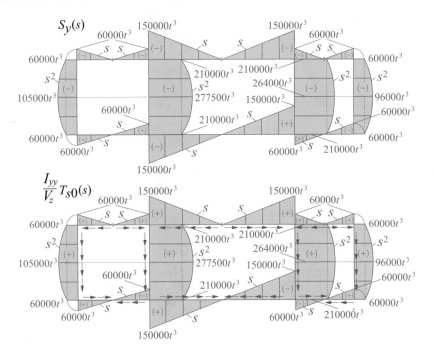

Fig. 6.18 Static moment $S_y(s)$ (top) and shear flow $T_{s0}(s)$ (bottom)

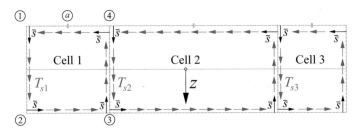

Fig. 6.19 Unit shear flows T_{si} ($i = 1, 2, 3$) and circumferential axes $\bar{s}$

In detail, with the help of the integral tables in Fig. 5.8 we have:

$$\frac{1}{t_{a-1}} \int_{\bar{s}_a}^{\bar{s}_1} S_y(s)\mathrm{d}\bar{s} = \frac{1}{0.8t} \cdot \frac{1}{2} \cdot 250t \cdot 1 \cdot \left(-60000t^3\right) = -9.375 \cdot 10^6 t^3,$$

$$\frac{1}{t_{1-2}} \int_{\bar{s}_1}^{\bar{s}_2} S_y(s)\mathrm{d}\bar{s} = \frac{1}{1.0t} \left[1 \cdot 600t \cdot 1 \cdot \left(-60000t^3\right) \right.$$
$$\left. + \frac{2}{3} \cdot 600t \cdot 1 \cdot \left(-45000t^3\right)\right] = -5.4 \cdot 10^7 t^3,$$

$$\frac{1}{t_{2-3}} \int_{\bar{s}_2}^{\bar{s}_3} S_y(s) \mathrm{d}\bar{s} = \frac{1}{0.8t} \left[\frac{1}{2} \cdot 250t \cdot 1 \cdot \left(-60000t^3 \right) \right.$$

$$\left. + \frac{1}{2} \cdot 250t \cdot 1 \cdot 60000t^3 \right] = 0,$$

$$-\frac{1}{t_{3-4}} \int_{\bar{s}_3}^{\bar{s}_4} S_y(s) \mathrm{d}\bar{s} = -\frac{1}{1.5t} \left[1 \cdot 600t \cdot 1 \cdot \left(-210000t^3 \right) \right.$$

$$\left. + \frac{2}{3} \cdot 600t \cdot 1 \cdot \left(-67500t^3 \right) \right] = 1.02 \cdot 10^8 t^3,$$

$$-\frac{1}{t_{4-a}} \int_{\bar{s}_4}^{\bar{s}_a} S_y(s) \mathrm{d}\bar{s} = -\frac{1}{0.8t} \cdot \frac{1}{2} \cdot 250t \cdot 1 \cdot \left(-60000t^3 \right) = 9.375 \cdot 10^6 t^3. \qquad (6.64)$$

Thus:

$$\Delta u_{10} = -4.8 \cdot 10^7 \frac{V_z t^3}{G I_{yy}}. \qquad (6.65)$$

The displacement discontinuities Δu_{20} and Δu_{30} result in a similar manner:

$$\Delta u_{20} = -2.1 \cdot 10^7 \frac{V_z t^3}{G I_{yy}}, \quad \Delta u_{30} = 6.0 \cdot 10^7 \frac{V_z t^3}{G I_{yy}}. \qquad (6.66)$$

For Δu_{11} we have:

$$\Delta u_{11} = \frac{1}{G} \oint \frac{\mathrm{d}s}{t(s)} = \frac{1}{G} \left(\frac{500t}{0.8t} + \frac{600t}{1.0t} + \frac{500t}{0.8t} + \frac{600t}{1.5t} \right) = \frac{2250}{G}. \qquad (6.67)$$

In an analogous manner:

$$\Delta u_{22} = \frac{2900}{G}, \quad \Delta u_{33} = \frac{2050}{G}. \qquad (6.68)$$

Finally, the displacement discontinuities Δu_{ij} ($i, j = 1, 2, 3$ with $i \neq j$) have to be considered. For Δu_{12} we get by integration over the common wall of cells 1 and 2:

$$\Delta u_{12} = -\frac{1}{G} \oint \frac{\mathrm{d}s}{t(s)} = -\frac{1}{G} \cdot \frac{600t}{1.5t} = -\frac{400}{G}. \qquad (6.69)$$

The quantity Δu_{23} results similarly as:

$$\Delta u_{23} = -\frac{500}{G}. \qquad (6.70)$$

The displacement discontinuity Δu_{13} is zero (cells 1 and 3 do not share a common wall):

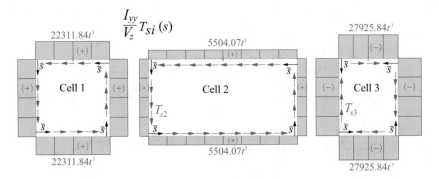

Fig. 6.20 Shear flows T_{si} ($i = 1, 2, 3$), drawn as functions of the circumferential coordinates $\bar{s}$

$$\Delta u_{13} = 0. \tag{6.71}$$

The following linear equation system follows in order to determine the unknown constant shear flows T_{si} ($i = 1, 2, 3$):

$$2250 T_{s1} - 400 T_{s2} + 0 = 4.8 \cdot 10^7 \cdot \frac{V_z t^3}{I_{yy}},$$

$$-400 T_{s1} + 2900 T_{s2} - 500 T_{s3} = 2.1 \cdot 10^7 \cdot \frac{V_z t^3}{I_{yy}},$$

$$0 - 500 T_{s2} + 2050 T_{s3} = -6.0 \cdot 10^7 \cdot \frac{V_z t^3}{I_{yy}}. \tag{6.72}$$

Solving for T_{si} yields (see Fig. 6.20):

$$T_{s1} = 22311.84 \frac{V_z t^3}{I_{yy}}, \quad T_{s2} = 5504.07 \frac{V_z t^3}{I_{yy}}, \quad T_{s3} = -27925.84 \frac{V_z t^3}{I_{yy}}. \tag{6.73}$$

With the statically indeterminate shear flows T_{si} ($i = 1, 2, 3$) available, the final shear flow of the closed cross-section can be determined as already shown with (6.58). The end result for the example considered here is shown in Fig. 6.21. Here the final distribution of $T_s(s)$ was plotted as a function of the circumferential axes s. It should be noted here that e.g. on the vertically running web of the cross-section under consideration between cell 1 and cell 2 (as well as between cell 2 and cell 3) not only the shear flow T_{s0} acts, but also the two statically indeterminate shear flows T_{s1} and T_{s2}, which is reflected when applying the calculation rule (6.58).

The above statements are valid for the case that a load exists exclusively in the form of a transverse force V_z. If the two shear forces V_y and V_z occur simultaneously, then it is advisable to use a somewhat more generalized form of the calculation which we will briefly outline below. The starting point for the considerations is the system of equations (6.56). We now replace in the definition

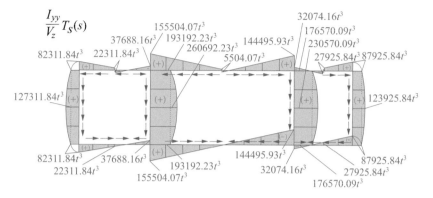

Fig. 6.21 Shear flow $T_s(s)$ of the closed cross-section, drawn as function of the circumferential axes s

$$\Delta u_{i0} = \oint \frac{T_{so}(s)}{Gt(s)}\,ds \tag{6.74}$$

the shear flow $T_{s0}(s)$ with (6.15), so that:

$$\Delta u_{i0} = -\frac{V_z}{GI_{yy}}\oint \frac{S_y(s)}{t(s)}\,ds - \frac{V_y}{GI_{zz}}\oint \frac{S_z(s)}{t(s)}\,ds. \tag{6.75}$$

With the abbreviations

$$\delta_{iz} = \oint \frac{S_y(s)}{t(s)}\,ds, \quad \delta_{iy} = \oint \frac{S_z(s)}{t(s)}\,ds \tag{6.76}$$

the compatibility equations (6.56) assume the following form:

$$\begin{bmatrix} \Delta u_{11} & \Delta u_{12} & \Delta u_{13} & \cdots & \Delta u_{1n} \\ \Delta u_{12} & \Delta u_{22} & \Delta u_{23} & \cdots & \Delta u_{2n} \\ \Delta u_{13} & \Delta u_{23} & \Delta u_{33} & \cdots & \Delta u_{3n} \\ \vdots & \vdots & \vdots & \ddots & \vdots \\ \Delta u_{1n} & \Delta u_{2n} & \Delta u_{3n} & \cdots & \Delta u_{nn} \end{bmatrix} \begin{pmatrix} T_{s1} \\ T_{s2} \\ T_{s3} \\ \vdots \\ T_{sn} \end{pmatrix} = \begin{pmatrix} \dfrac{V_z}{GI_{yy}}\delta_{1z} + \dfrac{V_y}{GI_{zz}}\delta_{1y} \\[2mm] \dfrac{V_z}{GI_{yy}}\delta_{2z} + \dfrac{V_y}{GI_{zz}}\delta_{2y} \\[2mm] \dfrac{V_z}{GI_{yy}}\delta_{3z} + \dfrac{V_y}{GI_{zz}}\delta_{3y} \\[2mm] \vdots \\[2mm] \dfrac{V_z}{GI_{yy}}\delta_{nz} + \dfrac{V_y}{GI_{zz}}\delta_{ny} \end{pmatrix}. \tag{6.77}$$

We now use $\frac{V_y}{I_{zz}} = 1$ and $\frac{V_z}{I_{yy}} = 0$. Then with (6.77):

$$
\begin{bmatrix}
\Delta u_{11} & \Delta u_{12} & \Delta u_{13} & \cdots & \Delta u_{1n} \\
\Delta u_{12} & \Delta u_{22} & \Delta u_{23} & \cdots & \Delta u_{2n} \\
\Delta u_{13} & \Delta u_{23} & \Delta u_{33} & \cdots & \Delta u_{3n} \\
\vdots & \vdots & \vdots & \ddots & \vdots \\
\Delta u_{1n} & \Delta u_{2n} & \Delta u_{3n} & \cdots & \Delta u_{nn}
\end{bmatrix}
\begin{pmatrix}
T_{s1y} \\ T_{s2y} \\ T_{s3y} \\ \vdots \\ T_{sny}
\end{pmatrix}
=
\begin{pmatrix}
\delta_{1y} \\ \delta_{2y} \\ \delta_{3y} \\ \vdots \\ \delta_{ny}
\end{pmatrix}.
\tag{6.78}
$$

The statically indeterminate shear flows T_{siy} can be determined from this system of equations.

We also set $\frac{V_y}{I_{zz}} = 0$ and $\frac{V_z}{I_{yy}} = 1$ so that:

$$
\begin{bmatrix}
\Delta u_{11} & \Delta u_{12} & \Delta u_{13} & \cdots & \Delta u_{1n} \\
\Delta u_{12} & \Delta u_{22} & \Delta u_{23} & \cdots & \Delta u_{2n} \\
\Delta u_{13} & \Delta u_{23} & \Delta u_{33} & \cdots & \Delta u_{3n} \\
\vdots & \vdots & \vdots & \ddots & \vdots \\
\Delta u_{1n} & \Delta u_{2n} & \Delta u_{3n} & \cdots & \Delta u_{nn}
\end{bmatrix}
\begin{pmatrix}
T_{s1z} \\ T_{s2z} \\ T_{s3z} \\ \vdots \\ T_{snz}
\end{pmatrix}
=
\begin{pmatrix}
\delta_{1z} \\ \delta_{2z} \\ \delta_{3z} \\ \vdots \\ \delta_{nz}
\end{pmatrix}.
\tag{6.79}
$$

The shear flows T_{siz} can be determined from this system of equations. The shear flows T_{siy} and T_{siz} are the fictitious shear flows of the cell i of the closed cross-section that are triggered by the two transverse forces V_y and V_z. The statically indeterminate shear flow T_{si} of cell i then results as:

$$
T_{si} = \frac{V_y}{I_{zz}} T_{siy} + \frac{V_z}{I_{yy}} T_{siz},
\tag{6.80}
$$

and the shear flow of the closed cross-section with n cells then follows in cell i as:

$$
T_s(s) = T_{s0}(s) + T_{si}.
\tag{6.81}
$$

6.3.3   Mixed Cross-Sections

In addition to exclusively open and exclusively closed cross-sections, there can also be cross-sections that are composed of open and closed components (Fig. 6.22). The calculation of the shear flow of such a cross-section can also be carried out with a statically indeterminate calculation, although it should be noted here that the unit shear flows to be applied in the individual cells are only taken into account in the closed cross-section parts in the calculation. This also applies to the application of the compatibility conditions (6.56). The shear flow calculation is then performed separately for the open and closed cross-sectional parts. In the open part, the shear flow $T_s(s)$ results as follows:

$$
T_s(s) = T_{s0}(s),
\tag{6.82}
$$

Fig. 6.22 Mixed multi-cell
cross-section

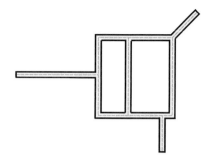

whereas in the closed parts a summation is carried out:

$$T_s(s) = T_{s0}(s) + T_{s1} + T_{s2} + \dots + T_{sn}. \tag{6.83}$$

6.3.4 Use of Symmetry Properties

In the case of closed cross-sections which consist of one cell or also of several
cells and in which certain symmetry properties are present, it is useful to use these
symmetry properties for the purpose of a simpler execution of the calculations. This
is briefly discussed here for the case that a multi-cell cross-section is subjected to the
transverse force V_z and the $z-$axis is an axis of symmetry. In this case there is a zero
crossing of the shear flow at those points where the $z-$axis intersects the skeleton
line of the cross section. An example of this is the cross-section given in Fig. 6.11
or 6.12. In such cases it is advisable to place the starting point s_A of the integration
at precisely these points of symmetry. The shear flow $T_s(s = s_A)$ then becomes zero
here.

6.4 Shear Center

6.4.1 Open Cross-Sections

There are a number of typical lightweight cross-sections where the application of
the transverse forces in the center of gravity S leads to (typically unplanned) torsion
of the beam. There is a certain point in which transverse shear forces must act if
torsion is to be avoided. We refer to this point as the so-called shear center M, the
determination of which is discussed in this section. As an introductory example we
consider the C-cross-section shown in Fig. 6.23, which is subjected to the transverse
shear force V_z. We define the (arbitrary) reference system $\bar{y}, \bar{z}$ at the point shown in

Fig. 6.23 and first determine the coordinates $\bar{y}_S$, $\bar{z}_S$ of the center of gravity S with the aid of the integral table of Fig. 5.8:

$$\bar{y}_S = \frac{\int_A \bar{y} \, dA}{\int_A dA} = \frac{2 \cdot \frac{1}{2} \cdot b \cdot 1 \cdot b \cdot t}{t \, (b \cdot 1 \cdot 1 + h \cdot 1 \cdot 1 + b \cdot 1 \cdot 1)} = \frac{b^2}{2b + h},$$

$$\bar{z}_S = \frac{\int_A \bar{z} \, dA}{\int_A dA} = \frac{t \left(\frac{1}{2} \cdot h \cdot 1 \cdot h + b \cdot 1 \cdot h \right)}{t \, (b \cdot 1 \cdot 1 + h \cdot 1 \cdot 1 + b \cdot 1 \cdot 1)} = \frac{h}{2}. \tag{6.84}$$

We want to assume that the cross-section height h is exactly twice the flange width, i.e. $h = 2b$. Then:

$$\bar{y}_S = \frac{b}{4}, \quad \bar{z}_S = \frac{h}{2} = b. \tag{6.85}$$

The moments of inertia $I_{\bar{y}\bar{y}}$, $I_{\bar{z}\bar{z}}$ and the deviation moment $I_{\bar{y}\bar{z}}$ result as:

$$I_{\bar{y}\bar{y}} = \int_A \bar{z}^2 dA = \frac{20tb^3}{3},$$

$$I_{\bar{z}\bar{z}} = \int_A \bar{y}^2 dA = \frac{2tb^3}{3},$$

$$I_{\bar{y}\bar{z}} = \int_A \bar{y}\bar{z} \, dA = tb^3. \tag{6.86}$$

With the help of Steiner's theorem we can relate the second-order area integrals to the center of gravity S of the cross-section and obtain:

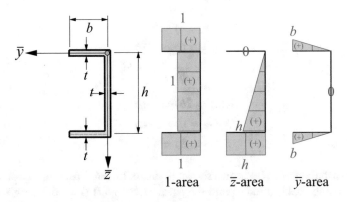

Fig. 6.23 C-cross-section with the axes $\bar{y}$, $\bar{z}$

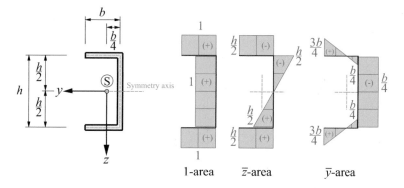

Fig. 6.24 C-cross-section with the principal axes y, z

$$I_{\hat{y}\hat{y}} = I_{\bar{y}\bar{y}} - \bar{z}_S^2 A = \frac{8tb^3}{3},$$

$$I_{\hat{z}\hat{z}} = I_{\bar{z}\bar{z}} - \bar{y}_S^2 A = \frac{5tb^3}{12},$$

$$I_{\hat{y}\hat{z}} = I_{\bar{y}\bar{z}} - \bar{y}_S \bar{z}_S A = 0. \tag{6.87}$$

Since the deviation moment $I_{\hat{y}\hat{z}}$ in the center of gravity coordinate system is zero, the reference system $\hat{y}$ and $\hat{z}$ is already the principal axis system so that we can use the designations y, z. This also results directly from the symmetry properties of the cross-section. The principal axis system together with the $y-$ and $z-$areas is shown in Fig. 6.24. The calculation of the static moment $S_y(s)$ is done section by section as follows. We introduce the circumferential local reference axes s_1, s_2 and s_3 as shown in Fig. 6.25, left, and calculate the necessary values for $S_y(s)$ in each segment. We obtain:

$$S_y(s_1 = 0) = 0,$$
$$S_y(s_1 = b) = -tb^2,$$
$$S_y(s_2 = 0) = -tb^2,$$
$$S_y(s_2 = b) = -\frac{3tb^2}{2},$$
$$S_y(s_2 = 2b) = -tb^2,$$
$$S_y(s_3 = 0) = -tb^2,$$
$$S_y(s_3 = b) = 0. \tag{6.88}$$

The distribution of $S_y(s)$ is also shown in Fig. 6.25, left. From this, both the shear flow $T_s(s)$ (Fig. 6.25, middle) and the shear stress $\tau_s(s)$ (Fig. 6.25, right) can be determined.

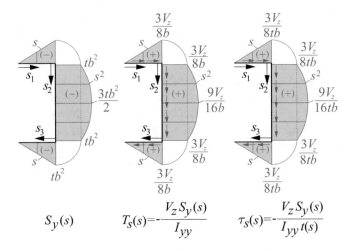

Fig. 6.25 Static moment $S_y(s)$, shear flow $T_s(s)$ and shear stress $\tau_s(s)$ for a C-cross-section under the transverse shear force V_z

Fig. 6.26 Resultant forces for the C-cross-section under the transverse shear force V_z

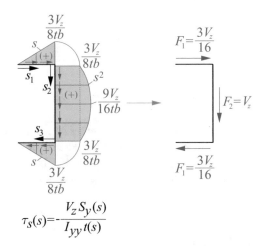

We now want to show that in the event that the transverse force acts in the center of gravity S, a (usually unwanted) torsional effect also occurs. For this, we first determine the forces resulting from the shear stresses in the two flanges and in the web (see Fig. 6.26). With the help of the integral table of Fig. 5.8 we get:

$$F_1 = \frac{1}{2} \cdot b \cdot t \cdot \frac{3V_z}{8tb} = \frac{3V_z}{16},$$

$$F_2 = 2b \cdot t \cdot \frac{3V_z}{8tb} + \frac{2}{3} \cdot 2b \cdot t \cdot \frac{3V_z}{16tb} = V_z. \tag{6.89}$$

This result allows two important conclusions. On the one hand, it can be seen that the vertical resulting force F_2 gives exactly the value of the acting transverse shear force V_z. Due to the requirement for equilibrium, this is a mandatory result and at the same time also a suitable way to check the correctness of the calculations. Furthermore, the two resultant forces in the flanges are identical but act in opposite directions, which is a mandatory result due to the balance of forces (in this example, no transverse force V_y is applied). In addition, the situation given in Fig. 6.26 shows that the resultant forces F_1 and F_2 generate a torsional moment M_x around the center of gravity S as follows:

$$M_x = -2 \cdot F_1 \cdot b - F_2 \cdot \frac{b}{4} = -\frac{5V_z b}{8}. \tag{6.90}$$

It can therefore be concluded that when the transverse shear force is applied in the center of gravity S, in addition to a bending effect there is also a torsional effect. In order to avoid this, the shear force must act in the so-called shear center M. If we define the coordinates of the shear center M as y_M and z_M, then we make the following requirement:

$$M_x = -\frac{5V_z b}{8} \equiv V_z y_M. \tag{6.91}$$

The y−coordinate y_M of the shear center M is thus:

$$y_M = -\frac{5b}{8}. \tag{6.92}$$

The position of the shear center M is shown in Fig. 6.27. It is apparently found outside the actual cross-section. Hence, if the additional torsional effect is to be avoided under the transverse force V_z, then a suitable design solution is required to position the force application point at y_M. The shear center coordinate z_M does not require calculation at this point. It can be shown that if there is an axis of symmetry for a considered cross-section, then the shear center M is always to be found on this axis of symmetry, as is also clear from Fig. 6.27. This issue will be discussed again at a later point.

In the following we want to develop a general method for determining the shear center M at arbitrary open and thin-walled cross-sections. The shear center M is the point in which transverse forces V_z and V_y must act so that a torsional effect does not occur in addition to bending of the beam under consideration. This can be expressed in such a way that it is required that the circumferential unit shear flows according to (6.28) do not cause a moment about the $x-$ axis, i.e.:

$$\oint T_y r_{tM} ds = 0, \quad \oint T_z r_{tM} ds = 0. \tag{6.93}$$

Fig. 6.27 Location of the
shear center M for a
C-cross-section

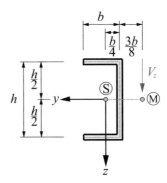

Fig. 6.28 Determination of
the shear center M for an
arbitrary cross-section

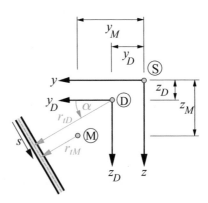

Herein, r_{tM} is the distance to the shear center M perpendicular to the tangent
of the skeleton line at each point. The ring integrals in (6.93) indicate that the prescribed integrations are to be performed over the entire circumference of the cross
section under consideration. However, since the shear center is usually unknown at
the beginning of such a calculation, first an arbitrary reference point is chosen that is
denoted as D. This arbitrary reference point does not necessarily have to be selected
to be identical to the center of gravity S, even though this might be an obvious choice.
In this reference point D we introduce the reference system y_D, z_D. Fig. 6.28 clarifies
the previous definitions. We denote the angle between the y_D−axis and the distance
r_{tD} as α.

The following relationship between the two distances r_{tM} and r_{tD} can be derived
from Fig. 6.28:

$$r_{tM} = r_{tD} - (y_M - y_D) \cos \alpha - (z_M - z_D) \sin \alpha. \qquad (6.94)$$

In addition, we can conclude:

$$\sin \alpha = -\frac{dy}{ds}, \quad \cos \alpha = \frac{dz}{ds}, \qquad (6.95)$$

so that the following connection results from the first requirement in (6.93):

$$\oint T_y r_{tM} ds = \oint T_y r_{tD} ds - (y_M - y_D) \oint T_y \frac{dz}{ds} ds + (z_M - z_D) \oint T_y \frac{dy}{ds} ds = 0.$$

$$(6.96)$$

At this point we carry out a partial integration for the second term and get:

$$- (y_M - y_D) \oint T_y \frac{dz}{ds} ds = - (y_M - y_D) T_y z \Big|_{s=0}^{s=s_e} + (y_M - y_D) \oint \frac{dT_y}{ds} z ds.$$

$$(6.97)$$

Herein s_e is the end point of the integration over the circumference of the cross section. However, since the unit shear flow at the points $s = 0$ and $s = s_e$ disappears (an open cross section is considered), the middle term in (6.97) can be omitted. With regard to the last term in (6.97) it follows from the definition of the unit shear flow $T_y = - \oint y t ds$ that

$$\frac{dT_y}{ds} = -yt \tag{6.98}$$

holds. Hence for the third term in (6.97):

$$(y_M - y_D) \oint \frac{dT_y}{ds} z ds = - (y_M - y_D) \oint yzt ds. \tag{6.99}$$

The expression $\oint yzt ds$ corresponds exactly to the deviation moment I_{yz}, which in the case of the principal axes y, z must be zero by definition.

In the same way, we carry out a partial integration for the third term, and after a few transformations we obtain:

$$(z_M - z_D) \oint T_y \frac{dy}{ds} ds = \underbrace{(z_M - z_D) T_y y \Big|_{s=0}^{s=s_e}}_{=0} - (z_M - z_D) \oint \underbrace{\frac{dT_y}{ds}}_{=-yt} y ds = (z_M - z_D) I_{zz}.$$

$$(6.100)$$

Hence:

$$\oint T_y r_{tD} ds + (z_M - z_D) I_{zz} = 0. \tag{6.101}$$

We can rearrange this expression immediately for the coordinate z_M of the shear center M and achieve:

$$z_M = z_D - \frac{1}{I_{zz}} \oint T_y r_{tD} ds. \tag{6.102}$$

Analogously we have for y_M:

$$y_M = y_D + \frac{1}{I_{yy}} \oint T_z r_{tD} ds. \tag{6.103}$$

Equations (6.102) and (6.103) can be simplified considerably in the case of thin-walled cross-sections that are composed of n straight segments (Fig. 6.29). Then:

$$y_M = y_D + \frac{1}{I_{yy}} \sum_{i_1}^{i=n} F_{z,i} r_{tD,i},$$

$$z_M = z_D - \frac{1}{I_{zz}} \sum_{i_1}^{i=n} F_{y,i} r_{tD,i}. \tag{6.104}$$

Here $r_{tD,i}$ is the vertical distance of the segment i to the reference point D. The quantities $F_{z,i}$ and $F_{y,i}$ are the resultants of the unit shear flows T_z and T_y in segment i which are tangential to the respective skeleton line.

We want to illustrate the calculation method for determining the shear center M of open thin-walled cross-sections using the already considered C-cross-section. Fig. 6.25, left, already contains the static moment S_y which according to (6.28) can be interpreted as the negative unit shear flow T_z. We now want to determine the shear center coordinate y_M according to (6.104) and choose the position of the reference point D in the center of gravity S (Fig. 6.24). This results in the distances $r_{tD,i}$ as:

$$r_{tD,1} = b, \quad r_{tD,2} = \frac{b}{4}, \quad r_{tD,3} = b. \tag{6.105}$$

The moment of inertia I_{yy} can be taken from (6.87) with the value $\frac{8tb^3}{3}$, the coordinate y_D is identically zero. The resultants $F_{z,i}$ amount to:

$$F_{z,1} = \frac{1}{2}tb^3, \quad F_{z,2} = \frac{8tb^3}{3}, \quad F_{z,3} = \frac{1}{2}tb^3. \tag{6.106}$$

Fig. 6.29 Determination of the shear center M at a thin-walled cross-section consisting of straight segments

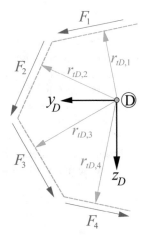

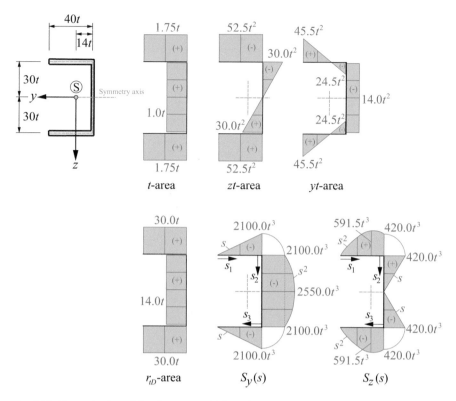

Fig. 6.30 Determination of the shear center M for a C-cross-section

They all point in the positive s−direction. Evaluation of (6.4.1) then yields:

$$y_M = -\frac{3}{8tb^3}\left(\frac{tb^4}{2} + \frac{2tb^4}{3} + \frac{tb^4}{2}\right) = -\frac{5b}{8}. \tag{6.107}$$

Obviously this result is in agreement with (6.92). The negative sign follows from the direction of rotation of the resultants with reference to $D = S$.

Another example is given with the C-cross-section shown in Fig. 6.30. The static moments $S_y(s)$ and $S_z(s)$ are shown in Fig. 6.30, bottom. Using the characteristic areas already given in Fig. 6.30, the position of the shear center M can be determined for this example as $y_M = -31.5t$, whereby the reference point D was placed in the center of gravity. Due to the symmetry properties of the cross-section, $z_M = 0$ results.

From the considerations made so far, some generally applicable rules regarding the shear center M of thin-walled profiles can be derived:

- If a cross section exhibits one axis of symmetry, then the shear center M is always located on this axis of symmetry.
- If the cross-section under consideration is point symmetric or doubly symmetric, then the shear center M is always located in the center of gravity S.
- If the cross-section under consideration consists of a number of thin-walled straight segments which all intersect at one point, then the shear center M lies exactly at the intersection point of the segments.
- The shear center is a load-independent quantity and depends exclusively on the geometry of the cross-section.

Figure 6.31 illustrates the position of the shear center M for a selection of thin-walled cross-sections.

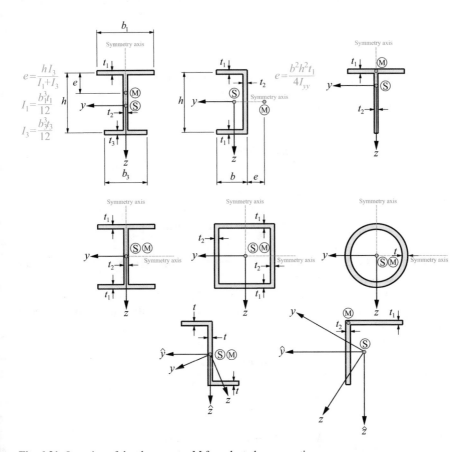

Fig. 6.31 Location of the shear center M for selected cross-sections

6.4.2 Closed Cross-Sections

The determination of the position of the shear center for closed cross-sections is performed as already shown for open cross-sections. However, it should be noted here that the unit shear flows must be determined on the closed profile and require a statically indeterminate calculation. Based on the calculation rule for the statically indeterminate shear flow of the cell i

$$T_s(s) = T_{s0}(s) + T_{si} \tag{6.108}$$

or

$$T_s(s) = \frac{Q_y}{I_{zz}} \left(-S_z(s) + T_{siy} \right) + \frac{V_z}{I_{yy}} \left(-S_y(s) + T_{siz} \right) \tag{6.109}$$

with $\frac{Q_y}{I_{zz}} = 1$ and $\frac{V_z}{I_{yy}} = 1$ using the unit shear flows $T_z(s)$ and $T_y(s)$ in cell i:

$$T_z(s) = -S_y(s) + T_{siz}, \quad T_y(s) = -S_z(s) + T_{siy}. \tag{6.110}$$

Insertion into (6.103) then yields:

$$
\begin{aligned}
y_M &= y_D + \frac{1}{I_{yy}} \oint T_z r_{tD} \mathrm{d}s \\
&= y_D + \frac{1}{I_{yy}} \left[-\oint S_y(s) r_{tD} \mathrm{d}s + T_{s1z} \oint r_{tD1} \mathrm{d}s \right. \\
&\quad \left. + T_{s2z} \oint r_{tD2} \mathrm{d}s + \dots + T_{snz} \oint r_{tDn} \mathrm{d}s \right],
\end{aligned} \tag{6.111}
$$

where we have made use of the fact that the shear flows T_{siz} are constant in every cell i and can therefore be written in front of the respective ring integral.

We also make use of the fact that the ring integral of cell i over the distance r_{tDi} corresponds exactly to double the area $A_{m,i}$ of cell i inscribed within the skeleton line:

$$\oint r_{tDi} \mathrm{d}s = 2A_{m,i}. \tag{6.112}$$

Then from (6.111) we obtain:

$$y_M = y_D + \frac{1}{I_{yy}} \left[-\oint S_y(s) r_{tD} \mathrm{d}s + 2T_{s1z} A_{m,1} + 2T_{s2z} A_{m,2} + \dots + 2T_{snz} A_{m,n} \right] \tag{6.113}$$

or:

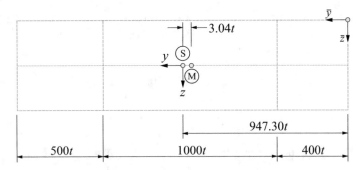

Fig. 6.32 Center of gravity S and shear center M of a three-cell closed cross-section

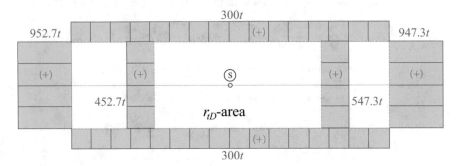

Fig. 6.33 r_{tD}-area for a three-cell closed cross-section

$$y_M = y_D - \frac{1}{I_{yy}} \oint S_y(s)r_{tD}ds + \frac{1}{I_{yy}} \left[2T_{s1z}A_{m,1} + 2T_{s2z}A_{m,2} + \ldots + 2T_{snz}A_{m,n}\right].$$
(6.114)

Analogously from (6.102):

$$z_M = z_D + \frac{1}{I_{zz}} \oint S_z(s)r_{tD}ds - \frac{1}{I_{zz}} \left[2T_{s1y}A_{m,1} + 2T_{s2y}A_{m,2} + \ldots + 2T_{sny}A_{m,n}\right].$$
(6.115)

The procedure is illustrated at the example of the three-cell cross-section given in Fig. 6.16. For the calculation, the center of gravity S of the cross section is first determined. The reference system $\bar{y}$ and $\bar{z}$ is applied as shown in Fig. 6.32. We obtain:

$$\bar{y}_S = \frac{S_{\bar{z}}}{A} = 947.30t, \quad \bar{z}_S = \frac{S_{\bar{y}}}{A} = 300t,$$
(6.116)

where the result $\bar{z} = 300t$ can also be deduced directly from the symmetry properties of the cross-section under consideration. Due to the simple symmetry of the cross-section, we can use the designations y and z directly for the centroid axes, these are already the principal axes. When determining the shear center M, we start from the center of gravity S of the cross-section, so that $y_D = 0$ and $z_D = 0$.

For the calculation of the shear center coordinates y_M and z_M the distances r_{tD} are required, which are shown in Fig. 6.33. The areas $A_{m,i}$ inscribed in the individual cells within the skeleton line result as:

$$A_{m,1} = 600000t^2, \quad A_{m,2} = 1200000t^2, \quad A_{m,3} = 480000t^2. \quad (6.117)$$

The expressions $\frac{2T_{siz}A_{m,i}}{I_{yy}}$ in (6.114) then result in (see also Fig. 6.20):

$$\frac{2T_{s1z}A_{m,1}}{I_{yy}} = 33.05t, \quad \frac{2T_{s2z}A_{m,2}}{I_{yy}} = 16.31t, \quad \frac{2T_{s3z}A_{m,3}}{I_{yy}} = -33.10t. \quad (6.118)$$

By superimposing the $S_y(s)$−area with the associated 1-area it also follows:

$$
\begin{aligned}
\frac{1}{I_{yy}} \left(-\oint S_y(s) r_{tD} ds \right) &= \frac{600t}{4.05 \cdot 10^8 t^4} \left[1 \cdot 1 \cdot 60000t^3 \cdot 952.7t + \frac{2}{3} \cdot 1 \cdot 45000t^3 \cdot 952.7t \right. \\
&\quad + 1 \cdot 1 \cdot 210000t^3 \cdot 452.7t + \frac{2}{3} \cdot 1 \cdot 67500t^3 \cdot 452.7t \\
&\quad - 1 \cdot 1 \cdot 210000t^3 \cdot 547.3t - \frac{2}{3} \cdot 1 \cdot 54000t^3 \cdot 547.3t \\
&\quad \left. - 1 \cdot 1 \cdot 60000t^3 \cdot 947.3t - \frac{2}{3} \cdot 1 \cdot 36000t^3 \cdot 947.3t \right] \\
&= -19.3t. \quad (6.119)
\end{aligned}
$$

Use was made here of the fact that the portions of the horizontal flanges cancel out due to the symmetry of the cross-section. The different signs of the individual terms result from the direction of rotation of the respective components with regard to the x−axis.

Based on (6.113) or (6.114) we eventually obtain:

$$y_M = -3,04t. \quad (6.120)$$

Due to the symmetry of the cross-section, $z_M = 0$ can be deduced directly without further calculation. The position of the shear center is also indicated in Fig. 6.32.

Bibliography

Bauchau OA, Craig JI (2009) Structural analysis: with applications to aerospace structures, Springer. Dordrecht et al, The Netherlands

Czerwenka G, Schnell W (1970) Einführung in die Rechenmethoden des Leichtbaus, Band 1, Bibliographisches Institut, Mannheim et al., Germany

Franke W, Friemann H (2005) Schub und Torsion in geraden Stäben, 3rd edn. Vieweg + Teubner, Wiesbaden, Germany

Gross D, Hauger W, Schröder J (2014) Technische Mechanik 2: Elastostatik, 12th edn. Springer Berlin et al, Germany

Hanswille G (1995) Vorlesungen über Stahl- und Verbundbau. Bergische Universität Wuppertal, Germany, Department of Civil Engineering

Kindmann R, Frickel J (2017) Elastische und plastische Querschnittstragfähigkeit, Online edition 2017, https://www.kindmann.de, last retrieved March 2020

Linke M, Nast E (2015) Festigkeitslehre für den Leichtbau. Springer Berlin et al, Germany

Megson THG (1999) Aircraft structures for engineering students, 3rd edn. Arnold, London, UK

Petersen C (1997) Stahlbau, 3rd edn. Vieweg Braunschweig et al, Germany

Roik KH (1978) Vorlesungen über Stahlbau. Ernst & Sohn Berlin et al, Germany

Shama M (2010) Torsion and shear stresses in ships. Springer Heidelberg et al, Germany

Wiedemann J (2007) Leichtbau 1: Elemente. Springer Berlin et al, Germany

Wlassow WS (1964) Dünnwandige elastische Stäbe, vol I. VEB Verlag für Bauwesen, Berlin, Germany

Wlassow WS (1965) Dünnwandige elastische Stäbe, vol II. VEB Verlag für Bauwesen, Berlin, Germany

Chapter 7
St. Venant Torsion

7.1 Introduction

The contents of Chaps. 5 and 6 focused on beams the loads of which cause normal forces, transverse shear forces and bending moments. However, in addition to such load cases, torsion may occur as well which constitutes the decisive load case for many lightweight construction applications and therefore must be taken into account in the design and dimensioning of lightweight structures accordingly. The consideration of the torsion of straight beams is the subject of this chapter (the so-called St. Venant's torsion[1]) and the following Chap. 8 (the so-called warping torsion). The main difference between these two theories is that in the framework of St. Venant torsion it is assumed that due to the twisting of the cross-section shear stresses in the cross-sectional plane occur, but no cross-sectional warping and thus no normal stresses in the direction of the longitudinal beam axis (the so-called warping stresses) arise. That the latter can very well occur and can have a decisive influence on the design becomes clear from consideration of warping torsion. We shall therefore consider both theories in detail in this and the following chapter.

Both St. Venant's torsion and warping torsion have in common that they are caused by torsional moments M_T, which can result e.g. due to an eccentric load of a beam (Fig. 7.1). Consider a beam which is supported at both ends against deflection as well as against twisting of the cross section and which is subjected to the eccentric single force F (eccentricity e). This is statically equivalent to shifting the force F to the beam axis and simultaneously considering a moment acting around the longitudinal axis of the beam, i.e. the torsional moment $M_T = Fe$. The former leads to beam bending under the single force F, in which bending moments M_y and transverse forces V_z (assuming principal axes) and the associated stresses σ_{xx} and τ_s result. This can be treated with the analysis methods as described in the previous chapters. On the other hand, a torsional moment in the form of M_T and the associated torsional

---

[1] Adhémar Jean Claude Barré de Saint-Venant, 1797–1886, French scientist.

© The Author(s), under exclusive license to Springer Nature Switzerland AG 2021 249
C. Mittelstedt, *Structural Mechanics in Lightweight Engineering*,
https://doi.org/10.1007/978-3-030-75193-7_7

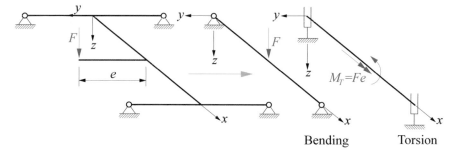

Fig. 7.1 Beam under eccentric load (left) and the resulting analysis models: bending (center), idealized as a simply supported beam, and torsion (right), idealized as a beam with fork bearings

Fig. 7.2 Beam with fork restraints subjected to the distributed torsional moment m_T

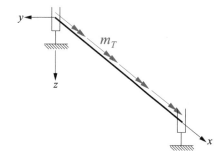

section moment $M_x(x)$ which we will deal with in this chapter also results. Of course, torsional moments can also occur in a distributed manner which we will refer to as m_T (Fig. 7.2) and which also cause the internal torsional moment $M_x(x)$.

We assume that bending and torsion can be treated completely independently of each other. In which cases this decoupling is permissible will be discussed in detail in Chap. 8. The special features of a so-called fork bearing, as indicated in the two Figs. 7.1 and 7.2, will be discussed later. For the present purpose it is sufficient to note that a fork bearing prevents a rotation of the cross section about the longitudinal axis x.

7.2 Assumptions and Constitutive Law

We begin the explanations with the treatment of St. Venant torsion, which, besides geometric and material linearity, is based on the following assumptions:

- Only beams with a straight longitudinal axis are treated.
- Torsion will result in a twisting of the beam and a rotation of the cross-section. We assume that the axis of rotation is not forced by any design measures and results freely.

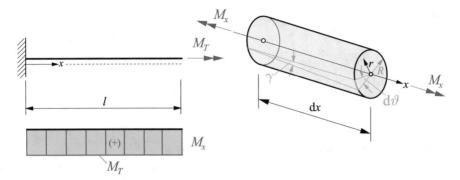

Fig. 7.3 Beam with circular cross section under torsional moment M_T and corresponding internal torsional moment M_x (left), infinitesimal element (right)

- We assume that the torsional load only causes a rotation $\vartheta(x)$ about the longitudinal axis x.
- We also assume that, as already mentioned above, torsion and bending can be treated independently of each other and that there is no coupling. We will discuss this in detail in Chap. 8.
- Furthermore, we assume that the cross section of the beam will be twisted due to torsion and that this is expressed by the rotation $\vartheta(x)$ about the longitudinal axis x, but that the cross section will completely retain its shape and dimensions in the deformed state. This assumption is often referred to as the so-called hypothesis of straightness of the radii.
- We assume warping-free cross-sections. We will define this term in more detail later on, but it is already mentioned here that warping-free cross-sections do not show any longitudinal displacements u due to the rotation ϑ around the beam axis x. As a result, the torsional moment M_T and the distributed torsional moment m_T may act at any point. Cross-sections that are not free of warping can only be treated strictly speaking within the framework of the torsion theory presented here if torsional moments act on the ends of the beam exclusively in the form of single moments and if the warping of the cross-section can occur freely and unhindered.

We first consider the elementary case of a bar with a circular cross section (first solved by Coulomb[2]), which is under the end moment M_T which causes the constant torsional moment M_x in the bar (see Fig. 7.3, left). We investigate an infinitesimal section element of length $\mathrm{d}x$ cut out of the member (Fig. 7.3, right). The shear strain γ of the lateral surface and the twist increment $\mathrm{d}\vartheta$ of the cross-section are also indicated. The radius of the cross section is R, and the radial coordinate, measured from the center of gravity of the cross section, is denoted as r. The shear strain γ and the cross-sectional rotation ϑ are related as follows:

$$\gamma \mathrm{d}x = r \mathrm{d}\vartheta. \tag{7.1}$$

---

[2]Charles Augustin de Coulomb, 1736–1806, French physicist.

Fig. 7.4 Infinitesimal
element dA of the circular
section (left), shear stress
distribution $\tau(r)$ (right)

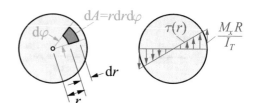

This can be solved for the shear strain γ as follows:

$$\gamma = r\frac{\mathrm{d}\vartheta}{\mathrm{d}x} = r\vartheta'. \tag{7.2}$$

Herein we call the first derivative of the rotation ϑ with respect to x the so-called twist ϑ' of the bar.

Once the shear strain γ is known, we can determine the shear stress distribution in the cross-section using the constitutive law $\tau = G\gamma$, where G is the shear modulus of the material:

$$\tau = G\gamma = Gr\vartheta'. \tag{7.3}$$

The shear stress is thus linearly distributed over the radial coordinate r, becomes zero at the center of gravity of the section and reaches its maximum value at the edge wherein this edge value needs to be determined.

If the shear stress distribution is given, we can determine the resulting torsional moment M_x:

$$M_x = \int_A \tau r \,\mathrm{d}A = \int_A Gr^2\vartheta'\,\mathrm{d}A = G\vartheta'\int_A r^2\,\mathrm{d}A. \tag{7.4}$$

We can decompose the surface integral with $\mathrm{d}A = r\,\mathrm{d}r\,\mathrm{d}\varphi$ (Fig. 7.4, left) into the two partial integrals as follows:

$$M_x = G\vartheta'\int_0^{2\pi}\int_0^R r^3\,\mathrm{d}r\,\mathrm{d}\varphi = G\vartheta'\frac{R^4}{4}\int_0^{2\pi}\mathrm{d}\varphi = G\vartheta'\frac{\pi R^4}{2}. \tag{7.5}$$

Introducing the torsional moment of inertia I_T according to

$$I_T = \frac{\pi R^4}{2}. \tag{7.6}$$

Then:

$$M_x = G I_T \vartheta'. \tag{7.7}$$

We refer to the quantity $G I_T$ as the so-called torsional stiffness, similar to the extensional stiffness EA of the rod or the bending stiffnesses EI_{yy} and EI_{zz} of the beam. This is the constitutive law for the St. Venant Torsion. With this we have an

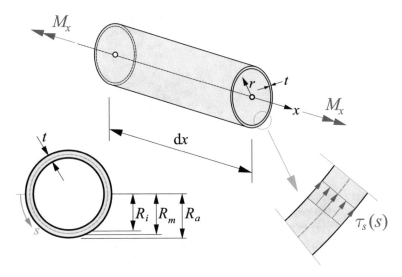

Fig. 7.5 Beam segment with circular cylindrical cross-section under torsional moment M_x (top), dimensions (bottom left), shear stress distribution (bottom right)

expression for (7.3) as follows:

$$\tau = Gr\vartheta' = \frac{M_x}{I_T}r. \tag{7.8}$$

This results in the maximum value of the torsional stress τ at the cross-section edge as (Fig. 7.4, right):

$$\tau = \frac{M_x R}{I_T}. \tag{7.9}$$

Eq. (7.8), in conjunction with Fig. 7.4, right, shows that the course of the shear stress τ is linear over r. This means that inside the solid cross section there is only a very poor utilization of the material, so that solid cross sections, depending on the application, are not the first choice for torsional loading when lightweight construction and design is the focus of the investigations.

Furthermore, we want to consider a beam with a circular cylindrical cross-section of constant thickness t (Fig. 7.5), assuming that the thickness t is sufficiently small compared to the radii R_i and R_a. The inner surface of the circular cylinder is located at $r = R_i$, the outer surface at $r = R_a$. The wall thickness t of the circular cylinder cross-section is halved by the skeleton line at the position $r = R_m$ at each position of the circumferential coordinate s.

As we already know from the investigation of the solid cross-section, the shear stress is distributed linearly over the radial coordinate r according to (7.8). However, if we assume that the thickness t is sufficiently small, especially in comparison to R_m, then an obvious approximation is to assume the shear stresses τ as constant over

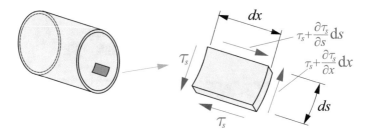

Fig. 7.6 Infinitesimal element of the beam with circular cylinder cross section

the thickness t of the section (Fig. 7.5, bottom right). As usual for such thin-walled cross-sections, we again introduce a local circumferential reference axis s so that the designation $\tau_s(s)$ can be used for the shear stress.

We now look at an infinitesimal element that we cut out of the cylinder and examine the shear stresses τ_s (Fig. 7.6). We form the sum of all forces in $x-$-direction at this element and obtain:

$$\tau t_s dx - \left(\tau_s + \frac{\partial \tau_s}{\partial s} ds \right) t dx = 0, \tag{7.10}$$

from which:

$$\frac{\partial \tau_s}{\partial s} = 0. \tag{7.11}$$

This equation states that τ_s does not undergo any change along the circumferential coordinate s and thus remains constant over the entire circumference of the cross-section. The torsional moment M_x is calculated as:

$$M_x = \int_A \tau r_s dA = \tau_s \int_A r dA. \tag{7.12}$$

Since the radius with $r = R_m$ is constant in the present case, it can be drawn directly before the integral and we get:

$$M_x = R_m \tau_s \int_A dA. \tag{7.13}$$

With the element $dA = R_m t d\varphi$ this results in:

$$M_x = \tau_s R_m^2 t \int_0^{2\pi} d\varphi = 2\pi R_m^2 t \tau_s. \tag{7.14}$$

We can solve this for τ_s and obtain:

$$\tau_s = \frac{M_x}{2\pi R_m^2 t}. \tag{7.15}$$

Since τ_s is assumed to be constant over the thickness of the cross section, this is also the maximum value relevant for design and justification:

$$\tau_{max} = \frac{M_x}{W_T}, \tag{7.16}$$

where W_T is the torsional resistance moment:

$$W_T = 2\pi R_m^2 t. \tag{7.17}$$

Furthermore, we use the relation (7.2) and the constitutive law $\tau = G\gamma$ and have:

$$\tau_s = Gr\vartheta', \tag{7.18}$$

which leads to the following expression after solving for ϑ' (with $r = R_m$) and insertion of (7.15):

$$\vartheta' = \frac{\tau_s}{GR_m} = \frac{M_x}{2\pi R_m^3 Gt}. \tag{7.19}$$

With the torsional moment of inertia

$$I_T = 2\pi R_m^3 t \tag{7.20}$$

this yields:

$$\vartheta' = \frac{M_x}{GI_T}. \tag{7.21}$$

In the form

$$M_x = GI_T\vartheta' \tag{7.22}$$

this again results in the constitutive law of St. Venant Torsion, as already shown with (7.7).

If one is interested in the mutual rotation $\Delta\vartheta$ of the two bar ends, the twist ϑ' must be integrated over the bar length:

$$\Delta\vartheta = \int_0^l \vartheta' dx = \int_0^l \frac{M_x}{GI_T} dx. \tag{7.23}$$

If the torsional moment M_x and the torsional moment of inertia I_T are both constant over the beam length, then the mutual rotation of the bar end sections results as:

$$\Delta\vartheta = \frac{M_x l}{GI_T}. \tag{7.24}$$

If one is interested in designing a circular cylinder cross-section that is as stiff as possible, the torsional stiffness $G I_T$ must be maximized. This will be done primarily by controlling the radius R_m which enters the equation to the third power in I_T.

7.3 Arbitrary Thin-Walled Hollow Cross-Sections

We now generalize the considerations and consider a thin-walled closed cross-section of arbitrary shape. For this purpose, we consider the situation of Fig. 7.7. The thickness t of the cross-section can be a function of the circumferential coordinate s, i.e. $t = t(s)$. The axis around which the cross-section rotates is called the axis of rotation, in Fig. 7.7 denoted as D. We first introduce the shear flow T_s (see also Chap. 6):

$$T_s = \tau_s(s) t(s). \tag{7.25}$$

Thus, the shear flow is the resultant of the shear stress $\tau_s(s)$ over the thickness $t(s)$ of the cross-section and has the unit of a force per unit length. On the basis of the free body diagram of Fig. 7.7 we can make the following statement by means of the equilibrium of forces in $x-$ direction:

$$\frac{\partial T_s}{\partial s} = 0. \tag{7.26}$$

This means that the shear flow T_s is constant at every point s on the skeleton line of the cross-section and thus assumes the same value everywhere. Note, however, that this does not apply to the shear stress $\tau_s(s) = \frac{T(s)}{t(s)}$. This is shown in Fig. 7.7: The shear stress assumes high values wherever the cross-section thickness becomes

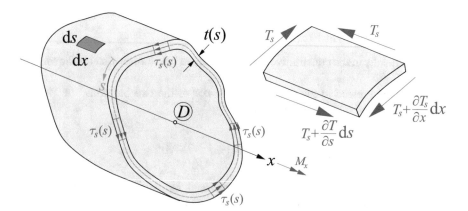

Fig. 7.7 Arbitrary thin-walled hollow cross-section

Fig. 7.8 Free body diagram
of a section ds

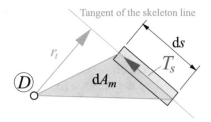

small. The maximum of the shear stress is then found at the location of the smallest
wall thickness t_{min}.

We now want to establish a connection between the shear flow T_s and the torsional
moment M_x. For this purpose, we look at the free body diagram in Fig. 7.8 wherein
a section element of the cross-section with the length ds and the shear flow T_s is
displayed. With respect to the rotation axis, the shear flow has the lever arm r_t, where
r_t represents the distance of the section element from the rotation axis, measured from
the tangent to the skeleton line. The distance of the section element from the rotation
axis is $r(s)$. The infinitesimal contribution dM_x of the shear flow to the torsional
moment M_x is:

$$dM_x = T_s ds r_t. \tag{7.27}$$

Summed up over the entire cross-section, this results in the torsional moment M_x:

$$M_x = \oint T_s r_t ds. \tag{7.28}$$

With the relation

$$dA_m = \frac{1}{2} r_t ds \tag{7.29}$$

we have for ds:

$$ds = 2 \frac{dA_m}{r_t}. \tag{7.30}$$

Insertion into (7.28) results in:

$$M_x = 2T_s \oint_{A_m} dA_m \tag{7.31}$$

or

$$M_x = 2T_s A_m. \tag{7.32}$$

Therein, A_m is the area of the cross-section enclosed by the skeleton line. Thus
we can determine the shear flow T_s due to the torsional moment M_x as follows:

$$T_s = \frac{M_x}{2A_m}. \tag{7.33}$$

The shear flow T_s due to a torsional moment M_x is thus calculated for any thin-walled closed cross-section as the torsional moment divided by twice the area A_m enclosed by the skeleton line. This is the so-called first Bredt formula[3]. The shear stress τ_s can thus be calculated as:

$$\tau_s(s) = \frac{T_s}{t(s)} = \frac{M_x}{2A_m t(s)}, \tag{7.34}$$

where it becomes clear that τ_s is a function of the circumferential coordinate s in contrast to T_s. It assumes its maximum at the position of the cross-section with the smallest wall thickness $t = t_{min}$:

$$\tau_{max} = \frac{M_x}{2A_m t_{min}}. \tag{7.35}$$

Thus the smallest wall thickness t_{min} is decisive for the design of a hollow section under torsional load. Conversely, this also means that all other areas with larger wall thicknesses are poorly utilized. For the purpose of a lightweight design, the wall thickness should therefore be kept constant over the circumference of such hollow sections. A discussion of Eq. (7.34) also shows that the shear stress $\tau_s(s)$ depends linearly on the applied torsional moment M_x, but also on the enclosed area A_m and the wall thickness $t(s)$. This means that the shear stress $\tau_s(s)$ can be reduced under a given moment M_x by increasing the area A_m and the wall thickness $t(s)$, where the increase of the wall thickness $t(s)$ is counterproductive with respect to lightweight design applications.

Introducing the torsional resistance moment W_T according to

$$W_T = 2A_m t_{min} \tag{7.36}$$

then we can write the following for (7.35):

$$\tau_{max} = \frac{M_x}{W_T}. \tag{7.37}$$

In order to apply the constitutive law (7.22), we still need an expression for the torsional moment of inertia I_T of the thin-walled closed cross-section under consideration. For this purpose, we consider the cross-section rotated around the axis of rotation D. Again, we assume that the cross-section retains its shape and its original dimensions during this rotation and that only a pure rigid body rotation takes place. We consider the displacements of the point P according to Fig. 7.9, where P is

[3]Rudolf Bredt, 1842–1900, German engineer.

Fig. 7.9 Displacement of a point P on the skeleton line of the cross-section

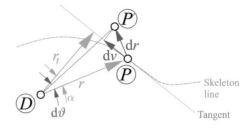

shifted by dr perpendicular to the radius r. Projected onto the tangent of the skeleton line this is the displacement dv. Since we assume small angles we can write:

$$dr = r d\vartheta. \tag{7.38}$$

Furthermore:

$$\cos\alpha = \frac{dv}{dr} = \frac{dv}{r d\vartheta}. \tag{7.39}$$

We also find that:

$$\cos\alpha = \frac{r_t}{r}. \tag{7.40}$$

Equating (7.39) and (7.40) then yields:

$$dv = r_t d\vartheta. \tag{7.41}$$

Differentiating (7.41) with respect to x (under the assumption that r_t is independent of x in a prismatic beam) yields

$$\frac{dv}{dx} = r_t \frac{d\vartheta}{dx} = r_t d\vartheta'. \tag{7.42}$$

We also consider the shear strain of an infinitesimal section element of the beam (Fig. 7.10). The shear strain is composed of the two angles α and β, whereby these can be calculated under the assumption of small angles:

$$\alpha = \frac{dv}{dx}, \quad \beta = \frac{du}{ds}, \tag{7.43}$$

so that

$$\gamma = \alpha + \beta = \frac{dv}{dx} + \frac{du}{ds}. \tag{7.44}$$

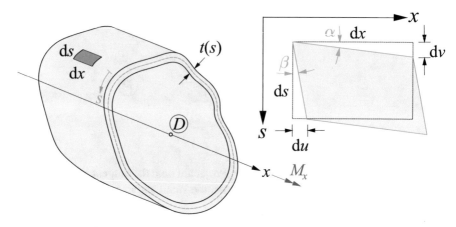

Fig. 7.10 Shear strain of an infinitesimal section element

With Hooke's law $\tau = G\gamma$ we have:

$$\gamma = \frac{\tau_s}{G} = \frac{dv}{dx} + \frac{du}{ds},\tag{7.45}$$

respectively with the shear flow $T_s = \tau_s(s)t(s)$ and Eq. (7.41):

$$\frac{T_s}{Gt(s)} = r_t\vartheta' + \frac{du}{ds}.\tag{7.46}$$

The term $\frac{du}{ds}$ represents the change of the longitudinal displacement u of the cross-section over the circumferential coordinate s and thus the warping of the cross-section.

We now form the integral of Eq. (7.46) over the circumference:

$$\oint \frac{T_s}{Gt(s)}ds = \oint r_t\vartheta'ds + \oint \frac{du}{ds}ds.\tag{7.47}$$

We first look at the second integral on the right side. The circumferential integral can be understood as an integral from a starting point $s = s_A$ to the end point $s = s_E$, so that:

$$\oint \frac{du}{ds}ds = \int_{s_A}^{s_E} \frac{du}{ds}ds = u(s = s_E) - u(s = s_A).\tag{7.48}$$

However, since we currently assume a closed cross-section and thus the points $s = s_E$ and $s = s_A$ coincide, the value of this integral is zero:

$$\oint \frac{du}{ds}ds = 0.\tag{7.49}$$

Then we have:

$$\frac{T_s}{G} \oint \frac{1}{t(s)} ds = \vartheta' \oint r_t ds. \tag{7.50}$$

The integral of r_t over the circumferential coordinate, which appears here on the right side, yields the value $2A_m$:

$$\frac{T_s}{G} \oint \frac{1}{t(s)} ds = 2A_m \vartheta'. \tag{7.51}$$

With the constitutive law (7.21) we obtain:

$$\frac{T_s}{G} \oint \frac{1}{t(s)} ds = 2A_m \frac{M_x}{GI_T}. \tag{7.52}$$

Furthermore, we know from (7.32) that $M_x = 2T_s A_m$ applies, so that:

$$T_s \oint \frac{1}{t(s)} ds = 4A_m^2 \frac{T_s}{I_T}, \tag{7.53}$$

or solved for the torsional moment of inertia I_T:

$$I_T = \frac{4A_m^2}{\oint \dfrac{ds}{t(s)}}. \tag{7.54}$$

This is the so-called second Bredt formula.

In many cases in lightweight engineering applications, one has to deal with cross-sections consisting of n straight segments of lengths l_i whose wall thicknesses t_i are constant. The integration prescribed in (7.54) can then be specified as:

$$I_T = \frac{4A_m^2}{\displaystyle\sum_{i=1}^{n} \int_0^{l_i} \dfrac{ds}{t_i}} = \frac{4A_m^2}{\displaystyle\sum_{i=1}^{n} \dfrac{l_i}{t_i}}. \tag{7.55}$$

For the special case of a cross-section whose wall thickness t is constant, this changes to the following form:

$$I_T = \frac{4A_m^2 t}{U}, \tag{7.56}$$

where U is the circumference of the cross-section marked by the skeleton line.

A lightweight construction specific discussion shows that the torsional stiffness GI_T of a bar with a given material can only be controlled by the torsional moment of inertia I_T. According to (7.54) and (7.56), the area A_m enclosed by the skeleton line is decisive and is to be maximized accordingly. If we are dealing with cross sections

Fig. 7.11 Thin-walled box
cross-section

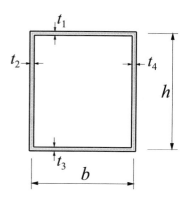

of constant thickness t, then the discussion of the ratio $\frac{A_m^2}{U}$ shows that the circular
cross section represents an optimum.

An elementary example is the cross section shown in Fig. 7.11 (width b, height h,
wall thicknesses t_1, t_2, t_3, t_4 as indicated). For this cross-section the torsional moment
of inertia I_T with (7.54) or (7.55) with $A_m = bh$ results as:

$$I_T = \frac{4b^2h^2}{\dfrac{b}{t_1} + \dfrac{h}{t_2} + \dfrac{b}{t_3} + \dfrac{h}{t_4}}. \tag{7.57}$$

If we now look at the case $t_1 = t, t_2 = 2t, t_3 = 5t, t_4 = 2t$ and $b = 100t, h = 200t$
we get:

$$I_T = \frac{4(20000t^2)^2}{\dfrac{100t}{t} + \dfrac{200t}{2t} + \dfrac{100t}{5t} + \dfrac{200t}{2t}} = 5 \cdot 10^6 t^4. \tag{7.58}$$

For the special case that all wall thicknesses have the uniform value t, (7.56) can
be used to calculate I_T, and with $U = 600t$ results in:

$$I_T = \frac{4(20000t^2)^2 t}{600t} = \frac{8}{3} \cdot 10^6 t^4. \tag{7.59}$$

7.4   Open Thin-Walled Cross Sections

Another important basic case which will be of importance in the following explanations concerns thin-walled open profiles, whereby we will first deal here with cross
sections consisting of segments of constant thickness t. Examples are thin-walled
beams with the common I, Z, C, L, or T profiles that are used in many technical
applications (Fig. 7.12).

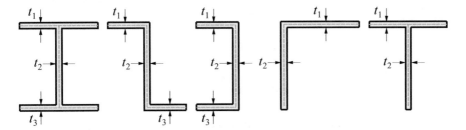

Fig. 7.12 Thin-walled open beam profiles, consisting of segments of constant thickness

Fig. 7.13 Cross-section
segment and its idealization

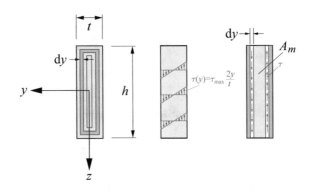

We consider Fig. 7.13, left, and look at a segment of such a thin-walled cross-section. The segment is a rectangular cross section with the height h and the thickness t, where we assume $t << h$. We can divide this thin section into a number of hollow sections, as shown in Fig. 7.13, on the left, which, when combined, result in the segment under consideration. The wall thickness of such a hollow section is dy. We assume that the thickness t is sufficiently small so that we can assume (similar to the full circle section) that the shear stress due to torsion is linearly distributed over the thickness of the section, but runs exclusively in the z–direction, with the maximum value τ_{max} at the section edge, as shown in Fig. 7.13, middle. The redistribution that would occur at the top and bottom of the segment is neglected if the cross-section is sufficiently thin, so that we assume a number of partial cross-sections, which can be calculated completely over the height h as shown in Fig. 7.13, right, and in which the horizontal sections in y–direction are completely neglected. Assuming sufficiently small values dy, we can assume that the shear stress is constant in each of these sections (Fig. 7.13, right). The area A_m enclosed by such a partial section can then be specified as $A_m = 2yh$. From the 1st Bredt formula (7.33) we obtain with the partial shear flow $dT_s = \tau(y)dy$ the following proportional torsional moment, which is carried by the considered partial cross-section:

$$dM_x = 2A_m dT_s, \tag{7.60}$$

or with

$$\tau(y) = \tau_{max} \frac{2y}{t} \tag{7.61}$$

and $A_m = 2yh$:

$$dM_x = \frac{8y^2 h \tau_{max}}{t} dy. \tag{7.62}$$

Integrated over the whole cross-section from $y = 0$ to $y = \frac{t}{2}$ leads to:

$$M_x = \frac{1}{3}\tau_{max} h t^2. \tag{7.63}$$

With the torsional resistance moment

$$W_T = \frac{1}{3} h t^2 \tag{7.64}$$

the maximum shear stress can be written as:

$$\tau_{max} = \frac{M_x}{W_T}. \tag{7.65}$$

In a similar way, an expression for the second order torsional moment of inertia can be derived for the cross-section. We use the second Bredt formula (7.54) for the partial cross-section of Fig. 7.13 and obtain for its partial moment of inertia dI_T:

$$dI_T = 8hy^2 dy, \tag{7.66}$$

which leads to I_T after integration over the entire cross-section as follows:

$$I_T = \int_0^{\frac{t}{2}} I_T dy = \frac{1}{3} h t^3. \tag{7.67}$$

The value $I_T = \frac{1}{3} h t^3$ thus describes the torsional moment of inertia for a single segment of a thin-walled open cross-section. How I_T can be determined for entire sections, such as the the profiles shown in Fig. 7.12, will be explained at a later stage. However, it can already be stated at this point that the distribution of the shear stresses due to torsion differs fundamentally for open and closed profiles (Fig. 7.14). Again, we compare the arbitrary cross-section of Fig. 7.7 (shown again in Fig. 7.14, left) with the case that this cross-section was additionally converted into an open cross-section by a longitudinal cut (Fig. 7.14, right). Contrary to what we have already determined for the closed cross-section, the shear stresses $\tau_s(s)$ are not constant in the open cross-section but rather linear over the wall thickness and reach their maximum values at the wall surfaces. Consequently, they reach their maximum in the open cross-section exactly where the wall thickness reaches its maximum (i.e.

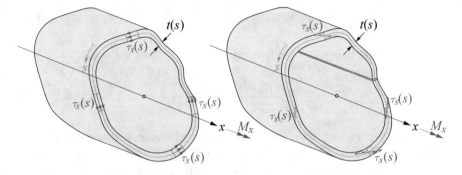

Fig. 7.14 Shear stress distribution in a closed and in a similar open cross-section

at $t = t_{max}$) and not in the area of the smallest wall thickness $t = t_{min}$ as in the case of the closed cross-section. Thus, the material utilization is not optimal for the open cross-section as opposed to the closed cross-section.

7.5   Comparison of Closed and Open Cross-Sections

In this section we want to compare the very general conclusions that we have worked out so far for closed and open cross-sections and to draw further conclusions for practical lightweight engineering applications. For this purpose, we consider the circular ring profile given in Fig. 7.15 (constant thickness t, radius R_m with respect to the skeleton line) and want to assume the profile to be closed, while for comparison purposes we also consider a similar open cross-section.

First we calculate the two torsional moments of inertia I_T. For the closed cross-section we get with (7.56):

$$I_T = \frac{4A_m^2 t}{U} = \frac{4\left(\pi R_m^2\right)^2 t}{2\pi R_m} = 2\pi R_m^3 t. \tag{7.68}$$

At the open cross-section, however, we have with (7.67):

Fig. 7.15 Discussion of a closed and an open circular ring profile

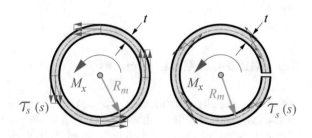

$$I_T = \frac{1}{3}ht^3 = \frac{1}{3}(2\pi R_m)\,t^3 = \frac{2}{3}\pi R_m t^3. \tag{7.69}$$

If we put these two values for I_T in relation to each other, we obtain:

$$\frac{2\pi R_m^3 t}{\frac{2}{3}\pi R_m t^3} = 3\left(\frac{R_m}{t}\right)^2. \tag{7.70}$$

The torsional moment of inertia I_T of the closed section is therefore higher by a factor of $3\left(\frac{R_m}{t}\right)^2$ than the torsional moment of inertia I_T of the open section. If we keep in mind that the radius R_m is significantly larger than the wall thickness t in the thin-walled cross sections considered here, then it can be concluded that the discrepancy between the two torsional moments of inertia is significant and can be several orders of magnitude. It can be concluded that open cross sections have an extremely low torsional stiffness compared to similar closed cross sections, which has to be considered appropriately in the design and analysis of lightweight structures.

We now want to carry out an analogous investigation for the shear stresses occurring in the closed and the open thin-walled circular ring cross section. The maximum shear stress $\tau_{\max}$ is calculated in both cases as $\tau_{\max} = \frac{M_x}{W_T}$, whereby the following torsional resistance moment W_T results for the closed cross-section:

$$W_T = 2A_m t_{\min} = 2\pi R_m^2 t. \tag{7.71}$$

At the open cross-section we get with (7.64) for W_T:

$$W_T = \frac{1}{3}ht^2 = \frac{2}{3}\pi R_m t^2. \tag{7.72}$$

The ratio of the maximum shear stresses at the closed and at the open cross-section then results as:

$$\frac{\tau_{\max,g}}{\tau_{\max,o}} = \frac{\frac{2}{3}\pi R_m t^2}{2\pi R_m^2 t} = \frac{1}{3}\frac{t}{R_m}. \tag{7.73}$$

Thus, the maximum shear stresses in the open cross-section are greater than in the closed cross-section by the factor $3\frac{R_m}{t}$.

An important conclusion to be drawn is that open sections are not suitable for carrying significant torsional loads and that closed sections should be used whenever possible.

7.6 Multi-cell Cross Sections

Another practically relevant case is when the considered cross-section is composed of a series of closed cells, as shown in Fig. 7.16 for an exemplary two-cell box section. This corresponds to a parallel connection of the two cells 1 and 2, whose torsional

Fig. 7.16 Two-cell closed box cross section

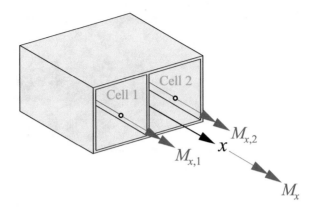

stiffnesses will add up to constitute the total torsional stiffness of the multi-cell cross-section. The equation (7.33) is also valid here for the two cells 1 and 2, so that we can calculate the torsional moment $M_{x,1}$ and the torsional moment $M_{x,2}$ from the shear flows $T_{s,1}$ and $T_{s,2}$ of the two cells as follows:

$$M_{x,1} = 2A_{m,1}T_{s,1}, \quad M_{x,2} = 2A_{m,2}T_{s,2}, \tag{7.74}$$

with the areas $A_{m,1}$ and $A_{m,2}$ enclosed by the two cells. The two proportional torsional moments can be added together to the total torsional moment M_x, whereby this also applies to n cells in general:

$$M_x = \sum_{i=1}^{n} M_{x,i} = 2\sum_{i=1}^{n} A_{m,i}T_{s,i}. \tag{7.75}$$

Here, too, we can make the connection (7.52) for both cells:

$$\oint \frac{T_{s,1}}{Gt_1(s)}\,ds = 2A_{m,1}\vartheta_1', \quad \oint \frac{T_{s,2}}{Gt_2(s)}\,ds = 2A_{m,2}\vartheta_2'. \tag{7.76}$$

The twists ϑ_1' and ϑ_2' are equal to the twist ϑ' of the total cross-section:

$$\vartheta_1' = \vartheta_2' = \vartheta'. \tag{7.77}$$

Assuming that the shear modulus G is identically and homogeneously distributed in each cell and that the shear flows are again constant over the circumference of the single cells, we obtain from (7.76)

$$\frac{T_{s,1}}{G} \oint \frac{ds}{t_1(s)} = 2A_{m,1}\vartheta', \quad \frac{T_{s,2}}{G} \oint \frac{ds}{t_2(s)} = 2A_{m,2}\vartheta'. \tag{7.78}$$

Fig. 7.17 Shear flows in a
two-cell closed hollow
cross-section

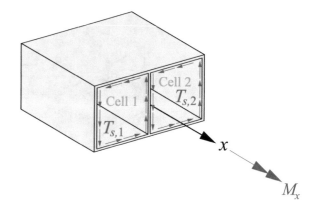

Fig. 7.17 Shear flows in a
two-cell closed hollow
cross-section

Solving for A_m yields:

$$2A_{m,1} = \frac{T_{s,1}}{G\vartheta'} \oint \frac{ds}{t_1(s)}, \quad 2A_{m,2} = \frac{T_{s,2}}{G\vartheta'} \oint \frac{ds}{t_2(s)}. \tag{7.79}$$

At this point it is appropriate to introduce the so-called torsion functions ψ_1, ψ_2 as follows

$$\psi_1 = \frac{T_{s,1}}{G\vartheta'}, \quad \psi_2 = \frac{T_{s,2}}{G\vartheta'}, \tag{7.80}$$

so that:

$$2A_{m,1} = \psi_1 \oint \frac{ds}{t_1(s)}, \quad 2A_{m,2} = \psi_2 \oint \frac{ds}{t_2(s)}. \tag{7.81}$$

Compatibility between cells 1 and 2 is achieved by considering that in those cross-sectional parts that are common to both cells, the difference of the shear flows $T_{s,1}$ and $T_{s,2}$ acts. This is shown in Fig. 7.17 and can be expressed by the following compatibility conditions:

$$\psi_1 \oint_1 \frac{ds}{t(s)} - \psi_2 \oint_{1,2} \frac{ds}{t(s)} = 2A_{m,1},$$
$$-\psi_1 \oint_{1,2} \frac{ds}{t(s)} + \psi_2 \oint_2 \frac{ds}{t(s)} = 2A_{m,2}. \tag{7.82}$$

The integral $\oint_{1,2} \frac{ds}{t(s)}$ extends over those parts of the cross section which are common to both cells 1 and 2. Thus, (7.82) is a linear system of equations from which the two torsion functions ψ_1 and ψ_2 can be determined.

The shown compatibility conditions can also be generalized for cross-sections with an arbitrary number of n closed cells:

$$
\begin{bmatrix}
\oint_1 \dfrac{ds}{t(s)} & -\oint_{1,2} \dfrac{ds}{t(s)} & -\oint_{1,3} \dfrac{ds}{t(s)} & \cdots & -\oint_{1,n} \dfrac{ds}{t(s)} \\[2mm]
-\oint_{2,1} \dfrac{ds}{t(s)} & \oint_2 \dfrac{ds}{t(s)} & -\oint_{2,3} \dfrac{ds}{t(s)} & \cdots & -\oint_{2,n} \dfrac{ds}{t(s)} \\[2mm]
-\oint_{3,1} \dfrac{ds}{t(s)} & -\oint_{3,2} \dfrac{ds}{t(s)} & \oint_3 \dfrac{ds}{t(s)} & \cdots & -\oint_{3,n} \dfrac{ds}{t(s)} \\[2mm]
\vdots & \vdots & \vdots & \ddots & \vdots \\[2mm]
-\oint_{n,1} \dfrac{ds}{t(s)} & -\oint_{n,2} \dfrac{ds}{t(s)} & -\oint_{n,3} \dfrac{ds}{t(s)} & \cdots & \oint_n \dfrac{ds}{t(s)}
\end{bmatrix}
\begin{pmatrix} \psi_1 \\ \psi_2 \\ \psi_3 \\ \vdots \\ \psi_n \end{pmatrix}
=
\begin{pmatrix} 2A_{m,1} \\ 2A_{m,2} \\ 2A_{m,3} \\ \vdots \\ 2A_{m,n} \end{pmatrix},
$$

$$(7.83)$$

or in symbolic form:

$$\underline{\underline{A}}\underline{\psi} = -\underline{a}. \tag{7.84}$$

The coefficient matrix $\underline{\underline{A}}$ is always symmetrical. The similarity to the system of equations (6.56) for the determination of statically indeterminate shear flows due to transverse shear forces at multicellular beam cross-sections is obvious.

With the constitutive law (7.21) we are able to determine an expression for the torsional moment of inertia I_T. From

$$\vartheta' = \frac{M_x}{GI_T} \tag{7.85}$$

it follows with (7.75) for a two-cell cross-section:

$$I_T = \frac{M_x}{G\vartheta'} = \frac{1}{G\vartheta'} \left(2A_{m,1}T_{s,1} + 2A_{m,2}T_{s,2} \right) = 2A_{m,1}\psi_1 + 2A_{m,2}\psi_2. \tag{7.86}$$

In the case of a cross-section with n cells we get

$$I_T = 2 \sum_{i=1}^{n} A_{m,i}\psi_i. \tag{7.87}$$

We now use the constitutive law (7.21) as well as (7.80) to determine the shear flows in the individual cells. We obtain for cells 1 and 2 of a two-cell cross section:

$$\vartheta' = \frac{M_x}{GI_T} = \frac{T_{s,1}}{G\psi_1} = \frac{T_{s,2}}{G\psi_2}, \tag{7.88}$$

from which we can calculate the shear flows of cells 1 and 2:

$$T_{s,1} = \frac{M_x}{I_T}\psi_1, \quad T_{s,2} = \frac{M_x}{I_T}\psi_2, \tag{7.89}$$

or in a more general form for a n−cell cross-section for cell i:

$$T_{s,i} = \frac{M_x}{I_T} \psi_i. \tag{7.90}$$

The derived relationships are illustrated below with an example. We again consider the cross section of Fig. 6.16 (see Chap. 6) which is shown here again for better clarity. We want to determine the torsional moment of inertia I_T and the shear flow due to torsion in the cells of the three-cell cross-section (Fig. 7.18).

The equation system to be solved here for the torsion functions ψ_1, ψ_2, ψ_3 currently reads:

$$\begin{bmatrix} \oint_1 \dfrac{ds}{t(s)} & -\oint_{1,2} \dfrac{ds}{t(s)} & -\oint_{1,3} \dfrac{ds}{t(s)} \\[2ex] -\oint_{2,1} \dfrac{ds}{t(s)} & \oint_2 \dfrac{ds}{t(s)} & -\oint_{2,3} \dfrac{ds}{t(s)} \\[2ex] -\oint_{3,1} \dfrac{ds}{t(s)} & -\oint_{3,2} \dfrac{ds}{t(s)} & \oint_3 \dfrac{ds}{t(s)} \end{bmatrix} \begin{pmatrix} \psi_1 \\ \psi_2 \\ \psi_3 \end{pmatrix} = \begin{pmatrix} 2A_{m,1} \\ 2A_{m,2} \\ 2A_{m,3} \end{pmatrix}. \tag{7.91}$$

The required areas of the cells, measured along the skeleton line, result here as

$$A_{m,1} = 300000t^2, \quad A_{m,2} = 600000t^2, \quad A_{m,3} = 240000t^2. \tag{7.92}$$

The individual integral terms required in (7.91) are

$$\oint_1 \frac{ds}{t(s)} = \frac{600t}{t} + \frac{500t}{0.8t} + \frac{600t}{1.5t} + \frac{500t}{0.8t} = 2250,$$

$$\oint_2 \frac{ds}{t(s)} = \frac{600t}{1.5t} + \frac{1000t}{t} + \frac{600t}{1.2t} + \frac{1000t}{t} = 2900,$$

$$\oint_3 \frac{ds}{t(s)} = \frac{600t}{1.2t} + \frac{400t}{t} + \frac{600t}{0.8t} + \frac{400t}{t} = 2050,$$

$$\oint_{1,2} \frac{ds}{t(s)} = \oint_{2,1} \frac{ds}{t(s)} = \frac{600t}{1.5t} = 400,$$

$$\oint_{1,3} \frac{ds}{t(s)} = \oint_{3,1} \frac{ds}{t(s)} = 0,$$

$$\oint_{2,3} \frac{ds}{t(s)} = \oint_{3,2} \frac{ds}{t(s)} = \frac{600t}{1.2t} = 500. \tag{7.93}$$

Then we have for (7.91):

$$\begin{bmatrix} 2250 & -400 & 0 \\ -400 & 2900 & -500 \\ 0 & -500 & 2050 \end{bmatrix} \begin{pmatrix} \psi_1 \\ \psi_2 \\ \psi_3 \end{pmatrix} = \begin{pmatrix} 600000t^2 \\ 1200000t^2 \\ 480000t^2 \end{pmatrix}. \tag{7.94}$$

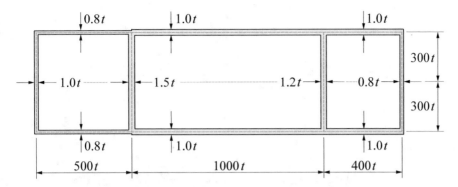

Fig. 7.18 Three-cell cross-section

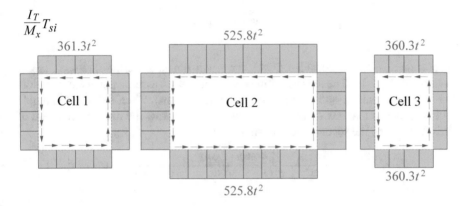

Fig. 7.19 Shear flows due to torsion in the three cells

This yields the torsion functions ψ_1, ψ_2, ψ_3 as:

$$\psi_1 = 361.3t^2, \quad \psi_2 = 525.8t^2, \quad \psi_3 = 360.3t^2. \tag{7.95}$$

The torsional moment of inertia can then be calculated using (7.87) as:

$$I_T = 2\sum_{i=1}^{3} A_{m,i}\psi_i = 1.02 \cdot 10^9 t^4. \tag{7.96}$$

The shear flows in the cells then follow from (7.90) as shown in Fig. 7.19. It is important to note that the differences of the shear flows act on the common walls of cells 1 and 2 as well as 2 and 3, as shown in Fig. 7.20.

The torsional moment of inertia I_t can also be narrowed down concerning its upper and lower limit without performing a complex calculation. For this purpose,

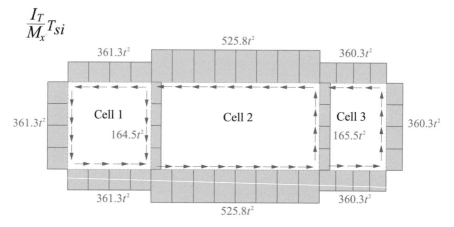

Fig. 7.20 Shear flow due to torsion in the three-cell cross-section

two scenarios can be considered which we will illustrate on the three-cell cross-section of Fig. 7.18. The lower limiting case occurs if we assume that the two middle bars between cells 1 and 2 and 2 and 3 are not present. Then the given example changes to a simple box section for which we can estimate the torsional moment of inertia I_T using the second Bredt formula (7.54):

$$I_T > \frac{4A_m^2}{\oint \dfrac{ds}{t(s)}} = \frac{4\,(600t \cdot 1900t)^2}{\dfrac{600t}{t} + 2 \cdot \dfrac{500t}{0.8t} + 2 \cdot \dfrac{1000t}{t} + 2 \cdot \dfrac{400t}{t} + \dfrac{600t}{0.8t}} = 9.63 \cdot 10^8 t^4. \quad (7.97)$$

An upper limit for I_T can be estimated by assuming that the middle bars between cells 1 and 2 and 2 and 3 have infinite wall thicknesses. Using the system of equations (7.91), we find that the secondary diagonal terms all assume zero values, so that:

$$\begin{bmatrix} \oint_1 \dfrac{ds}{t(s)} & 0 & 0 \\[2mm] 0 & \oint_2 \dfrac{ds}{t(s)} & 0 \\[2mm] 0 & 0 & \oint_3 \dfrac{ds}{t(s)} \end{bmatrix} \begin{pmatrix} \psi_1 \\ \psi_2 \\ \psi_3 \end{pmatrix} = \begin{pmatrix} 2A_{m,1} \\ 2A_{m,2} \\ 2A_{m,3} \end{pmatrix}. \quad (7.98)$$

Thus, the system of equations breaks down into three independent equations so that the torsion functions ψ_1, ψ_2, ψ_3 can be directly determined as ($i = 1, 2, 3$):

$$\psi_i = \frac{2A_{m,i}}{\oint_i \dfrac{ds}{t}}. \quad (7.99)$$

When calculating the ring integrals, however, we have to keep in mind that those fractions from the intermediate bars, which are assumed to be infinitely thick, result in zero values. We can estimate the upper limit for I_T by means of (7.87), so that:

$$I_T < 2 \sum_{i=1}^{3} A_{m,i} \psi_i = 4 \sum_{i=1}^{3} \frac{A_{m,i}^2}{\oint_i \frac{ds}{t}}$$

$$= 4 \left[\frac{\left(300000t^2\right)^2}{\dfrac{600t}{t} + 2 \cdot \dfrac{500t}{0.8t}} + \frac{\left(600000t^2\right)^2}{2 \cdot \dfrac{1000t}{t}} + \frac{\left(240000t^2\right)^2}{\dfrac{600t}{0.8t} + 2 \cdot \dfrac{400t}{t}} \right] = 1.06 \cdot 10^9 t^4.$$

$$(7.100)$$

A comparison with the exact result of $I_T = 1.02 \cdot 10^9 t^4$ shows that the estimation of the upper and lower limits gives very plausible results, the exact value lies approximately in the middle of the two calculated limit values.

7.7 Assembled Cross-Sections

As a last elementary case we want to consider thin-walled cross sections which are composed of a number of different segments (see also Chap. 6, Fig. 6.22), where both open and closed parts are possible. Since we assume that all cross-section parts are exposed to the same rotation ϑ and twist ϑ', we can write the following for segment i:

$$\vartheta_i' = \frac{M_{x,i}}{GI_{T,i}} = \vartheta'. \tag{7.101}$$

This also applies analogously to the complete cross-section:

$$\vartheta' = \frac{M_x}{GI_T}, \tag{7.102}$$

respectively

$$M_x = GI_T \vartheta'. \tag{7.103}$$

The total moment M_x again consists of the partial moments of the individual segments:

$$M_x = \sum_{i=1}^{n} M_{x,i} = \sum_{i=1}^{n} \vartheta' GI_{T,i} = \vartheta' \sum_{i=1}^{n} GI_{T,i}. \tag{7.104}$$

Fig. 7.21 Shear stresses due
to torsion in the I-beam

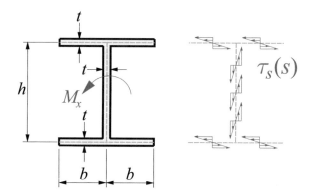

Equating (7.103) and (7.104) yields:

$$GI_T\vartheta' = \vartheta' \sum_{i=1}^{n} GI_{T,i},\tag{7.105}$$

or

$$GI_T = \sum_{i=1}^{n} GI_{T,i}.\tag{7.106}$$

Apparently, the torsional stiffness of a cross-section composed of several segments can be determined by adding up the contributions of the individual parts.

For a cross-section like the one shown in Fig. 7.21, left, the torsional moment of inertia I_T can then be calculated simply from the sum of the contributions of the individual segments:

$$I_T = \sum_{i=1}^{n} I_{T,i} = \frac{1}{3}\sum_{i=1}^{n} l_i t_i^3,\tag{7.107}$$

if l_i is the length of segment i and t_i is its wall thickness. For the given example we have:

$$I_T = \frac{1}{3}\left(4b + h\right)t^3.\tag{7.108}$$

The shear stresses $\tau_s(s)$ occurring at this profile are shown in Fig. 7.21, right.

7.8 Effective Wall Thicknesses

7.8.1 Truss Girders

In addition to full-wall box girders, there are also constructions in which a beam contains both full-walled components and truss structures. It can be shown that such wall sections composed of truss components are capable of transferring shear stresses resulting from torsion. An exemplary structural situation is shown in Fig. 7.22. A feasible way to analyze such structures with reasonable effort and with the means of analysis provided so far is to treat the truss wall mathematically as a solid wall to which a so-called effective or ideal wall thickness t_{eff} is assigned. It is therefore our main task in this section to determine this effective wall thickness t_{eff}.

Figure 7.23 shows the idealization applied in this regard. The cell width of the truss is a, the height of the considered truss wall is b. The inclination angle α of the diagonal bars (length d) is given with the uniform value α. Due to the circumferential shear flow T_s the truss wall is also stressed by this shear flow. We assume that the shear flow T_s runs linearly over the upper and lower chord and is introduced into the diagonal members at the nodes. This results in a linear distribution of the normal force between the two maximum values $\pm T_s a$ in the horizontal bars, as shown in Fig. 7.23, top. In addition the vertical resulting shear force $T_s b$ also acts in the truss plane. It is applied to the diagonal members with the amount $D = \frac{T_s b}{\sin \alpha}$. The vertical bars remain unloaded in this scenario. With $\sin \alpha = \frac{b}{d}$, the bar force in the diagonals results as $D = T_s d$.

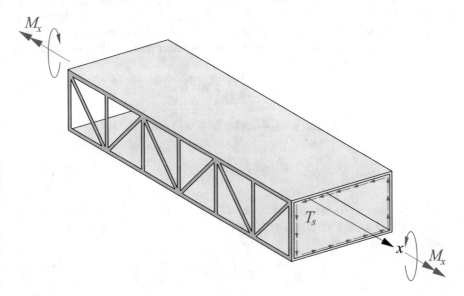

Fig. 7.22 Box girder under torsion with truss structure

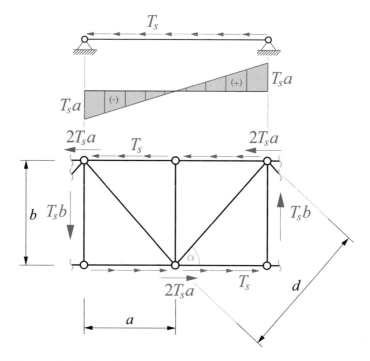

Fig. 7.23 Truss wall of the box beam

The determination of the effective wall thickness t_{eff} can now be carried out by comparing the internal work / strain energies W_i of the truss wall and a solid wall of thickness t_{eff}. It is assumed here that for the latter case the stress and strain quantities involved are exclusively the shear stress τ and the shear strain γ. This results in the following for the solid wall:

$$W_i = \frac{1}{2} \int_V \tau\gamma dV. \tag{7.109}$$

With $dV = t_{eff}dxdy$ and $\tau = \frac{T_s}{t_{eff}}$ and $\gamma = \frac{\tau}{G}$ we have:

$$W_i = \frac{T_s^2 ab}{2Gt_{eff}}. \tag{7.110}$$

For the truss wall, the contributions of the individual bars to the strain energy must be added up. For a bar of length l with the extensional stiffness EA the general result is

$$W_i = \frac{1}{2} \int_0^l \frac{N^2}{EA} dx. \tag{7.111}$$

If the normal force N is constant over the bar length l, then:

$$W_i = \frac{N^2 l}{2EA}. \tag{7.112}$$

If, on the other hand, a member with a linearly variable normal force N is considered, as is the case here for the upper and lower chord, then we obtain for a constantly distributed shear flow T_s and a member length of $2a$:

$$W_i = \frac{T_s^2 a^3}{6EA}. \tag{7.113}$$

Applied to the given situation of Fig. 7.23 the following strain energy results:

$$W_i = \frac{T_s^2}{2} \left[\frac{d^3}{E_d A_d} + \frac{a^3}{3} \left(\frac{1}{E_o A_o} + \frac{1}{E_u A_u} \right) \right]. \tag{7.114}$$

Herein, E_d, E_o, E_u are the moduli of elasticity of the diagonal and the upper and lower chord, A_d, A_o, A_u are the assigned cross-sectional areas. In the case that $E_d = E_o = E_u = E$ applies, we obtain:

$$W_i = \frac{T_s^2}{2E} \left[\frac{d^3}{A_d} + \frac{a^3}{3} \left(\frac{1}{A_o} + \frac{1}{A_u} \right) \right]. \tag{7.115}$$

Equating (7.110) and (7.115) results in the following expression for the effective wall thickness t_{eff} of the truss wall considered here:

$$t_{eff} = \frac{E}{G} \frac{ab}{\dfrac{d^3}{A_d} + \dfrac{a^3}{3} \left(\dfrac{1}{A_o} + \dfrac{1}{A_u} \right)}. \tag{7.116}$$

Figure 7.24 contains some further exemplary truss structures (see also Kollbrunner and Basler (1966) or Petersen (1997)).

For the truss wall of Fig. 7.24, top, the effective wall thickness is calculated as:

$$t_{eff} = \frac{E}{G} \frac{ab}{\dfrac{2d^3}{A_d} + \dfrac{b^3}{4A_v} + \dfrac{a^3}{12} \left(\dfrac{1}{A_o} + \dfrac{1}{A_u} \right)}. \tag{7.117}$$

The effective wall thickness of the truss given in Fig. 7.24, center, can be determined as:

$$t_{eff} = \frac{E}{G} \frac{ab}{\dfrac{d^3}{2A_d} + \dfrac{a^3}{12} \left(\dfrac{1}{A_o} + \dfrac{1}{A_u} \right)}. \tag{7.118}$$

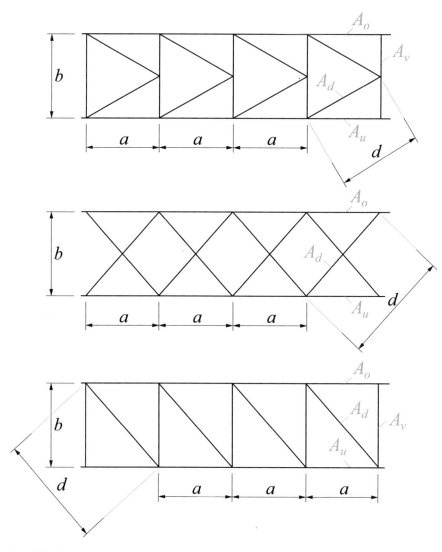

Fig. 7.24 Exemplary truss structures

For the truss structure shown in Fig. 7.24, bottom, we obtain:

$$t_{id} = \frac{E}{G} \frac{ab}{\dfrac{d^3}{A_d} + \dfrac{b^3}{A_v} + \dfrac{a^3}{12}\left(\dfrac{1}{A_o} + \dfrac{1}{A_u}\right)}. \tag{7.119}$$

7.8.2 Stiffened Box Beams

In many technical applications, thin-walled box beams are applied which are reinforced by stiffeners at regular intervals. The arrangement of stiffeners can serve many reasons. On the one hand, stiffeners are applied to introduce concentrated loads. On the other hand, stiffeners contribute significantly to the stability resistance of thin-walled lightweight structures. An exemplary stiffened box beam is shown in Fig. 7.25. If the stiffeners are longitudinal stiffeners with closed cross-sections, the determination of the torsional stiffness and the shear stresses would have to be performed according to the rules for multi-cell closed cross-sections. However, this procedure becomes very complex if a large number of stiffeners are present. Furthermore, it is shown that the contribution of the stiffeners to the torsional stiffness and to the load transfer of the torsional shear stresses is small if the dimensions of the stiffener cross-sections are significantly smaller than the dimensions of the component that is stiffened by them. It is therefore advisable to work with an effective wall thickness t_{eff} also in case of such stiffened cross-sections which is the content of the present section.

We consider the situation of a representative unit cell of a closed-profile stiffener as shown in Fig. 7.26. The wall thickness of the box beam is t_0 and the stiffener has a wall thickness of t_1. The stiffener is formed by two webs (length b_2) and a flange (length b_1). The area of the wall enclosed by the stiffener has the length b_0 and the two unstiffened areas of the wall have the length $\bar{b}$. We introduce the local axis s_0 for the box wall area within the stiffener cross section. Analogous to this, we assign the local circumferential axis s_1 to the stiffener cross-section, where $0 \leq s_1 \leq l_1$ ($l_1 = b_1 + 2b_2$).

The circumferential shear flow T_s will be divided into stiffener and wall with the components T_{s0} and T_{s1}, where T_{s0} and T_{s1} are a priori unknown. However, from the equilibrium of forces at the junction between the wall and the stiffener we immediately get the following condition:

$$T_s = T_{s0} + T_{s1}. \tag{7.120}$$

Fig. 7.25 Box beam stiffened at its upper and lower walls by longitudinal stiffeners

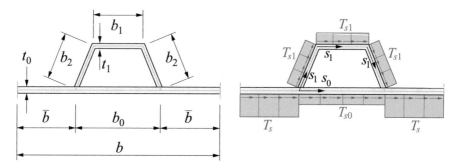

Fig. 7.26 Representative unit cell of the closed-profile stiffener

Fig. 7.27 Displacement
compatibility

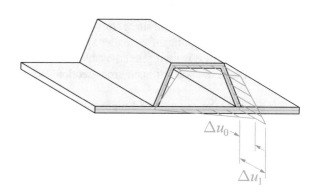

We can set up a compatibility condition for the given situation in such a way that
we require that the displacements u are identical due to the shear strain of both the
stiffener section at position $s_1 = l_1$ and the wall within the stiffener section at $s_0 = b_0$
(Fig. 7.27).

With

$$\Delta u = \int_s \gamma \, \mathrm{d}s = \frac{T_s}{G} \int_s \frac{\mathrm{d}s}{t} \tag{7.121}$$

we obtain:

$$\Delta u_0 = \frac{T_{s0} b_0}{G t_0}, \quad \Delta u_1 = \frac{T_{s1} l_1}{G t_1}. \tag{7.122}$$

From the requirement $\Delta u_0 = \Delta u_1$:

$$\frac{T_{s0} b_0}{G t_0} - \frac{T_{s1} l_1}{G t_1} = 0. \tag{7.123}$$

Together with (7.120) there are two equations from which the two partial shear
flows T_{s0} and T_{s1} can be determined:

$$T_{s0} = \frac{T_s}{1 + \dfrac{t_1 b_0}{t_0 l_1}}, \quad T_{s1} = \frac{T_s}{1 + \dfrac{t_0 l_1}{t_1 b_0}}. \tag{7.124}$$

The effective wall thickness t_{eff} can then be determined from the requirement that both the displacement Δu of the stiffened wall and the displacement of the wall with the effective wall thickness t_{eff} are identical:

$$\frac{2 T_s \bar{b}}{G t_0} + \frac{T_{s0} b_0}{G t_0} = \frac{T_s b}{G t_{eff}}. \tag{7.125}$$

Solving for t_{eff} yields:

$$t_{eff} = \frac{\left(b_0 + 2\bar{b}\right) t_0}{2\bar{b} + \dfrac{b_0}{1 + \dfrac{t_1 b_0}{t_0 l_1}}}. \tag{7.126}$$

7.9 Determination of Internal Forces

The constitutive law of St. Venant torsion is given with (7.22):

$$M_x = G I_T \vartheta'. \tag{7.127}$$

Using the free body diagram of Fig. 7.28, which shows an infinitesimal sectional element from a beam under torsion, we can determine the equilibrium between the torsional moment M_x and the distributed external torsional moment m_T. With $dM_x = \frac{dM_x}{dx} dx$ we obtain:

$$\frac{dM_x}{dx} = M_x' = -m_T. \tag{7.128}$$

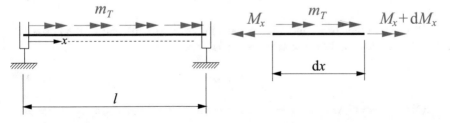

Fig. 7.28 Free body diagram of an infinitesimal sectional element of a beam under torsion

Fig. 7.29 Typical boundary conditions for a beam under torsion at the position $x = 0$: Fork restraint (left), free end (middle), free end with applied torsional moment (right)

The first derivative of the torsional moment M_x thus results in the negative distributed torsional load m_T. With the constitutive law (7.127) this yields:

$$(GI_T \vartheta')' = -m_T, \tag{7.129}$$

or for prismatic bars with a constant torsional stiffness GI_T:

$$GI_T \vartheta'' = -m_T. \tag{7.130}$$

The twist ϑ' or the torsional moment M_x as well as the cross-sectional rotation ϑ can be determined by single respectively twofold integration of the applied load m_T:

$$(GI_T \vartheta')' = -m_T,$$
$$GI_T \vartheta' = M_x = -\int m_T \, dx + C_1,$$
$$GI_T \vartheta = -\int \int m_T \, dx \, dx + C_1 x + C_2. \tag{7.131}$$

Thus, all relations are available to determine the torsional moment M_x as well as the twist ϑ' and the rotation ϑ for a given torsional load m_T. The constants C_1 and C_2 are determined from the given boundary conditions, a selection of which we will briefly discuss below (Fig. 7.29).

If there is a fork restraint at position $x = 0$ (Fig. 7.29, left), then the cross section at this position is prevented from rotating about the longitudinal axis. Thus:

$$\vartheta(x = 0) = 0. \tag{7.132}$$

If, on the other hand, there is no restraint at all and the end of the bar is free (Fig. 7.29, middle), the torsional moment M_x has to disappear:

$$M_x(x = 0) = 0. \tag{7.133}$$

Considering (7.127) this means that the twist ϑ' must also disappear at $x = 0$:

$$\vartheta'(x = 0) = 0. \tag{7.134}$$

Fig. 7.30 Beam under
uniform torsional load m_T
(top), resulting state
variables $\vartheta(x)$ and $M_x(x)$
(bottom)

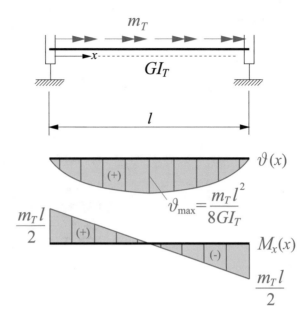

If there is a free bar end at which a torsional moment $M_{x,0}$ is introduced, then the
moment M_x assumes the value $M_{x,0}$ at this point:

$$M_x(x = 0) = M_{x,0}. \tag{7.135}$$

We discuss the general procedure using an elementary example. Let us consider
the beam shown in Fig. 7.30 under the constantly distributed torsional load m_T. The
beam also has a constant torsional stiffness GI_T. We use the solution (7.131) to
determine the course of the torsional moment M_x and the rotation ϑ of the cross-
section. The boundary conditions to be considered here are as follows:

$$\vartheta(x = 0) = 0, \quad \vartheta(x = l) = 0. \tag{7.136}$$

From the first condition in (7.136) we get with (7.131):

$$C_2 = 0. \tag{7.137}$$

After elementary transformations the second condition in (7.136) results in:

$$C_1 = \frac{1}{2}m_T l. \tag{7.138}$$

Fig. 7.31 Beam under
torsion with fork restraints
on both sides under the
single moment $M_{T,0}$ (top),
distribution of the torsional
moment $M_x(x)$ (bottom)

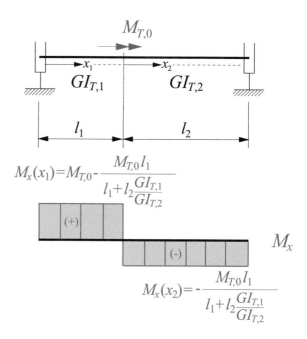

Thus the rotation of the cross section can be represented as follows:

$$\vartheta(x) = \frac{m_T x}{2GI_T}(l - x). \tag{7.139}$$

We can describe the torsional moment M_x as follows:

$$M_x(x) = m_T \left(\frac{l}{2} - x\right). \tag{7.140}$$

A graphical representation of the state variables is shown in Fig. 7.30. Thus, for a constant load m_T the torsional moment M_x is linear and the angle ϑ is quadratic over x.

Furthermore, we consider the situation shown in Fig. 7.31 where a beam under the single moment $M_{T,0}$ is given. The beam has a constant torsional stiffness $GI_{T,1}$ or $GI_{T,2}$ in each section, wherein the two beam sections have the lengths l_1 and l_2. We want to determine the torsional moment M_x.

We want to approach the task by integrating the constitutive law of St. Venant torsion by means of the solution given with (7.131). Since the moment distribution M_x will be discontinuous due to the point of action of the torsional moment $M_{T,0}$, we introduce the two longitudinal axes x_1 and x_2 as given in Fig. 7.31. We then obtain from (7.131) for the section $0 \le x_1 \le l_1$:

$$GI_{T,1}\vartheta_1'' = -m_{T,1},$$
$$GI_{T,1}\vartheta_1' = M_x = -m_{T,1}x_1 + C_1,$$
$$GI_{T,1}\vartheta_1 = -\frac{1}{2}m_{T,1}x_1^2 + C_1x_1 + C_2. \tag{7.141}$$

For the section with $0 \leq x_2 \leq l_2$ we have:

$$GI_{T,2}\vartheta_2'' = -m_{T,2},$$
$$GI_{T,2}\vartheta_2' = M_x = -m_{T,2}x_2 + C_3,$$
$$GI_{T,2}\vartheta_2 = -\frac{1}{2}m_{T,2}x_2^2 + C_3x_2 + C_4. \tag{7.142}$$

In the Eqs. (7.141) and (7.142) the terms $m_{T,1}$ and $m_{T,2}$ are both zero.

We now use the boundary and transition conditions of the given system to determine the constants $C_1, ..., C_4$. These are:

$$\vartheta_1(x_1 = 0) = 0,$$
$$\vartheta_1(x_1 = l_1) = \vartheta_2(x_2 = 0),$$
$$M_{x,1}(x_1 = l_1) = M_{x,2}(x_2 = 0) + M_{T,0},$$
$$\vartheta_2(x_2 = l_2) = 0. \tag{7.143}$$

From the resulting system of equations we obtain the constants $C_1, ..., C_4$ after a short calculation as follows:

$$C_1 = M_{T,0}\left(1 - \frac{l_1}{l_1 + l_2\dfrac{GI_{T,1}}{GI_{T,2}}}\right), \quad C_2 = 0,$$

$$C_3 = -\frac{M_{T,0}l_1}{l_1 + l_2\dfrac{GI_{T,1}}{GI_{T,2}}}, \quad C_4 = \frac{M_{T,0}l_1l_2}{l_1 + l_2\dfrac{GI_{T,1}}{GI_{T,2}}}. \tag{7.144}$$

From this the moment distribution M_x can be determined:

$$M_{x,1} = M_{T,0}\left(1 - \frac{l_1}{l_1 + l_2\dfrac{GI_{T,1}}{GI_{T,2}}}\right), \quad M_{x,2} = -\frac{M_{T,0}l_1}{l_1 + l_2\dfrac{GI_{T,1}}{GI_{T,2}}}. \tag{7.145}$$

The distribution of M_x is also shown in Fig. 7.31.

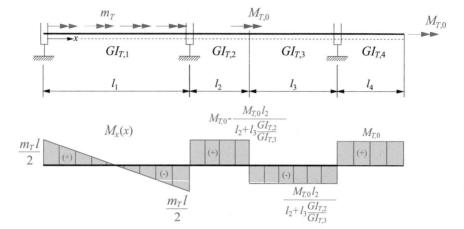

Fig. 7.32 Continuous beam supported by fork bearings (top), moment line M_x (bottom)

A remark has to be made about the determination of the state variables within the framework of St. Venant's torsion theory on continuous beams supported by fork bearings (see Fig. 7.32). The analysis rule (7.131) then results in the fact that both the distribution of the rotation ϑ and the torsional moment M_x end at each fork bearing and thus these state variables can be determined separately in each section of the beam. Warping stresses resulting from cross-sectional warping in case of cross-sections that are not free of warping as well as the resulting so-called warping moments, which are not considered in the torsion theory presented here and which typically extend beyond fork supports, cannot be treated within the theoretical framework outlined in this chapter. This then requires the treatment of the so-called warping torsion to which the following Chap. 8 is dedicated.

Bibliography

Bauchau OA, Craig JI (2009) Structural analysis: with applications to aerospace structures. Springer. Dordrecht et al, The Netherlands

Czerwenka G, Schnell W (1970) Einführung in die Rechenmethoden des Leichtbaus, Band 1. Bibliographisches Institut, Mannheim et al., Germany

Franke W, Friemann H (2005) Schub und Torsion in geraden Stäben, 3rd edn. Vieweg + Teubner, Wiesbaden, Germany

Hanswille G (1995) Vorlesungen über Stahl- und Verbundbau. Bergische Universität Wuppertal, Germany

Kindmann R, Frickel J (2017) Elastische und plastische Querschnittstragfähigkeit, Online edition 2017, https://kindmann.de. Accessed March 2020

Kollbrunner CF, Hajdin N (1963) Die St.-Venantsche Torsion, Mitteilungen der Technischen Kommission der Schweizer Stahlbau-Vereinigung, Heft 26, Verlag der Schweizer Stahlbau-Vereinigung, Zürich, Switzerland

Kollbrunner CF, Basler K (1963) Torsionsmomente und Stabverdrehung bei St.-Venantscher Torsion, Mitteilungen der Technischen Kommission der Schweizer Stahlbau-Vereinigung, Heft 27, Verlag der Schweizer Stahlbau-Vereinigung, Zürich, Switzerland

Kollbrunner CF, Basler K (1966) Torsion. Springer. Berlin et al, Germany

Linke M, Nast E (2015) Festigkeitslehre für den Leichtbau, Springer. Berlin et al, Germany

Megson THG (1999) Aircraft structures for engineering students, 3rd edn. Arnold, London, UK

Petersen C (1997) Stahlbau, Third edition, Vieweg. Braunschweig et al, Germany

Roik KH, Carl J, Lindner J (1972) Biegetorsionsprobleme gerader dünnwandiger Stäbe, Ernst & Sohn. Berlin et al, Germany

Roik KH (1978) Vorlesungen über Stahlbau, Ernst & Sohn. Berlin et al, Germany

Shama M (2010) Torsion and shear stresses in ships. Springer. Heidelberg et al, Germany

Wiedemann J (2007) Leichtbau 1: Elemente. Springer. Berlin et al, Germany

Wlassow WS (1964) Dünnwandige elastische Stäbe, vol I. VEB Verlag für Bauwesen, Berlin, Germany

Wlassow WS (1965) Dünnwandige elastische Stäbe, vol II. VEB Verlag für Bauwesen, Berlin, Germany

Chapter 8
Warping Torsion

8.1 Introduction

In the previous Chap. 7 we have dealt with St. Venant's theory of torsion of thin-walled cross-sections. In particular, we have assumed warping-free cross-sections and assumed that the cross-sections would remain flat under torsional load, respectively we have assumed cross-sections in which any warping would not be impeded. In this chapter, we want to find out when the limitations of St. Venant's torsion theory are permissible in application and in which cases we have to deal with the torsion problem of thin-walled beams by means of a more advanced theory. The generic term for this is the so-called warping torsion or the so-called first-order bending-torsion-theory. It can be shown that for many examples the assumptions of the St. Venant torsion are not fulfilled, practically relevant cross-sections are generally not free of warping so that due to the load and/or boundary conditions warping is restrained and corresponding additional stress states can be caused. This requires appropriate consideration in engineering practice.

To motivate warping torsion, we consider the cantilever beam shown in Fig. 8.1 which is loaded at the free end by the torsional moment M_T. The considered I-section has the flange width b, the web height h and the constant wall thickness t. Due to the applied torsional moment M_T, the inner moment M_x is constant over the entire length of the beam with $M_x = M_T$. In Fig. 8.2 the torsional moment M_x is divided into statically equivalent forces with $M_x = Fh$ which results in the rotation ϑ of the cross-section about its axis of rotation (not yet specified in detail here, but in this elementary case located in the center of gravity) which we denote as D. The force F in the upper and lower flange of the beam causes transverse shear forces and moments and thus also a bending deflection v of the flanges in $y-$direction and thus also a longitudinal displacement u as shown in Fig. 8.3. Since the beam is clamped at its left end, the moments in the flanges also produce bending stresses which we refer to as warping stresses $\sigma_{\omega\omega}$. The resulting warping of the cross-section is shown in Fig. 8.3. The indexing of $\sigma_{\omega\omega}$ indicates the cause, namely the warping of the cross-section which we will introduce in detail later and which will be denoted as ω. However, the

© The Author(s), under exclusive license to Springer Nature Switzerland AG 2021
C. Mittelstedt, *Structural Mechanics in Lightweight Engineering*,
https://doi.org/10.1007/978-3-030-75193-7_8

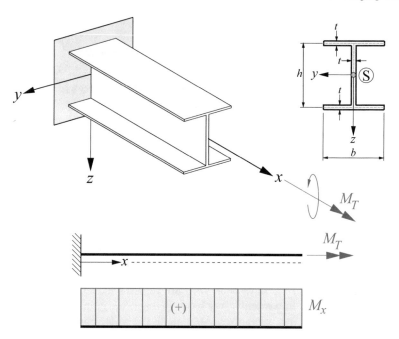

Fig. 8.1 Clamped I-beam under torsional moment M_T.

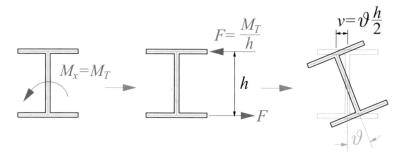

Fig. 8.2 Transformation of the torsional moment M_x into a force pair F, resulting rotation and displacements

development of warping stresses as a result of restrained warping does not only occur at clamped ends as considered here, but can also be the result of load introductions or strong changes in the cross-sectional properties.

In addition to the newly introduced warping stresses our simple example furthermore shows shear stresses due to warping torsion which are shown in Fig. 8.4. In order to be able to clearly distinguish between St. Venant torsion and warping torsion, in the further course of this chapter we will refer to the corresponding torsional moments M_x as M_{xp} (for primary torsion, i.e. St. Venant torsion) and M_{xs} (for secondary torsion, i.e. warping torsion). As already shown in Chap. 7 the torsional

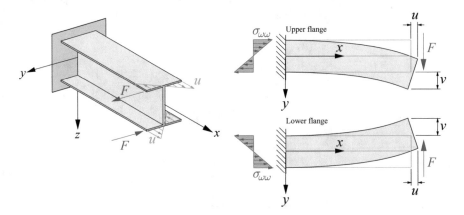

Fig. 8.3 Warping of the cross-section and warping stresses $\sigma_{\omega\omega}$

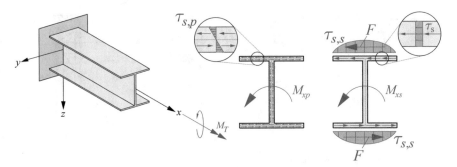

Fig. 8.4 Shear stresses due to primary torsion (M_{xp}) and secondary torsion (M_{xs})

moment M_{xp} results in primary shear stresses which are linearly distributed over the wall thickness of the section. In addition to these primary shear stresses, however, there are also the secondary shear stresses $\tau_{s,s}(s)$ due to the shear forces F acting in the flanges, which are constant over the wall thickness of the flange and show a parabolic distribution over the flange width. It can thus be concluded that the applied torsional moment M_T is transferred by St. Venant torsion and by warping torsion. The total torsional moment M_x thus results from the sum of the two components M_{xp} and $M_{xs} = Fh$ from primary and secondary torsion:

$$M_x = M_{xp} + M_{xs}. \tag{8.1}$$

The secondary torsional moment M_{xs} is thus the resultant of the shear stresses $\tau_{s,s}$ with lever arm on the rotation axis D. The warping stresses resulting from the flange bending moment M_{Fl} are shown in Fig. 8.5. In the context of warping torsion we also introduce a further sectional quantity, namely the so-called warping moment M_ω which we will discuss in detail later on.

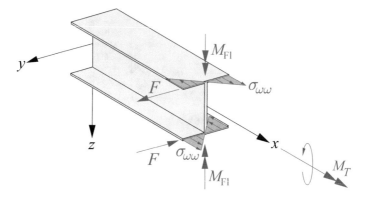

Fig. 8.5 Warping stresses due to secondary torsion M_{xs}

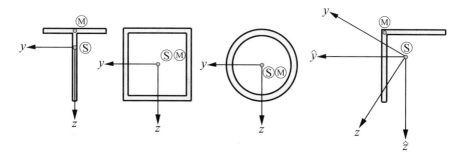

Fig. 8.6 Warping-free cross-sections

If we reconsider the assumptions of St. Venant's theory of torsion according to Chap. 7, we find that this theory is only applicable if the treated cross-section is free of warping, i.e. if we can assume that the cross-section remains flat, or if the warping of the cross-section can develop without restraints and is not hindered by given support conditions. However, the simple example discussed above already shows that in practical applications only in very rare cases we can consider cross-sections to be free of warping which makes it necessary to consider given torsion problems as warping torsion problems. As an anticipation of the further explanations in this chapter, the Figs. 8.6 and 8.7 show selected cross-sections which are warping-free (Fig. 8.6) or which exhibit warping under torsional loading (Fig. 8.7). As a consequence, when considering warping torsion, we have to drop the assumption of the validity of the hypothesis that cross-sections remain flat under torsion. We will show, however, that we can still maintain this hypothesis at least for single segments (i.e. flanges, webs) of the considered beam cross-sections. As we will also show, warping torsion becomes decisive especially for open-profile cross-sections.

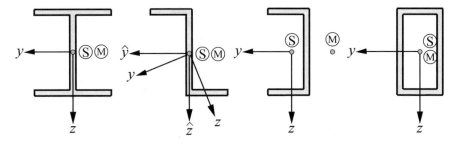

Fig. 8.7 Cross-sections with warping under torsional load

8.2 Warping of Open Cross-Sections

In this section we will deal with the determination of warping that will occur due to torsional loading of a beam cross-section. For this purpose, we define an initially completely arbitrary axis of rotation D as shown in Fig. 8.8. The axis system with its origin in D is designated $\bar{x}$, $\bar{y}$, $\bar{z}$, the resulting displacement quantities are $\bar{u}$, $\bar{v}$, $\bar{w}$ and the rotation angle $\bar{\vartheta}$. Figure 8.8 also shows the secondary shear stresses $\tau_{s,s}$ and the resulting shear flow $T_{s,s} = \tau_{s,s}(s)t(s)$, where $\tau_{s,s}$ is constant over the wall thickness of the section. In addition, a circumferential coordinate s is introduced as indicated. We refer here to a C cross-section, since this kind of cross-section is particularly suitable for illustrating warping torsion, but the relationships shown below are generally valid for any cross-section.

We now consider a section element with the edge lengths $\mathrm{d}\bar{x}$ and $\mathrm{d}s$ as given in Fig. 8.8 which has the distance r from the axis of rotation D. The distance r_t is measured orthogonally to the skeleton line (Fig. 8.9, left). Between r and r_t there is

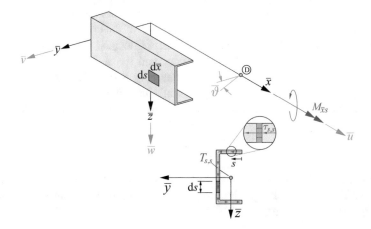

Fig. 8.8 Cross-section, axis of rotation, reference axes and displacements

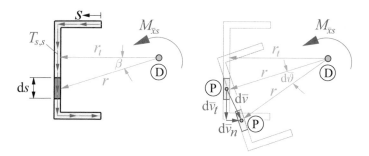

Fig. 8.9 Rotation of the cross-section

the angle β. Due to the secondary torsional moment $M_{\bar{x}s}$ the cross-section suffers the torsion $d\bar{\vartheta}$ as shown in Fig. 8.9, right. The point P thus is shifted by the displacement $d\bar{v}$, which is perpendicular to the radial coordinate r. We decompose $d\bar{v}$ into the component $d\bar{v}_t$ tangential to the skeleton line and the component $d\bar{v}_n$ orthogonal to the skeleton line.

The displacement $d\bar{v}$ results as:

$$d\bar{v} = r d\bar{\vartheta}. \tag{8.2}$$

Thus we can write the tangential component $d\bar{v}_t$ as:

$$d\bar{v}_t = d\bar{v} \cos \beta = r d\bar{\vartheta} \cos \beta = r_t d\bar{\vartheta}. \tag{8.3}$$

To obtain an expression for the longitudinal displacement $\bar{u}$, we consider the section element with the edge lengths ds and $d\bar{x}$ again, as shown in Fig. 8.10. Here we make use of the assumption that although we have to drop the hypothesis that the cross-sections remain flat within the scope of warping torsion, we can still maintain this hypothesis at least for the individual segments of the considered cross-section. Due to the displacement $\bar{v}_t$, the section element is tilted and thus also longitudinally displaced by $\bar{u}$, but no shear strains γ occur and the section element retains its rectangular shape even in the deformed state (Fig. 8.10). Furthermore, this is in accordance with the requirements of St. Venant's torsion theory (linear shear stresses due to primary torsion with zero values at the location of the skeleton line). We obtain:

$$-\frac{d\bar{u}}{ds} = \frac{d\bar{v}_t}{d\bar{x}}. \tag{8.4}$$

The negative sign on the left side of Eq. (8.4) results from the fact that the displacement $d\bar{v}_t$ causes a longitudinal displacement running counter to the $\bar{x}-$axis. Rearranged for $d\bar{u}$ we have:

Fig. 8.10 Determination of the longitudinal displacement u

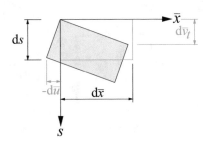

$$d\bar{u} = -\frac{d\bar{v}_t}{d\bar{x}}ds = -r_t\frac{d\bar{\vartheta}}{d\bar{x}}ds = -r_t\bar{\vartheta}'ds. \tag{8.5}$$

If this expression is integrated along the skeleton line s, then the displacement $\bar{u}$ at any position s can be determined:

$$\bar{u}(s) = -\int_{s_A}^{s} r_t\frac{d\bar{\vartheta}}{d\bar{x}}ds + \bar{u}_A, \tag{8.6}$$

where the expression $\bar{u}_A$ is added at the position $s = s_A$ as initial value of the displacement $\bar{u}$. Since we have also assumed prismatic beams with constant properties along the longitudinal axis of the beam, we can draw $\frac{d\bar{\vartheta}}{d\bar{x}}$ in front of the integral:

$$\bar{u}(s) = -\frac{d\bar{\vartheta}}{d\bar{x}}\int_{s_A}^{s} r_t ds + \bar{u}_A = -\bar{\vartheta}'\int_{s_A}^{s} r_t ds + \bar{u}_A. \tag{8.7}$$

This is a formulation for the warping or longitudinal displacement of the cross-section at any point s under torsion. We can also write:

$$\bar{u}(s) = -\bar{\vartheta}'(\bar{x})\bar{\omega}_{D,s}(\bar{y},\bar{z}) + \bar{u}_A = -\bar{\vartheta}'(\bar{x})\left(\bar{\omega}_{D,s}(\bar{y},\bar{z}) - \bar{\omega}_{D,A}\right) = -\bar{\vartheta}'(\bar{x})\bar{\omega}_D(\bar{y},\bar{z}), \tag{8.8}$$

where the integral term in (8.7) is denoted as $\bar{\omega}_{D,s}(\bar{y},\bar{z})$ and the boundary term as $\bar{\omega}_{D,A}$. The indexing with the letter D is explained by the fact that we currently still refer all calculations to the (initially not further specified) axis of rotation D. The expression

$$\bar{\omega}_{D,s}(\bar{y},\bar{z}) = \int_{s_A}^{s} r_t ds \tag{8.9}$$

is also called unit warping and is independent of the longitudinal axis $\bar{x}$, thus constituting a cross-sectional property. Thus, an expression for the longitudinal displacement $\bar{u}$ is found that we split into a portion $\bar{\vartheta}'(x)$ which depends only on $\bar{x}$, and a portion $\bar{\omega}_D(\bar{y},\bar{z})$ which depends only on the geometry of the cross-section. The unit warping therefore describes the warping of the cross-section for $\bar{\vartheta}'(\bar{x}) = -1.0$. Taking a closer look at the integral term (8.9) it can be noted that it describes the double area

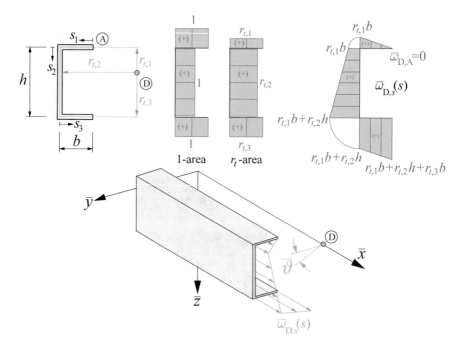

Fig. 8.11 Determination of the unit warping at the C-Cross-Section

A_m which is swept over by the radial coordinate between the points s_A and s:

$$\int_{s_A}^{s} r_t \mathrm{d}s = 2A_m. \tag{8.10}$$

We now take a closer look at the thin-walled C-cross-section treated so far (Fig. 8.11), and we determine the unit warping $\bar{\omega}_{D,s}(\bar{y}, \bar{z}) = \bar{\omega}_{D,s}(s)$. The height and width (measured along the skeleton line) are h and b, respectively, and the rotation axis D has the distances $r_{t,1}$, $r_{t,2}$ and $r_{t,3}$ to the respective parts of the cross-section. The circumferential coordinate s starts at the free end of the upper flange (point A) and covers the entire cross-section in the displayed directions. We now want to determine the unit warping and evaluate the expression (8.9) for the different relevant locations. We interpret the integral in (8.9) as a superposition of the 1-area with the r_t−area. Since the location $s = s_A$ describes the beginning of the integration, the initial warping $\bar{\omega}_{D,A} = \bar{\omega}_D(s = s_A)$ disappears:

$$\bar{\omega}_{D,A} = 0. \tag{8.11}$$

With the aid of the integral table contained in Fig. 5.8 we have for $s_1 = b$:

$$\bar{\omega}_{D,s}(s_1 = b) = \int_{s_A}^{b} 1 \cdot r_{t,1} ds = b \cdot 1 \cdot r_{t,1} = r_{t,1} b. \tag{8.12}$$

In the same way we obtain at $s_2 = h$:

$$\bar{\omega}_{D,s}(s_2 = h) = r_{t,1} b + \int_{0}^{h} 1 \cdot r_{t,2} ds = r_{t,1} b + r_{t,2} h. \tag{8.13}$$

Lastly we have at $s_3 = b$:

$$\bar{\omega}_{D,s}(s_3 = b) = r_{t,1} b + r_{t,2} h + \int_{0}^{b} 1 \cdot r_{t,3} ds = r_{t,1} b + r_{t,2} h + r_{t,3} b. \tag{8.14}$$

The distribution of the unit warping $\bar{\omega}_{D,s}(s)$ is shown in Fig. 8.11, right and bottom. Since the direction of integration always follows the local $s-$ axis, all ordinates of the unit warping shown here are positive.

8.3 Warping of Closed Cross-Sections

8.3.1 Single Cell Cross-Sections

The determination of the unit warping of closed cross-sections is performed analogous to the determination of the shear flow due to transverse shear forces in closed cross-sections (see Chap. 6). Here, however, we need to consider that the shear strains due to primary torsion are no longer negligible, the circumferential Bredt shear flow leads to shear strains which can no longer be neglected. Therefore, we consider the shear strains γ_{xs} which occur due to the shear flow. The problem with the evaluation of the Eq. (8.7) is, quite analogous to the calculation of the shear flows due to transverse shear forces in closed sections (Chap. 6), the a priori unknown boundary value of the displacement u_A so that we again resort to a statically indeterminate calculation. To illustrate the procedure, let us consider the box section shown in Fig. 8.12 whereby we want to avoid marking the individual quantities with an overline in this subsection for better readability. However, it has to be clearly emphasized that the explanations at this point all refer to any rotation axis D and any coordinate system x, y, z. We now cut the cross-section at an arbitrary point and determine the unit warping of this open cross-section, which we will denote as $\omega_{D,0}$. We obtain:

$$\omega_{D,0}(s) = \int_{0}^{s} r_t ds. \tag{8.15}$$

This results in the displacement $u_0(s)$ as:

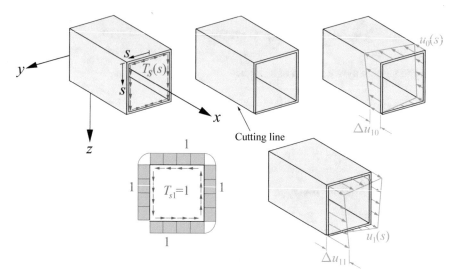

Fig. 8.12 Determination of the unit warping at the example of a closed box cross-section

$$u_0(s) = -\vartheta'(x)\omega_{D,0}(s). \tag{8.16}$$

The resultant displacement doscontinuity Δu_{10} (see Fig. 8.12, top right) is then calculated as:

$$\Delta u_{10} = -\vartheta'(x)\oint r_t ds = -2\vartheta'(x)A_m, \tag{8.17}$$

where A_m is the area enclosed by the cross-section.

Next we apply the constant circumferential Bredt shear flow T_{s1} (Fig. 8.12, bottom left) and determine the resulting displacement discontinuity Δu_{11} by considering the here no longer negligible shear strain γ_{xs} of the cross-section and first determine the displacement $u_1(s)$ using $du = \gamma_{xs}$ (see Fig. 8.13) at the position s:

$$u_1(s) = \int_0^s \gamma_{xs} ds. \tag{8.18}$$

With Hooke's Law $\tau = G\gamma_{xs}$ and $\gamma_{xs} = \frac{\tau}{G} = \frac{T}{Gt}Gt$ we obtain:

$$u_1(s) = \int_0^s \frac{\tau}{G} ds = \frac{T_{s1}}{G} \int_0^s \frac{ds}{t}. \tag{8.19}$$

At this point we can again use the torsion function ψ with $T_s = \psi G\vartheta'$ (see Chap. 7, Eq. 7.80) and obtain:

$$u_1(s) = \vartheta' \int_0^s \frac{\psi}{t} ds. \tag{8.20}$$

Fig. 8.13 Shear strain γ_{xs}

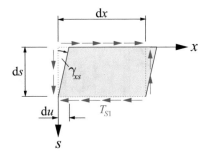

The displacement discontinuity Δu_{11} is calculated with $T_{s1} = 1$ as the ring integral over the whole cross-section:

$$\Delta u_{11} = \frac{1}{G} \oint \frac{ds}{t}, \tag{8.21}$$

respectively using the torsion function ψ:

$$\Delta u_{11} = \vartheta' \oint \frac{\psi}{t} ds. \tag{8.22}$$

We apply the following compatibility requirement as already employed with (6.42):

$$\Delta u_{10} + T_{s1} \Delta u_{11} = 0, \tag{8.23}$$

or in the following form:

$$-2\vartheta' A_m + \vartheta' \oint \frac{\psi}{t} ds = 0, \tag{8.24}$$

which can be solved for the torsion function ψ as follows:

$$\psi = \frac{2A_m}{\oint \dfrac{ds}{t}}. \tag{8.25}$$

From this the circumferential shear flow T_{s1} can be determined as $T_{s1} = G\psi\vartheta'$. The final warping or longitudinal displacement of the box cross-section is then calculated by adding the two displacements u_0 and u_1:

$$u(s) = -\vartheta' \left[\int_0^s r_t ds - \psi \int_0^s \frac{ds}{t} \right], \tag{8.26}$$

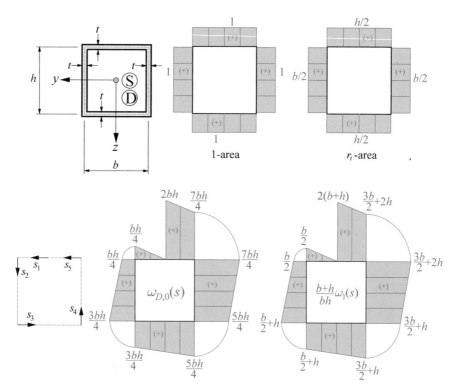

Fig. 8.14 Determination of warping $\omega_0(s)$ and $\omega_1(s)$ at the example of a closed box cross-section

so that:

$$u(s) = -\vartheta'(x)\left(\omega_{D,0}(s) - \omega_1(s)\right) = -\vartheta'\omega_D(s), \tag{8.27}$$

with:

$$\omega_{D,0}(s) = \int_0^s r_t \mathrm{d}s, \quad \omega_1(s) = \psi \int_0^s \frac{\mathrm{d}s}{t}. \tag{8.28}$$

We demonstrate the procedure at the example of a rectangular box cross-section as shown in Fig. 8.14. The cross-section has a constant wall thickness t, width b and height h.

We open the cross-section in its symmetry axis at the upper flange and assign local circumferential axes s_i ($i = 1, 2, 3, 4, 5$) as shown. We first determine the unit warping $\omega_{D,0}(s)$ of the open cross-section and obtain:

$$w_{D,0}(s_1 = 0) = 0,$$

$$w_{D,0}\left(s_1 = \frac{b}{2}\right) = \frac{h}{2}\frac{b}{2} = \frac{bh}{4},$$

$$w_{D,0}(s_2 = 0) = \frac{bh}{4},$$

$$w_{D,0}(s_2 = h) = \frac{bh}{4} + \frac{b}{2}h = \frac{3bh}{4},$$

$$w_{D,0}(s_3 = 0) = \frac{3bh}{4},$$

$$w_{D,0}(s_3 = b) = \frac{3bh}{4} + \frac{h}{2}b = \frac{5bh}{4},$$

$$w_{D,0}(s_4 = 0) = \frac{5bh}{4},$$

$$w_{D,0}(s_4 = h) = \frac{5bh}{4} + \frac{b}{2}h = \frac{7bh}{4},$$

$$w_{D,0}(s_5 = 0) = \frac{7bh}{4},$$

$$w_{D,0}\left(s_5 = \frac{b}{2}\right) = \frac{7bh}{4} + \frac{h}{2}\frac{b}{2} = 2bh. \tag{8.29}$$

The torsion function ψ can be calculated with (8.25) as:

$$\psi = \frac{2A_m}{\oint \dfrac{ds}{t}} = \frac{2bh}{\dfrac{2b}{t} + \dfrac{2h}{t}} = \frac{bht}{b+h}. \tag{8.30}$$

The warping w_1 due to the Bredt shear flow T_{s1} is then calculated with (8.28) as:

$$w_1(s) = \psi \int_0^s \frac{ds}{t} = \frac{bht}{b+h} \int_0^s \frac{ds}{t}, \tag{8.31}$$

and considering $t = \text{const.}$:

$$w_1(s) = \frac{bh}{b+h} \int_0^s ds. \tag{8.32}$$

The calculation of $w_{D,0}(s)$ and $w_1(s)$ according to (8.28) then yields the results as shown in Fig. 8.14, bottom. The superposition of these results leads to the final warping $w_D(s)$ of the closed box cross-section as given in Fig. 8.15. Note that the sign of the warping values depends on the ratio between b and h. If $b > h$ applies, the signs of $w_D(s)$ are as shown in Fig. 8.15. However, if $b < h$, then the signs are reversed. The special case of the square box section with $b = h$, on the other hand, is free of warping (i.e. $w_D(s) = 0$), as can be concluded from Fig. 8.15, and as was already shown in Fig. 8.6.

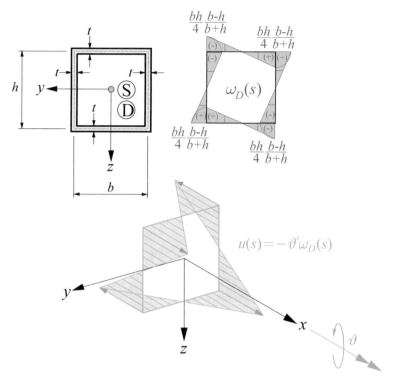

Fig. 8.15 Warping $\omega_D(s)$ (top) and longitudinal displacement (bottom) of the closed box cross-section

8.3.2   Multi-cell Cross-Sections

In the case of cross-sections consisting of n closed cells, the compatibility condition (8.23) or (8.24) has to be extended to n corresponding compatibility equations:

$$
\begin{bmatrix}
\oint_1 \dfrac{ds}{t(s)} & -\oint_{1,2} \dfrac{ds}{t(s)} & -\oint_{1,3} \dfrac{ds}{t(s)} & \cdots & -\oint_{1,n} \dfrac{ds}{t(s)} \\[2mm]
-\oint_{2,1} \dfrac{ds}{t(s)} & \oint_2 \dfrac{ds}{t(s)} & -\oint_{2,3} \dfrac{ds}{t(s)} & \cdots & -\oint_{2,n} \dfrac{ds}{t(s)} \\[2mm]
-\oint_{3,1} \dfrac{ds}{t(s)} & -\oint_{3,2} \dfrac{ds}{t(s)} & \oint_3 \dfrac{ds}{t(s)} & \cdots & -\oint_{3,n} \dfrac{ds}{t(s)} \\[1mm]
\vdots & \vdots & \vdots & \ddots & \vdots \\[1mm]
-\oint_{n,1} \dfrac{ds}{t(s)} & -\oint_{n,2} \dfrac{ds}{t(s)} & -\oint_{n,3} \dfrac{ds}{t(s)} & \cdots & \oint_n \dfrac{ds}{t(s)}
\end{bmatrix}
\begin{pmatrix} \psi_1 \\ \psi_2 \\ \psi_3 \\ \vdots \\ \psi_n \end{pmatrix}
=
\begin{pmatrix} 2A_{m,1} \\ 2A_{m,2} \\ 2A_{m,3} \\ \vdots \\ 2A_{m,n} \end{pmatrix}.
$$

$$(8.33)$$

This system of equations is formally completely identical to Eq. (7.83), Chap. 7.

8.4 Unit Warping with Respect to the Shear Center

Up to now we have considered the unit warping $\bar{\omega}_{D,s}(s)$ with respect to any point D. In this section we will now refer to a certain point M, where the term M already indicates that we will later use the shear center M as a reference point (Fig. 8.16). At this point, however, M is still an arbitrary point with the distances $\bar{y}_M$ and $\bar{z}_M$ from the rotational axis D. The center of gravity of the cross-section is denoted as S. The warping $\bar{\omega}_D(s)$ with respect to the center of rotation D results from (8.9):

$$\bar{\omega}_D(s) = \bar{\omega}_{D,s}(s) - \bar{\omega}_{D,A} = \int r_{tD}\,ds - \bar{\omega}_{D,A}. \tag{8.34}$$

where r_{tD} is the radial distance to the skeleton line of the respective cross-sectional segments measured from D. We now transform r_t from D to the point M (Fig. 8.17):

$$r_{tM} = r_{tD} - \bar{z}_M \sin \beta - \bar{y}_M \cos \beta, \tag{8.35}$$

with the angle β according to Fig. 8.17. We can thus write the unit warping $\bar{\omega}_M$ with respect to M as:

Fig. 8.16 Transformation of the axis of rotation D to the point M

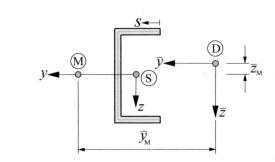

Fig. 8.17 Determination of the radial distance r_t with respect to M

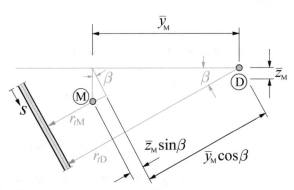

$$\bar{\omega}_{M,s} = \int r_{tM}\,\mathrm{d}s = \int r_{tD}\,\mathrm{d}s - \bar{z}_M \int \sin\beta\,\mathrm{d}s - \bar{y}_M \int \cos\beta\,\mathrm{d}s. \qquad (8.36)$$

Use of

$$\sin\beta = -\frac{\mathrm{d}\bar{y}}{\mathrm{d}s}, \quad \cos\beta = \frac{\mathrm{d}\bar{z}}{\mathrm{d}s} \qquad (8.37)$$

yields:

$$\bar{\omega}_{M,s} = \int r_{tD}\,\mathrm{d}s + \bar{z}_M \int \mathrm{d}\bar{y} - \bar{y}_M \int \mathrm{d}\bar{z}. \qquad (8.38)$$

or:

$$\bar{\omega}_M = \bar{\omega}_{D,s} - \bar{\omega}_{D,A} + \bar{z}_M\,(\bar{y}(s) - \bar{y}_A) - \bar{y}_M\,(\bar{z}(s) - \bar{z}_A). \qquad (8.39)$$

8.5   The First-Order Bending-Torsion Problem

We now want to investigate to what extent an interaction of the warping torsion considered here with a rod and beam effect exists. This topic of investigation is often referred to as the so-called first-order bending-torsion problem.

The longitudinal displacement $\bar{u}$ of a point at the position s with a cross-sectional rotation $\bar{\vartheta}$ around any axis of rotation D results, as already considered, as follows:

$$\bar{u}(s) = -\bar{\vartheta}'(\bar{x})\bar{\omega}_D(s). \qquad (8.40)$$

From this the longitudinal strain $\varepsilon_{\bar{x}\bar{x}}$ can be determined:

$$\varepsilon_{\bar{x}\bar{x}} = \frac{\mathrm{d}\bar{u}}{\mathrm{d}\bar{x}} = -\bar{\vartheta}''(\bar{x})\bar{\omega}_D(s). \qquad (8.41)$$

By means of Hooke's Law $\sigma_{\bar{x}\bar{x}} = E\varepsilon_{\bar{x}\bar{x}}$ the warping normal stress $\sigma_{\bar{\omega}\bar{\omega}} = \sigma_{\bar{x}\bar{x}}$ is obtained as:

$$\sigma_{\bar{\omega}\bar{\omega}} = E\varepsilon_{\bar{x}\bar{x}} = -E\bar{\vartheta}''(\bar{x})\omega_D(s). \qquad (8.42)$$

We now consider the case where the beam under consideration is not only subjected to torsional loading and thus to a warping $\bar{\omega}_D$, but also has to endure the longitudinal displacement $\bar{u}$ as well as the two curvatures $\bar{w}''$ and $\bar{v}''$ with respect to the $\bar{y}$−axis and the $\bar{z}$−axis, i.e. a simultaneous rod and biaxial bending effect is present. Then the stress calculation is as follows:

$$\sigma_{\bar{x}\bar{x}} = E\left[\bar{u}'(\bar{x}) - \bar{w}''(\bar{x})\bar{z} - \bar{v}''(\bar{x})\bar{y} - \bar{\vartheta}''(\bar{x})\bar{\omega}_D(s)\right]. \qquad (8.43)$$

In addition to the already known internal forces and moments $N_{\bar{x}}$, $M_{\bar{y}}$ and $M_{\bar{z}}$ there is another internal moment due to the warping of the cross-section, namely the so-called warping moment $M_{\bar{\omega}}$. The internal forces and moments result from the integration of the stress over the cross-section as follows:

$$N_{\bar{x}} = \int_A \sigma_{\bar{x}\bar{x}} dA, \quad M_{\bar{y}} = \int_A \sigma_{\bar{x}\bar{x}} \bar{z} dA, \quad M_{\bar{z}} = -\int_A \sigma_{\bar{x}\bar{x}} \bar{y} dA, \quad M_{\bar{\omega}} = \int_A \sigma_{\bar{x}\bar{x}} \bar{\omega}_{D,s} dA.$$

(8.44)

Performing the integrations using (8.43) then yields:

$$N_{\bar{x}} = E\left[\bar{u}' \int_A dA - \bar{w}'' \int_A \bar{z} dA - \bar{v}'' \int_A \bar{y} dA - \bar{\vartheta}'' \int_A \bar{\omega}_{D,s} dA\right],$$

$$M_{\bar{y}} = E\left[\bar{u}' \int_A \bar{z} dA - \bar{w}'' \int_A \bar{z}^2 dA - \bar{v}'' \int_A \bar{y}\bar{z} dA - \bar{\vartheta}'' \int_A \bar{\omega}_{D,s} \bar{z} dA\right],$$

$$-M_{\bar{z}} = E\left[\bar{u}' \int_A \bar{y} dA - \bar{w}'' \int_A \bar{y}\bar{z} dA - \bar{v}'' \int_A \bar{y}^2 dA - \bar{\vartheta}'' \int_A \bar{\omega}_{D,s} \bar{y} dA\right],$$

$$M_{\bar{\omega}} = E\left[\bar{u}' \int_A \bar{\omega}_{D,s} dA - \bar{w}'' \int_A \bar{\omega}_{D,s} \bar{z} dA - \bar{v}'' \int_A \bar{\omega}_{D,s} \bar{y} dA - \bar{\vartheta}'' \int_A \bar{\omega}_{D,s}^2 dA\right].$$

(8.45)

In a vector-matrox notation we have:

$$\begin{pmatrix} N_{\bar{x}} \\ M_{\bar{y}} \\ -M_{\bar{z}} \\ M_{\bar{\omega}} \end{pmatrix} = E \begin{bmatrix} \int_A dA & \int_A \bar{z} dA & \int_A \bar{y} dA & \int_A \bar{\omega}_{D,s} dA \\ \int_A \bar{z} dA & \int_A \bar{z}^2 dA & \int_A \bar{y}\bar{z} dA & \int_A \bar{\omega}_{D,s} \bar{z} dA \\ \int_A \bar{y} dA & \int_A \bar{y}\bar{z} dA & \int_A \bar{y}^2 dA & \int_A \bar{\omega}_{D,s} \bar{y} dA \\ \int_A \bar{\omega}_{D,s} dA & \int_A \bar{\omega}_{D,s} \bar{z} dA & \int_A \bar{\omega}_{D,s} \bar{y} dA & \int_A \bar{\omega}_{D,s}^2 dA \end{bmatrix} \begin{pmatrix} \bar{u}' \\ -\bar{w}'' \\ -\bar{v}'' \\ -\bar{\vartheta}'' \end{pmatrix}.$$

(8.46)

This is the constitutive law of the first-order bending-torsion problem, formulated for an arbitrary reference system $\bar{x}, \bar{y}, \bar{z}$. Apparently, it can be seen that, in addition to some already known area integrals, further quantities are added related to the warping $\bar{\omega}_{D,s}$. The set of equations (8.46) shows that not only, as already discussed in Chap. 5, a coupling between beam and rod effect can exist, but also a coupling between beam and rod effect on the one hand and warping torsion on the other hand can occur.

Besides the cross-sectional area A

$$A = \int_A dA$$

(8.47)

a number of first-order area integrals comes into play at this point. Beside the two already known static moments $S_{\bar{y}}$ and $S_{\bar{z}}$

$$S_{\bar{y}} = \int_A \bar{z} \mathrm{d}A, \quad S_{\bar{z}} = \int_A \bar{y} \mathrm{d}A \tag{8.48}$$

this is the newly added area integral $S_{\bar{\omega}}$:

$$S_{\bar{\omega}} = \int_A \bar{\omega}_{D,s} \mathrm{d}A, \tag{8.49}$$

which can be interpreted as a static moment regarding the warping $\bar{\omega}_{D,s}$. In contrast to $S_{\bar{y}}$ and $S_{\bar{z}}$, the unit of this quantity results as a length to the fourth power.

The second-order area integrals occurring here are the moments of inertia $I_{\bar{y}\bar{y}}$ and $I_{\bar{z}\bar{z}}$ and the deviation moment $I_{\bar{y}\bar{z}}$ already known from Chap. 5:

$$I_{\bar{y}\bar{y}} = \int_A \bar{z}^2 \mathrm{d}A, \quad I_{\bar{z}\bar{z}} = \int_A \bar{y}^2 \mathrm{d}A, \quad I_{\bar{y}\bar{z}} = \int_A \bar{y}\bar{z} \mathrm{d}A \tag{8.50}$$

In addition, we have the so-called warping moment of inertia $I_{\bar{\omega}\bar{\omega}}$ (in the unit of a length in sixth power) as well as the two area integrals which can be interpreted as deviation moments with respect to warping:

$$I_{\bar{y}\bar{\omega}} = \int_A \bar{\omega}_{D,s} \bar{y} \mathrm{d}A, \quad I_{\bar{z}\bar{\omega}} = \int_A \bar{\omega}_{D,s} \bar{z} \mathrm{d}A, \quad I_{\bar{\omega}\bar{\omega}} = \int_A \bar{\omega}_{D,s}^2 \mathrm{d}A. \tag{8.51}$$

The two area integrals $I_{\bar{y}\bar{\omega}}$ and result in the unit of a length in fifth power.

The system of equations (8.46) can be written in a more compact manner with the area values thus defined:

$$\begin{pmatrix} N_{\bar{x}} \\ M_{\bar{y}} \\ -M_{\bar{z}} \\ M_{\bar{\omega}} \end{pmatrix} = E \begin{bmatrix} A & S_{\bar{y}} & S_{\bar{z}} & S_{\bar{\omega}} \\ S_{\bar{y}} & I_{\bar{y}\bar{y}} & I_{\bar{y}\bar{z}} & I_{\bar{z}\bar{\omega}} \\ S_{\bar{z}} & I_{\bar{y}\bar{z}} & I_{\bar{z}\bar{z}} & I_{\bar{y}\bar{\omega}} \\ S_{\bar{\omega}} & I_{\bar{z}\bar{\omega}} & I_{\bar{y}\bar{\omega}} & I_{\bar{\omega}\bar{\omega}} \end{bmatrix} \begin{pmatrix} \bar{u}' \\ -\bar{w}'' \\ -\bar{v}'' \\ -\bar{\vartheta}'' \end{pmatrix}. \tag{8.52}$$

8.6 Cross-Sectional Normalizations

8.6.1 First Cross-Sectional Normalization: Center of Gravity S

Analogous to the explanations in Chap. 5 a partial decoupling of the correlations (8.52) can be achieved if the reference system is moved to the center of gravity S of the given cross-section. The following relationships then exist between the original system $\bar{x}$, $\bar{y}$, $\bar{z}$ and the system $\hat{x}$, $\hat{y}$, $\hat{z}$ in the center of gravity (see also Fig. 5.11):

$$\hat{x} = \bar{x}, \quad \hat{y} = \bar{y} - \bar{y}_S, \quad \hat{z} = \bar{z} - \bar{z}_S. \tag{8.53}$$

As a result the two static moments $S_{\hat{y}}$ and $S_{\hat{z}}$ become zero. Furthermore, we shift the initial value of the integration so that the static moment $S_{\hat{\omega}}$ becomes zero. The following then applies to the warping with respect to the coordinate system located in the center of gravity:

$$\hat{\omega}_D = \bar{\omega}_{D,s} - \bar{\omega}_{D,A}. \tag{8.54}$$

The position of the center of gravity results from the demand (5.30) for the disappearance of the static moments $S_{\hat{z}}$ and $S_{\hat{y}}$, so that (cf. (5.31)):

$$\bar{y}_S = \frac{\int_A \bar{y} \, dA}{\int_A dA} = \frac{S_{\bar{z}}}{A}, \quad \bar{z}_S = \frac{\int_A \bar{z} \, dA}{\int_A dA} = \frac{S_{\bar{y}}}{A}. \tag{8.55}$$

For the warping coordinate $\bar{\omega}_{D,A}$ follows analogously:

$$\bar{\omega}_{D,A} = \frac{\int_A \bar{\omega}_{D,s} \, dA}{\int_A dA} = \frac{S_{\bar{\omega}}}{A}. \tag{8.56}$$

Note that only the starting point A of the integration has changed during this transformation, but not the position of the rotation axis D. The consequence of this transformation is that a partial decoupling of (8.46) or (8.52) results as follows:

$$\begin{pmatrix} N_{\hat{x}} \\ M_{\hat{y}} \\ -M_{\hat{z}} \\ M_{\hat{\omega}} \end{pmatrix} = E \begin{bmatrix} A & 0 & 0 & 0 \\ 0 & I_{\hat{y}\hat{y}} & I_{\hat{y}\hat{z}} & I_{\hat{z}\hat{\omega}} \\ 0 & I_{\hat{y}\hat{z}} & I_{\hat{z}\hat{z}} & I_{\hat{y}\hat{\omega}} \\ 0 & I_{\hat{z}\hat{\omega}} & I_{\hat{y}\hat{\omega}} & I_{\hat{\omega}\hat{\omega}} \end{bmatrix} \begin{pmatrix} \hat{u}' \\ -\hat{w}'' \\ -\hat{v}'' \\ -\hat{\vartheta}'' \end{pmatrix}. \tag{8.57}$$

The transformed area integrals $I_{\hat{y}\hat{\omega}}$, $I_{\hat{z}\hat{\omega}}$, $I_{\hat{\omega}\hat{\omega}}$ appearing here result analogously to Steiner's theorem (cf. Eq. (5.34), Chap. 5) as:

$$I_{\hat{y}\hat{\omega}} = I_{\bar{y}\bar{\omega}} - \bar{y}_S \bar{\omega}_{D,A} A, \quad I_{\hat{z}\hat{\omega}} = I_{\bar{z}\bar{\omega}} - \bar{z}_S \bar{\omega}_{D,A} A, \quad I_{\hat{\omega}\hat{\omega}} = I_{\bar{\omega}\bar{\omega}} - \bar{\omega}_{D,A} S_{\bar{\omega}}. \tag{8.58}$$

8.6.2 Second Cross-Sectional Normalization: Principal Axes

Another essential decoupling of the constitutive law (8.57) is achieved by transforming the reference system $\hat{x}$, $\hat{y}$, $\hat{z}$ that is located in the center of gravity by rotation into the principal axis system x, y, z. In addition to the requirement for the disappearance of the deviation moment I_{yz}, already known from Chap. 5, the two area integrals $I_{y\omega}$ and $I_{z\omega}$ also disappear in the principal axis system. The position of the shear center M can be determined from the latter requirement, which is equivalent to the requirement that the rotational axis D is moved into the shear center M. The unit warping related to the shear center M follows directly from (8.39) with $\bar{y}_A = 0$ and $\bar{z}_A = 0$:

$$\omega_M = \hat{\omega}_D + \bar{z}_M \hat{y} - \bar{y}_M \hat{z}. \tag{8.59}$$

The integrals $I_{\hat{y}\hat{\omega}}^M$ and $I_{\hat{z}\hat{\omega}}^M$ related to the shear center M can be specified as:

$$I_{\hat{y}\hat{\omega}}^M = \int_A \hat{y}\hat{\omega}_M \mathrm{d}A$$

$$= \int_A \hat{y}\hat{\omega}_D \mathrm{d}A + \bar{z}_M \int_A \hat{y}^2 \mathrm{d}A - \bar{y}_M \int_A \hat{y}\hat{z} \mathrm{d}A,$$

$$I_{\hat{y}\hat{\omega}}^M = \int_A \hat{z}\hat{\omega}_D \mathrm{d}A + \bar{z}_M \int_A \hat{y}\hat{z} \mathrm{d}A - \bar{y}_M \int_A \hat{z}^2 \mathrm{d}A. \tag{8.60}$$

Setting these two expressions to zero results in a system of equations to determine the shear center coordinates:

$$\int_A \hat{y}\hat{\omega}_D \mathrm{d}A + \bar{z}_M \int_A \hat{y}^2 \mathrm{d}A - \bar{y}_M \int_A \hat{y}\hat{z} \mathrm{d}A = 0,$$

$$\int_A \hat{z}\hat{\omega}_D \mathrm{d}A + \bar{z}_M \int_A \hat{y}\hat{z} \mathrm{d}A - \bar{y}_M \int_A \hat{z}^2 \mathrm{d}A = 0. \tag{8.61}$$

This yields:

$$\bar{y}_M = \frac{I_{\hat{z}\hat{\omega}} I_{\hat{z}\hat{z}} - I_{\hat{y}\hat{\omega}} I_{\hat{y}\hat{z}}}{I_{\hat{y}\hat{y}} I_{\hat{z}\hat{z}} - I_{\hat{y}\hat{z}}^2}, \quad \bar{z}_M = \frac{I_{\hat{z}\hat{\omega}} I_{\hat{y}\hat{z}} - I_{\hat{y}\hat{\omega}} I_{\hat{y}\hat{y}}}{I_{\hat{y}\hat{y}} I_{\hat{z}\hat{z}} - I_{\hat{y}\hat{z}}^2}. \tag{8.62}$$

The warping moment of inertia $I_{\omega\omega}$ related to the shear center (an additional index M is omitted at this point, both cross-sectional normalizations are assumed) is calculated as:

$$I_{\omega\omega} = \int_A \omega_M^2 \mathrm{d}A = \int_A \left(\hat{\omega}_D + \bar{z}_M \hat{y} - \bar{y}_M \hat{z}\right)^2 \mathrm{d}A. \tag{8.63}$$

Inserting (8.62) results after some transformations:

$$I_{\omega\omega} = I_{\hat{\omega}\hat{\omega}} + \bar{z}_M I_{\hat{y}\hat{\omega}} - \bar{y}_M I_{\hat{z}\hat{\omega}}. \tag{8.64}$$

The second cross-sectional normalization leads to a complete decoupling of rod extension, beam bending and torsion. The constitutive law of the first-order bending-torsion changes to:

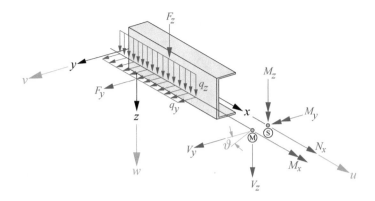

Fig. 8.18 Reference of the state variables to the center of gravity axis S and the shear center axis M

$$
\begin{pmatrix} N_x \\ M_y \\ -M_z \\ M_\omega \end{pmatrix} = E \begin{bmatrix} A & 0 & 0 & 0 \\ 0 & I_{yy} & 0 & 0 \\ 0 & 0 & I_{zz} & 0 \\ 0 & 0 & 0 & I_{\omega\omega} \end{bmatrix} \begin{pmatrix} u' \\ -w'' \\ -v'' \\ -\vartheta'' \end{pmatrix},
\tag{8.65}
$$

i.e. the following constitutive relations remain:

$$
N_x = E A u', \quad M_y = -E I_{yy} w'', \quad M_z = E I_{zz} v'', \quad M_\omega = -E I_{\omega\omega} \vartheta''. \tag{8.66}
$$

The prerequisite for this decoupling is that the bending moments have been transformed to the principal axes and that the normal force acts at the center of gravity S of the cross-section. Furthermore, shear loads, transverse forces and torsion moments are related to the shear center. The shear center axis is the natural axis of rotation of the cross-section since transverse shear forces and transverse loads do not cause a twisting of the cross-section when acting in the shear center. This is summarized in Fig. 8.18. Thus, if the reference axes are present as mentioned above, both the beam problem and the two bending problems and also the torsion problem can be considered separately.

8.7 Example

The previous remarks are illustrated with an example, based on Roik et al. (1972), Sect. 4.5, p. 111. We consider the unsymmetrical cross-section as shown in Fig. 8.19. The reference system $\bar{y}$ and $\bar{z}$ selected here is positioned as indicated. The rotation axis D and the starting point A of the integration are identical to the coordinate origin. The wall thickness distribution $t(s)$ is given as shown in Fig. 8.19. From the given dimensions the $\bar{y}$−line and the $\bar{z}$−line can be determined as shown in Fig. 8.19,

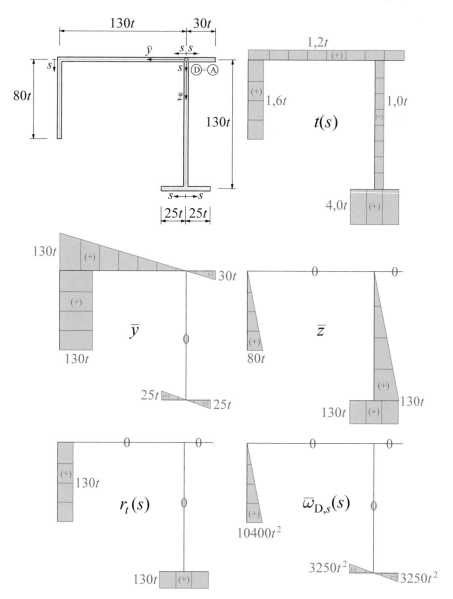

Fig. 8.19 Considered cross-section

middle. With the distance r_t of the rotation axis, the unit warping $\bar{\omega}_{D,s}(s)$ can be determined using (8.9). It is shown in Fig. 8.19, bottom right.

With the characteristic areas available in this way, all surface integrals required here can be calculated. The results are given here without detailing the calculation itself which results from simple superposition of the respective characteristic curves.

For the quantities related to bending the following results are obtained:

$$A = \int_A dA = \int_s t\,ds = 650t^2,$$

$$S_{\bar{z}} = \int_A \bar{y}\,dA = \int_s \bar{y}t\,ds = 26240t^3,$$

$$S_{\bar{y}} = \int_A \bar{z}\,dA = \int_s \bar{z}t\,ds = 39570t^3,$$

$$I_{\bar{y}\bar{y}} = \int_A \bar{z}^2\,dA = \int_s \bar{z}^2 t\,ds = 4385400t^4,$$

$$I_{\bar{z}\bar{z}} = \int_A \bar{y}^2\,dA = \int_s \bar{y}^2 t\,ds = \frac{9283400}{3}t^4,$$

$$I_{\bar{y}\bar{z}} = \int_A \bar{y}\bar{z}\,dA = \int_s \bar{y}\bar{z}t\,ds = 665600t^4. \tag{8.67}$$

For the quantities related to warping we have:

$$S_{\bar{\omega}} = \int_A \bar{\omega}_{D,s}\,dA = \int_s \bar{\omega}_{D,s}t\,ds = 665600t^4,$$

$$I_{\bar{y}\bar{\omega}} = \int_A \bar{y}\bar{\omega}_{D,s}\,dA = \int_s \bar{y}\bar{\omega}_{D,s}t\,ds = \frac{243334000}{3}t^5,$$

$$I_{\bar{z}\bar{\omega}} = \int_A \bar{z}\bar{\omega}_{D,s}\,dA = \int_s \bar{z}\bar{\omega}_{D,s}t\,ds = \frac{106496000}{3}t^5,$$

$$I_{\bar{\omega}\bar{\omega}} = \int_A \bar{\omega}_{D,s}^2\,dA = \int_s \bar{\omega}_{D,s}^2 t\,ds = \frac{15956980000}{3}t^5. \tag{8.68}$$

The first cross-sectional normalization concerns the translation of the reference system $\bar{x}, \bar{y}, \bar{z}$ into the system $\hat{x}, \hat{y}, \hat{z}$ located in the center of gravity, as well as the transformation of the starting point A of the integration of the unit warping. With (8.55) and (8.56) the position of the center of gravity $\bar{y}_S, \bar{z}_S$ and the initial warping $\bar{\omega}_{D,A}$ result as:

$$\bar{y}_S = \frac{S_{\bar{z}}}{A} = \frac{26240t^3}{650t^2} = 40.37t,$$

$$\bar{z}_S = \frac{S_{\bar{y}}}{A} = \frac{39570t^3}{650t^2} = 60.88t,$$

$$\bar{\omega}_{D,A} = \frac{S_{\bar{\omega}}}{A} = \frac{665600t^4}{650t^2} = 1024t^2. \tag{8.69}$$

With (8.53) and (8.54) the characteristic areas $\hat{y}, \hat{z}$ and $\hat{\omega}_D$ can then be drawn (Fig. 8.20).

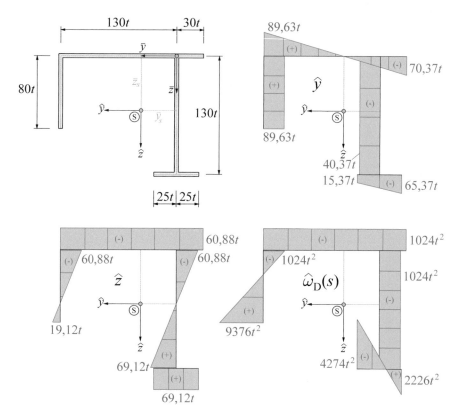

Fig. 8.20 Characteristic areas $\hat{y}$, $\hat{z}$ and $\hat{\omega}_D$

The transformed area integrals then result as follows:

$$I_{\hat{y}\hat{y}} = I_{\bar{y}\bar{y}} - \bar{z}_s^2 A = 1976256.64t^4,$$
$$I_{\hat{z}\hat{z}} = I_{\bar{z}\bar{z}} - \bar{y}_s^2 A = 2035137.68t^4,$$
$$I_{\hat{y}\hat{z}} = I_{\bar{y}\bar{z}} - \bar{y}_s \bar{z}_s A = -931921.64t^4,$$
$$I_{\hat{y}\hat{\omega}} = I_{\bar{y}\bar{\omega}} - \bar{y}_s \bar{\omega}_{D,A} A = 54241061.33t^5,$$
$$I_{\hat{z}\hat{\omega}} = I_{\bar{z}\bar{\omega}} - \bar{z}_s \bar{\omega}_{D,A} A = -5023061.33t^5,$$
$$I_{\hat{\omega}\hat{\omega}} = I_{\bar{\omega}\bar{\omega}} - \bar{\omega}_{D,A} S_{\bar{\omega}} = 4637418933.33t^6. \tag{8.70}$$

The second cross-sectional normalization has the objective of determining the principal axes x, y, z of the cross-section and of relating the warping to the natural axis of rotation, i.e. the shear center M. The principal axis angle φ_0 results from (5.53):

$$\tan 2\varphi_0 = \frac{2I_{\hat{y}\hat{z}}}{I_{\hat{z}\hat{z}} - I_{\hat{y}\hat{y}}} = -31.65, \tag{8.71}$$

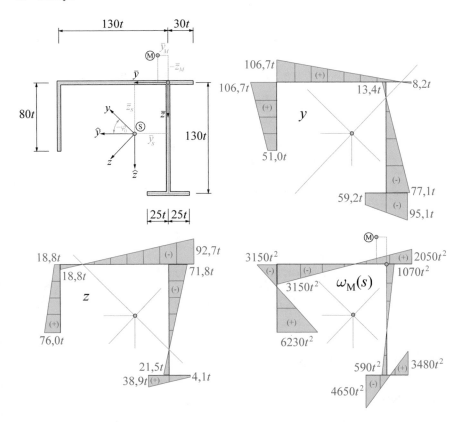

Fig. 8.21 Characteristic areas y, z and ω_M

so that:

$$\varphi_0 = -44.10°. \tag{8.72}$$

With $\sin \varphi_0 = -0.6959$ and $\cos \varphi_0 = 0.7181$ we obtain the transformed coordinates y and z according to (5.48):

$$y = \hat{y} \cos \varphi + \hat{z} \sin \varphi = 0.7181\hat{y} - 0.6959\hat{z},$$
$$z = -\hat{y} \sin \varphi + \hat{z} \cos \varphi = 0.6959\hat{y} + 0.7181\hat{z}. \tag{8.73}$$

The resulting areas y and z are shown in Fig. 8.21. The principal moments of inertia I_1 and I_2 follow using (5.56):

$$I_{1,2} = \frac{I_{\hat{y}\hat{y}} + I_{\hat{z}\hat{z}}}{2} \pm \sqrt{\left(\frac{I_{\hat{z}\hat{z}} - I_{\hat{y}\hat{y}}}{2}\right)^2 + I_{\hat{y}\hat{z}}^2}, \tag{8.74}$$

so that:

$$I_1 = 2938110t^4, \quad I_2 = 1073570t^4. \tag{8.75}$$

The shear center coordinates $\bar{y}_M$ and $\bar{z}_M$ are determined from (8.62) as follows:

$$\bar{y}_M = \frac{I_{\hat{z}\hat{\omega}}I_{\hat{z}\hat{z}} - I_{\hat{y}\hat{\omega}}I_{\hat{y}\hat{z}}}{I_{\hat{y}\hat{y}}I_{\hat{z}\hat{z}} - I_{\hat{y}\hat{z}}^2} = 12.78t,$$

$$\bar{z}_M = \frac{I_{\hat{z}\hat{\omega}}I_{\hat{y}\hat{z}} - I_{\hat{y}\hat{\omega}}I_{\hat{y}\hat{y}}}{I_{\hat{y}\hat{y}}I_{\hat{z}\hat{z}} - I_{\hat{y}\hat{z}}^2} = -32.50t. \tag{8.76}$$

The warping ω_M with respect to M is calculated with (8.59) as:

$$\omega_M = \hat{\omega}_D + \bar{z}_M\hat{y} - \bar{y}_M\hat{z}. \tag{8.77}$$

Its distribution is also shown in Fig. 8.21.
Finally, the warping moment of inertia $I_{\omega\omega}$ is determined:

$$I_{\omega\omega} = I_{\hat{\omega}\hat{\omega}} + \bar{z}_M I_{\hat{y}\hat{\omega}} - \bar{y}_M I_{\hat{z}\hat{\omega}} = 2938779163.84t^6. \tag{8.78}$$

8.8 Selected Basic Cases

8.8.1 Double Symmetrical I-Cross-Section

We consider the unit warping and the cross-sectional normalization for the doubly symmetric I-section shown in Fig. 8.22 with width b, height h and wall thickness t.

The warping $\omega_M(s)$ and the warping moment of inertia $I_{\omega\omega}$ are to be determined. Since this profile is a double symmetrical cross-section, a cross-sectional normalization is not necessary. The center of gravity and the shear center coincide here and the warping ω_M related to the shear center can be determined directly.

In Fig. 8.22 the $r_{tM}-$ area is shown. Apparently r_{tM} is zero in the web of the beam. The warping ω_M then results from the integration of r_{tM} via the circumferential coordinate s:

$$\omega_M(s) = \int_s r_{tM}\mathrm{d}s. \tag{8.79}$$

The maximum warping value results at the free ends of the cross-section:

$$\left|\omega_{M,\mathrm{max}}\right| = r_{tM}\int_0^{\frac{b}{2}}\mathrm{d}s = \frac{h}{2}\cdot\frac{b}{2} = \frac{bh}{4}. \tag{8.80}$$

The warping is linear within the individual cross-section segments, as shown in Fig. 8.22. Obviously, in a double-symmetrical I-section, only the flanges show

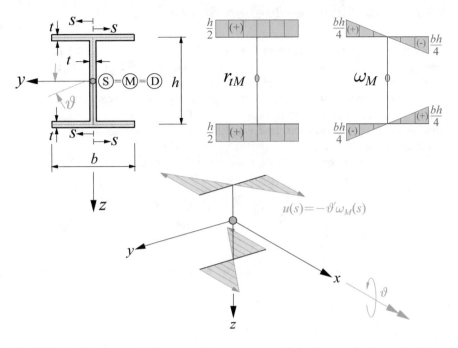

Fig. 8.22 Double symmetrical I-cross-section, r_{tM}–area and warping ω_M (top); warping figure (bottom)

warping, but not the web, as we have already assumed in the introductory part of this chapter.

Once the distribution of the warping of the cross-section is known, the warping moment of inertia $I_{\omega\omega}$ of the cross-section can be determined:

$$I_{\omega\omega} = \int_A \omega_M^2 \, dA, \tag{8.81}$$

which yields:

$$I_{\omega\omega} = \frac{h^2 b^3 t}{24}, \tag{8.82}$$

in the unit of a length unit in sixth power. With the moment of inertia $I_{zz} = \frac{tb^3}{6}$ we get:

$$I_{\omega\omega} = I_{zz} \frac{h^2}{4}. \tag{8.83}$$

8.8.2 Single Symmetrical I-Cross-Section

Furthermore, the single symmetrical cross-section as shown in Fig. 8.23 is considered. The dimensions and wall thicknesses are given as indicated in Fig. 8.23, and the starting point of the considerations is the intersection of the upper flange and the web. The necessary characteristic areas $\bar{y}$, r_{tD} and $\bar{\omega}_{D,s}$ are shown in Fig. 8.23. Due to the symmetry properties of the cross-section, the coordinates $\bar{y}_s$ and $\bar{y}_M$ become zero, as do the deviation moment $I_{\bar{y}\bar{z}}$ and the two area integrals $S_{\bar{\omega}}$ and $I_{\bar{z}\bar{\omega}}$. As a result the initial value $\bar{\omega}_{D,A}$ of the curvature also disappears.

We first calculate the following cross-sectional integrals:

$$I_{\bar{z}\bar{z}} = \int_A \bar{y}^2 dA = \frac{1}{12} t_o b_o^3 + \frac{1}{12} t_u b_u^3 = I_{zz,o} + I_{zz,u},$$

$$I_{\bar{y}\bar{\omega}} = \int_A \bar{y}\bar{\omega} dA = -\frac{1}{12} t_u b_u^3 h = -I_{zz,u} h,$$

$$I_{\bar{\omega}\bar{\omega}} = \int_A \bar{\omega}^2 dA = \frac{1}{12} t_u b_u^3 h^2 = I_{zz,u} h^2, \tag{8.84}$$

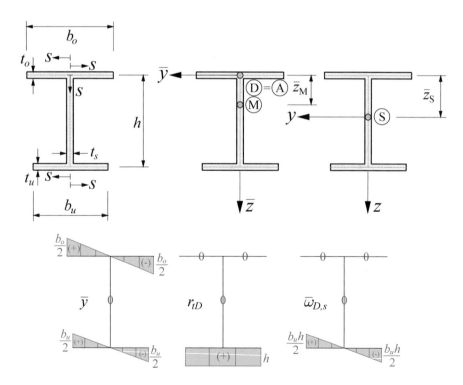

Fig. 8.23 Single symmetrical I-cross-section (top), $\bar{y}$−area, r_{tD}−area and unit warping $\bar{\omega}_{D,s}$ (bottom)

with $I_{zz,o} = \frac{1}{12}t_o b_o^3$ and $I_{zz,u} = \frac{1}{12}t_u b_u^3$. Due to the symmetry properties of the cross-section and as a consequence $\bar{y}_S = 0$ and $\bar{\omega}_{D,A} = 0$, the above area integrals are identical to those in the center of gravity system $\hat{y}$, $\hat{z}$, i.e.:

$$I_{\hat{z}\hat{z}} = I_{zz,o} + I_{zz,u}, \quad I_{\hat{y}\hat{\omega}} = -I_{zz,u}h, \quad I_{\hat{\omega}\hat{\omega}} = I_{zz,u}h^2. \tag{8.85}$$

The coordinate $\bar{y}_M$ of the shear center can be directly concluded as $\bar{y}_M = 0$. With (8.62) and with $I_{\hat{z}\hat{\omega}} = 0$ and $I_{\hat{y}\hat{z}} = 0$ the coordinate $\bar{z}_M$ follows as:

$$\bar{z}_M = -\frac{I_{\hat{y}\hat{\omega}}}{I_{\hat{z}\hat{z}}} = \frac{I_{zz,u}h}{I_{zz,o} + I_{zz,u}}. \tag{8.86}$$

The warping moment of inertia $I_{\omega\omega}$ then follows with $\bar{y}_M = 0$:

$$I_{\omega\omega} = I_{\hat{\omega}\hat{\omega}} + \bar{z}_M I_{\hat{y}\hat{\omega}} = \frac{I_{zz,o} I_{zz,u} h^2}{I_{zz,o} + I_{zz,u}}. \tag{8.87}$$

The warping ω_M with respect to the shear center M then results with $\bar{y}_M = 0$ as

$$\omega_M = \hat{\omega}_D + \bar{z}_M \hat{y}. \tag{8.88}$$

The distribution of ω_M over the simply symmetrical I-section is shown in Fig. 8.24. The ordinates ω_1 and ω_2 are as follows:

$$\omega_1 = \frac{b_o h}{2} \frac{I_{zz,u}}{I_{zz,o} + I_{zz,u}}, \quad \omega_2 = \frac{b_u h}{2} \frac{I_{zz,o}}{I_{zz,o} + I_{zz,u}}. \tag{8.89}$$

Fig. 8.24 Warping ω_M of the simple symmetrical I-cross-section

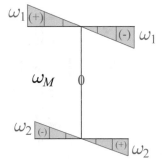

8.8.3 C-Cross-Section

Another important basic case is the C-cross-section as given in Fig. 8.25. The cross-section has the height h, the width b and section-wise constant wall thicknesses t_w and t_f. The position of the center of gravity S is known, and we are looking for the position of the shear center M, the warping moment of inertia $I_{\omega\omega}$ and the warping ω_M. The r_{tD}−area and the $\bar{z}$−area are already shown in Fig. 8.25. The axis of rotation D is to be set as shown in Fig. 8.25, the unit warping $\bar{\omega}_{D,s}$ is shown in Fig. 8.25, right. It is calculated according to (8.8) and (8.9) with the unit warping $\omega_{D,s}$ shown in Fig. 8.25 in the flanges with the boundary values $\pm\frac{bh}{2}$. Note that due to the orientation of the local circumferential axes s and the direction of rotation of the angle ϑ shown in Fig. 8.25, a positive unit warping occurs in the upper flange, whereas negative values occur in the lower flange.

Due to the symmetry of the cross-section, $\bar{z}_M = 0$ and $I_{\bar{y}\bar{z}} = I_{\hat{y}\hat{z}} = 0$ can be concluded. Furthermore, $\bar{z} = \hat{z}$ can be assumed immediately. The static moment $S_{\bar{\omega}}$ follows as zero, so that $\bar{\omega}_{D,A} = 0$ and thus $\hat{\omega}_D = \bar{\omega}_{D,s}$ hold. From this it follows immediately that $I_{\hat{z}\hat{\omega}} = I_{\bar{z}\bar{\omega}}$ and $I_{\hat{y}\hat{y}} = I_{\bar{y}\bar{y}}$.

We now calculate the area moment $I_{\hat{z}\hat{\omega}}$ as:

$$I_{\hat{z}\hat{\omega}} = \int_A \bar{z}\bar{\omega}_{D,s}\,dA = \int_s \bar{z}\bar{\omega}_{D,s}t\,ds = -\frac{t_f b^2 h^2}{4}. \tag{8.90}$$

To determine the position of the shear center M the area moment of inertia $I_{\hat{y}\hat{y}}$ is also required:

$$I_{\hat{y}\hat{y}} = \int_A \hat{z}^2\,dA = \int_s \hat{z}^2 t\,ds = \frac{t_f b h^2}{2} + \frac{t_w h^3}{12}. \tag{8.91}$$

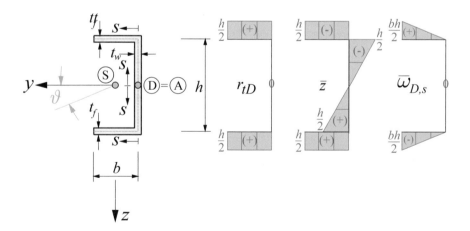

Fig. 8.25 C-cross-section, r_{tD}−area, $\bar{z}$−area and unit warping $\bar{\omega}_{D,s}$

Due to the symmetry properties of the cross-section, these are also the area values $I_{z\omega}$ and I_{yy}. The position of the shear center M can then be determined with $I_{\hat{y}\hat{z}} = 0$ as:

$$\bar{y}_M = \frac{I_{\hat{z}\hat{\omega}}}{I_{\hat{y}\hat{y}}} = -\frac{t_f b^2 h^2}{4 I_{\hat{y}\hat{y}}} = -\frac{t_f b^2 h^2}{4 I_{yy}}. \tag{8.92}$$

Obviously, the shear center M lies outside of the profile, as already mentioned in Chap. 6, Fig. 6.31. If the special case $2b = h$ as well as $t_f = t_w = t$ applies, then we get:

$$\bar{y}_M = -\frac{3b}{8}. \tag{8.93}$$

The comparison with Fig. 6.27 shows that the determination of the shear center M with the calculation method presented here yields identical results with the procedure described in Chap. 6.

The warping moment of inertia $I_{\bar{\omega}\bar{\omega}} = I_{\hat{\omega}\hat{\omega}}$ with respect to the center of rotation D results as:

$$I_{\hat{\omega}\hat{\omega}} = \frac{t_f b^3 h^2}{6}. \tag{8.94}$$

We obtain the warping moment of inertia $I_{\omega\omega}$ with $\bar{z}_M = 0$ as:

$$I_{\omega\omega} = I_{\hat{\omega}\hat{\omega}} - \bar{y}_M I_{\hat{z}\hat{\omega}} = \frac{t_f b^3 h^2}{6} \left(1 - \frac{3}{4} \frac{1}{1 + \frac{t_w h^3}{6 t_f b h^2}} \right). \tag{8.95}$$

Finally, we want to calculate the warping coordinates ω_M with respect to the shear center M, which currently result as follows:

$$\omega_M = \hat{\omega}_D - \bar{y}_M \hat{z}. \tag{8.96}$$

The resulting warping ω_M is shown in Fig. 8.26. The warping coordinates $\omega_1, \omega_2, \omega_3, \omega_4$ are as follows:

$$\omega_1 = -\omega_4 = \frac{bh}{2} \left(1 - \frac{1}{2} \frac{1}{1 + \frac{t_w h^3}{6 t_f b h^2}} \right), \quad -\omega_2 = \omega_3 = \frac{1}{4} \frac{bh}{1 + \frac{t_w h}{6 t_f b}}. \tag{8.97}$$

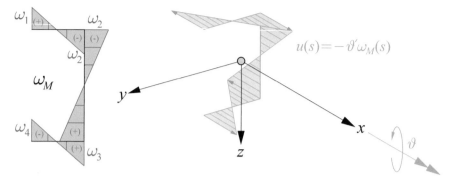

Fig. 8.26 Warping ω_M and longitudinal displacement u of the C-section

8.8.4   Z-Cross-Section

The final basic case is the Z-cross-section shown in Fig. 8.27. For this cross-section the center of gravity S and the shear center M are identical so that the shear center is selected as starting point A of the integration. Figure 8.27 also shows the unit warping $\bar\omega_{D,s}$. Obviously, $S_{\bar\omega} = 0$ is not zero so that a normalization of the warping surface has to be performed. For $S_{\bar\omega}$ we have:

$$S_{\bar\omega} = \int_A \bar\omega \, dA = \int_s \bar\omega t \, ds = \frac{t_f b^2 h}{2}. \tag{8.98}$$

The initial warping $\bar\omega_{D,A}$ results as:

$$\bar\omega_{D,A} = \frac{S_{\bar\omega}}{A} = \frac{bh}{2} \frac{1}{2 + \dfrac{t_w h}{t_f b}}. \tag{8.99}$$

The warping ω_M can then be determined as shown in Fig. 8.27, right. The warping coordinates ω_1, ω_2, ω_3, ω_4 result as:

$$\omega_1 = \omega_4 = \frac{bh}{2} \frac{1 + \dfrac{t_w h}{t_f b}}{2 + \dfrac{t_w h}{t_f b}}, \qquad \omega_2 = \omega_3 = \frac{bh}{2} \frac{1}{2 + \dfrac{t_w h}{t_f b}}. \tag{8.100}$$

The warping moment of inertia $I_{\omega\omega}$ follows as:

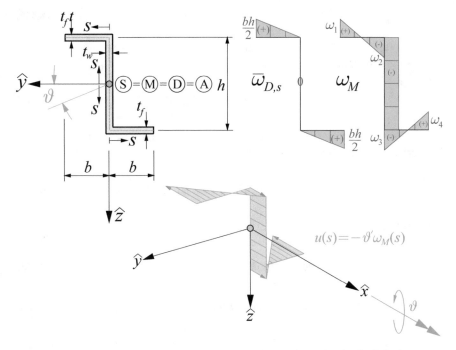

Fig. 8.27 Z-cross-section, unit warping $\bar{\omega}_{D,s}$ and warping ω_M (top); longitudinal displacement u (bottom)

$$I_{\omega\omega} = \frac{t_f b^3 h^2}{6} \cdot \frac{\dfrac{1}{2} + \dfrac{t_w h}{t_f b}}{2 + \dfrac{t_w h}{t_f b}}. \tag{8.101}$$

8.9 Determination of Internal Moments

8.9.1 Differential Equation of Warping Torsion

In this section we want to discuss which differential equation describes the present warping torsion problem and how to determine the internal moments that arise due to torsion of a beam. For this purpose, we will again consider the situation of the I-beam under torsional load as shown in Figs. 8.1, 8.2, 8.3, 8.4 and 8.5. The inner torsional moment M_x is equivalent to the force pair $F = \frac{M_{xs}}{h}$, which generates both the flange bending moment M_{Fl} and the resulting flange transverse shear force V_{Fl} in the two flanges if we consider the two flanges as beams under bending about the $z-$ axis. For the flange bending moment M_{Fl} we have:

$$M_{Fl}(x) = -EI_{zz}v''(x),$$
(8.102)

where I_{zz} represents the moment of inertia of the flange cross-section with respect to bending about the $z-$axis. From Fig. 8.2 we can conclude $v = \vartheta\frac{h}{2}$ so that the following applies:

$$M_{Fl}(x) = -EI_{zz}\vartheta''(x)\frac{h}{2}.$$
(8.103)

The transverse shear force V_{Fl} acting in the flanges can then be calculated as the first derivative of the flange moment M_{Fl}:

$$V_{Fl}(x) = -EI_{zz}v'''(x) = -EI_{zz}\vartheta'''(x)\frac{h}{2}.$$
(8.104)

From this the secondary torsional moment M_{xs} can be determined as:

$$M_{xs} = V_{Fl}h = -EI_{zz}\vartheta'''(x)\frac{h^2}{2}.$$
(8.105)

From Eq. (8.83) we can conclude that the product $I_{zz}\frac{h^2}{2}$ corresponds exactly to the warping moment of inertia $I_{\omega\omega}$, so that with $I_{zz} = \frac{tb^3}{12}$ we have:

$$M_{xs} = -EI_{\omega\omega}\vartheta'''(x).$$
(8.106)

The two flange moments M_{Fl}, multiplied by the distance h, are then interpreted as the so-called warping moment M_ω:

$$M_\omega = M_{Fl}h = -EI_{\omega\omega}\vartheta''(x).$$
(8.107)

According to Eq. (8.1) the total torsional moment M_x consists of M_{xp} from St. Venant torsion and M_{xs} as a consequence of secondary torsion. Hence:

$$M_x = M_{xp} + M_{xs} = GI_T\vartheta'(x) - EI_{\omega\omega}\vartheta'''(x).$$
(8.108)

Considering the equilibrium condition (7.128) from Chap. 7

$$M_x' = -m_T,$$
(8.109)

we obtain the differential equation of warping torsion as follows for the case of a prismatic beam with constant cross-section and elastic properties:

$$EI_{\omega\omega}\vartheta''''(x) - GI_T\vartheta''(x) = m_T.$$
(8.110)

This equation has been derived at this point using the example of the I-beam, but in the form given here it also applies to any other prismatic beam. It has the following

Fig. 8.28 Typical boundary conditions of a beam under torsion at the position $x = 0$: fork restraint (left), clamping (middle), free end (right)

general solution:

$$\vartheta(x) = \frac{C_1}{\lambda^2} \sinh(\lambda x) + \frac{C_2}{\lambda^2} \cosh(\lambda x) + C_3 x + C_4 + F(x). \qquad (8.111)$$

Here $F(x)$ is the term resulting from the particular solution of (8.110). For the special case that m_T is a constant torsional load, this results in:

$$\vartheta(x) = \frac{C_1}{\lambda^2} \sinh(\lambda x) + \frac{C_2}{\lambda^2} \cosh(\lambda x) + C_3 x + C_4 - \frac{m_T x^2}{2 G I_T}. \qquad (8.112)$$

The quantity λ is the so called decay factor, the meaning of which we will discuss later. It is defined as:

$$\lambda = \sqrt{\frac{G I_T}{E I_{\omega\omega}}}. \qquad (8.113)$$

Given that state variables are related to the center of gravity axis S and the shear center axis M, as shown in Fig. 8.18, the bending-torsion problem considered here is described by the following set of constitutive equations:

$$N = E A u', \quad M_y = -E I_{yy} w'', \quad M_z = E I_{zz} v'', \quad E I_{\omega\omega} \vartheta''''(x) - G I_T \vartheta''(x) = m_T, \qquad (8.114)$$

or:

$$E A u'' = -n, \quad E I_{yy} w'''' = q_z, \quad E I_{zz} v'''' = q_y, \quad E I_{\omega\omega} \vartheta''''(x) - G I_T \vartheta''(x) = m_T. \qquad (8.115)$$

The solution (8.111) of the differential equation of warping torsion (8.110) requires the definition of boundary conditions to determine the constants C_1, C_2, C_3 and C_4. For this purpose, we take another look at the conditions already described in Chap. 7, Fig. 7.29. If the case of the fork restraint (Fig. 8.28, left) is given, then the rotation ϑ is identically zero at this location. In addition, the second derivative of ϑ disappears at this point since the warping moment M_ω becomes zero:

$$\vartheta = 0, \quad \vartheta'' = 0. \qquad (8.116)$$

By a fork restraint we thus understand a type of bearing where torsional moments can be absorbed, but no rotation ϑ is possible, and where the warping of the cross-section can occur unhindered.

If the beam under consideration is fully clamped (Fig. 8.28, middle), then both the rotation ϑ and its first derivative ϑ' are zero at this location:

$$\vartheta = 0, \quad \vartheta' = 0. \tag{8.117}$$

If, on the other hand, there is a free unloaded end (Fig. 8.28, right), then the warping moment M_ω and thus the second derivative ϑ'' of the rotation ϑ disappears there. Furthermore, M_x becomes zero at a free end:

$$\vartheta'' = 0, \quad G I_T \vartheta'(x) - E I_{\omega\omega} \vartheta'''(x) = 0. \tag{8.118}$$

8.9.2 Selected Basic Cases

In the following, we want to consider a beam with fork restraints at both ends (Fig. 8.29) under constant load m_T (see also Fig. 7.30, Chap. 7). The boundary conditions are as follows:

$$\vartheta(x = 0) = 0, \quad \vartheta(x = l) = 0, \quad \vartheta''(x = 0) = 0, \quad \vartheta''(x = l) = 0. \tag{8.119}$$

With the general solution (8.112) we obtain the following constants C_1, C_2, C_3 and C_4:

$$C_1 = \frac{m_T}{G I_T} \frac{1 - \cosh \lambda l}{\sinh \lambda l}, \quad C_2 = \frac{m_T}{G I_T}, \quad C_3 = \frac{m_T l}{2 G I_T}, \quad C_4 = -\frac{m_T}{\lambda^2 G I_T}. \tag{8.120}$$

The rotation $\vartheta(x)$ can then be written as:

$$\vartheta(x) = \frac{m_T}{\lambda^2 G I_T} \left[\frac{1 - \cosh \lambda l}{\sinh \lambda l} \sinh \lambda x + \cosh \lambda x - 1 + \frac{\lambda^2 x}{2} (l - x) \right]. \tag{8.121}$$

From this the torsional moments M_{xp} and M_{xs} due to primary and secondary torsion as well as the warping moment M_ω can be determined. For M_{xp} we obtain:

$$M_{xp} = G I_T \vartheta' = \frac{m_T}{\lambda} \left[\frac{1 - \cosh \lambda l}{\sinh \lambda l} \cosh \lambda x + \sinh \lambda x + \frac{\lambda}{2} (l - 2x) \right]. \tag{8.122}$$

The secondary torsional moment M_{xs} reads:

$$M_{xs} = -E I_{\omega\omega} \vartheta'''(x) = -\frac{m_T}{\lambda} \left[\frac{1 - \cosh \lambda l}{\sinh \lambda l} \cosh \lambda x + \sinh \lambda x \right]. \tag{8.123}$$

Fig. 8.29 Beam under uniform load m_T (top), state variables (bottom)

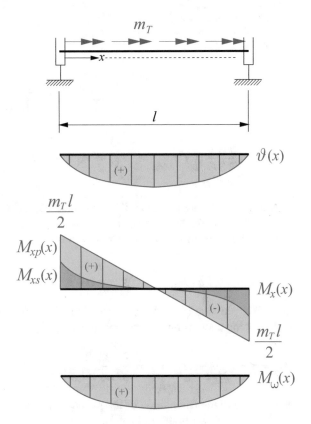

The warping moment M_ω results as:

$$M_\omega = -EI_{\omega\omega}\vartheta''(x) = -\frac{m_T}{\lambda^2}\left[\frac{1 - \cosh \lambda l}{\sinh \lambda l}\sinh \lambda x + \cosh \lambda x - 1\right]. \quad (8.124)$$

The comparison with the example from Chap. 7, Fig. 7.30, shows that not only the primary torsional moment M_{xp} and the rotation ϑ show significantly different distributions when considering this problem within the framework of warping torsion, but also the other state variables M_{xs} and M_ω have to be considered. A qualitative representation of all state variables can be found in Fig. 8.29, bottom. Obviously, the distribution of the two torsional moments M_{xp} and M_{xs} is non-linear over x, but the sum of these two quantities results in a function linear over x with the boundary values as already shown in Fig. 7.30. If, in addition, we have the case that the cross-section is free of warping with $EI_{\omega\omega} = 0$, then it is straightforward to verify that the state variables M_{xs} and M_ω, which are specific for warping torsion, become zero and that both the primary torsional moment M_{xp} and the rotation ϑ assume the values according to St. Venant torsion. Then:

Fig. 8.30 Moments M_{xp} and M_{xs} of the clamped beam under torsional moment M_T

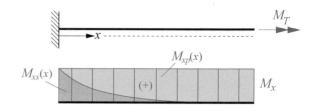

$$\vartheta(x) = \frac{m_T x}{2GI_T}(l - x), \quad M_{xp}(x) = m_T \left(\frac{l}{2} - x\right), \quad M_{xs}(x) = 0, \quad M_\omega = 0.$$
$$(8.125)$$

A further example is the cantilever beam under a single torsional moment M_T at the free end (see Fig. 8.1). The beam has the length l and the two constant properties GI_T and $EI_{\omega\omega}$. With the boundary conditions

$$\vartheta(x = 0) = 0, \quad \vartheta'(x = 0) = 0, \quad \vartheta''(x = l) = 0, \quad GI_T\vartheta'(x = l) - EI_{\omega\omega}\vartheta'''(x = l) = M_T$$
$$(8.126)$$

and the expression (8.111) for the rotation $\vartheta(x)$ ($F(x) = 0$, since $m_T = 0$) we get the following expressions for the constants C_1, C_2, C_3 and C_4:

$$C_1 = -\frac{M_T}{\sqrt{GI_T EI_{\omega\omega}}}, \quad C_2 = \frac{M_T \tanh \lambda l}{\sqrt{GI_T EI_{\omega\omega}}}, \quad C_3 = \frac{M_T}{GI_T}, \quad C_4 = -\frac{M_T \tanh \lambda l}{GI_T \lambda}.$$
$$(8.127)$$

This allows the calculation of the primary and secondary torsional moment M_{xp} and M_{xs} and the warping moment M_ω:

$$M_{xp} = GI_T\vartheta' = -M_T \left(\cosh \lambda x - \sinh \lambda x \tanh \lambda l - 1\right),$$
$$M_{xs} = -EI_{\omega\omega}\vartheta'''(x) = M_T \left(\cosh \lambda x - \sinh \lambda x \tanh \lambda l\right),$$
$$M_\omega = -EI_{\omega\omega}\vartheta''(x) = \frac{M_T}{\lambda} \left(\sinh \lambda x - \cosh \lambda x \tanh \lambda l\right). \quad (8.128)$$

We can verify the calculation by considering that the sum of M_{xp} and M_{xs} must result in the acting torsional moment $M_x = M_T$. A discussion of the resulting moments in (8.128) also shows that the proportion M_{xs} due to warping torsion decays with increasing distance from the clamping point at $x = 0$ wherein M_{xs} decays more rapidly with increasing torsional stiffness GI_T compared to the warping stiffness $EI_{\omega\omega}$, i.e. the greater λ is. The same applies to the warping stresses $\sigma_{\omega\omega}$ resulting from M_ω (see Eq. (8.137)). A qualitative representation of the distribution of the moments M_{xp} and M_{xs} is shown in Fig. 8.30. Apparently, the effect of warping torsion clearly dominates at the clamping point, while the proportion of primary torsion increases with increasing values of x.

As another example, a prismatic bar is shown in Fig. 8.31 (length l) which is loaded at one point by the single torsional moment $M_{T,0}$ (partial bar lengths l_1 and l_2). The stiffnesses GI_T and $EI_{\omega\omega}$ are constant over the entire length of the beam. Two local axes x_1 and x_2 are introduced as indicated. The moment distribution M_x

Fig. 8.31 Beam with fork restraints at both ends under single moment $M_{T,0}$ (top), distribution of the torsional moment $M_x(x)$ (bottom)

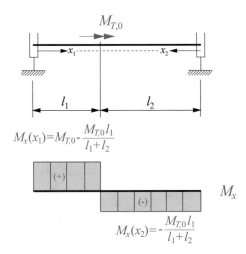

$$M_x(x_1) = M_{T,0} - \frac{M_{T,0} l_1}{l_1 + l_2}$$

$$M_x(x_2) = -\frac{M_{T,0} l_1}{l_1 + l_2}$$

according to St. Venant's theory is also shown here and can be derived from the result of Fig. 7.31 for the special case $GI_{T,1} = GI_{T,2}$.

The boundary and transition conditions associated with the solution (8.111) of the differential equation of warping torsion (8.110) are as follows. The following applies at the left fork bearing:

$$\vartheta_1(x_1 = 0) = 0, \quad \vartheta_1''(x_1 = 0) = 0. \tag{8.129}$$

Analogously we have at the right fork bearing:

$$\vartheta_2(x_2 = 0) = 0, \quad \vartheta_2''(x_2 = 0) = 0. \tag{8.130}$$

At $x_1 = l_1$ and $x_2 = l_2$ the following transition conditions must be observed:

$$\vartheta_1(x_1 = l_1) = \vartheta_2(x_2 = l_2),$$
$$\vartheta_1'(x_1 = l_1) = \vartheta_2'(x_2 = l_2),$$
$$\vartheta_1''(x_1 = l_1) = \vartheta_2''(x_2 = l_2),$$
$$EI_{\omega\omega}\left(\vartheta_1'''(x_1 = l_1) - \vartheta_2'''(x_2 = l_2)\right) + M_{T,0} = 0. \tag{8.131}$$

Hence, exactly as many conditions are available as the evaluation of (8.111) requires for the two beam segments 1 and 2. The presentation of the concrete determination of the constants is omitted here; only the final result is reported at this point (see e.g. Roik et al. (1972)). For the rotation ϑ in the two segments we obtain:

$$EI_{\omega\omega}\vartheta_1(x_1) = \frac{M_{T,0}}{\lambda^3}\left(\frac{l_2 \lambda x_1}{l} - \frac{\sinh \lambda l_2}{\sinh \lambda l}\sinh \lambda x_1\right),$$

$$EI_{\omega\omega}\vartheta_2(x_2) = \frac{M_{T,0}}{\lambda^3}\left(\frac{l_1 \lambda x_2}{l} - \frac{\sinh \lambda l_1}{\sinh \lambda l}\sinh \lambda x_2\right). \tag{8.132}$$

The primary torsional moment M_{xp} results in:

$$M_{xp,1} = \lambda^2 E I_{\omega\omega} \vartheta_1'(x_1) = M_{T,0} \left(\frac{l_2}{l} - \frac{\sinh \lambda l_2}{\sinh \lambda l} \cosh \lambda x_1 \right),$$

$$M_{xp,2} = \lambda^2 E I_{\omega\omega} \vartheta_2'(x_2) = M_{T,0} \left(-\frac{l_1}{l} + \frac{\sinh \lambda l_1}{\sinh \lambda l} \cosh \lambda x_2 \right). \quad (8.133)$$

The warping moment M_ω follows as:

$$M_{\omega,1} = -E I_{\omega\omega} \vartheta_1''(x_1) = \frac{M_{T,0}}{\lambda} \frac{\sinh \lambda l_2}{\sinh \lambda l} \sinh \lambda x_1,$$

$$M_{\omega,2} = -E I_{\omega\omega} \vartheta_2''(x_2) = \frac{M_{T,0}}{\lambda} \frac{\sinh \lambda l_1}{\sinh \lambda l} \sinh \lambda x_2. \quad (8.134)$$

The secondary torsional moment M_{xs} is obtained in the following form:

$$M_{xs,1} = -E I_{\omega\omega} \vartheta_1'''(x_1) = M_{T,0} \frac{\sinh \lambda l_2}{\sinh \lambda l} \cosh \lambda x_1,$$

$$M_{xs,2} = -E I_{\omega\omega} \vartheta_2'''(x_2) = -M_{T,0} \frac{\sinh \lambda l_1}{\sinh \lambda l} \cosh \lambda x_2. \quad (8.135)$$

A qualitative graphical representation of the state variables discussed here is shown in Fig. 8.32.

Further elementary basic cases can be taken from the special literature (see selection at the end of this chapter).

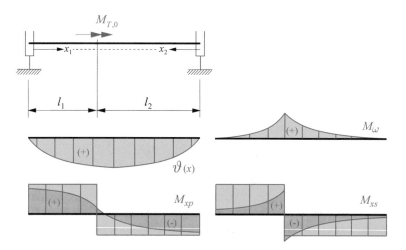

Fig. 8.32 Beam under torsion with fork restraints on both sides under the single moment $M_{T,0}$ (top), qualitative representation of the state variables (bottom)

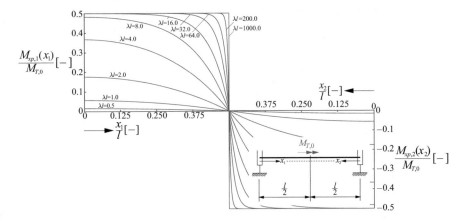

Fig. 8.33 Primary torsional moment M_{xp} of the beam under torsion with fork restraints on both sides under the single torsional moment $M_{T,0}$ for different values λl

Based on the example of Fig. 8.32 the influence of the decay factor λ can be discussed. We will examine the special case $l_1 = l_2 = \frac{l}{2}$ and consider the internal moments M_{xp}, M_{xs} and M_ω for different values for λl (Figs. 8.33, 8.34 and 8.35). The concentration of both the secondary torsional moment M_{xs} and the warping moment M_ω at the introduction point of the torsional moment $M_{T,0}$ is clearly evident. As a general conclusion it can be stated that with increasing values for λl and thus an increasing torsional stiffness GI_T with a constant warping stiffness $EI_{\omega\omega}$ the moments M_{xs} and M_ω decay faster with increasing distance than with small values λl. For very small values for λ the share of the secondary torsional moment M_{xs} in the torsional moment M_x dominates, whereas for large values for λ the problem converges closer and closer to St. Venant torsion.

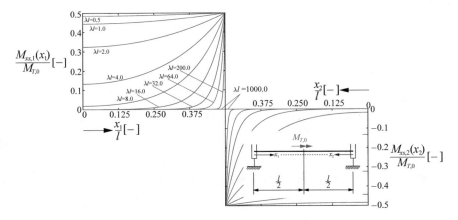

Fig. 8.34 Secondary torsional moment M_{xs} of the beam under torsion with fork restraints on both sides under the single torsional moment $M_{T,0}$ for different values λl

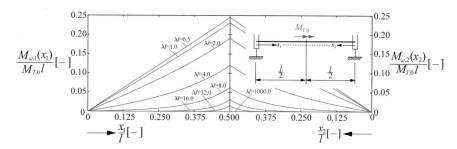

Fig. 8.35 Warping moment M_ω of the beam under torsion with fork restraints on both sides under the single torsional moment $M_{T,0}$ for different values λl

8.9.3 Influence of Normal Forces

For the calculation of the warping moment it is important to note that this can also be caused by normal forces if the normal forces act in cross-sectional points where the warping ω is not zero. This can be seen quite clearly in the I-section of Fig. 8.36 under eccentric normal force F.

The normal force F can be divided into four force groups as shown in Fig. 8.37. Obviously these four groups of forces result in a normal force N, two bending moments M_y and M_z, and a warping moment $M_\omega = F\omega$. Considering the warping of Fig. 8.36 with the maximum value $\omega_{\text{max}} = \frac{bh}{4}$, then the corresponding warping moment M_ω results as:

$$M_\omega = F\omega = \frac{Fbh}{4}. \tag{8.136}$$

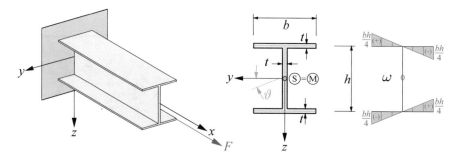

Fig. 8.36 I-beam under eccentric normal force F (left), cross-section and warping ω (right)

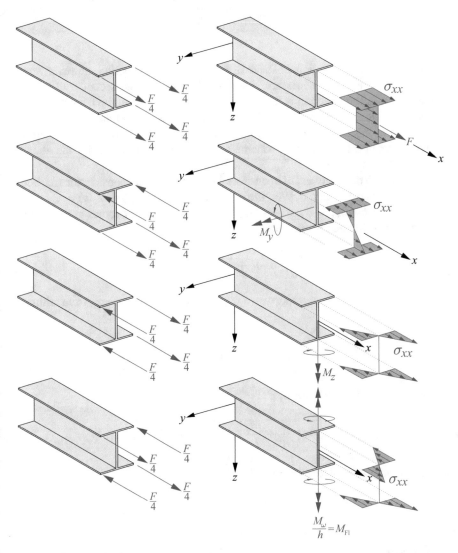

Fig. 8.37 Force groups, internal forces and moments, stress distributions on the I-beam under eccentric normal force F

8.10 Stress Analysis

If the reference axes have been selected as shown in Fig. 8.18, then the rod problem, the two bending problems and also the torsion problem are uncoupled. The calculation of the normal stress σ_{xx} can then be performed as shown below:

$$\sigma_{xx} = \frac{N}{A} + \frac{M_y}{I_{yy}} z - \frac{M_z}{I_{zz}} y + \frac{M_\omega}{I_{\omega\omega}} \omega. \tag{8.137}$$

Regarding the calculation of shear stresses in the cross-section, however, some further explanations are necessary. As shown in Fig. 8.4, the secondary torsional moment M_{xs} at the I-section is translated into shear stresses in the flanges and it is possible to calculate these stresses like the shear stresses $\tau_s(s)$ due to a transverse shear force at thin-walled cross-sections (see Chap. 6). We now want to develop a generally applicable calculation scheme for any cross-section besides the already considered I-section.

We start with the treatment of open cross-sections and again consider the free body diagram of Fig. 6.1, Chap. 6. With the circumferential shear flow $T_s(s) = \tau_s(s)t(s)$ and the force flow $n_{xx} = \sigma_{xx}t$ resulting from σ_{xx}, the sum of forces at the section element in $x-$direction results in:

$$\frac{dn_{xx}}{dx} dx ds + \frac{dT_s}{ds} ds dx = 0. \tag{8.138}$$

Dividing by dx and integrating over s from the starting point $s = s_A$ to an arbitrary point s leads to:

$$\int_{s_A}^{s} \frac{dT_s}{ds} ds = T_s(s) - T_s(s = s_A) = -\int_{s_A}^{s} \frac{dn_{xx}}{dx} ds. \tag{8.139}$$

With (8.137) we obtain for n_{xx}:

$$n_{xx} = \left(\frac{N}{A} + \frac{M_y}{I_{yy}} z - \frac{M_z}{I_{zz}} y + \frac{M_\omega}{I_{\omega\omega}} \omega \right) t, \tag{8.140}$$

and differentiating with respect to x yields:

$$\frac{dn_{xx}}{dx} = n'_{xx} = \left(\frac{N'}{A} + \frac{M'_y}{I_{yy}} z - \frac{M'_z}{I_{zz}} y + \frac{M'_\omega}{I_{\omega\omega}} \omega \right) t. \tag{8.141}$$

If we again assume that the normal force N in the considered beam is constant (i.e. $N' = 0$ applies), and if we use the relations $M'_y = V_z$, $M'_z = -V_y$ and $M'_\omega = M_{xs}$, then we obtain:

$$T_s(s) - T_s(s = s_A) = -\frac{V_z}{I_{yy}} \int_{s_A}^{s} t z ds - \frac{V_y}{I_{zz}} \int_{s_A}^{s} t y ds - \frac{M_{xs}}{I_{\omega\omega}} \int_{s_A}^{s} t \omega ds. \tag{8.142}$$

If an open cross-section is considered, then the start of the integration is placed at a free end of the cross-section so that $T_s(s = s_A) = 0$ applies. Introducing the area integrals

$$S_y(s) = \int_{S_A}^s tz\,ds, \quad S_z(s) = \int_{S_A}^s ty\,ds, \quad S_\omega(s) = \int_{S_A}^s t\omega\,ds \qquad (8.143)$$

yields:

$$T_s(s) = -\frac{V_z}{I_{yy}} S_y(s) - \frac{V_y}{I_{zz}} S_z(s) - \frac{M_{xs}}{I_{\omega\omega}} S_\omega(s). \qquad (8.144)$$

This is an expression for the determination of the shear flow $T_s(s)$ due to transverse shear forces and secondary torsional loading for an open thin-walled cross-section. The resulting shear stresses then result from this shear flow plus the share from primary / St. Venant torsion:

$$\tau_s(s) = -\frac{V_z S_y(s)}{I_{yy} t(s)} - \frac{V_y S_z(s)}{I_{zz} t(s)} - \frac{M_{xs} S_\omega(s)}{I_{\omega\omega} t(s)} + \frac{M_{xp}}{I_T} t(s). \qquad (8.145)$$

If, however, a closed cross-section is present, the calculation can be carried out in a similar way as described in Chap. 6. We refer here to the procedure as shown in Fig. 6.9 for a closed box section. The cross-section is converted into an open cross-section by introducing a section, and the shear flow $T_{s0}(s)$ resulting from the secondary torsional moment M_{xs} can be determined from (8.144) as:

$$T_{s0}(s) = -\frac{M_{xs} S_\omega(s)}{I_{\omega\omega}}. \qquad (8.146)$$

The resulting displacement discontinuity Δu_{10} then results as (see also Chap. 6, Eq. (6.40)):

$$\Delta u_{10} = \oint \frac{T_{s0}(s)}{Gt(s)}\,ds = -\frac{M_{xs}}{GI_{\omega\omega}} \oint \frac{S_\omega(s)}{t(s)}\,ds. \qquad (8.147)$$

Furthermore, a constant unit shear flow $T_{s1} = 1$ is applied and the resulting displacement discontinuity Δu_{11} is considered (see also Chap. 6, Eq. (6.41)):

$$\Delta u_{11} = \frac{1}{G} \oint \frac{ds}{t(s)}. \qquad (8.148)$$

The compatibility equation for determining the shear flow T_{s1} then results as in Chap. 6, Eq. (6.42)

$$\Delta u_{10} + T_{s1} \Delta u_{11} = 0. \qquad (8.149)$$

Thus:

$$T_{s1} = -\frac{\Delta u_{10}}{\Delta u_{11}} = \frac{M_{xs}}{I_{\omega\omega}} \frac{\displaystyle\oint \frac{S_\omega(s)}{t(s)}\,ds}{\displaystyle\oint \frac{ds}{t(s)}}. \qquad (8.150)$$

Hence the shear flow $T_s(s)$ can be calculated as:

$$T_s(s) = T_{s0}(s) + T_{s1} = -\frac{M_{xs}}{I_{\omega\omega}}\left[S_\omega(s) - \frac{\oint \frac{S_\omega(s)}{t(s)}ds}{\oint \frac{ds}{t(s)}}\right]. \tag{8.151}$$

The shear flow resulting from primary and secondary torsion can then be determined by the superposition of (7.33) and (8.151).

The procedure for multi-cellular cross-sections can be derived analogously, similarly to Chap. 6. This remains without further explanation at this point.

8.11 Comparison of Open and Closed Cross-Sections

We conclude this chapter with the comparative consideration of two cross-sections under torsional loading (see Fig. 8.38). This example is based on a study by Hanswille (1995). A bar of length l is given, which is supported by fork restraints on both sides and loaded by the torsional moment $M_{T,0}$ in the center point. Two different cross-sections are considered, namely a double symmetrical I-section on the one hand, and a closed box section on the other hand. Dimensions and wall thicknesses are given as shown in Fig. 8.38.

We first determine some relevant cross-sectional values, namely the maximum warping ω, the warping moment of inertia $I_{\omega\omega}$, the torsional moment of inertia I_T, and the decay factor λ. For beam A we get:

$$|\omega_{\max}| = \frac{bh}{4} = 87500 t^2,$$

$$I_{\omega\omega} = \frac{h^2 b^3 t}{24} = 38.28125 \cdot 10^{12} t^6,$$

$$I_T = \frac{1}{3}\sum_{i=1}^{n} l_i t_i^3 = 7.425 \cdot 10^6 t^4,$$

$$\lambda = \sqrt{\frac{G I_T}{E I_{\omega\omega}}} = 2.73 \cdot 10^{-4}\frac{1}{t}. \tag{8.152}$$

For beam B we obtain:

Fig. 8.38 Beam under torsion (top), investigated cross-sections (bottom)

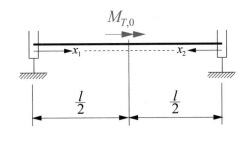

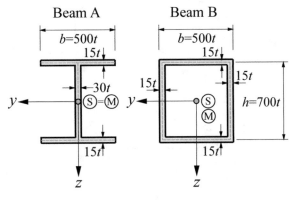

$$|\omega_{max}| = \left|\frac{bh}{4}\frac{b-h}{b+h}\right| = \frac{43750}{3}t^2,$$

$$I_{\omega\omega} = \frac{b^2h^2t_F}{8}\frac{(b-h)^2}{b+h} = 7.65625 \cdot 10^{12}t^6,$$

$$I_T = \frac{4A_m^2}{U} = 3.0625 \cdot 10^9 t^4,$$

$$\lambda = \sqrt{\frac{GI_T}{EI_{\omega\omega}}} = 0.0124\frac{1}{t}. \qquad (8.153)$$

The distributions of the internal moments M_{xp}, M_{xs} and $M_{\omega\omega}$ as well as of the rotation ϑ can be derived as follows. For the primary torsional moment M_{xp} we have:

$$M_{xp,1}(x_1) = M_{T,0}\left(\frac{1}{2} - \frac{\sinh\left(\frac{\lambda l}{2}\right)}{\sinh\lambda l}\cosh\lambda x_1\right),$$

$$M_{xp,2}(x_2) = M_{T,0}\left(-\frac{1}{2} + \frac{\sinh\left(\frac{\lambda l}{2}\right)}{\sinh\lambda l}\cosh\lambda x_2\right). \qquad (8.154)$$

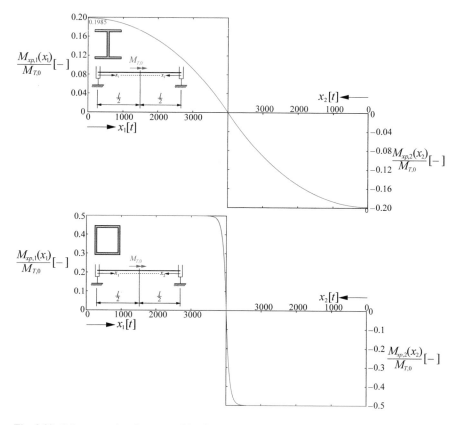

Fig. 8.39 Primary torsional moment M_{xp} for beam A and B

The distribution of M_{xp} over the length of the beam is shown in Fig. 8.39. For the secondary torsional moment M_{xs} we have:

$$M_{xs,1}(x_1) = M_{T,0} \frac{\sinh\left(\dfrac{\lambda l}{2}\right)}{\sinh \lambda l} \cosh \lambda x_1,$$

$$M_{xs,2}(x_2) = -M_{T,0} \frac{\sinh\left(\dfrac{\lambda l}{2}\right)}{\sinh \lambda l} \cosh \lambda x_2. \tag{8.155}$$

A graphic representation is given in Fig. 8.40.

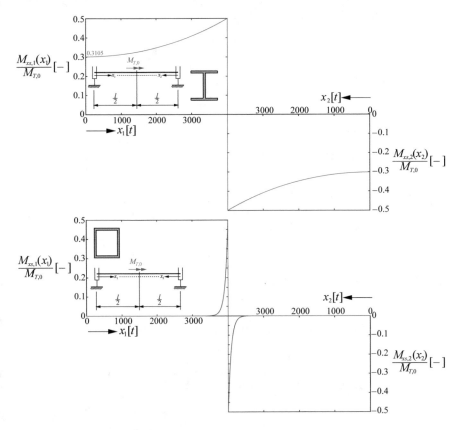

Fig. 8.40 Secondary torsional moment M_{xs} for beam A and B

The warping moment M_ω can be given as follows:

$$M_{\omega,1}(x_1) = \frac{M_{T,0}}{\lambda} \frac{\sinh\left(\dfrac{\lambda l}{2}\right)}{\sinh \lambda l} \sinh \lambda x_1,$$

$$M_{\omega,2}(x_2) = \frac{M_{T,0}}{\lambda} \frac{\sinh\left(\dfrac{\lambda l}{2}\right)}{\sinh \lambda l} \sinh \lambda x_2. \tag{8.156}$$

The distribution of M_ω over the beam length is given in Fig. 8.41.

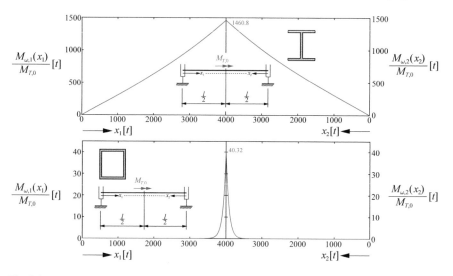

Fig. 8.41 Warping moment M_ω for beam A and B

The rotation ϑ of the cross-section of the beam can be represented as:

$$\frac{E\vartheta_1(x_1)}{M_{T,0}} = \frac{1}{\lambda^3 I_{\omega\omega}} \left(\frac{\lambda x_1}{2} - \frac{\sinh\left(\dfrac{\lambda l}{2}\right)}{\sinh \lambda l} \sinh \lambda x_1 \right),$$

$$\frac{E\vartheta_2(x_2)}{M_{T,0}} = \frac{1}{\lambda^3 I_{\omega\omega}} \left(\frac{\lambda x_2}{2} - \frac{\sinh\left(\dfrac{\lambda l}{2}\right)}{\sinh \lambda l} \sinh \lambda x_2 \right). \qquad (8.157)$$

A graphical representation is shown in Fig. 8.42.

If the warping moment M_ω is given, the normal stresses due to warping for both beams in the center (where the maximum warping moment due to the moment introduction occurs in both cases) can be determined as:

$$\sigma_{xx} = \frac{M_\omega}{I_{\omega\omega}} \omega. \qquad (8.158)$$

For beam A we obtain:

$$\left|\sigma_{xx,\max}\right| = 3.3389 \cdot 10^{-6} \frac{M_{T,0}}{t^3}. \qquad (8.159)$$

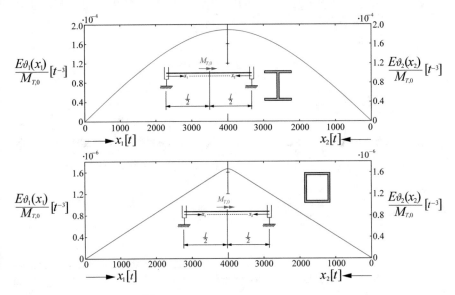

Fig. 8.42 Relative rotation $\frac{E\vartheta}{M_{T,0}}$ for beam A and B

For beam B:

$$\left|\sigma_{xx,\max}\right| = 7.68 \cdot 10^{-8}\frac{M_{T,0}}{t^3}. \tag{8.160}$$

A graphical representation of the stress distributions for both beams is shown in Fig. 8.43.

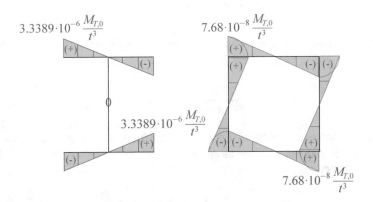

Fig. 8.43 Warping normal stresses $\sigma_{\omega\omega} = \sigma_{xx}$ for beam A and B

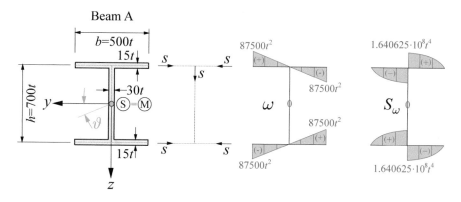

Fig. 8.44 Determination of the static moment $S_\omega(s)$ for beam A

The results show that considerable normal stresses occur at the open cross-section, whereas the normal stresses at the closed cross-section are two orders of magnitude smaller.

Furthermore, the shear stresses due to torsion are determined for the two considered cross-sections. For beam A, the following applies due to superposition of primary and secondary torsion:

$$\tau_s(s) = -\frac{M_{xs}S_\omega(s)}{I_{\omega\omega}t(s)} + \frac{M_{xp}}{I_T}t(s). \qquad (8.161)$$

The static moment S_ω necessary for the secondary component results as (see also Fig. 8.44)

$$S_\omega(s) = \int_A \omega \, dA = \int_s \omega t \, ds. \qquad (8.162)$$

The maximum shear stresses due to torsion at midspan are calculated with $M_{xp} = 0$ and $M_{xs} = \frac{1}{2}M_{T,0}$ as (Fig. 8.45, left):

$$\max \tau_s = -\frac{M_{xs}S_\omega(s)}{I_{\omega\omega}t(s)} = 1.4286 \cdot 10^{-7}\frac{M_{T,0}}{t^3}. \qquad (8.163)$$

At the left support we get the following components of the shear stress τ_s (Fig. 8.45, right). With the primary torsional moment $M_{xp} = 0.1985 M_{T,0}$ we obtain in the flange of the cross-section:

$$\max \tau_s = \frac{M_{xp}}{I_T}t(s) = 4.0101 \cdot 10^{-7}\frac{M_{T,0}}{t^3}. \qquad (8.164)$$

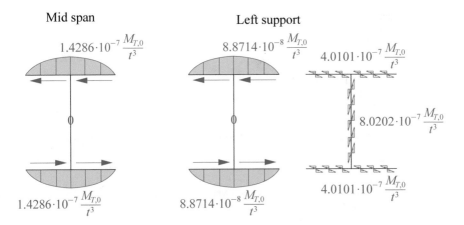

Fig. 8.45 Shear stresses τ_s due to torsion for beam A at midspan (left) and at the left supported end (right)

In the web we have:

$$\max \tau_s = 8.0202 \cdot 10^{-7} \frac{M_{T,0}}{t^3}. \tag{8.165}$$

Due to secondary torsion with $M_{xs} = 0.3105 \cdot M_{T,0}$ we obtain:

$$\max \tau_s = -\frac{M_{xs} S_\omega(s)}{I_{\omega\omega} t(s)} = 8.8714 \cdot 10^{-8} \frac{M_{T,0}}{t^3}. \tag{8.166}$$

The shear stresses for beam B (Fig. 8.46) as a consequence of the primary and secondary torsional moment M_{xp} and M_{xs} result as (cf. Eq (8.151) and (7.33)):

$$T_s(s) = -\frac{M_{xs}}{I_{\omega\omega}} \left[S_\omega(s) - \frac{\oint \dfrac{S_\omega(s)}{t(s)} ds}{\oint \dfrac{ds}{t(s)}} \right] + \frac{M_{xp}}{2 A_m}. \tag{8.167}$$

To determine the static moment S_ω, the cross-section is opened at the marked position and both the warping ω and the static moment S_ω are determined. The results of this calculation step are shown in Fig. 8.46. The ring integrals occurring in (8.167) currently result as follows:

$$\oint \frac{S_\omega(s)}{t(s)} ds = 1166666666.67 t^4, \quad \oint \frac{ds}{t(s)} = 160. \tag{8.168}$$

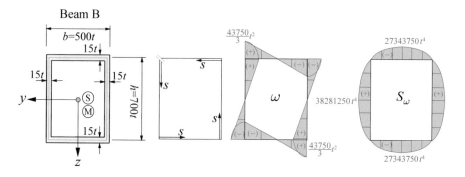

Fig. 8.46 Warping ω and static moment S_ω of the box beam

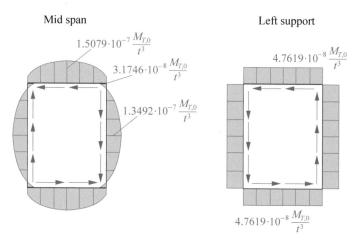

Fig. 8.47 Shear stresses τ_s due to torsion for beam B at midspan (left) and at the left supported end (right)

Hence we have for the shear flow due to M_{xs}:

$$T_s(s) = -\frac{M_{xs}}{I_{\omega\omega}}\left[S_\omega(s) - 7291666.67t^4\right]. \tag{8.169}$$

The resulting shear stress $\tau_s(s) = \frac{T_s(s)}{t(s)}$ at midspan with $M_{xs} = 0.5M_{T,0}$ is shown in Fig. 8.47, left.

At the left supported end with $M_{xp} = 0.5M_{T,0}$ and $M_{xs} = 0$ the constant circumferential shear stress τ_s over the wall thickness results as

$$T_s = \frac{M_{xp}}{2A_m t} = 4.7619 \cdot 10^{-8}\frac{M_{T,0}}{t^3}. \tag{8.170}$$

It is shown in Fig. 8.47, right.

Bibliography

Czerwenka G, Schnell W (1970) Einführung in die Rechenmethoden des Leichtbaus, Band 1, Bibliographisches Institut, Mannheim et al., Germany

Franke W, Friemann H (2005) Schub und Torsion in geraden Stäben, 3rd edn. Vieweg + Teubner, Wiesbaden, Germany

Hanswille G (1995) Vorlesungen über Stahl- und Verbundbau. Bergische Universität Wuppertal, Germany, Fachbereich Bauingenieurwesen

Kindmann R, Frickel J (2017) Elastische und plastische Querschnittstragfähigkeit, online edition 2017, https://kindmann.de. Accessed March 2020

Kollbrunner CF, Basler K (1964) Sektorielle Größen und Spannungen bei offenen, dünnwandigen Querschnitten, Mitteilungen der Technischen Kommission der Schweizer Stahlbau-Vereinigung, Heft 28. Verlag der Schweizer Stahlbau-Vereinigung, Zürich, Switzerland

Kollbrunner CF, Hajdin N (1964) Wölbkrafttorsion dünnwandiger Stäbe mit offenem Profil Teil I, Mitteilungen der Technischen Kommission der Schweizer Stahlbau-Vereinigung, Heft 29. Verlag der Schweizer Stahlbau-Vereinigung, Zürich, Switzerland

Kollbrunner CF, Hajdin N (1965) Wölbkrafttorsion dünnwandiger Stäbe mit offenem Profil Teil II, Mitteilungen der Technischen Kommission der Schweizer Stahlbau-Vereinigung, Heft 30. Verlag der Schweizer Stahlbau-Vereinigung, Zürich, Switzerland

Kollbrunner CF, Basler K (1966) Torsion. Springer. Berlin et al, Germany

Linke M, Nast E (2015) Festigkeitslehre für den Leichtbau. Springer. Berlin et al, Germany

Murray NW (1986) Introduction to the theory of thin-walled structures. Oxford University Press, Oxford, UK

Petersen C (1997) Stahlbau, Third edition, Vieweg. Braunschweig et al, Germany

Roik KH, Carl J, Lindner J (1972) Biegetorsionsprobleme gerader dünnwandiger Stäbe, Ernst & Sohn. Berlin et al, Germany

Roik KH (1978) Vorlesungen über Stahlbau, Ernst & Sohn. Berlin et al, Germany

Schnell W, Czerwenka G (1970) Einführung in die Rechenmethoden des Leichtbaus, Band 2. Bibliographisches Institut, Mannheim et al., Germany

Shama M (2010) Torsion and shear stresses in ships. Springer. Heidelberg et al, Germany

Wiedemann J (2007) Leichtbau 1: Elemente. Springer. Berlin et al, Germany

Wlassow WS (1964) Dünnwandige elastische Stäbe, vol I. VEB Verlag für Bauwesen, Berlin, Germany

Wlassow WS (1965) Dünnwandige elastische Stäbe, vol II. VEB Verlag für Bauwesen, Berlin, Germany

Part III
Energy Methods

Chapter 9
Work and Energy

9.1 Introduction

So far we have been dealing with the consideration of beam and rod structures and their exact-analytical treatment. However, in practical application, calculation tasks often arise in which exact solutions of given beam and rod problems are no longer possible or only possible with unproportional effort and one has to use one of the many available analysis and approximation methods of structural mechanics, which in many cases are based on work and energy considerations. Such work and energy methods are discussed in detail in this part of this book and their practical application is highlighted. Before this, however, a closer look at the concept of work and energy in the sense of structural mechanics is necessary which is the subject of the present chapter. In all our considerations we assume conservative forces, i.e. forces where the work W performed by them depends only on the starting and end point of their motion.

9.2 Work and Energy

9.2.1 Fundamentals

Figure 9.1 shows a particle that is subjected to a force F and that is shifted by the displacement u in the direction of the force F.

The work W performed by the force F is then calculated as:

$$W = Fu. \qquad (9.1)$$

This simple result is due to the fact that we tacitly assumed that the force F does not change over the distance u and that the displacement u takes place exactly in

© The Author(s), under exclusive license to Springer Nature Switzerland AG 2021   347
C. Mittelstedt, *Structural Mechanics in Lightweight Engineering*,
https://doi.org/10.1007/978-3-030-75193-7_9

Fig. 9.1 Shift of a particle under the force F by the displacement u in the direction of the force

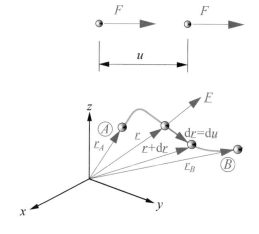

Fig. 9.2 Particle under the force $\underline{F}$ on a spatial trajectory between points A and B

the positive direction of the force F. We now want to extend the considerations to the general case as shown in Fig. 9.2. A particle under the force $\underline{F} = (F_x, F_y, F_z)^T$ is given, which moves on an arbitrary spatial trajectory with the position vector $\underline{r} = (r_x, r_y, r_z)^T$ from point A to point B. The location vectors of points A and B are $\underline{r}_A$ and $\underline{r}_B$, respectively. The force $\underline{F}$ is location-dependent, i.e. $\underline{F}$ can show different amounts and directions of action at different points of the path curve. The work increment dW which is performed by $\underline{F} = (F_x, F_y, F_z)^T$ along a displacement increment $d\underline{u} = (du, dv, dw)^T$ can be written as:

$$dW = \underline{F}\,d\underline{u} = F_x du + F_y dv + F_z dw. \tag{9.2}$$

This is the scalar product of $\underline{F}$ and $d\underline{u}$. The total work done by $\underline{F}$ on the trajectory between point A and point B is the sum of all work increments or the integral over the scalar product $\underline{F}\,d\underline{u}$:

$$W = \int_A^B \underline{F}\,d\underline{u}. \tag{9.3}$$

Hence, work is a scalar quantity. It is expressed in Newtonmeters[1] [Nm] or in Joule [J],[2] where the relation 1 Nm = 1 J holds. Joule is the SI-unit for work and energy:

$$1\,J = 1\frac{kg\,m^2}{s^2}. \tag{9.4}$$

Accordingly, a work of 10 J is performed when an object with the weight of 1 kg is lifted by 1 m in the gravity field of the earth. For simplicity we have assumed here that the acceleration due to gravity is $10\frac{m}{s^2}$.

---

[1] Isaac Newton, 1642–1726, English scientist.

[2] James Prescott Joule, 1818–1889, English physicist.

Fig. 9.3 Change in length u of a spring loaded by a force F with stiffness k, corresponding work

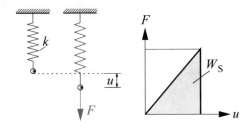

In case of a moment $\underline{M} = (M_x, M_y, M_z)^T$ which is subjected to the rotation $\underline{\varphi} = (\varphi_x, \varphi_y, \varphi_z)^T$ the work increment dW can be specified as follows:

$$dW = \underline{M}d\underline{\varphi} = M_x d\varphi_x + M_y d\varphi_y + M_z d\varphi_z. \tag{9.5}$$

The total work done between a start and an end rotation φ_A und φ_B is analogous to (9.3):

$$W = \int_{\varphi_A}^{\varphi_B} \underline{M}d\underline{\varphi}. \tag{9.6}$$

An elementary example is the work W_S performed when extending an elastic spring (see Fig. 9.3, left). A linear-elastic spring with the spring stiffness k is given which is tensioned by the force F. Let us consider the work W_S to be performed when the spring is extended from the unloaded state to an elongation u. Since the linear relation $F = ku$ is assumed to be valid here, the performed work W_S can be written as:

$$W_S = \int_0^u F du = \int_0^u ku du = \frac{1}{2}ku^2 = \frac{1}{2}Fu. \tag{9.7}$$

The work performed thus corresponds to the area below the line in Fig. 9.3, right, in the shown force-displacement diagram.

9.2.2 Internal and External Work

For an arbitrary solid it is necessary to distinguish between different kinds of work. In the context of this chapter the distinction between the external work or the work of the applied forces W_a on the one hand and the internal work W_i on the other hand is important. The external work is the work which is performed by the applied force quantities (single forces and moments, line loads, area and volume loads) imposed on a solid body due to the resulting deformations. Using the example of the linear spring just discussed, the work W_S performed according to (9.7) is an external work W_a.

The external force quantities lead to internal force quantities or a state of stress, as already discussed in detail in the previous chapters at the examples of rods and beams. The internal work is therefore the work performed by the internal forces along the displacements of the individual body points. We can illustrate this quite simply by the example of the elastic spring already discussed. The externally acting force F will cause an internal spring force of the same amount. The inner work increment dW_i is the product of the inner force F and the displacement increment du, i.e.:

$$dW_i = F\,du. \tag{9.8}$$

The total internal work W_i then results as the sum of all work increments dW_i:

$$W_i = \int_0^u F\,du. \tag{9.9}$$

In the case of a linear-elastic spring with the constitutive relation $F = ku$ we get:

$$W_i = \int_0^u ku\,du = \frac{1}{2}ku^2 = \frac{1}{2}Fu. \tag{9.10}$$

Apparently $W_i = W_a$ applies, i.e. the internal work is identical to the external work. We will define the term internal work in more detail later on for various other structural elements as relevant in lightweight engineering. The term strain energy or internal energy is also in use here, a term which we will discuss more precisely and in a more general form at a later point. If there is an elastic and, as a special case, even linear-elastic structure, as with the spring already considered, then the internal work which is a consequence of the external work is completely stored in the body as internal or inner energy and can be completely recovered when the body is relieved. This explains the concept of internal energy or strain energy.

We now want to extend the considerations to a solid, i.e. a sum of an infinite number of particles. The body under consideration is assumed to be linearly elastic. The internal work W_i as a consequence of the state of stress and the associated form changes will be discussed in detail later. The external work W_a of the applied loads in the form of single loads, single moments, line loads, area loads or volume loads is performed when the considered solid deforms under the load.

Let V be the volume of the considered solid of Fig. 9.4, and let the load be given in the form of m forces $\underline{P}_i$ $(i = 1, 2, ..., m)$ and n moments $\underline{M}_j$ $(j = 1, 2, ..., n)$ as well as the volume load $\underline{f}$ and the surface load $\underline{t}$ which acts on the partial surface ∂V_t. Then the external work W_a can be written as follows:

$$W_a = \frac{1}{2}\int_V \underline{f}\underline{u}\,dV + \frac{1}{2}\int_{\partial V_t} \underline{t}\underline{u}\,dS + \frac{1}{2}\sum_{i=1}^m \underline{P}_i\underline{u}_i + \frac{1}{2}\sum_{j=1}^n \underline{M}_j\underline{\varphi}_j. \tag{9.11}$$

Fig. 9.4 Three-dimensional solid under volume load $\underline{f}$, surface load $\underline{t}$, forces $\underline{P}_i$ and moments $\underline{M}_j$

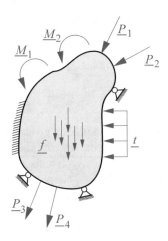

9.2.3 Principle of Work and Energy

The principle of work and energy in the framework of elastostatics states that the external work W_a, which is performed by the given applied loads acting on a solid, is completely converted into internal work W_i. The prerequisite for this is that the system is conservative. Thus:

$$W_i = W_a. \tag{9.12}$$

It also follows that the work done by the applied external loads W_a is stored as internal energy in the elastic solid and can be fully recovered when the body is relieved. The principle of work and energy applies to any closed elastic system, and it also states that no energy can be lost in such an elastic system. We will go into detail later on concerning the internal or inner energy or the so-called strain energy. It should be noted that in the literature a number of symbols, e.g. Π_i, W_i, U, are used for the strain energy.

In order to define the concept of the energy fundamentally, we state the following:

Energy is the ability to perform work.

Since we want to refer here exclusively to static problems and exclude dynamic effects, kinetic energy does not play a role in the further course of the explanations. However, the so-called potential energy is important for us at this point. A simple example would be a point mass m which is raised against the gravity field of the earth by the measure h. The energy required for lifting the point mass is $W = mgh$ and is also the potential energy supplied to the mass point (Fig. 9.5). Hence, by supplying the potential energy the mass point is enabled to perform work. Analogously, this also applies to the elastic spring already discussed, which also performs an internal work by the tensioning process or by the external work performed, and to which an internal energy was supplied which can be recovered and converted into work. The

Fig. 9.5 Potential energy at the example of a point mass

spring has thus been given the ability to do work itself again through the tensioning process and the internal energy stored in it. The potential energy in the two examples mentioned above is thus characterized by the changed position of the mass point and the changed shape of the spring, respectively, and can be used for work. Often the potential energy is simply called potential. The amount of potential energy must therefore be the same as the amount of work that was previously done and was necessary to obtain the potential energy. This circumstance will be discussed in detail later on.

It is obvious that it is impossible to realize a mechanical system in which the potential energy generated by the work performed can be greater than the work itself. Such a so-called perpetuum mobile is therefore a physically impossible structure. However, energy can easily be transferred from one form to another. For example, the potential energy of a mass m raised by the height h can be transferred into kinetic energy by releasing it in the gravitational field of the earth. From this, the energy conservation law or the energy conservation principle can be motivated as follows:

If a frictionless mechanical and self-contained system without external influences is considered, its total energy is the same at any given time. In this case, energy can neither be lost nor generated, it can only pass from one form to another.

9.3 Strain Energy and Complementary Strain Energy

9.3.1 The Rod

The first basic structural element is the rod shown in Fig. 9.6, which is under a tensile normal force N. The tensile member has the cross-sectional area A on which the normal stress $\sigma_{xx} = \frac{N}{A}$ is uniformly distributed. We assume elastic material behavior, but initially not necessarily linear elasticity. We now consider an infinitesimal section element of length dx and investigate its deformation behavior. The displacement at the negative section edge in the direction of the member has the value u, at the positive section edge it assumes the value $u + \frac{du}{dx}dx = u + u'dx = u + \varepsilon_{xx}dx$. The normal stress σ_{xx} performs a work increment along the infinitesimal strain $d\varepsilon_{xx}$ of the amount $\sigma_{xx}A d\varepsilon_{xx}dx$. With $dV = Adx$ the result is $\sigma_{xx}d\varepsilon_{xx}dV$. With the abbreviation $dU_0 = \sigma_{xx}d\varepsilon_{xx}$ we get $dU_0 dV$. Here dU_0 is the so-called incremental strain energy density, which is shown in the stress-strain diagram of Fig. 9.7, top left.

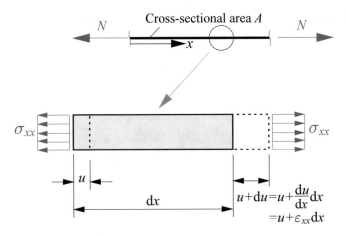

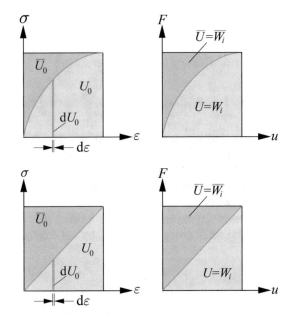

Fig. 9.6 Rod under tension (top), infinitesimal element (bottom) in the undeformed and the deformed state

Fig. 9.7 Stress-strain diagram with the strain energy density U_0 and the complementary strain energy density $\bar{U}_0$ (left), force-displacement diagram with strain energy $U = W_i$ and complementary strain energy $\bar{U} = \bar{W}_i$ (right) at the example of a rod under tensile load; elastic material behavior (top) and the special case of linear elasticity (bottom)

The strain energy density U_0 is then obtained by integrating dU_0 over the entire strain curve ε_{xx} up to the desired strain:

$$U_0 = \int_0^{\varepsilon_{xx}} \sigma_{xx} d\varepsilon_{xx}. \tag{9.13}$$

It represents the area below the line shown in Fig. 9.7, top left.

If the considered material is not only elastic, but if the special case of linear elasticity holds, the following expression for the normal stress σ_{xx} results from Hooke's law $\sigma_{xx} = E\varepsilon_{xx}$ and the kinematic relation $\varepsilon_{xx} = u'$:

$$\sigma_{xx} = E\varepsilon_{xx} = Eu'. \tag{9.14}$$

The normal force N of the rod results from the integration of the normal stress σ_{xx} over the cross-section A:

$$N = \int_A \sigma_{xx} \, dA = \int_A Eu' \, dA = Eu' \int_A dA = EAu', \tag{9.15}$$

so that:

$$u' = \frac{N}{EA}. \tag{9.16}$$

For the normal stress σ_{xx} the relation already known from Chap. 5 results:

$$\sigma_{xx} = Eu' = \frac{N}{A}. \tag{9.17}$$

For the strain energy density U_0 we obtain:

$$U_0 = \int_0^{\varepsilon_{xx}} \sigma_{xx} \, d\varepsilon_{xx} = E \int_0^{\varepsilon_{xx}} \varepsilon_{xx} \, d\varepsilon_{xx} = \frac{1}{2} E\varepsilon_{xx}^2 = \frac{1}{2}\sigma_{xx}\varepsilon_{xx} = \frac{1}{2}Eu'^2. \tag{9.18}$$

The term strain energy density is explained by the fact that it is the energy introduced into the rod as a result of the strain ε_{xx}. If there is an elastic rod whose state of deformation is completely reversible, then it is possible to recover this energy completely and convert it into work. The use of the term density indicates that here the strain energy is related to the volume of the sectional element.

The work increment dW_i of the infinitesimal element is thus obtained by multiplying the strain energy density U_0 by the volume dV:

$$dW_i = \int_0^{\varepsilon_{xx}} \sigma_{xx} \, d\varepsilon_{xx} \, dV = U_0 \, dV. \tag{9.19}$$

The internal energy or the strain energy W_i or U, respectively, stored in the rod results from the integration over the rod volume. With $dV = Adx$ we obtain

$$W_i = \int_0^l \int_0^{\varepsilon_{xx}} \sigma_{xx} \, d\varepsilon_{xx} \, Adx = A \int_0^l U_0 \, dx = U. \tag{9.20}$$

The strain energy is shown in the force-displacement diagram in Fig. 9.7, right, and is thus the area below the shown line.

In the case of linear elasticity W_i results as:

$$W_i = \int_V U_0 dx = \frac{1}{2} E \int_0^l \int_A u'^2 dA dx = \frac{1}{2} EA \int_0^l u'^2 dx. \tag{9.21}$$

Using $u'^2 = \varepsilon_{xx}^2 = \frac{\sigma_{xx}^2}{E^2} = \frac{N^2}{E^2 A^2}$ yields:

$$W_i = \frac{1}{2} EA \int_0^l \frac{N^2}{E^2 A^2} dx = \frac{1}{2} \int_0^l \frac{N^2}{EA} dx. \tag{9.22}$$

If, in addition, the normal force N and the extensional stiffness EA are constant, then we obtain:

$$W_i = \frac{N^2 l}{2EA}. \tag{9.23}$$

This is comparable to the linear elastic spring (constitutive law $F = ku$), where the strain energy is given by the value $\frac{1}{2} Fu = \frac{F^2}{2k}$.

Similarly, the complementary strain energy density can be defined:

$$\overline{U}_0 = \int_0^{\sigma_{xx}} \varepsilon_{xx} d\sigma_{xx}. \tag{9.24}$$

By means of Fig. 9.7, left, it can be concluded that

$$U_0 + \overline{U}_0 = \sigma_{xx} \varepsilon_{xx} \tag{9.25}$$

holds in every case. If the special case of linear elasticity is considered, then:

$$\overline{U}_0 = \frac{1}{2} \sigma_{xx} \varepsilon_{xx}. \tag{9.26}$$

This result can also be concluded immediately from Fig. 9.7, bottom left. In the case of linear elasticity, the stress-strain curve is a straight line and the area above the working line represents exactly the value for $\overline{U}_0$, leading to (9.26).

The corresponding complementary strain energy $\overline{W}_i$ or $\overline{U}$ can now be determined as:

$$d\overline{W}_i = \int_0^{\sigma_{xx}} \varepsilon_{xx} d\sigma_{xx} dV = \overline{U}_0 dV, \tag{9.27}$$

and after integration over the rod volume:

$$\overline{W}_i = \int_0^l \int_0^{\sigma_{xx}} \varepsilon_{xx} d\sigma_{xx} A dx = A \int_0^l \overline{U}_0 dx = \overline{U}. \tag{9.28}$$

If the special case of linear elasticity is present, then we have:

$$\overline{W}_i = \frac{1}{2} \int_0^l \frac{N^2}{EA} dx. \tag{9.29}$$

It can be seen that the expressions for W_i and $\overline{W}_i$ are identical when linear elasticity is considered (see also Fig. 9.7, bottom).

9.3.2 The Euler-Bernoulli Beam

We now want to extend the considerations to the Euler-Bernoulli beam (see also Chap. 5). For the necessary correlations for the calculation of such structures, we refer here explicitly to Chap. 5 and make the assumption that we are dealing with uniaxial bending (bending moment M and additionally the normal force N), i.e. that one of the two principal axes of the considered cross-section coincides with the z- axis, so that we can discard the indices of the bending moment $M_y = M$ and the transverse shear force $V_z = V$. The beam is subjected to the arbitrary continuous line load $q(x)$ in z-direction as well as the arbitrary but continuous axial line load $n(x)$. The beam has the variable, but continuous extensional stiffness $EA(x)$ and bending stiffness $EI(x)$. Since we want to keep the explanations as general as possible, no specific boundary conditions are defined at this point.

The equilibrium conditions for the present beam situation result in (see also Chap. 5, Eq. (5.95)):

$$N' = -n, \quad V' = -q, \quad M' = V. \tag{9.30}$$

The necessary area integrals are the moment of inertia I_{yy}, which we will refer to as I for the sake of simplicity, the static moment $S_y = S$ and the cross-sectional area A:

$$A = \int_A \mathrm{d}A, \quad S = \int_A z\mathrm{d}A, \quad I = \int_A z^2\mathrm{d}A. \tag{9.31}$$

In order to make this section self-contained, we now briefly go back to the basic equations of the Euler-Bernoulli beam. The displacement field of the Euler-Bernoulli beam in the case of uniaxial bending is obtained as follows (cf. Eqs. (5.6) and (5.7)):

$$u_P = u - z_P w', \quad w_P = w, \tag{9.32}$$

where P is an arbitrary point of the cross-section at the position z_P which does not necessarily have to lie on the beam's axis of gravity ($z_P = 0$). The quantities u and w are the displacements of the centroid axis in x- and z- direction, respectively.

Of all strains only the normal strain ε_{xx} remains here which we can calculate as follows (see Eq. (5.9))

$$\varepsilon_{xx} = \frac{\mathrm{d}u_P}{\mathrm{d}x} = u'_P = u' - z_P w''. \tag{9.33}$$

The resulting normal stresses σ_{xx} can then be determined from Hooke's Law $\sigma_{xx} = E\varepsilon_{xx}$ as (see Eq. (5.10) or (5.64)):

$$\sigma_{xx} = E\left(u' - zw''\right), \tag{9.34}$$

where we want to drop the indexing with respect to point P from this point on. The normal force N and bending moment M can then be calculated from this as follows. The normal force N results as:

$$N = \int_A \sigma_{xx} dA = E \int_A u' dA - E \int_A zw'' dA$$
$$= Eu' \int_A dA - Ew'' \int_A z dA. \tag{9.35}$$

The area integrals appearing here are the cross-sectional area A and the static moment S according to (9.31), whereby the latter must disappear in a principal axis system (cf. Chap. 5). Thus, we obtain:

$$N = EAu'. \tag{9.36}$$

Analogously we can treat and obtain the bending moment M:

$$M = \int_A \sigma_{xx} z dA = E \int_A u' z dA - E \int_A z^2 w'' dA$$
$$= Eu' \int_A z dA - Ew'' \int_A z^2 dA. \tag{9.37}$$

With $S = 0$ and the moment of inertia I we have:

$$M = -EIw''. \tag{9.38}$$

Inserting (9.36) and (9.38) in (9.34) yields the following expression for the normal stress σ_{xx} in the case of uniaxial bending (see also Eq. (5.65)):

$$\sigma_{xx} = \frac{N}{A} + \frac{M}{I}z. \tag{9.39}$$

The strain energy density U_0 can be written as:

$$U_0 = \int_0^{\varepsilon_{xx}} \sigma_{xx} d\varepsilon_{xx} = E \int_0^{\varepsilon_{xx}} \varepsilon_{xx} d\varepsilon_{xx} = \frac{1}{2}E\varepsilon_{xx}^2 = \frac{1}{2}E\left(u' - zw''\right)^2. \tag{9.40}$$

We determine the strain energy W_i by integration over the beam volume $dV = dAdx$ as follows:

$$W_i = \int_V U_0 dV = \frac{1}{2}E \int_0^l \int_A \left(u'^2 - 2zu'w'' + z^2 w''^2\right) dAdx, \tag{9.41}$$

or when considering (9.31) and $S = 0$:

$$W_i = \frac{1}{2}EA\int_0^l u'^2 dx + \frac{1}{2}EI\int_0^l w''^2 dx. \tag{9.42}$$

Analogously we can determine the complementary strain energy density as:

$$\overline{U}_0 = \int_0^{\sigma_{xx}} \varepsilon_{xx} d\sigma_{xx} + \int_0^{\tau_{xz}} \gamma_{xz} d\tau_{xz}. \tag{9.43}$$

It is important to note that the second term cannot be neglected a priori. Although the shear strain γ_{xz} becomes zero due to the kinematic assumptions for the Euler-Bernoulli beam, the shear stress τ_{xz} is required for reasons of equilibrium (see also Chaps. 5 and 6). From Hooke's law $\sigma_{xx} = E\varepsilon_{xx}$ and $\tau_{xz} = G\gamma_{xz}$ then results:

$$\overline{U}_0 = \int_0^{\sigma_{xx}} \frac{\sigma_{xx}}{E} d\sigma_{xx} + \int_0^{\tau_{xz}} \frac{\tau_{xz}}{G} d\tau_{xz} = \frac{\sigma_{xx}^2}{2E} + \frac{\tau_{xz}^2}{2G}. \tag{9.44}$$

From this we can determine the complementary strain energy $\overline{W}_i$ with $\tau_{xz} = \frac{VS}{Ib}$ e.g. for a rectangular cross section of width b by integration over the bar volume:

$$\begin{aligned}
\overline{W}_i &= \int_V \overline{U}_0 dV = \int_0^l \int_A \left(\frac{\sigma_{xx}^2}{2E} + \frac{\tau_{xz}^2}{2G} \right) dA dx \\
&= \int_0^l \int_A \left[\frac{1}{2E}\left(\frac{N}{A} + \frac{M}{I}z\right)^2 + \frac{1}{2G}\left(\frac{VS}{Ib}\right)^2 \right] dA dx.
\end{aligned} \tag{9.45}$$

Performing the integrations yields:

$$\overline{W}_i = \frac{1}{2}\int_0^l \frac{N^2}{EA} dx + \frac{1}{2}\int_0^l \frac{M^2}{EI} dx + \frac{1}{2}\int_0^l \frac{V^2}{GA_{eff}} dx, \tag{9.46}$$

wherein A_{eff} with

$$\frac{1}{A_{eff}} = \frac{1}{I^2}\int_A \frac{S^2}{b^2} dA \tag{9.47}$$

is the so-called effective area. The effective area is the area of a shear-loaded cross-section that is effectively available for load transfer. It can be expressed by the so-called shear correction factor K and the actual cross-sectional area A as follows:

$$A_{eff} = KA, \tag{9.48}$$

so that:

$$\overline{W}_i = \frac{1}{2}\int_0^l \frac{N^2}{EA} dx + \frac{1}{2}\int_0^l \frac{M^2}{EI} dx + \frac{1}{2}\int_0^l \frac{V^2}{KGA} dx. \tag{9.49}$$

Fig. 9.8 Simply supported
beam

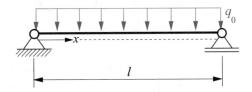

For the derivation and interpretation of the shear correction factor and the influence
of shear deformations in beam structures we explicitly refer to Chap. 16 of this book
where the beam theory according to Timoshenko for shear-deformable beams is
discussed in detail.

Often the term concerning the transverse shear force occurring in (9.49) is
neglected for slender beams with sufficient shear stiffness. For illustration we consider the simple example shown in Fig. 9.8. A simply supported straight beam (length
l, constant bending stiffness EI and effective shear stiffness KGA) under the uniform
load $q = q_0$ is considered. The complementary strain energy $\overline{W}_i$ is in this case:

$$\overline{W}_i = \overline{W}_{i,M} + \overline{W}_{i,V} = \frac{1}{2} \int_0^l \frac{M^2}{EI} dx + \frac{1}{2} \int_0^l \frac{V^2}{KGA} dx, \tag{9.50}$$

which, after performing the integrations of the transverse shear force and moment
areas, leads to the following expression for the example shown in Fig. 9.8:

$$\overline{W}_{i,M} + \overline{W}_{i,V} = \frac{q^2 l^5}{240EI} + \frac{q^2 l^3}{24kGA}. \tag{9.51}$$

In order to estimate the influence of the transverse shear force term we calculate the
quotient of $\overline{W}_{i,M}$ and $\overline{W}_{i,V}$ and obtain

$$\frac{\overline{W}_{i,M}}{\overline{W}_{i,V}} = \frac{KGA}{10EI} l^2. \tag{9.52}$$

For the sake of simplicity, a rectangular cross-section (width b, height h, shear correction factor $K = \frac{5}{6}$) is considered. After insertion we have:

$$\frac{\overline{W}_{i,M}}{\overline{W}_{i,V}} = \frac{G}{E} \left(\frac{l}{h} \right)^2. \tag{9.53}$$

If an isotropic material with the Poisson's ratio $\nu = 0.3$ is assumed, then the shear
modulus results from (2.156) as

$$G = \frac{E}{2(1+\nu)} = \frac{5}{13} E, \tag{9.54}$$

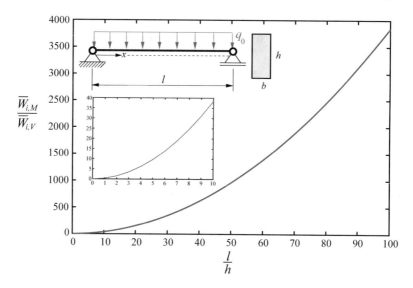

Fig. 9.9 Ratio $\frac{\overline{W}_{i,M}}{\overline{W}_{i,V}}$, plotted as function of the slenderness $\frac{l}{h}$

so that:

$$\frac{\overline{W}_{i,M}}{\overline{W}_{i,V}} = \frac{5}{13}\left(\frac{l}{h}\right)^2. \tag{9.55}$$

A graphical representation of $\frac{\overline{W}_{i,M}}{\overline{W}_{i,V}}$, plotted over the ratio $\frac{l}{h}$, is shown in Fig. 9.9.

From this graph it is easy to see that the transverse shear force term $\overline{W}_{i,V}$ loses importance with increasing slenderness $\frac{l}{h}$ compared to the bending moment term $\overline{W}_{i,M}$. Significant values for $\overline{W}_{i,V}$ are only reached at $\frac{l}{h}$ under conditions where the validity of the Euler-Bernoulli beam theory is questionable anyway. Consequently, transverse shear force influences are neglected when considering the complementary strain energy of slender beams in many applications of lightweight construction. Exceptions to this are compact beams, especially those with low shear stiffness, which is often the case with fibre composite materials used in lightweight engineering. The Chaps. 16 and 18 contain a number of further explanations in this regard.

9.3.3   Torsion

A straight bar of length l is given which is subjected to a torsional moment M_x. The strain energy increment $dW_{i,p}$ (the index p stands for primary torsion, see Chap. 7) is then:

$$dW_{i,p} = \frac{1}{2}M_x d\vartheta. \tag{9.56}$$

From the constitutive law (7.7) of St. Venant torsion $GI_T\vartheta' = M_x$ we obtain

$$d\vartheta = \frac{M_x dx}{GI_T},\tag{9.57}$$

so that from (9.56):

$$dW_{i,p} = \frac{1}{2}\frac{M_x^2}{GI_T}dx = \frac{1}{2}GI_T\left(\vartheta'\right)^2 dx.\tag{9.58}$$

By integration over the beam length l we then obtain the strain energy of the beam due to St. Venant torsion as:

$$W_{i,p} = \frac{1}{2}\int_0^l \frac{M_x^2}{GI_T}dx = \frac{1}{2}\int_0^l M_x\vartheta' dx = \frac{1}{2}\int_0^l GI_T\left(\vartheta'\right)^2 dx.\tag{9.59}$$

The derivation of the strain energy due to warping torsion (here the index s is used for the secondary torsion, cf. Chap. 8) can also be motivated again quite easily at the example of the I-beam (see Fig. 8.1 and the explanations there), where we first consider the strain energy of a flange with the deflection v and the bending stiffness EI_f. We have:

$$dW_{i,s} = \frac{1}{2}EI_f\left(v''\right)^2 dx.\tag{9.60}$$

With $v = \frac{h}{2}\vartheta$ and the warping stiffness $EI_{\omega\omega}$ of a double symmetric I-profile as $EI_{\omega\omega} = EI_f\frac{h^2}{2}$ we get:

$$dW_{i,s} = \frac{1}{4}EI_{\omega\omega}\left(\vartheta''\right)^2 dx.\tag{9.61}$$

This result is to be multiplied by the value 2 since such an I-beam of course has two flanges. Integration over the beam length l then yields the strain energy due to warping torsion as:

$$W_{i,s} = \frac{1}{2}\int_0^l EI_{\omega\omega}\left(\vartheta''\right)^2 dx = \frac{1}{2}\int_0^l \frac{M_\omega^2}{EI_{\omega\omega}}dx,\tag{9.62}$$

where M_ω is the warping moment. Although this expression was motivated at the example of an I-beam, it is valid for any thin-walled cross-sectional shape.

The total internal energy / strain energy W_i of a beam under torsion then results from the addition of (9.59) and (9.62):

$$W_i = W_{i,p} + W_{i,s} = \frac{1}{2} \int_0^l GI_T \left(\vartheta'\right)^2 \mathrm{d}x + \frac{1}{2} \int_0^l EI_{\omega\omega} \left(\vartheta''\right)^2 \mathrm{d}x$$

$$= \frac{1}{2} \int_0^l \frac{M_x^2}{GI_T} \mathrm{d}x + \frac{1}{2} \int_0^l \frac{M_\omega^2}{EI_{\omega\omega}} \mathrm{d}x. \qquad (9.63)$$

This expression corresponds to the complementary strain energy $\overline{W}_i$ because of the linear elasticity assumed here.

9.3.4 Combined Loading

If a beam structure is subjected to axial extension, bending about its two principal axes, transverse shear forces in both principal axis directions, and primary and secondary torsion, then the strain energy can be specified by superposition of the individual components (assuming that all cross-sectional normalizations have been performed):

$$W_i = \frac{1}{2} \int_0^l EAu'^2 \mathrm{d}x + \frac{1}{2} \int_0^l EI_{yy} w''^2 \mathrm{d}x + \frac{1}{2} \int_0^l EI_{zz} v''^2 \mathrm{d}x$$

$$+ \frac{1}{2} \int_0^l GI_T \left(\vartheta'\right)^2 \mathrm{d}x + \frac{1}{2} \int_0^l EI_{\omega\omega} \left(\vartheta''\right)^2 \mathrm{d}x. \qquad (9.64)$$

The complementary strain energy $\overline{W}_i$ is in this case:

$$\overline{W}_i = \frac{1}{2} \int_0^l \frac{N^2}{EA} \mathrm{d}x + \frac{1}{2} \int_0^l \frac{M_y^2}{EI_{yy}} \mathrm{d}x + \frac{1}{2} \int_0^l \frac{M_y^2}{EI_{yy}} \mathrm{d}x$$

$$+ \frac{1}{2} \int_0^l \frac{V_y^2}{K_y GA} \mathrm{d}x + \frac{1}{2} \int_0^l \frac{V_z^2}{K_z GA} \mathrm{d}x + \frac{1}{2} \int_0^l \frac{M_x^2}{GI_T} \mathrm{d}x + \frac{1}{2} \int_0^l \frac{M_\omega^2}{EI_{\omega\omega}} \mathrm{d}x, \qquad (9.65)$$

where K_y and K_z are the shear correction factors related to the two transverse shear forces V_y and V_z (see Chap. 16).

9.3.5 Generalization for the Continuum

We now want to generalize the relationships established so far for the three-dimensional continuum. For this purpose, we consider the infinitesimal three-dimensional sectional element shown in Fig. 9.10 and assume that only the normal stress σ_{xx} is present. The normal stress σ_{xx} performs the work increment $\sigma_{xx} \mathrm{d}y \mathrm{d}z \mathrm{d}\varepsilon_{xx} \mathrm{d}x$ along the infinitesimal strain $\mathrm{d}\varepsilon_{xx}$. With $\mathrm{d}x \mathrm{d}y \mathrm{d}z = \mathrm{d}V$ this expression changes to $\sigma_{xx} \mathrm{d}\varepsilon_{xx} \mathrm{d}V$ or $\mathrm{d}U_0 \mathrm{d}V$, so that we get the following expression for the work increment $\mathrm{d}W_i$:

Fig. 9.10 Infinitesimal
three-dimensional sectional
element

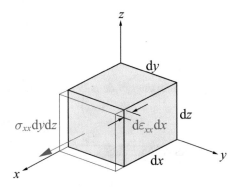

$$dW_i = \int_0^{\varepsilon_{xx}} \sigma_{xx} d\varepsilon_{xx} dV = U_0 dV = dU. \tag{9.66}$$

We now extend the consideration to the case where all normal stresses σ_{xx}, σ_{yy}, σ_{zz} and all shear stresses τ_{yz}, τ_{xz}, τ_{xy} are present. We then obtain:

$$dW_i = \int_0^{\varepsilon_{xx}} \sigma_{xx} d\varepsilon_{xx} dV + \int_0^{\varepsilon_{yy}} \sigma_{yy} d\varepsilon_{yy} dV + \int_0^{\varepsilon_{zz}} \sigma_{zz} d\varepsilon_{zz} dV$$
$$+ \int_0^{\gamma_{yz}} \tau_{yz} d\gamma_{yz} dV + \int_0^{\gamma_{xz}} \tau_{xz} d\gamma_{xz} dV + \int_0^{\gamma_{xy}} \tau_{xy} d\gamma_{xy} dV$$
$$= U_0 dV, \tag{9.67}$$

or, when using the index notation and employing Einstein's summation convention:

$$dW_i = \int_0^{\varepsilon_{ij}} \sigma_{ij} d\varepsilon_{ij} dV = U_0 dV = dU. \tag{9.68}$$

Similarly we obtain for $d\overline{W}_i$:

$$d\overline{W}_i = \int_0^{\sigma_{xx}} \varepsilon_{xx} d\sigma_{xx} dV + \int_0^{\sigma_{yy}} \varepsilon_{yy} d\sigma_{yy} dV + \int_0^{\sigma_{zz}} \varepsilon_{zz} d\sigma_{zz} dV$$
$$+ \int_0^{\tau_{yz}} \gamma_{yz} d\tau_{yz} dV + \int_0^{\tau_{xz}} \gamma_{xz} d\tau_{xz} dV + \int_0^{\tau_{xy}} \gamma_{xy} d\tau_{xy} dV$$
$$= \overline{U}_0 dV, \tag{9.69}$$

and in index notation:

$$d\overline{W}_i = \int_0^{\sigma_{ij}} \varepsilon_{ij} d\sigma_{ij} dV = \overline{U}_0 dV. \tag{9.70}$$

The quantities W_i respectively $\overline{W}_i$ then become:

$$W_i = \int_V U_0 dV = U,$$

$$\overline{W}_i = \int_V \overline{U}_0 dV = \overline{U}. \tag{9.71}$$

For the strain energy density U_0 the following relation applies:

$$dU_0 = \sigma_{ij} d\varepsilon_{ij}. \tag{9.72}$$

We can thus conclude:

$$\sigma_{ij} = \frac{\partial U_0}{\partial \varepsilon_{ij}}. \tag{9.73}$$

Thus, once the strain energy density is given, the stress component σ_{ij} can be determined by partial derivation of U_0 with respect to the assigned strain component ε_{ij}. Similarly we have for $\overline{U}_0$ with $d\overline{U}_0 = \varepsilon_{ij} d\sigma_{ij}$:

$$\varepsilon_{ij} = \frac{\partial \overline{U}_0}{\partial \sigma_{ij}}. \tag{9.74}$$

9.4 Application of the Principle of Work and Energy to Elastic Deformations

The principle of work and energy $W_i = W_a$ (Eq. (9.12)) can be used at this point for a rather simple problem class, namely the calculation of deformations of structures under point loads or single moments. For this purpose, we consider a truss structure consisting of two rods, as shown in Fig. 9.11. The two rods have identical extensional stiffnesses EA. We are looking for the vertical displacement w of the point of force application, i.e. the displacement of this point in the direction of the line of action of F. We use the principle of work and energy (9.12) to solve this problem. The normal forces in the members result in $N_1 = \frac{F}{\tan(\alpha)}$ and $N_2 = -\frac{F}{\sin(\alpha)}$. We then obtain the strain energy W_i of the truss as:

$$W_i = \frac{1}{2} \int_0^l \frac{N_1^2}{EA} dx + \frac{1}{2} \int_0^{\frac{l}{\cos(\alpha)}} \frac{N_2^2}{EA} dx, \tag{9.75}$$

and after inserting N_1 and N_2:

$$W_i = \frac{F^2 l}{2EA \tan^2(\alpha)} \left(1 + \frac{1}{\cos^3(\alpha)}\right). \tag{9.76}$$

The external work W_a results as:

Fig. 9.11 Truss under point load F

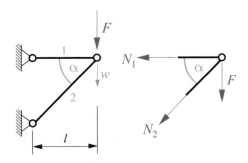

$$W_a = \frac{1}{2}Fw. \tag{9.77}$$

Imposing $W_i = W_a$ then yields:

$$\frac{F^2 l}{2EA \tan^2(\alpha)}\left(1 + \frac{1}{\cos^3(\alpha)}\right) = \frac{1}{2}Fw. \tag{9.78}$$

Solving for the displacement w results in:

$$w = \frac{Fl}{EA \tan^2(\alpha)}\left(1 + \frac{1}{\cos^3(\alpha)}\right). \tag{9.79}$$

We also demonstrate the application of the principle of work and energy on the cantilever beam of Fig. 9.12 which is loaded at its free end by a single force F. We want to determine the vertical displacement of the point of force application. First, we determine the distribution of the bending moment $M(x)$ over the length of the beam, as shown in Fig. 9.12. When using the integral table of Fig. 5.8 we obtain:

$$W_i = \frac{1}{2}\int_0^l \frac{M^2}{EI}dx = \frac{F^2 l^3}{6EI}. \tag{9.80}$$

For the external work W_a we have:

$$W_a = \frac{1}{2}Fw. \tag{9.81}$$

Imposing $W_i = W_a$ then yields:

$$w = \frac{Fl^3}{3EI}. \tag{9.82}$$

Finally, the rotation φ of the left support of the bar shown in Fig. 9.13 is determined. The beam is loaded at the left support by an edge moment M_0. The resulting moment line is also shown in Fig. 9.13. The strain energy of the beam is then obtained as:

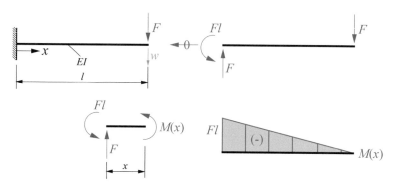

Fig. 9.12 Cantilever beam under point load F

Fig. 9.13 Beam under
single moment M_0

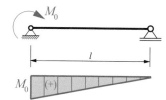

$$W_i = \frac{1}{2}\int_0^l \frac{M^2}{EI}\mathrm{d}x = \frac{M_0^2 l}{6EI}. \tag{9.83}$$

The external work W_a results in:

$$W_a = \frac{1}{2}M\varphi. \tag{9.84}$$

The theorem $W_i = W_a$ then yields the rotation φ as:

$$\varphi = \frac{M_0 l}{3EI}. \tag{9.85}$$

Obviously, the principle of work and energy is suitable for the simple calculation of deformations of structures under single forces and moments. However, this method is limited since in this way we can only calculate displacements or rotations at the points of application of the single forces and moments, and here only those deformations whose directions also correspond to the directions of action of the acting forces or moments.

9.5 General Principle of Work and Energy of Elastostatics

From the equilibrium conditions (2.45) and the kinematic equations (2.70) a general formulation of the principle of work and energy can be derived which establishes a relationship between any equilibrium group of stresses and forces (characterized by the superscript index (1)) and any kinematically permissible field of strains and displacements (characterized by the superscript index (2)). Note that in the following general formulation, the equilibrium group and the strain and displacement field do not necessarily have to be a consequence of each other.

We now consider an equilibrium group consisting of the three normal stresses $\sigma_{xx}^{(1)}$, $\sigma_{yy}^{(1)}$, $\sigma_{zz}^{(1)}$ and the three shear stresses $\tau_{xy}^{(1)}$, $\tau_{xz}^{(1)}$, $\tau_{yz}^{(1)}$. Furthermore, the volume forces $f_x^{(1)}$, $f_y^{(1)}$, $f_z^{(1)}$ as well as the stress vectors $t_x^{(1)}$, $t_y^{(1)}$, $t_z^{(1)}$ are given. The stress field satisfies the equilibrium conditions (2.45) identically. In addition, the boundary conditions $t_x = t_{x0}$, $t_y = t_{y0}$, $t_z = t_{z0}$ on ∂V_t are identically fulfilled. The strain and displacement field, on the other hand, is composed of the displacements $u^{(2)}$, $v^{(2)}$, $w^{(2)}$ and the strain components $\varepsilon_{xx}^{(2)}$, $\varepsilon_{yy}^{(2)}$, $\varepsilon_{zz}^{(2)}$, $\gamma_{yz}^{(2)}$, $\gamma_{xz}^{(2)}$, $\gamma_{xy}^{(2)}$. The displacements and strains are coupled via the kinematic relations (2.70). The displacement field also satisfies given displacement boundary conditions of the type $u = u_0$, $v = v_0$, $w = w_0$ on the corresponding partial surface ∂V_u.

We first consider the local equilibrium conditions (2.45):

$$\frac{\partial \sigma_{xx}^{(1)}}{\partial x} + \frac{\partial \tau_{xy}^{(1)}}{\partial y} + \frac{\partial \tau_{xz}^{(1)}}{\partial z} + f_x^{(1)} = 0,$$

$$\frac{\partial \tau_{xy}^{(1)}}{\partial x} + \frac{\partial \sigma_{yy}^{(1)}}{\partial y} + \frac{\partial \tau_{yz}^{(1)}}{\partial z} + f_y^{(1)} = 0,$$

$$\frac{\partial \tau_{xz}^{(1)}}{\partial x} + \frac{\partial \tau_{yz}^{(1)}}{\partial y} + \frac{\partial \sigma_{zz}^{(1)}}{\partial z} + f_z^{(1)} = 0. \qquad (9.86)$$

In the further explanations, we make use of the index notation and Einstein's summation convention, using the coordinates x_1, x_2, x_3 and the indices i and j (where $i, j = 1, 2, 3$). Thus:

$$\frac{\partial \sigma_{ij}^{(1)}}{\partial x_j} + f_i^{(1)} = 0.$$

We now multiply these equations with the displacements $u_i^{(2)}$ and perform an integration over the body volume V:

$$\int_V \frac{\partial \sigma_{ij}^{(1)}}{\partial x_j} u_i^{(2)} \, \mathrm{d}V + \int_V f_i^{(1)} u_i^{(2)} \, \mathrm{d}V = 0.$$

The integrand of the first integral is transformed using the product rule of differential calculus as follows:

$$\frac{\partial \sigma_{ij}^{(1)}}{\partial x_j} u_i^{(2)} = \frac{\partial}{\partial x_j} \left(\sigma_{ij}^{(1)} u_i^{(2)} \right) - \sigma_{ij}^{(1)} \frac{\partial u_i^{(2)}}{\partial x_j}. \tag{9.87}$$

We obtain:

$$\int_V \frac{\partial}{\partial x_j} \left(\sigma_{ij}^{(1)} u_i^{(2)} \right) dV - \int_V \sigma_{ij}^{(1)} \frac{\partial u_i^{(2)}}{\partial x_j} dV + \int_V f_i^{(1)} u_i^{(2)} dV = 0. \tag{9.88}$$

If we now apply the Gaussian integral theorem to the first integral and thus convert the volume integral over V into an integral on the boundary ∂V, we obtain:

$$\int_V \frac{\partial}{\partial x_j} \left(\sigma_{ij}^{(1)} u_i^{(2)} \right) dV = \int_{\partial V} \sigma_{ij}^{(1)} u_i^{(2)} n_j dS. \tag{9.89}$$

Therein, n_j is the component of the normal vector $\underline{n}$ in the direction of the coordinate x_j. The result is as follows:

$$\int_{\partial V} \sigma_{ij}^{(1)} u_i^{(2)} n_j dS - \int_V \sigma_{ij}^{(1)} \frac{\partial u_i^{(2)}}{\partial x_j} dV + \int_V f_i^{(1)} u_i^{(2)} dV = 0. \tag{9.90}$$

The relationship between the internal stresses and the external stress vector reads $\sigma_{ij}^{(1)} n_j = t_i^{(1)}$, where t_i is the i-th component of the stress vector $\underline{t}$. With $\sigma_{ij}^{(1)} \frac{\partial u_i^{(2)}}{\partial x_j} = \sigma_{ij}^{(1)} \varepsilon_{ij}^{(2)}$ we then obtain the general principle of work and energy of elastostatics as follows:

$$\int_V \sigma_{ij}^{(1)} \varepsilon_{ij}^{(2)} dV = \int_V f_i^{(1)} u_i^{(2)} dV + \int_{\partial V} t_i^{(1)} u_i^{(2)} dS. \tag{9.91}$$

The boundary or surface ∂V of the body can be divided into the areas ∂V_t and ∂V_u, whereby the stresses $\underline{t}_0$ are given on the section ∂V_t and the displacements $\underline{u}_0$ on the section ∂V_u. Then we get:

$$\int_V \sigma_{ij}^{(1)} \varepsilon_{ij}^{(2)} dV = \int_V f_i^{(1)} u_i^{(2)} dV + \int_{\partial V_t} t_{i0}^{(1)} u_i^{(2)} dS + \int_{\partial V_u} t_i^{(1)} u_{i0}^{(2)} dS. \tag{9.92}$$

If we use the reference system x, y, z again, we obtain:

$$\int_V \left(\sigma_{xx}^{(1)} \varepsilon_{xx}^{(2)} + \sigma_{yy}^{(1)} \varepsilon_{yy}^{(2)} + \sigma_{zz}^{(1)} \varepsilon_{zz}^{(2)} + \tau_{yz}^{(1)} \gamma_{yz}^{(2)} + \tau_{xz}^{(1)} \gamma_{xz}^{(2)} + \tau_{xy}^{(1)} \gamma_{xy}^{(2)} \right) dV$$

$$= \int_V \left(f_x^{(1)} u^{(2)} + f_y^{(1)} v^{(2)} + f_z^{(1)} w^{(2)} \right) dV$$

$$+ \int_{\partial V_t} \left(t_{x0}^{(1)} u^{(2)} + t_{y0}^{(1)} v^{(2)} + t_{z0}^{(1)} w^{(2)} \right) dS$$

$$+ \int_{\partial V_u} \left(t_x^{(1)} u_0^{(2)} + t_y^{(1)} v_0^{(2)} + t_z^{(1)} w_0^{(2)} \right) dS. \tag{9.93}$$

Obviously the integral on the left side represents the work performed by the stresses along the respective strain components, i.e. the internal work. The integral of the right side, on the other hand, contains the work performed by the volume forces as well as by the surface forces. It is important to note that the general theorem presented here is completely independent of the type of material behavior. It is therefore valid for any kind of material.

Bibliography

Becker W, Gross D (2002) Mechanik elastischer Körper und Strukturen. Springer, Berlin, Germany

Gross D, Hauger W, Wriggers P (2014) Technische Mechanik 4: Hydromechanik, Elemente der Höheren Mechanik, Numerische Methoden, 9th edn. Springer, Berlin, Germany

Kossira H (1996) Grundlagen des Leichtbaus. Springer, Berlin, Germany

Langhaar HL (2016) Energy methods in applied mechanics. Dover Publications, New York, USA

Linke M, Nast E (2015) Festigkeitslehre für den Leichtbau. Springer, Berlin, Germany

Mittelstedt C (2017) Energiemethoden der Elastostatik, Studienbereich Mechanik, Technische Universität Darmstadt, Germany

Reddy JN (2017) Energy principles and variational methods in applied mechanics, 3rd edn. Wiley, New York, USA

Tauchert TR (1974) Energy principles in structural mechanics. McGraw-Hill, New York, USA

Chapter 10
Principle of Virtual Displacements

10.1 Introduction

This chapter deals with the principle of virtual displacements. It will be shown that this is a very fundamental and important principle in the framework of elastostatics, and thus also for lightweight engineering, from which a number of practically relevant analysis methods can be developed. Applications which we will discuss in this chapter are the determination of forces and moments at statically determinate systems as well as the determination of influence lines for forces and moments, also at statically determinate systems. Furthermore, we will discuss in detail to what extent the principle of virtual displacements can be used for the derivation of differential equations and boundary conditions for problems in the framework of elastostatics.

10.2 Virtual Displacements and Virtual Works

In order to introduce the concept of virtual displacements we consider the beam shown in Fig. 10.1 which is subjected to a distributed line load $q(x)$. The beam has the length l and is simply supported at both ends.

Let $w(x)$ be the deflection of the beam in its state of static equilibrium. We now introduce an infinitesimal variation $\delta w(x)$ from the equilibrium state as shown. This variation $\delta w(x)$ is called virtual displacement. Virtual displacements have the following properties:

- Virtual displacements are assumed to be infinitesimal small.
- They are virtual quantities and do not exist in reality.
- It is required that they are in accordance with the given geometric boundary conditions.

C. Mittelstedt, *Structural Mechanics in Lightweight Engineering*,
https://doi.org/10.1007/978-3-030-75193-7_10

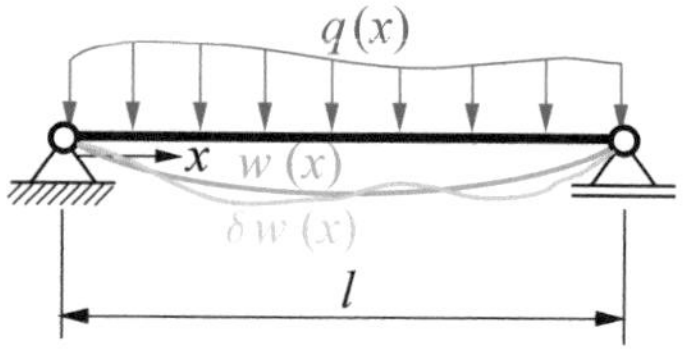

Fig. 10.1 Equilibrium
configuration $w(x)$ of a
beam and admissible virtual
displacement $\delta w(x)$

Fig. 10.2 Cantilever beam

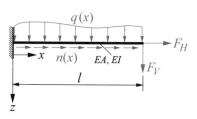

The last point means in the example shown in Fig. 10.1 that at the two supported ends the virtual displacements $\delta w(x)$ must disappear, i.e. $\delta w(x = 0) = 0$ and
$\delta w(x = l) = 0$. Any infinitesimal variation $\delta w(x)$ which fulfills these requirements
is thus admissible.

If a given static system is subjected to a virtual displacement, virtual work is
performed. Since these are virtual displacements from the equilibrium position, the
internal and external forces and stresses do not change. However, they perform virtual
work along the virtual displacements and thus also along the virtual strains. We
distinguish between virtual internal work δW_i and virtual external work δW_a. If,
for example, a solid body is given on which a single force F acts, then the virtual
external work δW_a is the product of the force F and the virtual displacement δu, i.e.
$\delta W_a = F \delta u$, or in vectorial notation $\delta W_a = \underline{F} \, \delta \underline{u}$.

We consider the cantilever beam shown in Fig. 10.2 which is loaded by the two
line loads $q(x)$ and $n(x)$ and at its free end by the two single forces F_H and F_V. We
assume bending in the xz-plane and further assume that y and z are the principal
axes of the beam cross-section. The geometric boundary conditions can currently be
described as follows:

$$w(x = 0) = 0, \quad w'(x = 0) = 0, \quad u(x = 0) = 0. \tag{10.1}$$

At the end of the cantilever beam the following dynamic boundary conditions apply:

$$V(x = l) = F_V, \quad N(x = l) = F_H, \quad M(x = l) = 0. \tag{10.2}$$

We now consider the two virtual displacements δu and δw, for which we assume that
they fulfill the geometric boundary conditions (10.1) identically, i.e. the requirements
$\delta u(x = 0) = 0$, $\delta w(x = 0) = 0$ and $\delta w'(x = 0) = 0$ are fulfilled.

We first discuss the virtual external work δW_a. The two single forces F_H and
F_V perform virtual work along the virtual displacements $\delta u(x = l)$ and $\delta w(x = l)$,
respectively. The line load $q(x)$, which we assume to be arbitrarily but continuously
distributed over the length of the beam, is multiplied by the virtual displacement

$\delta w(x)$ to determine its virtual work and integrated over the entire length of the beam. We proceed analogously with the line load $n(x)$. We then obtain δW_a as:

$$\delta W_a = \int_0^l q(x)\delta w\,dx + \int_0^l n(x)\delta u\,dx + F_V \delta w(x=l) + F_H \delta u(x=l). \quad (10.3)$$

The virtual internal work δW_i is obtained from the virtual strain energy density δU_0:

$$\delta W_i = \int_V \delta U_0\,dV = \int_V \sigma_{ij}\delta\varepsilon_{ij}\,dV = \int_0^l \int_A \sigma_{xx}\delta\varepsilon_{xx}\,dA\,dx. \quad (10.4)$$

The connection already known from Chap. 5 between the strain ε_{xx} and the displacements u and w

$$\varepsilon_{xx} = u' - zw'' \quad (10.5)$$

is used to represent the virtual strain $\delta\varepsilon_{xx}$ as follows:

$$\delta\varepsilon_{xx} = \delta u' - z\delta w''. \quad (10.6)$$

The virtual internal work δW_i is then:

$$\delta W_i = \int_0^l \int_A \sigma_{xx}\left(\delta u' - z\delta w''\right)dA\,dx. \quad (10.7)$$

Integration over the cross-sectional area A of the beam yields with $N = \int_A \sigma_{xx}\,dA$ and $M = \int_A \sigma_{xx}z\,dA$:

$$\delta W_i = \int_0^l \left(N\delta u' - M\delta w''\right)dx. \quad (10.8)$$

It should be noted that the previous relations are valid independently of the material behavior. However, if linear elasticity is considered and we assume Hooke's law to be valid, then we obtain with the constitutive relationships $N = EAu'$ and $M = -EIw''$:

$$\delta W_i = \int_0^l EAu'\delta u'\,dx + \int_0^l EIw''\delta w''\,dx. \quad (10.9)$$

A further example is given in Fig. 10.3 in the context of St. Venant's torsion theory (see Chap. 7). The virtual external work can be specified as:

$$\delta W_a = \int_0^l m_T\delta\vartheta\,dx + M_{T0}\delta\vartheta(x=l). \quad (10.10)$$

The internal virtual work follows as:

$$\delta W_i = \int_0^l M_x\delta\vartheta'\,dx = \int_0^l GI_T\vartheta'\delta\vartheta'\,dx. \quad (10.11)$$

Fig. 10.3 Cantilever beam
under torsional load

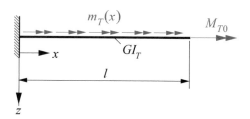

We now extend the considerations to an arbitrary three-dimensional solid (volume V) which is loaded by the volume forces f and m single forces $\underline{F}_i$. We subdivide the boundary ∂V into the two subareas ∂V_u and ∂V_t, whereby displacements are given on the subarea ∂V_u, i.e. the displacement vector $\underline{u}$ is given on ∂V_u with $\underline{u}_0$. On the other hand, the stress vector $\underline{t}$ is preset to $\underline{t}_0$ on the subsurface ∂V_t. If we now apply virtual displacements $\delta \underline{u}$ to a solid determined in this way, then these must disappear on the subsurface ∂V_u. The virtual external work index δW_a then results as:

$$\delta W_a = \int_V \underline{f}\,\delta \underline{u}\,dV + \int_{\partial V_t} \underline{t}_0 \delta \underline{u}\,dS + \sum_{i=1}^{m} \underline{F}_i \delta \underline{u}_i. \tag{10.12}$$

The virtual strain energy density δU_0 is obtained as:

$$\delta U_0 = \int_0^{\delta \varepsilon_{ij}} \sigma_{ij}\,d(\delta \varepsilon_{ij}). \tag{10.13}$$

Since all stress components remain unchanged during the virtual displacements, we can write immediately:

$$\delta U_0 = \sigma_{ij}\delta \varepsilon_{ij}. \tag{10.14}$$

The virtual strain energy or the virtual internal work δW_i is then obtained by integrating δU_0 over the volume of the solid under consideration:

$$\delta W_i = \int_V \delta U_0\,dV = \int_V \sigma_{ij}\delta \varepsilon_{ij}\,dV. \tag{10.15}$$

10.3 The Principle of Virtual Displacements

The principle of virtual displacements can be verbalized as follows:

A body is in equilibrium exactly when, at any permissible virtual displacement from the equilibrium position, the virtual internal work is equal to the virtual external work.

Thus:

$$\delta W_i = \delta W_a. \tag{10.16}$$

As we will show in detail later, the principle of virtual displacements always leads to an equilibrium statement. It has some very interesting and for our purposes essential applications which we will discuss in the following.

10.3.1 Determination of Forces and Moments in Statically Determinate Systems

A first elementary application of the principle of virtual displacements consists in the determination of forces and moments (e.g. support reactions or internal forces and moments) at statically determinate beam and rod structures. For motivation, we consider the simply supported beam shown in Fig. 10.4, top left. The beam is subjected to the uniformly distributed load q_0, the single load P_0 (acting exactly in the center of the beam) and the single moment M_0. The beam has the length l. The reaction force in the support B and the bending moment M at the position $x = l/2$ are to be determined using the principle of virtual displacements.

To determine the reaction force at point B, we cut away the support and thus release the reaction force B (Fig. 10.4, top right). In this way, the beam, which is actually supported in a statically determinate manner, becomes kinematically displaceable

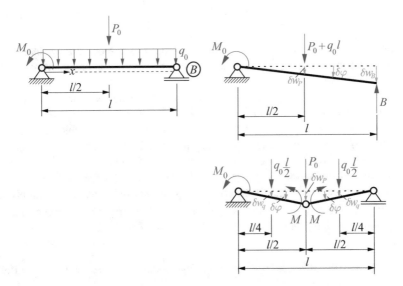

Fig. 10.4 Simply supported beam under line load q_0, single force P_0 and single moment M_0 (left), kinematics for determining the reaction force B (top right), kinematics for determining the bending moment M at position $x = l/2$ (bottom right)

(which means that there is a clearly defined degree of freedom from which a unique displacement figure can be constructed) and will undergo a virtual rotation $\delta\varphi$ about the support A. It is important to note that this is a pure rigid body rotation, so that the beam itself remains undeformed and retains its ideally straight configuration. As a consequence, in this case no internal virtual works δW_i occur due to the absence of virtual strains, and only the virtual external works δW_a are to be considered here. The single force P_0 and the resultant of the line load q_0 with the amount $q_0 l$ both act in the middle of the beam and pass through the virtual displacement δw_p. Since the virtual displacements are assumed to be infinitesimally small, we can write $\delta w_p = \delta\varphi\frac{l}{2}$. The reaction force B passes through the virtual displacement $\delta w_B = \delta\varphi l$. The moment M_0 performs a virtual work along the virtual rotation $\delta\varphi$. Note that this virtual work must be entered into the work balance with a negative sign, since the moment M_0 and the virtual displacement $\delta\varphi$ have opposite directions of rotation. Thus:

$$\delta W = \delta W_a = -M_0\delta\varphi + (P_0 + q_0 l)\,\delta\varphi\frac{l}{2} - B l\delta\varphi = 0, \tag{10.17}$$

or:

$$\left(-M_0 + (P_0 + q_0 l)\frac{l}{2} - B l\right)\delta\varphi = 0. \tag{10.18}$$

The trivial solution $\delta\varphi = 0$ is not useful here, as it would imply that no virtual displacements have been applied to the system. Therefore we set the expression in parentheses to zero which we can solve immediately for the reaction force B:

$$B = -\frac{M_0}{l} + \frac{P_0}{2} + \frac{q_0 l}{2}. \tag{10.19}$$

Obviously, this is the reaction force that we would also receive from elementary equilibrium considerations. Hence, the principle of virtual displacements obviously leads to a statement of equilibrium.

Analogously, we proceed with the determination of the bending moment M in the center of the beam at the position $x = \frac{l}{2}$. We release the moment by inserting a full joint and observe the resulting virtual displacement figure, which results from the virtual rotation $\delta\varphi$ of the two beam segments of length $\frac{l}{2}$ around the respective supported ends. Both the released bending moment M and the external moment M_0 then perform virtual work along these virtual rotations. Furthermore, the single force P_0 is displaced by the virtual displacement $\delta w_P = \frac{l}{2}\delta\varphi$, and the two resultants of the line load with the respective amount $q_0\frac{l}{2}$ on the left and right bar segment perform virtual work along the virtual displacement $\delta w_q = \frac{l}{4}\delta\varphi$. Thus:

$$\delta W = \delta W_a = -2M\delta\varphi - M_0\delta\varphi + 2q_0\frac{l}{2}\frac{l}{4}\delta\varphi + P_0\frac{l}{2}\delta\varphi = 0. \tag{10.20}$$

This can be solved immediately for the bending moment M at the position $x = \frac{l}{2}$:

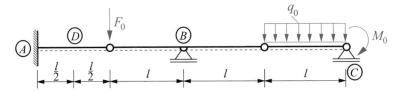

Fig. 10.5 Continuous beam under single force F_0, end moment M_0 and uniform line load q_0

$$M = -\frac{M_0}{2} + \frac{q_0 l^2}{8} + \frac{P_0 l}{4}. \tag{10.21}$$

This result can also be determined from elementary equilibrium considerations.

When determining forces and moments in statically determinate systems with the aid of the principle of virtual displacements, the system under consideration is displaced kinematically by releasing the quantity that is to be determined, whereby the resultant displacement figure can always be clearly and uniquely defined. This unique displacement figure is also denoted as the kinematic chain of the system under consideration.

Another example is the continuous beam shown in Fig. 10.5 under the single force F_0, the moment M_0 and the uniform line load q_0. The forces and moments C_V, B_V, A_V, M_A, M_B and M_D are to be determined with the help of the principle of virtual displacements.

All virtual displacement figures/kinematic chains necessary for the present calculation task are shown in Fig. 10.6. The calculation method is briefly explained here at the example of the support force C_V. The kinematic chain that is created after the release of this reaction force is shown in Fig. 10.6, top. The virtual work reads:

$$\delta W = \delta W_a = M_0 \delta \varphi + q_0 l \delta w_1 - C_V \delta w_2 = 0. \tag{10.22}$$

From the kinematic relations $\delta w_1 = \delta \varphi \frac{l}{2}$ and $\delta w_2 = \delta \varphi l$ it follows:

$$\left(M_0 + q_0 l \frac{l}{2} - C_V l \right) \delta \varphi = 0. \tag{10.23}$$

Setting the term enclosed in parentheses to zero then leads to the support force C_V as:

$$C_V = \frac{M_0}{l} + \frac{q_0 l}{2}. \tag{10.24}$$

Analogously, the other forces and moments can be determined, and we have:

$$B_V = q_0 l - 2\frac{M_0}{l},$$

$$A_V = F_0 - \frac{q_0 l}{2} + \frac{M_0}{l},$$

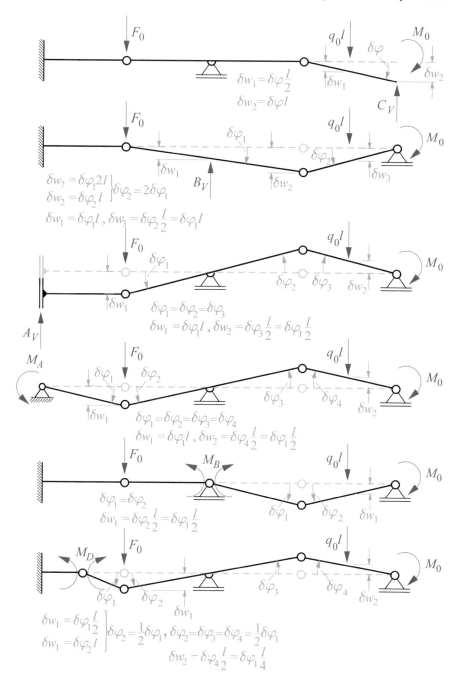

Fig. 10.6 Kinematic chains for the determination of the required forces and moments

$$M_A = F_0 l - \frac{q_0 l^2}{2} + M_0,$$

$$M_B = M_0 - \frac{q_0 l^2}{2},$$

$$M_D = -\frac{F_0 l}{2} + \frac{q_0 l^2}{4} - \frac{M_0}{2}. \tag{10.25}$$

10.3.2 Influence Lines for Forces and Moments in Statically Determinate Systems

A further elementary application of the principle of virtual displacements is the determination of influence lines for forces and moments of statically determinate systems. By an influence line we understand the graphical representation of a force or a moment of a static system in case the system is loaded by a moving load. From the influence line, the considered quantity can then be determined for any possible position of the moving load. As an introduction, let us consider a simply supported beam (Fig. 10.7) which is subjected to a moving unit load $F = 1$. We want to determine the influence line for the reaction force in support B (abbreviated as IL B), it is already shown in Fig. 10.7, bottom. This rather simple influence line IL B can be set up simply by using engineering intuition. In this case it is clear that the reaction force B takes the value zero if the moving load is exactly above the left support. In the same way it can be concluded immediately that B takes the value $B = 1$ if the moving load $F = 1$ is exactly above support B. Between these two values the influence line IL B is a straight line.

For more complex systems than the one shown in Fig. 10.7 we need a clear procedure for the determination of such influence lines. Here, we use the principle of virtual displacements and cut free the force or moment for which we want to determine the influence line. As a result, the system becomes kinematically displaceable and a clearly defined kinematic chain is established. The influence line that is to be determined then corresponds exactly to the kinematic chain that is created as we will demonstrate later on. Since the kinematic chain is composed of the individual beam segments which all remain undeformed in themselves, the influence lines for forces and moments in statically determinate systems are always composed of straight lines.

Fig. 10.7 Simply supported beam under rolling load (top) and influence line for the reaction force B (bottom)

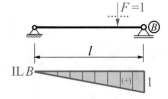

The procedure for the determination of influence lines for forces and moments in statically determinate systems is that the bond that is energetically related to the desired force or moment is removed and we apply the quantity thus released to the static system. Based on this, the kinematic chain of the system is determined, whereby we have to take into account that the displacement or rotation assigned to the desired force or moment assumes exactly the value -1, respectively the force quantity that has been released performs a negative work along the unit displacement or rotation. Then the desired influence line for the quantity under consideration corresponds exactly to the resulting kinematic chain. This is called Land's theorem.[1]

We prove Land's theorem by means of the situation shown in Fig. 10.8. Consider a beam on two supports which is loaded by a moving unit load $F = 1$. The moment M_S is to be determined at an arbitrary point x. For this purpose, we introduce a moment hinge at the position x, which releases the desired moment on both sides of the hinge. Assuming that the joint is rotated by the virtual angle $\delta\varphi$, the rolling load will undergo the virtual displacement δw. Thus, the virtual work balance can be established which in this case consist exclusively of virtual external works:

$$\delta W_a = M_S \delta\varphi + F\delta w = 0. \tag{10.26}$$

Since we have assumed the rolling load as a unit load, we get:

$$M_S \delta\varphi + \delta w = 0. \tag{10.27}$$

If we now define the virtual rotation with $\delta\varphi = -1$, as described above, we obtain:

$$-M_S + \delta w = 0, \tag{10.28}$$

Fig. 10.8 On Land's theorem

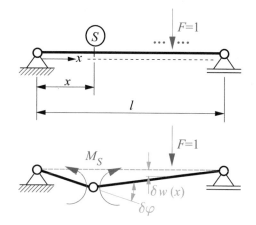

---

[1]Robert Land, 1857–1899, German civil engineer.

or

$$M_S = \delta w. \tag{10.29}$$

This proves Land's theorem.

As an elementary example consider the beam shown in Fig. 10.9 under the rolling unit load $F = 1$. The influence lines for the bending moment M_S and the transverse shear force V_S are to be determined at the marked position S.

In order to determine the influence line IL M_S, the bending moment M_S is released and the virtual angle is assumed as $\delta\varphi = 1$ in such a way that the released bending moment performs a negative virtual work. With the virtual joint displacement δw a relation between the angles $\delta\varphi_1$ and $\delta\varphi_2$ can be determined:

$$\delta w = \delta\varphi_1 l_1 = \delta\varphi_2 l_2. \tag{10.30}$$

Hence:

$$\delta\varphi_2 = \delta\varphi_1 \frac{l_1}{l_2}. \tag{10.31}$$

The sum of the partial angles has to result in $\delta\varphi = 1$, i.e. $\delta\varphi_1 + \delta\varphi_2 = 1$. Thus:

$$\delta\varphi_1 \left(1 + \frac{l_1}{l_2}\right) = 1. \tag{10.32}$$

We can solve this expression for $\delta\varphi_1$ as follows:

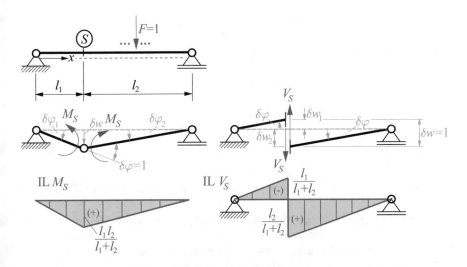

Fig. 10.9 Influence lines for a simply supported beam under rolling unit load $F = 1$

$$\delta\varphi_1 = \frac{l_2}{l_1 + l_2}. \tag{10.33}$$

For the partial angle $\delta\varphi_2$ we then have:

$$\delta\varphi_2 = \frac{l_1}{l_1 + l_2}. \tag{10.34}$$

The virtual displacement δw of the joint then results in:

$$\delta w = \frac{l_1 l_2}{l_1 + l_2}. \tag{10.35}$$

Due to Land's theorem, the desired influence line IL M_S then corresponds exactly to the kinematic chain that results when a moment joint is introduced. It is shown in Fig. 10.9, bottom left.

The influence line IL V_S is determined in the same way. A shear force joint is inserted at the position S for the calculation of IL V_S. This way, virtual transverse displacements are enabled at this point, but virtual angular rotations and longitudinal displacements are prevented. The system now experiences the virtual total displacement $\delta w = 1$, whereby the directions of the two partial displacements are selected in such a way that the released transverse shear force V_S performs a negative virtual work. The two beam segments both experience identical virtual rotation angles $\delta\varphi$ around their respective support points. The partial displacements are called δw_1 and δw_2, their sum corresponds exactly to the applied displacement $\delta w = 1$:

$$\delta w_1 + \delta w_2 = 1. \tag{10.36}$$

Furthermore, the following relations apply between δw_1, δw_2 and $\delta\varphi$:

$$\delta w_1 = \delta\varphi l_1, \quad \delta w_2 = \delta\varphi l_2. \tag{10.37}$$

Thus three equations are available for the three unknown virtual quantities. The result is:

$$\delta\varphi = \frac{1}{l_1 + l_2}, \quad \delta w_1 = \frac{l_1}{l_1 + l_2}, \quad \delta w_2 = \frac{l_2}{l_1 + l_2}. \tag{10.38}$$

The influence line IL V_S can thus be determined, and it is shown in Fig. 10.9, bottom right.

Another example is the continuous beam already shown in Figs. 10.5 and 10.6, for which different influence lines are to be determined under a moving load (Fig. 10.10). Figure 10.11 contains the resulting kinematic chains and associated influence lines. A detailed discussion of the solution is omitted at this point.

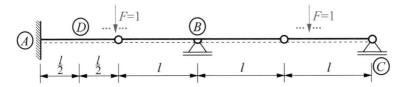

Fig. 10.10 Continuous beam under moving load

10.4   Pole Plans and Kinematic Chains

The systems discussed so far which have been investigated with the help of the principle of virtual displacements are of a quite simple form and are straightforward in their analysis. If, however, a system of higher complexity is given, then a clear procedure for determining the required kinematic chains must be found. In the case of the methodology that is presented in the following, we speak of the so-called kinematic method, and an important tool here is the preparation of a so-called pole plan which is indispensable for the determination of kinematic chains.

As an introduction, a rigid disk (see Fig. 10.12) of triangular shape is considered. The vertices of the disk are denoted as A, B and C. The disk is now rotated around the point P by the virtual angle $\delta\varphi$. The points A, B and C have the radii r_A, r_B, r_C to point P. Since the rotation of the disk is around the point P, this point is also called the (absolute) center of rotation or the main pole. Every virtual displacement of a point of the disk can be represented as a function of this virtual rotation, whereby the virtual displacement is always perpendicular to the respective radius r. For the example shown here the following applies:

$$\delta w_A = r_A \delta\varphi, \quad \delta w_B = r_B \delta\varphi, \quad \delta w_C = r_C \delta\varphi. \tag{10.39}$$

The absolute center of rotation of a rigid disk is called the main pole, whereas the relative center of rotation of two disks is called the relative center of rotation or the relative pole. Using the example of the rigid disk in Fig. 10.12 the point P is a main pole, no virtual displacement of any kind takes place here.

For further illustration we consider the situation shown in Fig. 10.13. Two rigid disks of arbitrary shape are given here which we want to denote as disk 1 and disk 2. Disk 1 is assumed to be supported at an arbitrary point and is also connected to disk 2 via a moment joint. Disk 2 is supported at an arbitrary point as indicated. Obviously, for the given system under the shown support conditions a kinematic chain can be determined. The following general rules apply:

- A support of a rigid disk that invokes two reaction forces that act perpendicularly to each other is always the main pole of this disk, no virtual displacement is possible. Any movement of this disk at any point will then be perpendicular to the radius between the main pole and the point under consideration. Consequently, the main

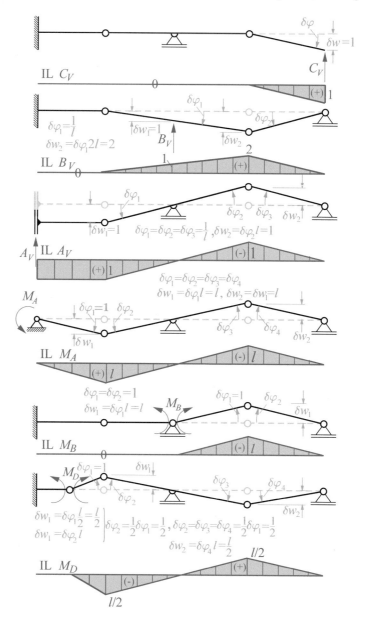

Fig. 10.11 Resulting kinematic chains and influence lines for the continuous beam

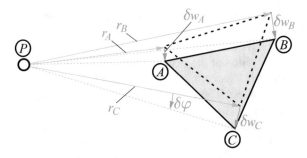

Fig. 10.12 Rigid disk (vertices A, B, C) under a virtual rotation $\delta\varphi$ around the main pole P

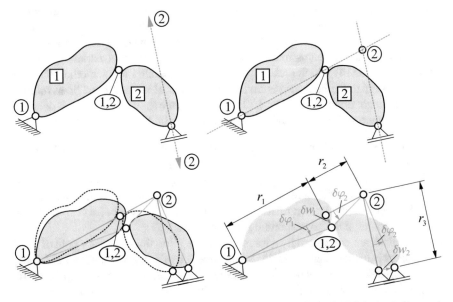

Fig. 10.13 Determination of the pole plan for a kinematic system of two rigid disks (top), kinematic chain (bottom)

    pole of disk 1 lies exactly in its support, marked by the number of the disk, enclosed in a circle.

- Furthermore, a moment joint between two disks always forms their common pole, secondary pole or relative pole—both disks undergo identical displacements at such a joint. Relative poles are marked with the numbers of the two disks connected by them and enclosed in an oval frame.
- In addition, the main pole of a supported disk where the kind of guided support invokes a single reaction force is always located on a line running through the corresponding support in the direction of the line of action of the corresponding reaction force. In Fig. 10.13 this is indicated by the line passing through the support of disk 2.

- Main pole and relative pole of two disks always lie on a straight line. Using the example of the Fig. 10.13, the main pole of disk 1 and the relative pole of the two disks 1 and 2 have already been determined. The main pole of disk 2 is now located at the intersection of the two drawn lines. The pole plan for the example shown in Fig. 10.13 is thus complete, and the resulting kinematic chain is shown in Fig. 10.13, bottom left. The corresponding kinematic relations can be seen in Fig. 10.13, bottom right.
- The poles of a movable system do not necessarily have to be defined unambiguously in their position. An example of poles lying at infinity is shown in Fig. 10.14. Shown here is a movable system of three beams. The two vertical beams 1 and 3 are supported as indicated and connected to beam 2 at the upper end by ideal moment joints. Thus, the main poles of beams 1 and 3 are located at their supported ends. The relative poles between beams 1 and 2 or 2 and 3 can be immediately located in the moment joints. With the rule that the main and relative poles of two connected disks or beams must always lie on a straight line (marked by dashed lines in Fig. 10.14) which presently, however, are parallel to each other, the only conclusion that can be drawn is that the main pole of beam 2 must lie at infinity. The kinematic chain of the Fig. 10.14 consequently shows that beam 2 does not undergo any rotation, but rather undergoes a pure rigid body displacement.

- The relative poles of three disks always lie on a straight line, as can be seen in Fig. 10.15. Here four rigid disks are shown which are connected by moment joints. The relative poles (1,2), (2,3), (3,4) and (1,4) are directly found in the joints. The position of the remaining relative poles can be concluded by drawing straight lines through the already known relative poles. The still unknown poles (1,3) and (2,4) are then found at the corresponding intersections.
- If two disks are connected by a transverse force joint, the secondary pole of these two disks is located perpendicular to the direction of motion of the joint at infinity (Fig. 10.16, left). For two disks connected by a normal force joint (Fig. 10.16, right), an analogous conclusion is drawn. Two disks whose relative pole lie at infinity always rotate by the same angle.

- Three disks that form a joint triangle always behave like a rigid disk and are therefore immovable in themselves (Fig. 10.17). The same applies analogously to three connected rods that form a triangle.
- Projection constancy (Fig. 10.18): All displacements and rotations considered in this section are virtual and infinitesimally small. The movable bar of the Fig. 10.18 shows that in the case of finite, but not infinitesimal displacements, a kinematic chain would result as given in Fig. 10.18, left. The moment joint in the middle of the beam would move on a circular path and as a consequence the right support would follow this movement and shift to the left. For the context of this chapter, i.e. virtual and thus infinitesimally small displacements, this circular line can be understood as a straight downward displacement. This means that displacements are always replaced by their tangential components perpendicular to the pole radius. In the

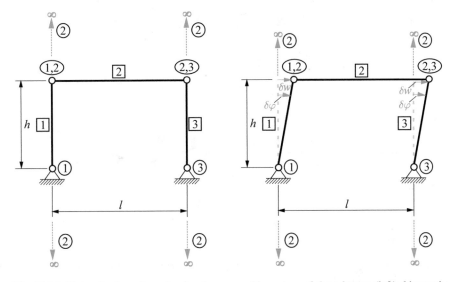

Fig. 10.14 Determination of a pole plan for a movable system of three beams (left), kinematic chain (right)

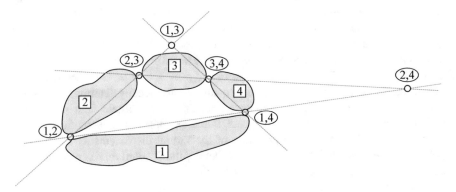

Fig. 10.15 Determination of relative poles

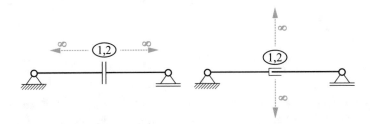

Fig. 10.16 Determination of relative poles for transverse force joints and normal force joints

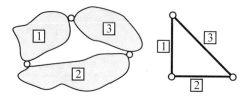

Fig. 10.17 Three disks connected by joints

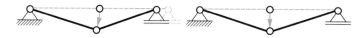

Fig. 10.18 Projection constancy

case of the Fig. 10.18 this also means that the horizontal displacement of the right support is not considered.

The above rules allow us to determine both main and relative poles and thus also the pole plans and kinematic chains of any arbitrary movable static system. Figure 10.19 contains a number of exemplary pole plans and kinematic chains of movable static systems. The kinematic method discussed here therefore allows us to determine both forces and moments and influence lines for force quantities at complex statically determinate systems by means of the principle of virtual displacements. When determining influence lines of force quantities at statically determinate systems it has to be considered that the influence lines have zero values under main poles and changes in slope under relative poles. Some examples for the solution of such problems with the help of the principle of the virtual displacements when using pole plans are presented in the following whereby the detailed representation of the calculation is omitted here for reasons of brevity.

Figure 10.20 shows a statically determinate frame for which the following force quantities are to be determined: Support reactions M_A, B, A_H; bending moments M_C, M_D; vertical joint force G_V. The corresponding pole plans and kinematic chains are shown in Fig. 10.21.

The result is:

$$M_A = Fh, \quad B = \frac{ql}{2}, \quad M_D = -\frac{Fh}{2}, \quad A_H = F, \quad M_C = \frac{ql^2}{8}, \quad G_V = \frac{ql}{2}.$$
$$(10.40)$$

For the truss structure given in Fig. 10.22 the forces in rods 1, 2, 3 and 9 are to be determined with the aid of the principle of virtual displacements. The respective necessary pole plans and kinematic chains are shown in Fig. 10.23.

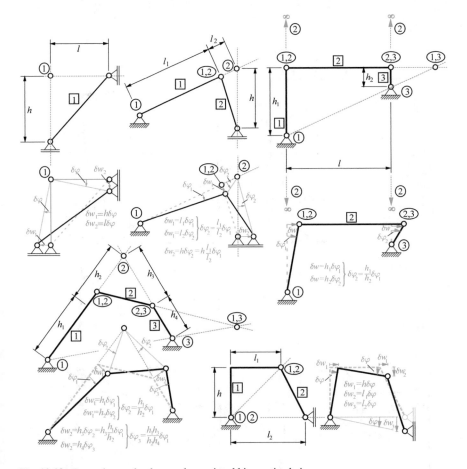

Fig. 10.19 Exemplary pole plans and associated kinematic chains

Fig. 10.20 Statically
determinate frame under
uniform load q and single
load F

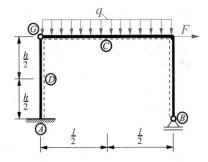

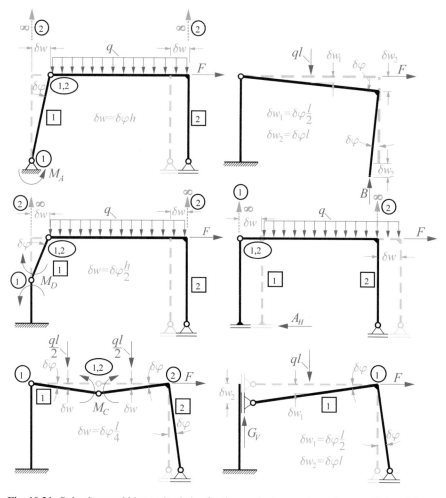

Fig. 10.21 Pole plans and kinematic chains for the required support reactions and internal forces and moments

Fig. 10.22 Truss under two point loads

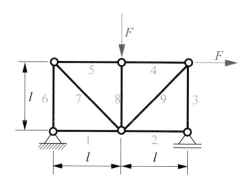

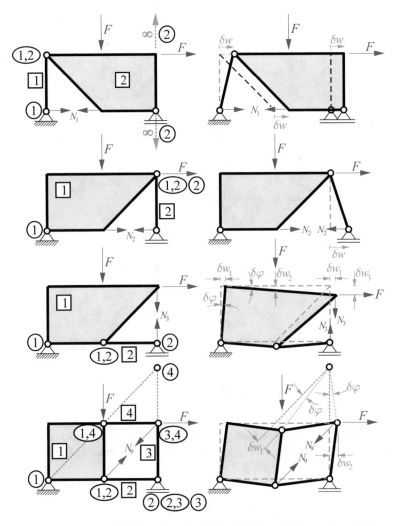

Fig. 10.23 Determination of the forces in rods 1, 2, 3 and 9 using the kinematic method

The result is:

$$N_1 = F, \quad N_2 = 0, \quad N_3 = -F, \quad N_9 = \sqrt{2}F. \tag{10.41}$$

The frame given in Fig. 10.20 is again considered whereby now a moving force is applied (see Fig. 10.24). We want to determine the influence lines for the reaction force B, the bending moment M_C and the joint force G_V, which are also shown in 10.24.

Finally, the truss shown in Fig. 10.22 is now considered under a moving load. The influence lines for the reaction force B and for the rods 1, 2, 3 and 9 are to

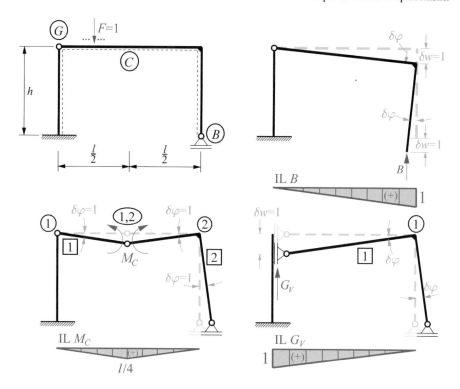

Fig. 10.24 Statically determinate frame; influence lines

be determined. The corresponding pole plans and kinematic chains as well as the influence lines IL B, IL N_1, IL N_2, IL N_3 and IL N_9 are shown in Fig. 10.25.

10.5 The Variational Operator δ

We have already used the variational operator δ several times, and in this section we want to consider this operator in some more detail. The variational operator δ symbolizes a very small change or the so-called variation of a certain quantity, e.g. a displacement u, whereby the quantity that undergoes a variation can depend on one or more variables. We speak of the so-called first variation δu, for example, if we apply a virtual displacement δu to a static system. This is very similar to the differential calculus if we speak of a change in the quantity dx with respect to x and want to express an infinitesimally small increase of x. Therefore, we can establish some general rules of calculation for the variational operator δ which are closely related to those of differential calculus and which we will apply frequently in the further course of this book. We will discuss these analysis rules here but will not

Fig. 10.25 Influence lines
for a truss under moving load

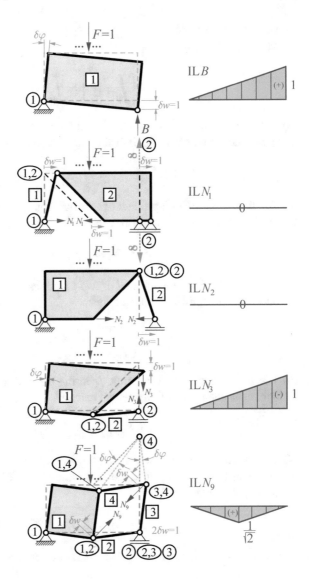

provide any detailed mathematical proofs. For this purpose the interested reader is
referred to the widely available literature on the calculus of variations. Chapter 11
of this book contains some more detailed information.

Let the functions $f_1, f_2, f_3, \ldots, f_n$ be dependent on variables such as the displacement u. The following calculation rules apply when using the variational operator δ:

(1) The order of variation and differentiation is interchangeable:

$$\delta(\nabla u) = \nabla(\delta u),\qquad(10.42)$$

wherein ∇ is the gradient: $\nabla = \left(\frac{\partial}{\partial x}, \frac{\partial}{\partial y}, \frac{\partial}{\partial z} \right)^T$.

(2) The order of integration and variation is interchangeable:

$$\delta \left[\int_V u \, dV \right] = \int_V \delta u \, dV. \tag{10.43}$$

(3) The first variation of the sum of several functions f_1, f_2, f_3, ..., f_n can be formed similar to the summation rule of differential calculus:

$$\delta \left[f_1 \pm f_2 \pm f_3 \pm ... \pm f_n \right] = \delta f_1 \pm \delta f_2 \pm \delta f_3 \pm ... \pm \delta f_n. \tag{10.44}$$

(4) The first variation of a product of two functions f_1, f_2 can be formed similar to the product rule of differential calculus:

$$\delta \left[f_1 f_2 \right] = f_2 \delta f_1 + f_1 \delta f_2. \tag{10.45}$$

(5) The first variation of a quotient of two functions f_1, f_2 can be formed similar to the quotient rule of differential calculus:

$$\delta \left[\frac{f_1}{f_2} \right] = \frac{\delta f_1}{f_2} - \frac{f_1 \delta f_2}{f_2^2}. \tag{10.46}$$

(6) The first variation of a function f_1 in the form of a power law can be formed similar to the chain rule of differential calculus:

$$\delta \left[f_1^n \right] = n f_1^{n-1} \delta f_1. \tag{10.47}$$

(7) If a function f is given which is a function of several dependent variables (e.g. the displacements u, v, w), then the total variation δf can be formed from the sum of the partial variations:

$$\delta f (u, v, w) = \delta_u f + \delta_v f + \delta_w f. \tag{10.48}$$

The operators δ_u, δ_v, δ_w are the partial variations with respect to u, v, w.

10.6 Formulation for a Continuum

We now want to derive a generally valid formulation of the principle of virtual displacements for a three-dimensional continuum and subsequently discuss and evaluate this formulation in detail for rods and beams. For now, we will switch back to the index notation in connection with Einstein's summation convention. We assume that the considered continuum is in a state of equilibrium and that both surface loads t_i

and volume loads f_i are present. As a consequence, the stress field σ_{ij} is invoked inside the solid. We now apply the virtual displacements δu_i which in turn cause the virtual strains $\delta\varepsilon_{ij}$.

We now employ the general principle of work and energy (see Section 9.5) and replace the quantities $\sigma_{ij}^{(1)}$, $f_i^{(1)}$, $t_i^{(1)}$ with the real quantities σ_{ij}, f_i, t_i. Analogously we replace the kinematic quantities $u_i^{(2)}$, $\varepsilon_{ij}^{(2)}$ by the virtual kinematic quantities δu_i, $\delta\varepsilon_{ij}$. It should be noted that no virtual displacements δu_i may occur on those areas ∂V_u on which displacements are specified. We can then write the principle of virtual displacements as:

$$\int_V \sigma_{ij}\delta\varepsilon_{ij}\,dV = \int_V f_i\delta u_i\,dV + \int_{\partial V_t} t_{i0}\delta u_i\,dS. \tag{10.49}$$

When using the reference system x, y, z we obtain:

$$\int_V \left(\sigma_{xx}\delta\varepsilon_{xx} + \sigma_{yy}\delta\varepsilon_{yy} + \sigma_{zz}\delta\varepsilon_{zz} + \tau_{yz}\delta\gamma_{yz} + \tau_{xz}\delta\gamma_{xz} + \tau_{xy}\delta\gamma_{xy}\right)dV$$
$$= \int_V \left(f_x\delta u + f_y\delta v + f_z\delta w\right)dV + \int_{\partial V_t}\left(t_{x0}\delta u + t_{y0}\delta v + t_{z0}\delta w\right)dS. \tag{10.50}$$

Apparently both the internal virtual works δW_i and the external virtual works δW_a appear which can be distinguished as follows:

$$\delta W_i = \int_V \sigma_{ij}\delta\varepsilon_{ij}\,dV,$$
$$\delta W_a = \int_V f_i\delta u_i\,dV + \int_{\partial V_t} t_{i0}\delta u_i\,dS. \tag{10.51}$$

The principle of virtual displacements is then, as discussed earlier:

$$\delta W_i = \delta W_a. \tag{10.52}$$

The principle of virtual displacements can thus be derived from the general principle of work and energy of elastostatics. This clearly shows that the principle of virtual displacements is not only a consequence of the equilibrium conditions, but is even completely equivalent to them. We can verbalize the principle of virtual displacements as follows:

A deformable body is in equilibrium if, for any permissible virtual displacement from the equilibrium position, the virtual inner work is equal to the virtual outer work.

Thus, the principle of virtual displacements is a statement of equilibrium formulated via corresponding virtual works. It applies to any kind of material law.

10.7 The Rod

We want to use the principle of virtual displacements to derive the equilibrium conditions and boundary conditions for a rod. For this purpose, we consider a tensile member with the length l and the constant extensional stiffness EA, as shown in Fig. 10.26. The rod is loaded by the constant line load $n(x)$ in axial direction and the single force F at the free end.

The equilibrium condition (see also Chap. 5) reads for this case:

$$N' = -n. \tag{10.53}$$

The boundary conditions can be concluded as follows. At the clamping point at $x = 0$ the rod displacement u must disappear, whereas at the free end of the bar at $x = l$ the normal force N must correspond to the acting tensile force F:

$$u(x = 0) = 0, \quad N(x = l) = F. \tag{10.54}$$

We now want to derive the Eqs. (10.53) and (10.54) with the help of the principle of virtual displacements and first consider the virtual inner work δW_i. Since we only have to consider the normal stress σ_{xx} and the virtual strain $\delta\varepsilon_{xx}$ for the member currently under consideration, this results in the following:

$$\delta W_i = \int_V \sigma_{xx} \delta\varepsilon_{xx} \mathrm{d}V. \tag{10.55}$$

We split the volume integral into an integral concerning the cross-sectional area and an integral with respect to the longitudinal axis x as follows:

$$\delta W_i = \int_0^l \sigma_{xx} A \delta u' \mathrm{d}x = \int_0^l N \delta u' \mathrm{d}x. \tag{10.56}$$

Furthermore, we perform a partial integration to reduce the degree of derivation regarding the virtual displacement δu by one:

$$\delta W_i = N \delta u \big|_0^l - \int_0^l N' \delta u \mathrm{d}x. \tag{10.57}$$

Fig. 10.26 Rod under line load $n(x)$ and single force F

The virtual external work δW_a is composed of the parts of the line load $n(x)$ and the single force F:

$$\delta W_a = \int_0^l n\delta u dx + F\delta u(x = l). \tag{10.58}$$

Thus we can state the principle of virtual displacements $\delta W_i = \delta W_a$ for the present case as:

$$N\delta u|_0^l - \int_0^l N'\delta u dx - \int_0^l n\delta u dx - F\delta u(x = l) = 0, \tag{10.59}$$

respectively with $N\delta u|_0^l = N\delta u(x = l) - N\delta u(x = 0)$:

$$-\int_0^l (N' + n)\delta u dx - N\delta u(x = 0) + (N - F)\,\delta u(x = l) = 0. \tag{10.60}$$

The variations of the displacement u which appear here are not only arbitrary (within the given boundary conditions), but also completely independent of each other. Thus, Eq. (10.60) can only be fulfilled if each of the terms appearing therein becomes zero.

First we take a closer look at the integral term in (10.60) which we require to become zero:

$$\int_0^l (N' + n)\,\delta u dx = 0. \tag{10.61}$$

On the one hand there is the possibility of the trivial solution $\delta u = 0$ which is not useful here. Furthermore, there is the possibility of setting term enclosed in parentheses to zero so that:

$$N' + n = 0. \tag{10.62}$$

Obviously this corresponds exactly to the equilibrium condition (10.53).

We now examine the second term appearing in (10.60). This term can be interpreted to mean that at the position $x = 0$ either the normal force N disappears or that the variation δu becomes zero.

$$\text{Either} \quad N(x = 0) = 0 \quad \text{or} \quad \delta u(x = 0) = 0. \tag{10.63}$$

While the first possibility corresponds to a free unloaded end, the second possibility indicates that the displacement u at $x = 0$ must assume a fixed value $u\,(x = 0) = u_0$ (e.g., $u_0 = 0$). In the present example, only the second possibility comes into consideration so that we obtain as boundary condition at $x = 0$:

$$u(x = 0) = u_0 = 0. \tag{10.64}$$

This is identical to the first boundary condition in (10.54).

We now look at the third expression appearing in (10.60). It can be concluded that the following possibilities exist:

$$\text{Either} \quad \delta u(x = l) = 0 \quad \text{or} \quad N(x = l) - F = 0. \tag{10.65}$$

Since no displacement u is prescribed at $x = l$, only the second possibility comes into consideration so that:

$$N(x = l) = F. \tag{10.66}$$

This corresponds to the second boundary condition in (10.54).

In summary we can state on the basis of this elementary example that the principle of virtual displacements not only provides the equilibrium conditions of a given system, but also unambiguously yields all potentially possible associated boundary conditions. It should be noted that the principle of virtual displacements always leads to statements formulated in the corresponding force quantities, i.e. forces and moments.

10.8   The Euler-Bernoulli Beam

We consider the cantilever beam shown in Fig. 10.2 and want to derive the equilibrium conditions as well as all potentially possible boundary conditions for this example using the principle of virtual displacements. The virtual external work can currently be described as (see also Eq. (10.3)):

$$\delta W_a = \int_0^l q(x)\delta w\,\mathrm{d}x + \int_0^l n(x)\delta u\,\mathrm{d}x + F_V\delta w(x = l) + F_H\delta u(x = l). \tag{10.67}$$

The virtual inner work δW_i is already given with Eq. (10.8):

$$\delta W_i = \int_0^l \left(N\delta u' - M\delta w''\right)\mathrm{d}x. \tag{10.68}$$

The principle of virtual displacements $\delta W_i = \delta W_a$ then reads:

$$\int_0^l \left(N\delta u' - M\delta w''\right)\mathrm{d}x - \int_0^l q(x)\delta w\,\mathrm{d}x - \int_0^l n(x)\delta u\,\mathrm{d}x - F_V\delta w(x = l) - F_H\delta u(x = l) = 0. \tag{10.69}$$

For the part of the inner virtual work δW_i we integrate the first term partially and obtain:

$$\int_0^l N\delta u'\,\mathrm{d}x = N\delta u\big|_0^l - \int_0^l N'\delta u\,\mathrm{d}x. \tag{10.70}$$

For the second term the partial integration is performed twice:

$$-\int_0^l M\delta w''\mathrm{d}x = -M\delta w'|_0^l + \int_0^l M'\delta w'\mathrm{d}x = -M\delta w'|_0^l + M'\delta w|_0^l - \int_0^l M''\delta w\mathrm{d}x.$$

$$(10.71)$$

The principle of virtual displacements thus reads:

$$-\int_0^l \left(N' + n\right)\delta u\mathrm{d}x - \int_0^l \left(M'' + q\right)\delta w\mathrm{d}x$$
$$- N\delta u(x = 0) + (N - F_H)\delta u(x = l) - M'\delta w(x = 0)$$
$$+ (M' - F_V)\delta w(x = l) + M\delta w'(x = 0) - M\delta w'(x = l) = 0.$$

$$(10.72)$$

The two integral terms result in the equilibrium conditions of the beam already known from Chap. 5 for the case of uniaxial bending:

$$N' = -n, \quad M'' = -q. \tag{10.73}$$

The boundary terms occurring in (10.72) can be interpreted as follows:

$$
\begin{array}{lll}
\text{Either} & N(x = 0) = 0 & \text{or} \quad \delta u(x = 0) = 0, \\
\text{either} & N(x = l) - F_H = 0 & \text{or} \quad \delta u(x = l) = 0, \\
\text{either} & M'(x = 0) = 0 & \text{or} \quad \delta w(x = 0) = 0, \\
\text{either} & M'(x = l) - F_V = 0 & \text{or} \quad \delta w(x = l) = 0, \\
\text{either} & M(x = 0) = 0 & \text{or} \quad \delta w'(x = 0) = 0, \\
\text{either} & M(x = l) = 0 & \text{or} \quad \delta w'(x = l) = 0. \quad (10.74)
\end{array}
$$

10.9   Beam Under Torsion

We now consider the clamped beam under torsion as given in Fig. 10.3. The virtual external and internal work δW_a and δW_i have already been formulated with (10.10) and (10.11):

$$\delta W_a = \int_0^l m_T\delta\vartheta\mathrm{d}x + M_{T0}\delta\vartheta(x = l),$$

$$\delta W_i = \int_0^l M_x\delta\vartheta'\mathrm{d}x. \tag{10.75}$$

We now use the principle of virtual displacements to derive the equilibrium condition and all potential boundary conditions for the given structural situation. For δW_i we perform a partial integration and obtain:

$$\int_0^l M_x \delta\vartheta' \mathrm{d}x = M_x \delta\vartheta\big|_0^l - \int_0^l M_x' \delta\vartheta \mathrm{d}x. \tag{10.76}$$

The principle of virtual displacements $\delta W_i = \delta W_a$ then results in:

$$-\int_0^l \left(M_x' + m_T\right) \delta\vartheta \mathrm{d}x - M_x \delta\vartheta(x = 0) + (M_x - M_{T0}) \delta\vartheta(x = l) = 0. \tag{10.77}$$

From this, the corresponding equilibrium condition can be concluded immediately as:

$$M_x' = -m_T. \tag{10.78}$$

This expression obviously matches (7.128). The two boundary terms can be interpreted as follows:

$$\text{Either} \quad M_x(x = 0) = 0 \quad \text{or} \quad \delta\vartheta(x = 0) = 0,$$
$$\text{either} \quad M_x(x = l) - M_{T0} = 0 \quad \text{or} \quad \delta\vartheta(x = l) = 0. \tag{10.79}$$

10.10 Beam Under Combined Loads

We now consider a beam of length l which is subjected to the distributed loads $n(x), q_y(x), q_z(x)$ and $m_T(x)$ (Fig. 10.27). These loads are arbitrary, but distributed continuously over the length l. The properties $EA(x), EI_{yyy}(x), EI_{zz}(x), GI_T(x), EI_{\omega\omega}(x)$ are arbitrarily but continuously distributed over x. We assume that all cross-sectional normalizations (see Chaps. 5 and 8) have been carried out. The strain energy is available with the expression (9.64) and is given here again for the sake of clarity:

Fig. 10.27 Beam under combined loads

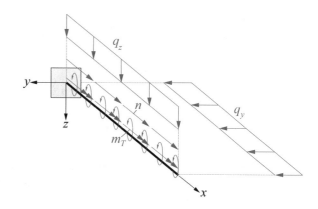

$$W_i = \frac{1}{2} \int_0^l E A u'^2 dx + \frac{1}{2} \int_0^l E I_{yy} w''^2 dx + \frac{1}{2} \int_0^l E I_{zz} v''^2 dx$$

$$+ \frac{1}{2} \int_0^l G I_T \left(\vartheta'\right)^2 dx + \frac{1}{2} \int_0^l E I_{\omega\omega} \left(\vartheta''\right)^2 dx, \tag{10.80}$$

or

$$W_i = \frac{1}{2} \int_0^l N u' dx - \frac{1}{2} \int_0^l M_y w'' dx + \frac{1}{2} \int_0^l M_z v'' dx$$

$$+ \frac{1}{2} \int_0^l M_x \vartheta' dx - \frac{1}{2} \int_0^l M_\omega \vartheta'' dx. \tag{10.81}$$

The virtual inner work δW_i then follows as:

$$\delta W_i = \int_0^l N \delta u' dx - \int_0^l M_y \delta w'' dx + \int_0^l M_z \delta v'' dx$$

$$+ \int_0^l M_x \delta \vartheta' dx - \int_0^l M_\omega \delta \vartheta'' dx. \tag{10.82}$$

The virtual external work δW_a results as follows:

$$\delta W_a = \int_0^l n \delta u dx + \int_0^l q_y \delta v dx + \int_0^l q_z \delta w dx + \int_0^l m_T \delta \vartheta dx. \tag{10.83}$$

Performing the necessary partial integrations and applying the principle of virtual displacements $\delta W_i = \delta W_a$ results in the following expression after some transformations and combining terms of equal variations $\delta u, \delta v, \delta w, \delta \vartheta$:

$$- \int_0^l \left(N' + n\right) \delta u dx + \int_0^l \left(M_z'' - q_y\right) \delta v dx - \int_0^l \left(M_y'' + q_z\right) \delta w dx$$

$$- \int_0^l \left(M_x' + M_\omega'' + m_T\right) \delta \vartheta dx + N \delta u \big|_0^l - M_y \delta w' \big|_0^l + M_y' \delta w \big|_0^l + M_z \delta v' \big|_0^l - M_z' \delta v \big|_0^l$$

$$- M_\omega \delta \vartheta' \big|_0^l + \left(M_x + M_\omega'\right) \delta \vartheta \big|_0^l = 0. \tag{10.84}$$

From the integral term $\int_0^l \left(N' + n\right) \delta u dx$ the following equilibrium condition results:

$$N' = -n, \tag{10.85}$$

which corresponds to the rod equilibrium already considered. Using the constitutive law $N = E A u'$ (5.117) the following result is obtained:

$$\left(E A u'\right)' = -n, \tag{10.86}$$

which corresponds to (5.120).

The same procedure can be used for the other integral expressions in (10.84). The second and third term result in:

$$M_y'' = -q_z, \quad M_z'' = q_y. \tag{10.87}$$

This corresponds to the two moment equilibrium conditions (5.95) and (5.98) already derived for the Euler-Bernoulli beam. Using the constitutive law (5.96) and (5.99) results in:

$$(EI_{yy}w'')'' = q_z, \quad (EI_{zz}v'')'' = q_y. \tag{10.88}$$

From the fourth integral expression we can conclude:

$$M_x' + M_\omega'' + m_T = 0. \tag{10.89}$$

If the constitutive law for primary torsion and secondary torsion is used here, the following results:

$$(EI_{\omega\omega}\vartheta'')'' - (GI_T\vartheta')' = m_T. \tag{10.90}$$

Except for the variability of the stiffnesses GI_T and $EI_{\omega\omega}$ assumed here, this expression corresponds to the differential equation (8.110) of warping torsion.

The boundary terms in (10.84) concerning $x = 0$ and $x = l$ are discussed in the following. The term

$$N\delta u|_0^l = 0 \tag{10.91}$$

corresponds to the expression already derived at the example of the rod and requires no further explanation at this point. The second term

$$M_y\delta w'\big|_0^l = 0 \tag{10.92}$$

leads to the conclusion that at $x = 0$ and $x = l$ either the bending moment $M_y = -EI_{yyy}w''$ disappears (which would be the case with a free unloaded beam end) or the variation $\delta w'$ is zero which is equivalent to w' assuming a fixed value. This would be the case, for example, if the end of the beam was clamped. Thus, the boundary term (10.92) corresponds to the respective cases shown in Fig. 5.27. The boundary term regarding the bending moment M_z

$$M_z\delta v'\big|_0^l = 0 \tag{10.93}$$

leads to analogous statements and requires no further interpretation.

The third boundary term in (10.84)

$$M'_y \delta w \big|_0^l = 0 \tag{10.94}$$

can be interpreted with $M'_y = V_z$ to the effect that at the positions $x = 0$ and $x = l$ either the transverse shear force $V_z = -(EI_{yyy}w'')'$ or the variation δw of the deflection w disappears. The latter condition would be equivalent to specifying the deflection w with a fixed value, e.g. with the value zero in the case of a fixed support (see also Fig. 5.27). An analogous expression regarding the transverse shear force V_y and the deflection v can also be found in (10.84) which does not require further discussion at this point.

The expression

$$M_\omega \delta \vartheta' \big|_0^l = 0 \tag{10.95}$$

results in the statement that at $x = 0$ and $x = l$ either the warping moment $M_\omega = -EI_{\omega\omega}\vartheta''$ becomes zero, or that the variation $\delta \vartheta'$ of the twist ϑ' disappears. From the term

$$\left(M_x + M'_\omega\right) \delta \vartheta \big|_0^l = 0 \tag{10.96}$$

it can be concluded that either the sum $M_x + M'_\omega = GI_T\vartheta' - (EI_\omega\vartheta'')'$ becomes zero or the variation $\delta\vartheta$ of the rotation ϑ disappears.

Bibliography

Becker W, Gross D (2002) Mechanik elastischer Körper und Strukturen. Springer, Berlin, Germany
Gross D, Hauger W, Wriggers P (2014) Technische Mechanik 4: Hydromechanik, Elemente der Höheren Mechanik, Numerische Methoden, 9th edn. Springer, Berlin, Germany
Hirschfeld K (2006) Baustatik, 7th edn. Springer, Berlin, Germany
Lanczos C (1986) The variational principles of mechanics, 4th edn. Dover, New York, USA
Langhaar HL (2016) Energy methods in applied mechanics. Dover, New York, USA
Meskouris K, Hake E (2009) Statik der Stabtragwerke: Einführung in die Tragwerkslehre, 2nd edn. Springer, Berlin, Germany
Mittelstedt C (2017) Energiemethoden der Elastostatik, Studienbereich Mechanik, Technische Universität Darmstadt, Germany
Oden JT, Reddy JN (1983) Variational methods in theoretical mechanics, 2nd edn. Springer, Berlin, Germany
Reddy JN (2017) Energy principles and variational methods in applied mechanics, 3rd edn. Wiley, New York, USA
Tauchert TR (1974) Energy principles in structural mechanics. McGraw-Hill, New York, USA

Chapter 11
Principle of Stationary Value of the Total Elastic Potential

11.1 Introduction

This chapter is dedicated to an energy principle which is very important for structural mechanics and thus also for lightweight engineering, namely the principle of the stationary value of the total elastic potential. It can be derived from the principle of virtual displacements and represents the basis for a range of energy-based approximation methods in lightweight engineering. However, before we turn to this principle, let us first address some selected important aspects of the closely related calculus of variations. This chapter concludes with two interesting applications, namely the first theorem of Castigliano and the theorem of Clapeyron.

11.2 Fundamentals of Calculus of Variations

A one-dimensional boundary value problem is given which is described by a differential equation and a set of given boundary conditions. The solution of this boundary value problem is the function $y(x)$. An integral expression $I(y(x))$ can now be specified for many boundary value problems. This expression assumes an extreme value if $y(x)$ is the solution for the given boundary value problem. In general, the following applies:

$$I(y) = \int_{x_1}^{x_2} F\left(y, y', y'', ..., x\right) \mathrm{d}x = \text{Extremum}. \tag{11.1}$$

We want to determine the function $y(x)$ in a way that the integral $I(y)$ takes an extreme value. This kind of problem is called a so-called variation problem and is closely related to the fundamental problem of differential calculus, to calculate for a given function $y(x)$ those values x where $y(x)$ has extreme values. The difference here is that in the case of a variation problem the function y itself is initially unknown

© The Author(s), under exclusive license to Springer Nature Switzerland AG 2021
C. Mittelstedt, *Structural Mechanics in Lightweight Engineering*,
https://doi.org/10.1007/978-3-030-75193-7_11

Fig. 11.1 Variation problem

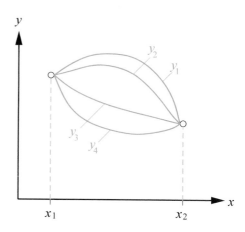

and, within the framework of given restrictions, e.g. between the values x_1 and x_2, is to be determined from the set of all permissible functions so that the integral $I(y)$ assumes an extreme value. This is shown in Fig. 11.1.

11.2.1 Functional with First Order Derivatives

We first consider the case that a given problem is solved by an ordinary differential equation of second order of the type

$$Ay'' + By' + Cy + D = R(x), \tag{11.2}$$

where $A, ..., D$ are constants and $R(x)$ is an arbitrary function. Thus an integral expression of the form

$$I(y) = \int_{x_1}^{x_2} F\left(y, y', x\right) dx \tag{11.3}$$

applies. We now want to determine the so-called functional $F\left(y, y', x\right)$ in such a way that $I(y)$ has a minimum if the function $y(x)$ satisfies the given differential equation and the boundary conditions of the form

$$a_1 y + a_2 y' = a_3 \quad \text{at} \quad x = x_1,$$
$$b_1 y + b_2 y' = b_3 \quad \text{at} \quad x = x_2 \tag{11.4}$$

Herein a_1, a_2, a_3 and b_1, b_2, b_3 are constant quantities.

Let $y(x)$ be the solution of the given boundary value problem. In addition, $\bar{y}$ is a function adjacent to y, i.e.:

$$\bar{y} = y + \delta y. \tag{11.5}$$

Fig. 11.2 Function y and an adjacent function $\overline{y}$ formed by variation $\overline{y} = y + \delta y$

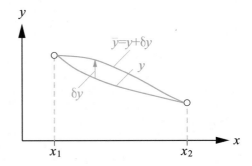

The function $\overline{y}$ is thus created by a variation δy of y, see also Fig. 11.2.

We now form the integral expressions $I(y)$ and $I(\overline{y})$ for both functions and determine their difference:

$$\Delta I = I(\overline{y}) - I(y) = \int_{x_1}^{x_2} F\left(\overline{y}, \overline{y}', x\right) dx - \int_{x_1}^{x_2} F\left(y, y', x\right) dx. \tag{11.6}$$

We develop the functional $F\left(\overline{y}, \overline{y}', x\right)$ as a Taylor series:

$$
\begin{aligned}
F\left(\overline{y}, \overline{y}', x\right) &= F\left(y + \delta y, y' + \delta y', x\right) \\
&= F\left(y, y', x\right) + \frac{\partial F}{\partial y}\delta y + \frac{\partial F}{\partial y'}\delta y' \\
&+ \frac{1}{2}\frac{\partial^2 F}{\partial y^2}(\delta y)^2 + \frac{\partial^2 F}{\partial y \partial y'}\delta y \delta y' + \frac{1}{2}\frac{\partial^2 F}{\partial (y')^2}(\delta y')^2 + \dots \tag{11.7}
\end{aligned}
$$

The difference of the integral expressions ΔI then results as:

$$
\begin{aligned}
\Delta I = \int_{x_1}^{x_2} \Bigg(&\frac{\partial F}{\partial y}\delta y + \frac{\partial F}{\partial y'}\delta y' \\
&+ \frac{1}{2}\frac{\partial^2 F}{\partial y^2}(\delta y)^2 + \frac{\partial^2 F}{\partial y \partial y'}\delta y \delta y' + \frac{1}{2}\frac{\partial^2 F}{\partial (y')^2}(\delta y')^2 + \dots \Bigg) dx. \tag{11.8}
\end{aligned}
$$

The first two terms appearing here

$$\frac{\partial F}{\partial y}\delta y + \frac{\partial F}{\partial y'}\delta y' \tag{11.9}$$

are called the first variation δI. All further terms of higher order are called second variation $\delta^2 I$. If we now assume that the adjacent function $\overline{y}$ of y is sufficiently close to y, then we can assume that the terms δy and $\delta y'$ are sufficiently small. Thus, the higher order terms are negligible and we have:

$$\Delta I = \int_{x_1}^{x_2} \left(\frac{\partial F}{\partial y} \delta y + \frac{\partial F}{\partial y'} \delta y' \right) dx. \tag{11.10}$$

We now assume that y is the solution of the given boundary value problem and that the integral expression $I(y)$ should assume a minimum. Then ΔI would be greater than or equal to zero for each allowable variation. Since $\Delta I \geq 0$ is valid for all $y \pm \delta y$, $\Delta I = 0$ must therefore be the necessary condition for the existence of an extreme value for I, i.e.:

$$\Delta I = \int_{x_1}^{x_2} \left(\frac{\partial F}{\partial y} \delta y + \frac{\partial F}{\partial y'} \delta y' \right) dx = 0. \tag{11.11}$$

The order of differentiation and variation is interchangeable. If y is now a continuous function with continuous first derivation, then:

$$\delta y' = \delta \frac{dy}{dx} = \frac{d}{dx} (\delta y). \tag{11.12}$$

This allows us to partially integrate (11.11):

$$\int_{x_1}^{x_2} \left[\frac{\partial F}{\partial y} - \frac{d}{dx} \left(\frac{\partial F}{\partial y'} \right) \right] \delta y \, dx + \left[\frac{\partial F}{\partial y'} \delta y \right]_{x_1}^{x_2} = 0. \tag{11.13}$$

This equation can only be fulfilled if the following holds:

$$\frac{\partial F}{\partial y} - \frac{d}{dx} \left(\frac{\partial F}{\partial y'} \right) = 0, \quad \text{for} \quad x_1 \leq x \leq x_2,$$

$$\frac{\partial F}{\partial y'} = 0 \quad \text{or} \quad \delta y = 0, \quad \text{for} \quad x = x_1, x = x_2. \tag{11.14}$$

The first equation is the so-called Eulerian differential equation. From the second equation the boundary conditions of the given problem can be read.

11.2.2 Functional with Second Order Derivatives

Furthermore we consider the case that the functional F contains derivatives of $y(x)$ up to y'':

$$F = F(y, y', y'', x). \tag{11.15}$$

Thus:

$$\Delta I = \int_{x_1}^{x_2} \left(\frac{\partial F}{\partial y} \delta y + \frac{\partial F}{\partial y'} \delta y' + \frac{\partial F}{\partial y''} \delta y'' \right) dx = 0. \tag{11.16}$$

Partial integration of the second term yields:

$$\int_{x_1}^{x_2} \frac{\partial F}{\partial y'} \delta y' dx = \left[\frac{\partial F}{\partial y'} \delta y\right]_{x_1}^{x_2} - \int_{x_1}^{x_2} \frac{d}{dx} \frac{\partial F}{\partial y'} \delta y dx. \tag{11.17}$$

The third term in (11.16) results in:

$$\int_{x_1}^{x_2} \frac{\partial F}{\partial y''} \delta y'' dx = \left[\frac{\partial F}{\partial y''} \delta y'\right]_{x_1}^{x_2} - \int_{x_1}^{x_2} \frac{d}{dx} \frac{\partial F}{\partial y''} \delta y' dx$$

$$= \left[\frac{\partial F}{\partial y''} \delta y'\right]_{x_1}^{x_2} - \left[\frac{d}{dx} \frac{\partial F}{\partial y''} \delta y\right]_{x_1}^{x_2} + \int_{x_1}^{x_2} \frac{d^2}{dx^2} \frac{\partial F}{\partial y''} \delta y dx. \tag{11.18}$$

We thus obtain:

$$\int_{x_1}^{x_2} \left(\frac{\partial F}{\partial y} - \frac{d}{dx} \frac{\partial F}{\partial y'} + \frac{d^2}{dx^2} \frac{\partial F}{\partial y''}\right) \delta y dx$$

$$+ \left[\frac{\partial F}{\partial y'} \delta y\right]_{x_1}^{x_2} + \left[\frac{\partial F}{\partial y''} \delta y'\right]_{x_1}^{x_2} - \left[\frac{d}{dx} \frac{\partial F}{\partial y''} \delta y\right]_{x_1}^{x_2} = 0. \tag{11.19}$$

The Eulerian differential equation results from the expression in parentheses in the integral term as

$$\frac{\partial F}{\partial y} - \frac{d}{dx} \frac{\partial F}{\partial y'} + \frac{d^2}{dx^2} \frac{\partial F}{\partial y''} = 0. \tag{11.20}$$

In the case of a functional F with derivatives up to degree two it is therefore a differential equation of fourth degree.

The associated boundary conditions result from (11.19) as follows. For $x = x_1$ and $x = x_2$ we get:

$$\frac{\partial F}{\partial y'} - \frac{d}{dx} \frac{\partial F}{\partial y''} = 0 \quad \text{or} \quad \delta y = 0,$$

$$\frac{\partial F}{\partial y''} = 0 \quad \text{or} \quad \delta y' = 0. \tag{11.21}$$

11.2.3 Functional with n-th Order Derivatives

We now consider the general case that the basic function F contains derivatives of $y(x)$ up to the degree n:

$$F = F\left(y, y', y'', y''', ..., y^{(n)}, x\right). \tag{11.22}$$

The difference ΔI results as:

11 Principle of Stationary Value ...

$$\Delta I = \int_{x_1}^{x_2} \left(\frac{\partial F}{\partial y} \delta y + \frac{\partial F}{\partial y'} \delta y' + \frac{\partial F}{\partial y''} \delta y'' + \ldots + \frac{\partial F}{\partial y^{(n)}} \delta y^{(n)} \right) = 0. \qquad (11.23)$$

We integrate (11.23) partially step by step as shown here for the term of the order k:

$$\int_{x_1}^{x_2} \frac{\partial F}{\partial y^{(k)}} \delta y^{(k)} dx = \int_{x_1}^{x_2} \frac{\partial F}{\partial y^{(k)}} \frac{d^{(k)}}{dx^{(k)}} \delta y dx$$

$$= \left[\frac{\partial F}{\partial y^{(k)}} \frac{d^{k-1}}{dx^{k-1}} \delta y \right]_{x_1}^{x_2} - \int_{x_1}^{x_2} \frac{d}{dx} \frac{\partial F}{\partial y^{(k)}} \frac{d^{k-1}}{dx^{k-1}} \delta y dx$$

$$= \left[\frac{\partial F}{\partial y^{(k)}} \frac{d^{k-1}}{dx^{k-1}} \delta y \right]_{x_1}^{x_2} - \left[\frac{d}{dx} \frac{\partial F}{\partial y^{(k)}} \frac{d^{k-2}}{dx^{k-2}} \delta y \right]_{x_1}^{x_2}$$

$$+ \int_{x_1}^{x_2} \frac{d^2}{dx^2} \frac{\partial F}{\partial y^{(k)}} \frac{d^{k-2}}{dx^{k-2}} \delta y dx. \qquad (11.24)$$

Performing the partial integrations k-fold results in:

$$\int_{x_1}^{x_2} \frac{\partial F}{\partial y^{(k)}} \delta y^{(k)} dx = \left[\frac{\partial F}{\partial y^{(k)}} \frac{d^{k-1}}{dx^{k-1}} \delta y \right]_{x_1}^{x_2} - \left[\frac{d}{dx} \frac{\partial F}{\partial y^{(k)}} \frac{d^{k-2}}{dx^{k-2}} \delta y \right]_{x_1}^{x_2}$$

$$+ \ldots - \ldots + (-1)^{k-1} \left[\frac{d^{k-1}}{dx^{k-1}} \frac{\partial F}{\partial y^{(k)}} \delta y \right]_{x_1}^{x_2} + (-1)^k \int_{x_1}^{x_2} \frac{d^k}{dx^k} \frac{\partial F}{\partial y^{(k)}} \delta y dx.$$

$$(11.25)$$

Carrying out the partial integrations for all terms in (11.23) leads to:

$$\Delta I = \int_{x_1}^{x_2} \left(\frac{\partial F}{\partial y} - \frac{d}{dx} \frac{\partial F}{\partial y'} + \frac{d^2}{dx^2} \frac{\partial F}{\partial y''} - \cdots + \cdots + (-1)^n \frac{d^n}{dx^n} \frac{\partial F}{\partial y^{(n)}} \right) \delta y dx$$

$$+ \left[\left(\frac{\partial F}{\partial y'} - \frac{d}{dx} \frac{\partial F}{\partial y''} + \cdots - \cdots + (-1)^{n-1} \frac{d^{n-1}}{dx^{n-1}} \frac{\partial F}{\partial y^{(n)}} \right) \delta y \right]_{x_1}^{x_2}$$

$$+ \left[\left(\frac{\partial F}{\partial y''} - \frac{d}{dx} \frac{\partial F}{\partial y'''} + \cdots - \cdots + (-1)^{n-2} \frac{d^{n-2}}{dx^{n-2}} \frac{\partial F}{\partial y^{(n)}} \right) \delta y' \right]_{x_1}^{x_2} + \cdots +$$

$$+ \left[\left(\frac{\partial F}{\partial y^{(n-1)}} - \frac{d}{dx} \frac{\partial F}{\partial y^{(n)}} \right) \delta y^{(n-2)} \right]_{x_1}^{x_2}$$

$$+ \left[\frac{\partial F}{\partial y^{(n)}} \delta y^{(n-1)} \right]_{x_1}^{x_2} = 0. \qquad (11.26)$$

The term enclosed in parentheses appearing in the integral contains the Eulerian differential equation describing the given problem:

$$\frac{\partial F}{\partial y} - \frac{d}{dx} \frac{\partial F}{\partial y'} + \frac{d^2}{dx^2} \frac{\partial F}{\partial y''} - \cdots + \cdots + (-1)^n \frac{d^n}{dx^n} \frac{\partial F}{\partial y^{(n)}} = 0. \qquad (11.27)$$

Apparently the variation problem leads to a differential equation of degree $2n$ if a functional F of order n is given.

The corresponding boundary conditions can be read directly from (11.26) and are as follows at $x = x_1$ and x_2:

$$\frac{\partial F}{\partial y'} - \frac{d}{dx}\frac{\partial F}{\partial y''} + \cdots - \cdots + (-1)^{n-1}\frac{d^{n-1}}{dx^{n-1}}\frac{\partial F}{\partial y^{(n)}} = 0 \quad \text{or} \quad \delta y = 0,$$

$$\frac{\partial F}{\partial y''} - \frac{d}{dx}\frac{\partial F}{\partial y'''} + \cdots - \cdots + (-1)^{n-2}\frac{d^{n-2}}{dx^{n-2}}\frac{\partial F}{\partial y^{(n)}} = 0 \quad \text{or} \quad \delta y' = 0,$$

$$\cdots$$

$$\frac{\partial F}{\partial y^{(n-1)}} - \frac{d}{dx}\frac{\partial F}{\partial y^{(n)}} = 0 \quad \text{or} \quad \delta y^{(n-2)} = 0,$$

$$\frac{\partial F}{\partial y^{(n)}} = 0 \quad \text{or} \quad \delta y^{(n-1)} = 0.$$

$$(11.28)$$

Boundary conditions in which for a basic function of degree n derivatives up to degree $n - 1$ are contained are called essential boundary conditions. In (11.29) these are the expressions

$$\delta y = 0,$$
$$\delta y' = 0,$$
$$\cdots$$
$$\delta y^{(n-2)} = 0,$$
$$\delta y^{(n-1)} = 0. \qquad (11.29)$$

The variation δy of the still to be determined function y must fulfill all essential boundary conditions, so that the variation problem leads to the solution of the given boundary value problem.

Those boundary conditions which contain derivatives of an order higher than $n - 1$ are called secondary boundary conditions. In (11.29) these are all remaining expressions.

11.2.4 Functional with n Functions with First Order Derivatives

We now extend our considerations to the case where the functional F contains the n functions $y_1, y_2, ..., y_n$, whereby we want to limit ourselves to the case where first order derivatives occur:

$$F = F\left(y_1, y_2, \cdots, y_n, y_1', y_2', \cdots, y_n', x\right). \qquad (11.30)$$

The difference ΔI then contains each unknown function $y_1, y_2, \ldots, y_n$:

$$
\Delta I = \int_{x_1}^{x_2} \left(\frac{\partial F}{\partial y_1} \delta y_1 + \frac{\partial F}{\partial y_2} \delta y_2 + \cdots + \frac{\partial F}{\partial y_n} \delta y_n \right.
$$
$$
\left. + \frac{\partial F}{\partial y_1'} \delta y_1' + \frac{\partial F}{\partial y_2'} \delta y_2' + \cdots + \frac{\partial F}{\partial y_n'} \delta y_n' \right) dx = 0. \tag{11.31}
$$

Partial integration yields:

$$
\int_{x_1}^{x_2} \left[\left(\frac{\partial F}{\partial y_1} - \frac{d}{dx} \frac{\partial F}{\partial y_1'} \right) \delta y_1 + \left(\frac{\partial F}{\partial y_2} - \frac{d}{dx} \frac{\partial F}{\partial y_2'} \right) \delta y_2 + \cdots \left(\frac{\partial F}{\partial y_n} - \frac{d}{dx} \frac{\partial F}{\partial y_n'} \right) \delta y_n \right] dx
$$
$$
+ \left[\frac{\partial F}{\partial y_1'} \delta y_1 \right]_{x_1}^{x_2} + \left[\frac{\partial F}{\partial y_2'} \delta y_2 \right]_{x_1}^{x_2} + \cdots + \left[\frac{\partial F}{\partial y_n'} \delta y_n \right]_{x_1}^{x_2} = 0.
$$

$$\tag{11.32}$$

The variations $\delta y_1, \delta y_2, \ldots, \delta y_n$ are independent of each other so that the following n Eulerian differential equations result:

$$
\frac{\partial F}{\partial y_1} - \frac{d}{dx} \frac{\partial F}{\partial y_1'} = 0,
$$
$$
\frac{\partial F}{\partial y_2} - \frac{d}{dx} \frac{\partial F}{\partial y_2'} = 0,
$$
$$
\cdots
$$
$$
\frac{\partial F}{\partial y_n} - \frac{d}{dx} \frac{\partial F}{\partial y_n'} = 0. \tag{11.33}
$$

The essential boundary conditions at $x = x_1$ and $x = x_2$ result as:

$$
\delta y_1 = 0, \quad \delta y_2 = 0, \quad \cdots \quad \delta y_n = 0. \tag{11.34}
$$

The secondary boundary conditions read:

$$
\frac{\partial F}{\partial y_1'} = 0, \quad \frac{\partial F}{\partial y_2'} = 0, \quad \cdots \quad \frac{\partial F}{\partial y_n'} = 0. \tag{11.35}
$$

We refrain from discussing further cases of functionals at this point and refer to the available relevant literature.

11.3 Principle of the Stationary Value of the Total Elastic Potential

After clarification of some necessary basics of the calculus of variations, we now consider the principle of the stationary value of the total elastic potential which is the basis of many important approximation methods of lightweight engineering and can be derived from the principle of virtual displacements (see Chap. 10). We restrict ourselves to the consideration of solids that exhibit elastic material behavior, but do not necessarily assume linear elasticity at this point.

Consider an arbitrary elastic body. Then the virtual internal work δW_i corresponds exactly to the virtual change $\delta \Pi_i$ of the total strain energy or the inner potential stored in the body. All energy stored in an elastic body can thus be recovered after unloading. In index notation we can write:

$$\delta W_i = \int_V \sigma_{ij} \delta \varepsilon_{ij} dV = \int_V \delta U_0 dV = \delta \Pi_i. \tag{11.36}$$

If the external forces (here the volume forces f_i and surface forces t_{i0}) have a potential Π_a, then we may assume that the virtual change $\delta \Pi_a$ of the external potential corresponds to the external virtual work δW_a. Here, however, a negative sign is to be used because although work was done, potential energy was released, which is equivalent to a loss of potential:

$$\delta W_a = \int_V f_i \delta u_i dV + \int_{\partial V_t} t_{i0} \delta u_i dS = -\delta \Pi_a. \tag{11.37}$$

The total elastic potential Π of the considered body consists of the internal potential Π_i and the external potential Π_a:

$$\Pi = \Pi_i + \Pi_a. \tag{11.38}$$

The principle of virtual displacements then changes into the principle of virtual total potential:

$$\delta \Pi = \delta (\Pi_i + \Pi_a) = 0. \tag{11.39}$$

This means that an elastic body is in a state of equilibrium exactly when, for all permissible variations δu_i and $\delta \varepsilon_{ij}$ of the displacements and strains, the change in the total potential is zero. Therefore, the first variation of the total potential must disappear in order to keep the considered elastic system in equilibrium. The disappearance of the first variation $\delta \Pi$ of the potential with any permissible virtual displacements and strains δu_i and $\delta \varepsilon_{ij}$ then also means that the total elastic potential Π has an extreme value in the state of equilibrium:

$$\Pi = \Pi_i + \Pi_a = \text{Extremum}. \tag{11.40}$$

This is also called the principle of the stationary value of the total elastic potential. This is obviously a variation task, as we have already considered in the previous section.

If linear elasticity is assumed, then the inner potential Π_i depends quadratically on the strains so that the extreme value required in (11.40) is a minimum:

$$\Pi = \Pi_i + \Pi_a = \text{Minimum.} \tag{11.41}$$

This is called the principle of minimum total elastic potential or the Green-Dirichlet minimum principle.

In the following we show the application of this principle to the derivation of equilibrium conditions and boundary conditions for rods and beams. The importance of this principle for the derivation of approximation methods for lightweight engineering is discussed in detail in Chaps. 13 and 14.

11.3.1 The Rod

We consider the rod shown in Fig. 11.3 (length l, constant extensional stiffness EA) which is loaded by the variable but continuous line load $n(x)$. We want to derive the equilibrium conditions and boundary conditions for the given situation and use the principle of the minimum of the total elastic potential $\Pi = \Pi_i + \Pi_a = \text{Minimum}$. The external potential results from the line load $n(x)$ as:

$$\Pi_a = -\int_0^l nu(x)\mathrm{d}x. \tag{11.42}$$

The inner potential Π_i results from the normal stress σ_{xx} and the strain ε_{xx}:

$$\Pi_i = \frac{1}{2}\int_V \sigma_{xx}\varepsilon_{xx}\mathrm{d}V = \frac{1}{2}\int_V \sigma\varepsilon\mathrm{d}V, \tag{11.43}$$

where we can omit the indexing of the individual state variables here. We split the volume integral into an integral over the cross-sectional area A and into an integral over the rod length l:

$$\Pi_i = \frac{1}{2}\int_A \int_0^l \sigma\varepsilon\mathrm{d}x\mathrm{d}A. \tag{11.44}$$

Fig. 11.3 Rod under line load $n(x)$

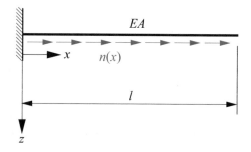

Integration over the area A then results in:

$$\Pi_i = \frac{1}{2} \int_0^l \sigma \varepsilon A \mathrm{d}x. \tag{11.45}$$

With Hooke's law $\sigma = E\varepsilon$ we get with $EA = \text{const.}$:

$$\Pi_i = \frac{EA}{2} \int_0^l \varepsilon^2 \mathrm{d}x. \tag{11.46}$$

With the kinematic equation $\varepsilon = \frac{\partial u}{\partial x} = u'$ then the following results:

$$\Pi_i = \frac{EA}{2} \int_0^l u'^2 \mathrm{d}x. \tag{11.47}$$

The total elastic potential Π can therefore be written as

$$\Pi = \frac{EA}{2} \int_0^l u'^2 \mathrm{d}x - n \int_0^l u(x)\mathrm{d}x, \tag{11.48}$$

where we want to assume from this point on that the line load $n(x)$ is constant over the length of the bar. We rearrange (11.48) as follows:

$$\Pi = \int_0^l \left(\frac{EA}{2} u'^2 - nu \right) \mathrm{d}x = \text{Minimum.} \tag{11.49}$$

This is obviously a variational problem with $y(x) = u(x)$:

$$I(y) = \int_0^l F\left(u, u', x\right) \mathrm{d}x = \text{Minimum.} \tag{11.50}$$

The integral term I thus represents the total elastic potential Π, and the functional F is the integrand in (11.49):

$$F = \frac{EA}{2} u'^2 - nu. \tag{11.51}$$

We now want to determine the Eulerian differential equation (from which we can determine the solution $y = u$) and the underlying boundary conditions at $x_1 = 0$ and $x_2 = l$. They can be determined from (11.14) as: follows:

$$\frac{\partial F}{\partial u} - \frac{\mathrm{d}}{\mathrm{d}x}\left(\frac{\partial F}{\partial u'} \right) = 0, \quad \text{for} \quad 0 \le x \le l,$$

$$\frac{\partial F}{\partial u'} = 0 \quad \text{or} \quad \delta u = 0, \quad \text{for} \quad x = 0, x = l. \tag{11.52}$$

Evaluation of (11.52) using (11.51) yields:

$$\frac{\partial F}{\partial u} = -n,$$
$$\frac{\partial F}{\partial u'} = E A u',$$
$$\frac{d}{dx}\left(\frac{\partial F}{\partial u'}\right) = E A u''. \tag{11.53}$$

The partial derivatives with respect to u or u' are to be performed as known from differential calculus. This results in the following Eulerian differential equation:

$$\frac{\partial F}{\partial u} - \frac{d}{dx}\left(\frac{\partial F}{\partial u'}\right) = -n - E A u'' = 0. \tag{11.54}$$

This can be rearranged as follows:

$$E A u'' = -n. \tag{11.55}$$

Obviously this is the differential equation of the rod which describes the behavior of the member under a constant line load n in the area $0 \le x \le l$ (see Chap. 5).

The boundary conditions result as:

$$\frac{\partial F}{\partial u'} = E A u' = 0 \quad \text{or} \quad \delta u = 0, \quad \text{for} \quad x = 0, x = l. \tag{11.56}$$

This means that at the edges $x = 0$ and $x = l$ either the normal force $N = E A u'$ disappears or the variation δu of the displacement u must become zero, which would correspond to a prescribed value $u = u_0$, for example the value $u_0 = 0$. For the situation discussed here, only the following two boundary conditions come into question:

$$u(x = 0) = 0, \quad E A u'(x = l) = N(x = l) = 0. \tag{11.57}$$

It can be shown quite simply that we get identical results if we appropriately transform the first variation of the total elastic potential:

$$\delta \Pi = \delta \left[\frac{E A}{2}\int_0^l u'^2 dx - n \int_0^l u(x) dx\right] = 0. \tag{11.58}$$

Since the order of variation and integration is interchangeable:

$$\delta \Pi = \frac{E A}{2}\int_0^l \delta\left(u'^2\right) dx - n \int_0^l \delta u dx = 0. \tag{11.59}$$

We rearrange the first integral term as follows:

$$\delta\left(u'^2\right) = 2u'\delta u'.\tag{11.60}$$

Thus:

$$\delta\Pi = EA\int_0^l u'\delta u'\mathrm{d}x - n\int_0^l \delta u\mathrm{d}x = 0.\tag{11.61}$$

We now partially integrate the first term:

$$EA\int_0^l u'\delta u'\mathrm{d}x = EAu'\delta u\big|_0^l - EA\int_0^l u''\delta u\mathrm{d}x.\tag{11.62}$$

Thus we have for $\delta\Pi$:

$$\delta\Pi = EAu'\delta u\big|_0^l - \int_0^l \left(EAu'' + n\right)\delta u\mathrm{d}x = 0.\tag{11.63}$$

For an arbitrary allowable variation δu this can only be fulfilled if the individual terms in (11.63) disappear for themselves. Thus we get from the integral term in (11.63):

$$EAu'' = -n.\tag{11.64}$$

This is obviously the already derived differential equation (11.55) of the rod. The term remaining in (11.63) then corresponds to the boundary conditions of the member already derived with (11.56) at $x = 0$ and $x = l$. Thus, the principle of the minimum of the total elastic potential supplies the equilibrium conditions of the considered structure as well as all possible boundary conditions. In contrast to the principle of virtual displacements, formulations in the sense of the displacement quantities (in this case the displacement u) result here and not in the corresponding force quantities.

11.3.2 The Euler-Bernoulli Beam

As another example we consider the beam shown in Fig. 11.4 with the bending stiffness $EI = EI(x)$ which is loaded by the line load $q(x)$. We consider bending in the xz-plane, i.e. we assume principal axes of the cross-section one of which coincides with the z-axis. The differential equation that describes the behavior of the beam within the framework of the Euler-Bernoulli beam theory is known as (see also Chap. 5):

$$\left(EIw''\right)'' - q = 0.\tag{11.65}$$

The boundary conditions here can be deduced from Fig. 11.4 as:

$$w(x = 0) = 0, \quad w'(x = 0) = 0, \quad M(x = l) = 0, \quad V(x = l) = 0.\tag{11.66}$$

Fig. 11.4 Beam under line
load $q(x)$

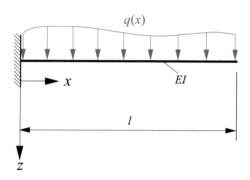

As in the previous example for the rod, we want to derive the beam differential
equation and the underlying boundary conditions from the principle of the minimum
of the total elastic potential.

The inner potential Π_i in the case of the Euler-Bernoulli beam for uniaxial bending
reads:

$$\Pi_i = \frac{1}{2} \int_V \sigma_{xx}\varepsilon_{xx}\mathrm{d}V = \frac{1}{2}\int_V \sigma\varepsilon\mathrm{d}V. \tag{11.67}$$

Here, too, we dispense with indexing the stresses and strains and use Hooke's law
$\sigma = E\varepsilon$ as a first step. Furthermore, we split the volume integral into an area integral
and an integral with respect to the longitudinal direction x:

$$\Pi_i = \frac{1}{2}\int_0^l \int_A E\varepsilon^2 \mathrm{d}A\mathrm{d}x. \tag{11.68}$$

With the strain ε as

$$\varepsilon = u' = -zw'' \tag{11.69}$$

we obtain:

$$\Pi_i = \frac{1}{2}\int_0^l \int_A E\left(-zw''\right)^2 \mathrm{d}A\mathrm{d}x = \frac{1}{2}\int_0^l \int_A E z^2 w''^2\mathrm{d}A\mathrm{d}x, \tag{11.70}$$

or

$$\Pi_i = \frac{1}{2}\int_0^l Ew''^2\mathrm{d}x \int_A z^2\mathrm{d}A. \tag{11.71}$$

With the moment of inertia $\int_A z^2\mathrm{d}A = I$ we get:

$$\Pi_i = \frac{1}{2}\int_0^l EI w''^2\mathrm{d}x. \tag{11.72}$$

For the external potential we have:

$$\Pi_a = - \int_0^l q w dx. \tag{11.73}$$

The total potential $\Pi = \Pi_i + \Pi_a$ is then:

$$\Pi = \int_0^l \left(\frac{1}{2} E I w''^2 - q w \right) dx. \tag{11.74}$$

We now demand that Π assumes a minimum if $w(x)$ is the exact solution of the given beam problem and all underlying constraints are fulfilled.

The functional $F = \frac{1}{2} E I w''^2 - q w$ occurring in this variation problem with the function $y = w$ describes a problem with derivatives up to degree $n = 2$, so that:

$$F = F\left(w, w', w'', x\right). \tag{11.75}$$

The corresponding Eulerian differential equation can be taken from (11.20):

$$\frac{\partial F}{\partial w} - \frac{d}{dx}\frac{\partial F}{\partial w'} + \frac{d^2}{dx^2}\frac{\partial F}{\partial w''} = 0. \tag{11.76}$$

For the individual terms we obtain:

$$\frac{\partial F}{\partial w} = -q,$$

$$-\frac{d}{dx}\frac{\partial F}{\partial w'} = 0,$$

$$\frac{\partial F}{\partial w''} = E I w'',$$

$$\frac{d^2}{dx^2}\frac{\partial F}{\partial w''} = (E I w'')''. \tag{11.77}$$

The Eulerian differential equation is then:

$$(E I w'')'' - q = 0. \tag{11.78}$$

Again, it is shown that the variation problem leads to the already known differential equation of beam bending. Since the functional is an equation of order $n = 2$, the result is a differential equation of degree $2n = 4$.

The corresponding boundary conditions can be read from (11.21). We get at $x = 0$ and $x = l$:

$$\frac{\partial F}{\partial w'} - \frac{d}{dx}\frac{\partial F}{\partial w''} = 0 \quad \text{or} \quad \delta w = 0,$$

$$\frac{\partial F}{\partial w''} = 0 \quad \text{or} \quad \delta w' = 0. \tag{11.79}$$

This yields:

$$(EIw'')' = 0 \quad \text{or} \quad \delta w = 0,$$
$$EIw'' = 0 \quad \text{or} \quad \delta w' = 0. \tag{11.80}$$

We can write in a rearranged form:

$$V = 0 \quad \text{or} \quad \delta w = 0,$$
$$M = 0 \quad \text{or} \quad \delta w' = 0. \tag{11.81}$$

This means that either the transverse shear force V must be zero at $x = 0$ and $x = l$ or the deflection w is prescribed with a fixed value $w = w_0$ (e.g. $w_0 = 0$). In addition, either the bending moment M must disappear at these points or the angle w' is prescribed with a fixed value. In the present case only the boundary conditions (11.66) are relevant for the problem under consideration.

Finally, we show that we encounter an identical set of equations if we consistently vary the potential formulation (11.72). With

$$\delta \Pi = \frac{1}{2} \int_0^l EI\delta(w'')^2 \mathrm{d}x - \int_0^l q\delta w \mathrm{d}x = 0 \tag{11.82}$$

and applying the chain rule to the first term we get:

$$\int_0^l EIw''\delta w'' \mathrm{d}x - \int_0^l q\delta w \mathrm{d}x = 0. \tag{11.83}$$

Partial integration of the first term in (11.83) yields:

$$EIw''\delta w'\big|_0^l - \int_0^l (EIw'')'\delta w' \mathrm{d}x - \int_0^l q\delta w \mathrm{d}x = 0. \tag{11.84}$$

Renewed partial integration leads to:

$$\int_0^l \left((EIw'')'' - q\right)\delta w \mathrm{d}x + (EIw'')'\delta w(x = 0) - EIw''\delta w'(x = 0)$$
$$+ EIw''\delta w'(x = l) - (EIw'')'\delta w(x = l) = 0. \tag{11.85}$$

With the constitutive law of the bar $EIw'' = -M$ or $(EIw'')' = -V$ this results in:

$$\int_0^l \left((EIw'')'' - q\right)\delta w \mathrm{d}x - V\delta w(x = 0) + M\delta w'(x = 0)$$
$$- M\delta w'(x = l) + V\delta w(x = l) = 0. \tag{11.86}$$

Again, this condition can be fulfilled with arbitrary admissible variations δw as well as $\delta w'$ only if each term disappears in (11.86) by itself. From the integral expression in (11.86) we then obtain, if we do not consider the trivial solution $\delta w = 0$ any further:

$$(EIw'')'' - q = 0. \tag{11.87}$$

Thus, as already shown with (11.78), the differential equation of the beam results. The remaining terms in (11.86) then represent all possible boundary conditions of the given problem. A renewed discussion is omitted at this point.

11.3.3 Beam Under Torsion

We consider the beam shown in Fig. 11.5 with the length l subjected to the distributed torsional moment $m_T(x)$. The member has arbitrary stiffnesses GI_T and $EI_{\omega\omega}$ which are continuously distributed over x. We derive the equilibrium condition and all potential boundary conditions with the help of the principle of the minimum of the elastic total potential. Again, we follow two paths, namely the application of the calculation rules of the calculus of variations (here Eq. (11.15)), and the direct variation of the resulting expressions.

The total elastic potential is currently:

$$\Pi = \Pi_i + \Pi_a = \frac{1}{2} \int_0^l GI_T \left(\vartheta'\right)^2 dx + \frac{1}{2} \int_0^l EI_{\omega\omega} \left(\vartheta''\right)^2 dx - \int_0^l m_T \vartheta dx. \tag{11.88}$$

The functional is therefore of the form

$$F = F\left(\vartheta, \vartheta', \vartheta'', x\right). \tag{11.89}$$

The corresponding Eulerian differential equation is available with (11.20) and can currently be written as:

$$\frac{\partial F}{\partial \vartheta} - \frac{d}{dx}\frac{\partial F}{\partial \vartheta'} + \frac{d^2}{dx^2}\frac{\partial F}{\partial \vartheta''} = 0. \tag{11.90}$$

Fig. 11.5 Beam subjected to a distributed torsional moment $m_T(x)$

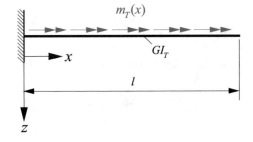

Evaluation of the individual terms yields:

$$\frac{\partial F}{\partial \vartheta} = -m_T,$$

$$\frac{\mathrm{d}}{\mathrm{d}x}\frac{\partial F}{\partial \vartheta'} = \left(GI_T\vartheta'\right)',$$

$$\frac{\mathrm{d}^2}{\mathrm{d}x^2}\frac{\partial F}{\partial \vartheta''} = \left(EI_{\omega\omega}\vartheta''\right)'', \qquad (11.91)$$

so that the following equilibrium condition results:

$$\left(EI_{\omega\omega}\vartheta''\right)'' - \left(GI_T\vartheta'\right)' = m_T. \qquad (11.92)$$

This expression agrees with (10.90).

The corresponding boundary conditions at $x = 0$ and $x = l$ follow from (11.21) as:

$$\frac{\partial F}{\partial \vartheta'} - \frac{\mathrm{d}}{\mathrm{d}x}\frac{\partial F}{\partial \vartheta''} = 0 \quad \text{or} \quad \delta\vartheta = 0,$$

$$\frac{\partial F}{\partial \vartheta''} = 0 \quad \text{or} \quad \delta\vartheta' = 0, \qquad (11.93)$$

which currently leads to the following expressions:

$$GI_T\vartheta' - \left(EI_{\omega\omega}\vartheta''\right)' = 0 \quad \text{or} \quad \delta\vartheta = 0,$$

$$EI_{\omega\omega}\vartheta'' = 0 \quad \text{or} \quad \delta\vartheta' = 0. \qquad (11.94)$$

Thus the boundary conditions (10.95) and (10.96) are again confirmed.

In the following we want to show that the equilibrium condition and all potential boundary conditions can also be obtained by consistently varying the total elastic potential Π. The first variation $\delta\Pi$ of the elastic total potential Π (11.88) must become zero by definition, which results in the following expression:

$$\delta\Pi = \int_0^l GI_T\vartheta'\delta\vartheta'\mathrm{d}x + \int_0^l EI_{\omega\omega}\vartheta''\delta\vartheta''\mathrm{d}x - \int_0^l m_T\delta\vartheta\mathrm{d}x = 0. \quad (11.95)$$

Partial integration of the first term results in:

$$\int_0^l GI_T\vartheta'\delta\vartheta'\mathrm{d}x = \left. GI_T\vartheta'\delta\vartheta\right|_0^l - \int_0^l \left(GI_T\vartheta'\right)'\delta\vartheta\mathrm{d}x. \qquad (11.96)$$

Twofold partial integration of the second term yields:

$$\int_0^l EI_{\omega\omega}\vartheta''\delta\vartheta''\mathrm{d}x = \left. EI_{\omega\omega}\vartheta''\delta\vartheta'\right|_0^l - \left. \left(EI_{\omega\omega}\vartheta''\right)'\delta\vartheta\right|_0^l + \int_0^l \left(EI_{\omega\omega}\vartheta''\right)''\delta\vartheta\mathrm{d}x. \quad (11.97)$$

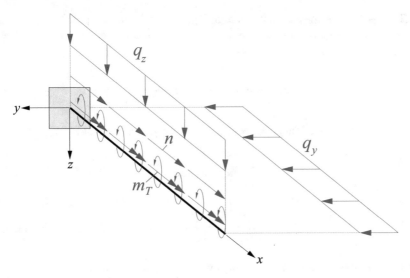

Fig. 11.6 Beam under combined loads

In summary we have:

$$\int_0^l \left[\left(EI_{\omega\omega}\vartheta'' \right)'' - \left(GI_T\vartheta' \right)' - m_T \right] \delta\vartheta \, dx$$
$$+ \left(GI_T\vartheta' - \left(EI_{\omega\omega}\vartheta'' \right)' \right) \delta\vartheta \Big|_0^l + EI_{\omega\omega}\vartheta'' \delta\vartheta' \Big|_0^l = 0. \qquad (11.98)$$

11.3.4 Beam Under Combined Load

Consider a beam under combined load as shown in Fig. 11.6. The governing equations and the underlying boundary conditions can be determined as already shown in the previous examples. Since all load cases shown here have already been discussed separately before, the superposition principle can be used for a combined load due to the geometric and material linearity assumed here. A further discussion is therefore not necessary at this point.

11.4 First Theorem of Castiglianio

A further application of the principle of the stationary value of the total elastic potential is the first theorem of Castigliano.[1] For this purpose, we consider a static system that is loaded exclusively by point loads. The total elastic potential $\Pi = \Pi_i + \Pi_a$ is present here in the form of discrete displacements, i.e. $\Pi_i = \Pi_i(u_i)$, $\Pi_a = \Pi_a(u_i)$. Thus, if there are only point loads, then we can immediately write Π_a

---

[1]Carlo Alberto Castigliano, 1847–1884, Italian civil engineer.

as $\Pi_a = -F_i u_i$ for the external potential Π_a, if F_i is the load at point i and u_i is the displacement of the point of action in the direction of the load. Then we can write the first variation $\delta\Pi$ of the total potential as:

$$\delta\Pi = \delta\left(\delta\Pi_i + \delta\Pi_a\right) = \frac{\partial\Pi_i}{\partial u_i}\delta u_i + \frac{\partial\Pi_a}{\partial u_i}\delta u_i = 0. \tag{11.99}$$

With $\Pi_a = -F_i u_i$ we immediately obtain $\frac{\partial\Pi_a}{\partial u_i} = -F_i$, and we have:

$$\frac{\partial\Pi_i}{\partial u_i}\delta u_i - F_i\delta u_i = 0, \tag{11.100}$$

or:

$$\left(\frac{\partial\Pi_i}{\partial u_i} - F_i\right)\delta u_i = 0. \tag{11.101}$$

If we ignore the trivial solution $\delta u_i = 0$ we get:

$$\frac{\partial\Pi_i}{\partial u_i} = F_i. \tag{11.102}$$

This is the first theorem of Castigliano. It states that the single force F_i can be determined from the partial derivative of the inner potential Π_i with respect to the associated displacement u_i, if the potential can be formulated as a function of the displacements u_i. If the system under consideration is loaded by single moments M_i, then Castigliano's first theorem reads:

$$\frac{\partial\Pi_i}{\partial\varphi_i} = M_i. \tag{11.103}$$

It is important to note that the name of this principle is not clear and uniform in the technical literature. As we will see in Chap. 12 there is another theorem named after Castigliano. There seems to be no agreement in the technical literature which of the two theorems is called the first theorem of Castigliano and which is called the second theorem.

11.5 Theorem of Clapeyron

Another theorem which can be concluded from the treated potential expressions Π_i and Π_a is the so-called theorem of Clapeyron.[2] It can be concluded from the general theorem of work and energy if linear-elastic material behavior is assumed and if it is also assumed that the external loads are dead loads and thus not dependent on

---

[2]Émile Clapeyron, 1799–1864, French physicist.

the deformation history. Furthermore, we assume that there are no displacements u_{i0} ($i = 1, 2, 3$) other than zero on the surface of the considered body. If we now use the actual stress, displacement and strain quantities in the general work and energy theorem, then we get the following result in index notation:

$$\int_V \sigma_{ij}\varepsilon_{ij}\mathrm{d}V = \int_V f_i u_i \mathrm{d}V + \int_{\partial V_t} t_{i0} u_i \mathrm{d}S. \qquad (11.104)$$

The integral expression on the left side represents twice the strain energy or inner potential stored in the body, i.e. $2\Pi_i$. On the right side, however, the work done by the dead loads during the deformation process is W_a. Since dead loads have a potential, $\Pi_a = -W_a$ applies, so that (11.104) represents the so-called theorem of Clapeyron:

$$2\Pi_i + \Pi_a = 0. \qquad (11.105)$$

Bibliography

Becker W, Gross D (2002) Mechanik elastischer Körper und Strukturen. Springer, Berlin, Germany

Gross D, Hauger W, Wriggers P (2014) Technische Mechanik 4: Hydromechanik, Elemente der Höheren Mechanik, Numerische Methoden, 9th edn. Springer, Berlin, Germany

Kossira H (1996) Grundlagen des Leichtbaus. Springer, Berlin, Germany

Lanczos C (1986) The variational principles of mechanics, 4th edn. Dover, New York, USA

Langhaar HL (2016) Energy methods in applied mechanics. Dover Publications, New York, USA

Mittelstedt C (2017) Energiemethoden der Elastostatik, Studienbereich Mechanik, Technische Universität Darmstadt, Germany

Oden JT, Reddy JN (1983) Variational methods in theoretical mechanics, 2nd edn. Springer, Berlin, Germany

Reddy JN (2017) Energy principles and variational methods in applied mechanics, 3rd edn. Wiley, New York, USA

Tauchert TR (1974) Energy principles in structural mechanics. McGraw-Hill, New York, USA

Washizu K (1982) Variational methods in elasticity and plasticity, 3rd edn. Pergamon Press, New York, USA

Chapter 12
Principle of Virtual Forces

12.1 Introduction

So far we have treated principles based on virtual displacements. However, it is also possible to derive work and energy principles based on the consideration of virtual forces. Such principles have a significant value for lightweight engineering practice and are presented in this chapter. We then formalize these considerations with an explanation of the so-called force method for the analysis of statically determinate and statically indeterminate systems, before concluding this chapter with the treatment of reciprocity theorems and some other interesting applications of the discussed principles. We first introduce the definition of the terms virtual forces and complementary virtual work.

12.2 Virtual Forces and Complementary Virtual Work

Consider a body in a state of equilibrium that is exposed to a virtual equilibrium group $\underline{F}$, which performs the complementary virtual work $\delta \overline{W}$ along the actual displacements:

$$\delta \overline{W} = \delta \underline{F} \underline{u}. \tag{12.1}$$

Analogous to the previous considerations we can then define a virtual external complementary work $\delta \overline{W}_a$, which is performed by virtual volume forces $\delta \underline{f}$, virtual surface loads $\delta \underline{t}$ and m virtual point forces $\delta \underline{F}_i$, and which we can formulate in a vector-matrix notation as follows:

$$\delta \overline{W}_a = \int_V \delta \underline{f} \underline{u} dV + \int_{\partial V_u} \delta \underline{t} \underline{u} dS + \sum_{i=1}^{m} \delta \underline{F}_i \underline{u}. \tag{12.2}$$

Fig. 12.1 Cantilever beam under virtual loads

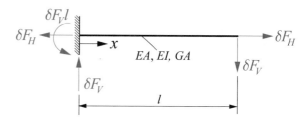

The virtual internal complementary work $\delta \overline{W}_i$ results from the virtual complementary strain energy density $\delta \overline{U}_0$, which is written in index notation as $\delta \overline{U}_0 = \varepsilon_{ij}\delta\sigma_{ij}$. Thus we obtain $\delta \overline{W}_i$ as:

$$\delta \overline{W}_i = \int_V \delta \overline{U}_0 \mathrm{d}V = \int_V \varepsilon_{ij}\delta\sigma_{ij}\mathrm{d}V. \tag{12.3}$$

We now consider the cantilever beam shown in Fig. 12.1 (length l, stiffnesses EA, EI, GA), which is loaded at its free end at $x = l$ by the virtual forces δF_V and δF_H. Due to the virtual forces the two reaction forces δF_H in horizontal direction and δF_V in vertical direction will result at the clamping point. In addition to this, the clamping moment with the value $\delta F_V l$ occurs. Obviously, this group of forces and moments forms an equilibrium group, it fulfills all equilibrium conditions. However, the virtual complementary external work $\delta \overline{W}_a$ contains only the two individual forces δF_H and δF_V at the end of the cantilever, but not the resulting support reactions - these do not perform any work due to the support situation (displacements and rotations are not possible at this point). Thus:

$$\delta \overline{W}_a = \delta F_V w(x = l) + \delta F_H u(x = l). \tag{12.4}$$

The virtual complementary inner work $\delta \overline{W}_i$ results from the virtual normal stress $\delta\sigma_{xx}$ and the virtual shear stress $\delta\tau_{xz}$ as well as from the corresponding real strain quantities:

$$\delta \overline{W}_i = \int_V (\varepsilon_{xx}\delta\sigma_{xx} + \gamma_{xz}\delta\tau_{xz})\,\mathrm{d}V. \tag{12.5}$$

We divide the normal strain ε_{xx} into the two parts ε_{xx}^0 and ε_{xx}^1, where ε_{xx}^0 is related to the center of gravity axis of the beam and thus to a stress invoked by the normal force N. The fraction ε_{xx}^1, on the other hand, refers to the linear variability of the strain over z and is thus related to the moment M. Furthermore, we decompose the volume integral into an integral over the cross-sectional area A and an integral over the longitudinal axis x:

$$\delta \overline{W}_i = \int_0^l \int_A \left[\left(\varepsilon_{xx}^0 + z\varepsilon_{xx}^1\right)\delta\sigma_{xx} + \gamma_{xz}\delta\tau_{xz}\right]\mathrm{d}A\mathrm{d}x. \tag{12.6}$$

The integration over the cross-sectional area A yields:

$$\delta \overline{W}_i = \int_0^l \left(\varepsilon_{xx}^0 \delta N + \varepsilon_{xx}^1 \delta M + \gamma_{xz} \delta V \right) \mathrm{d}x. \tag{12.7}$$

The quantities δN, δM and δV are the virtual internal forces and moment caused by the virtual external forces which perform virtual work along the real strains.

If we assume linear elasticity, then we obtain the following expressions for ε_{xx}^0 and ε_{xx}^1:

$$\varepsilon_{xx}^0 = \frac{N}{EA}, \quad \varepsilon_{xx}^1 = \frac{M}{EI}. \tag{12.8}$$

The shear strain γ_{xz} can be calculated with the help of Hooke's Law $\tau_{xz} = G\gamma_{xz}$ and the assumption of a constant shear stress distribution $\tau_{xz} = \frac{V}{KA}$ (where K is the shear correction factor, see Chap. 9 and the explanations on the Timoshenko beam theory in Chap. 16) can be substituted in (12.7), and it follows:

$$\delta \overline{W}_i = \int_0^l \left(\frac{N\delta N}{EA} + \frac{M\delta M}{EI} + \frac{V\delta V}{KGA} \right) \mathrm{d}x. \tag{12.9}$$

Since the Euler-Bernoulli beam theory excludes transverse shear strains γ_{xz}, as a consequence the shear stresses τ_{xz} and thus also their resultant, the transverse shear force V, cannot be calculated from constitutive equations. Consequently, the influence of the transverse shear force V is neglected from this point on and, as a consequence, the third term in (12.9) is no longer considered. This corresponds to the assumption $GA \to \infty$ which is consistent with the Euler-Bernoulli beam theory assumed here. This is also shown in Fig. 9.9 and the explanations there.

12.3 The Principle of Virtual Forces

Analogous to the already discussed principle of virtual displacements, the principle of virtual forces can also be derived from the general principle of work and energy of elastostatics. For this purpose we use the virtual force quantities $\delta\sigma_{ij}$, δf_i and δt_i. The kinematic field consists of the real displacements u_i and strains ε_{ij}. On those boundary areas ∂V_t of the body where stresses are given, the virtual surface forces δt_i must be zero. The virtual force quantities applied here have the following properties, analogous to the virtual displacements:

- The virtual forces are virtual and do not exist in reality.
- The virtual forces are infinitesimally small.
- The virtual forces must fulfill the equilibrium conditions and all given stress boundary conditions.

From the general principle of work and energy of elastostatics we then obtain

$$\int_V \varepsilon_{ij} \delta\sigma_{ij} \mathrm{d}V = \int_V u_i \delta f_i \mathrm{d}V + \int_{\partial V_u} u_{i0} \delta t_i \mathrm{d}S. \tag{12.10}$$

The first integral term appearing here is the so-called inner virtual complementary work $\delta \overline{W}_i$:

$$\delta \overline{W}_i = \int_V \varepsilon_{ij} \delta \sigma_{ij} \mathrm{d}V. \tag{12.11}$$

The two remaining terms represent the so-called external virtual complementary work $\delta \overline{W}_a$:

$$\delta \overline{W}_a = \int_V u_i \delta f_i \mathrm{d}V + \int_{\partial V_u} u_{i0} \delta t_i \mathrm{d}S. \tag{12.12}$$

The principle of virtual forces is thus:

$$\delta \overline{W}_i = \delta \overline{W}_a. \tag{12.13}$$

The principle of virtual forces is also called the principle of virtual complementary work. It is valid for any virtual force quantities which fulfill the given stress boundary conditions. Thus, it is not only a consequence of the kinematic compatibility of displacements and strains, but it is even completely equivalent to them. It can be verbalized as follows:

> A deformable body is in a kinematically compatible state of displacement and strain exactly when, in any equilibrium system of virtual forces, the inner complementary work is equal to the external complementary work.

The principle of virtual forces is valid for any material behavior.

12.4   The Unit Load Theorem

From the principle of virtual forces an extremely useful theorem, namely the so-called unit load theorem, can be derived. For motivation we consider an Euler-Bernoulli beam which is loaded by the two virtual forces δF_H and δF_V as well as the virtual single moment δM. The assigned kinematic quantities are the displacements u, w and the rotation φ. The principle of virtual forces is then as follows:

$$\int_0^l \left(\frac{N \delta N}{EA} + \frac{M \delta M}{EI} \right) \mathrm{d}x = u \delta F_H + w \delta F_V + \varphi \delta M. \tag{12.14}$$

If, on the other hand, a truss consisting of m rods is considered that is loaded exclusively by point forces at its nodes, the normal forces occurring in the rods are constant and we have:

$$\sum_{i=1}^m \frac{N_i \delta N_i l_i}{(EA)_i} = u \delta F_H + w \delta F_V. \tag{12.15}$$

Therein l_i and $(EA)_i$ are the length and the extensional stiffness of rod i.

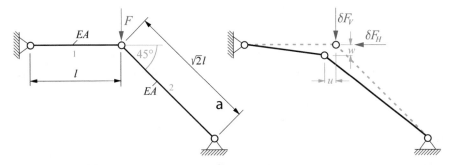

Fig. 12.2 Truss under single force F (left), deformed structure under virtual forces with real node displacements (right)

To illustrate the practical value of the unit load theorem, let us consider the truss shown in Fig. 12.2 that consists of two rods and which is subjected to the vertically acting single force F. We want to determine the two node displacements u and w as indicated. Both members have identical extensional stiffnesses EA. In order to determine the node displacements u and w, we now apply the two virtual forces δF_V and δF_H. The principle of virtual forces is then:

$$\sum_{i=1}^{m} \frac{N_i \delta N_i l_i}{(EA)_i} = \frac{N_1 \delta N_1 l}{EA} + \frac{N_2 \delta N_2 \sqrt{2}l}{EA} = \delta F_H u + \delta F_V w. \tag{12.16}$$

To determine the displacement u we set the virtual force δF_V to zero. We obtain:

$$\frac{N_1 \delta N_1 l}{EA} + \frac{N_2 \delta N_2 \sqrt{2}l}{EA} = \delta F_H u. \tag{12.17}$$

The virtual forces may be arbitrary within the requirements of the considered static system. Thus, for δF_H we can also simply apply a unit force $\delta F_H = 1$:

$$\frac{N_1 \delta N_1 l}{EA} + \frac{N_2 \delta N_2 \sqrt{2}l}{EA} = u. \tag{12.18}$$

As a result, the displacement u can be read directly from (12.18). The forces N_1 and N_2 are the real member forces due to the single force F, and δN_1 and δN_2 are the member forces due to the virtual unit force $\delta F_H = 1$. At the given example the following applies:

$$N_1 = -F, \quad N_2 = -\sqrt{2}F,$$
$$\delta N_1 = -1, \quad \delta N_2 = 0. \tag{12.19}$$

The displacement u can thus be determined immediately from (12.18) as:

$$u = \frac{-F \cdot (-1) \cdot l}{EA} = \frac{Fl}{EA}. \tag{12.20}$$

We can proceed in the same way to determine the displacement w. In the formulation (12.16) we set the virtual single force δF_H to zero and apply the virtual single force δF_V as a unit load $\delta F_V = 1$. We obtain:

$$\frac{N_1 \delta N_1 l}{EA} + \frac{N_2 \delta N_2 \sqrt{2} l}{EA} = w. \tag{12.21}$$

The member forces N_1 and N_2 due to the single force F are already given in (12.19). The virtual member forces δN_1 and δN_2 as a result of the virtual unit force $\delta F_V = 1$ result as:

$$\delta N_1 = -1, \quad \delta N_2 = -\sqrt{2}. \tag{12.22}$$

We then obtain the displacement w as:

$$w = (1 + 2\sqrt{2}) \frac{Fl}{EA}. \tag{12.23}$$

It turns out that the principle of virtual forces in combination with the unit load theorem allows to calculate deformations of static systems quite easily. Formalized this procedure is also denoted as the so-called force method, which we will discuss in detail later in this chapter.

Another example is the beam shown in Fig. 12.3 which has the length l and the constant bending stiffness EI. The beam is loaded by the single load F at midspan. The deflection w is to be determined by using the principle of virtual forces.

In the first step we determine the moment line of the beam due to the given single force F. It is shown in Fig. 12.3, left. In addition, we apply the virtual single force

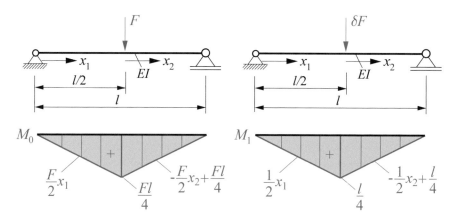

Fig. 12.3 Beam under point load F (top left), beam under virtual force $\delta F = 1$ (top right), moment diagrams (bottom)

δF to calculate the desired deflection w and determine the moment line for this load case as well. It is shown in Fig. 12.3, right. The principle of virtual forces is then:

$$\int_0^l \frac{M\delta M}{EI} dx = w\delta F. \tag{12.24}$$

Therein, M is the bending moment due to the given load F, and δM represents the moment distribution due to the virtual unit force $\delta F = 1$. Using a unit load $\delta F = 1$ results in:

$$\int_0^l \frac{M\delta M}{EI} dx = w. \tag{12.25}$$

The deflection w can then be easily calculated from (12.25). Note that due to the discontinuity of the two moment lines, the integration has to be split into two partial integrals. It is therefore advantageous to introduce two local axes x_1 and x_2 as shown in Fig. 12.3, so that the following applies:

$$w = \int_0^{\frac{l}{2}} \frac{M(x_1)\delta M(x_1)}{EI} dx_1 + \int_0^{\frac{l}{2}} \frac{M(x_2)\delta M(x_2)}{EI} dx_2. \tag{12.26}$$

In what follows we denote the moments due to the given loads as M_0, and the moments due to the virtual load with M_1:

$$w = \int_0^{\frac{l}{2}} \frac{M_0(x_1)M_1(x_1)}{EI} dx_1 + \int_0^{\frac{l}{2}} \frac{M_0(x_2)M_1(x_2)}{EI} dx_2. \tag{12.27}$$

Performing the necessary integrations then results in:

$$w = \int_0^{\frac{l}{2}} \frac{1}{EI} \frac{F}{2} x_1 \frac{1}{2} x_1 dx_1 + \int_0^{\frac{l}{2}} \frac{1}{EI} \left(-\frac{F}{2} x_2 + \frac{Fl}{4} \right) \left(-\frac{1}{2} x_2 + \frac{l}{4} \right) dx_2, \tag{12.28}$$

thus:

$$w = \frac{Fl^3}{48EI}. \tag{12.29}$$

At this point the great value of using the integral table as shown in Fig. 5.8 becomes apparent. The above result is obtained much faster if the integrations prescribed in (12.25) or (12.26) are carried out using the integral table instead of actually solving the integrals.

12.5 The Principle of the Stationary Value of the Elastic Complementary Potential

Analogous to the previous explanations, the so-called principle of the stationary value of the total complementary potential can be derived. We assume that an elastic structure is given, although we do not necessarily require linear elasticity at this point. The virtual inner complementary work $\delta\overline{W}_i$ then corresponds to the virtual change $\delta\overline{\Pi}_i$ of the inner complementary potential $\overline{\Pi}_i$, which we can represent in index notation as follows:

$$\delta\overline{W}_i = \int_V \varepsilon_{ij}\delta\sigma_{ij}dV = \delta\overline{\Pi}_i. \tag{12.30}$$

If there is a complementary potential for the displacements that occur, then this is identical to the negative virtual external complementary work. This is also referred to as external complementary potential $\overline{\Pi}_a$:

$$\delta\overline{W}_a = \int_V u_i\delta f_i dV + \int_{S_u} u_{i0}\delta t_i dS = -\overline{\Pi}_a. \tag{12.31}$$

The total complementary potential $\overline{\Pi}$ is then the sum of the inner and outer complementary potential: $\overline{\Pi} = \overline{\Pi}_i + \overline{\Pi}_a$. The principle of virtual forces then changes into the principle of the virtual total complementary potential as follows:

$$\delta\overline{\Pi} = \delta\left(\overline{\Pi}_i + \overline{\Pi}_a\right) = 0. \tag{12.32}$$

It is required that the variation $\delta\overline{\Pi}$ vanishes for any arbitrary virtual forces. From this it follows that the total complementary potential in the actual state of deformation assumes an extremal value:

$$\overline{\Pi} = \overline{\Pi}_i + \overline{\Pi}_a = \text{Extremum}. \tag{12.33}$$

We call this requirement the principle of the stationary value of the total complementary potential.

If, however, not only elastic material behavior is present, but if the material under consideration is even linearly elastic, then the internal complementary potential is a quadratic form of the stresses. As a consequence, the extreme value required in (12.33) is a minimum. In this case we speak of the principle of the minimum of the total complementary potential:

$$\overline{\Pi} = \overline{\Pi}_i + \overline{\Pi}_a = \text{Minimum}. \tag{12.34}$$

This is also known as the principle of Castigliano and Menabrea.

12.6 Second Theorem of Castigliano

From the principle of the stationary value of the elastic complementary potential, another important theorem can be derived, namely the second theorem of

Castigliano.[1] For this purpose we consider a system which is subjected to single forces F_i and moments M_j and for which we can express the complementary potential $\overline{\Pi} = \overline{\Pi}_i + \overline{\Pi}_a$ through these single forces and moments, whereby we will consider only point loads F_i for the moment:

$$\overline{\Pi}_i = \overline{\Pi}_i\,(F_i), \quad \overline{\Pi}_a = \overline{\Pi}_a\,(F_i). \tag{12.35}$$

The first variation of the complementary potential is then:

$$\delta\overline{\Pi}_i = \frac{\partial\overline{\Pi}_i}{\partial F_i}\delta F_i, \quad \delta\overline{\Pi}_a = \frac{\partial\overline{\Pi}_a}{\partial F_i}\delta F_i = -u_i\delta F_i. \tag{12.36}$$

Hence:

$$\delta\overline{\Pi} = \delta\overline{\Pi}_i + \delta\overline{\Pi}_a = \left(\frac{\partial\overline{\Pi}_i}{\partial F_i}\delta F_i - u_i\delta F_i\right) = \left(\frac{\partial\overline{\Pi}_i}{\partial F_i} - u_i\right)\delta F_i = 0. \tag{12.37}$$

Avoiding the trivial solution $\delta F_i = 0$ by setting the term in parentheses to zero then results in:

$$\frac{\partial\overline{\Pi}_i}{\partial F_i} - u_i = 0. \tag{12.38}$$

This is the 2nd theorem of Castigliano. It states that by partial derivation of the inner complementary potential $\overline{\Pi}_i$ with respect to the single force F_i, the displacement u_i at the point of force application in the direction of the force can be determined.

An elementary example is a linear spring with the spring stiffness k, where the spring law $F = ku$ is assumed to be valid. The inner complementary potential $\overline{\Pi}_i$ is then $\overline{\Pi}_i = \frac{1}{2}Fu = \frac{1}{2}\frac{F^2}{k}$. Derivation of $\overline{\Pi}_i$ with respect to F then returns $\frac{F}{k}$, which obviously corresponds exactly to the displacement u.

As a further example we consider the cantilever beam of the Fig. 12.4. The deflection of the point of force application is to be determined with the help of Castigliano's 2nd theorem. For this purpose we first determine the moment line for the given situation as shown in Fig. 12.4, bottom left. The inner complementary potential $\overline{\Pi}_i$ can then be written as:

$$\overline{\Pi}_i = \frac{1}{2}\int_0^l \frac{M^2}{EI}dx = \frac{1}{2}\int_0^l \frac{(Fx - Fl)^2}{EI}dx = \frac{F^2l^3}{6EI}. \tag{12.39}$$

The deflection w can then be determined as follows:

$$w = \frac{\partial\overline{\Pi}_i}{\partial F} = \frac{Fl^3}{3EI}. \tag{12.40}$$

---

[1]Carlo Alberto Castigliano, 1847–1884, Italian civil engineer.

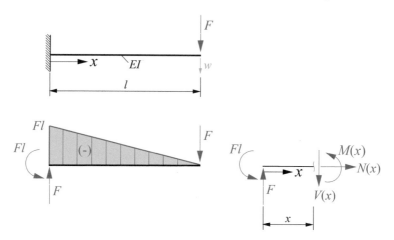

Fig. 12.4 Cantilever beam under point load F

12.7 Theorem of Menabrea

A further application of the second theorem of Castigliano is the treatment of statically indeterminate systems, whereby we then speak of the so-called theorem of Menabrea.[2] For explanation, we consider the statically indeterminate beam in Fig. 12.5 which has the length l and the constant bending stiffness EI. The load consists of a uniformly distributed line load q. At this point we want to determine the support force B using the second theorem of Castigliano, respectively the theorem of Menabrea.

In addition to the x-axis, another reference axis $\overline{x}$ is introduced as shown in Fig. 12.5. Now we cut the beam at an arbitrary position $\overline{x}$ as shown in Fig. 12.5, right, and determine the moment $M(\overline{x})$ as follows:

$$M(\overline{x}) = B\overline{x} - \frac{1}{2}q\overline{x}^2. \tag{12.41}$$

The inner complementary potential $\overline{\Pi}_i$ then results as follows:

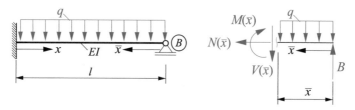

Fig. 12.5 Statically indeterminate beam under line load q

---

[2]Federico Luigi Conte di Menabrea, 1809–1896, Italian scientist.

$$\overline{\Pi}_i = \frac{1}{2}\int_0^l \frac{M(\overline{x})^2}{EI}dx = \frac{1}{2EI}\int_0^l \left(B\overline{x} - \frac{1}{2}q\overline{x}^2\right)^2 dx = \frac{1}{2EI}\left(\frac{1}{3}B^2l^3 - \frac{1}{4}Bql^4 + \frac{1}{20}q^2l^5\right).$$

(12.42)

The second theorem of Castigliano or the theorem of Menabrea then results in the deflection w_B at the support B, which as a result must obviously assume the value zero:

$$\frac{\partial\overline{\Pi}_i}{\partial B} = w_B = 0.$$

(12.43)

This yields the expression

$$\frac{2}{3}Bl^3 - \frac{1}{4}ql^4 = 0,$$

(12.44)

from which we can then determine the desired support force B:

$$B = \frac{3}{8}ql.$$

(12.45)

Thus the theorem of Menabrea states that the partial derivative of the inner complementary potential with respect to a statically indeterminate quantity must become zero. It can also be applied to multiply statically indeterminate systems as we can see from the example in Fig. 12.6.

The support reactions A_V and M_A are to be determined with the help of Menabrea's theorem for the triple statically indeterminate supported beam (length l, constant bending stiffness EI) under uniform load q. For this purpose, we cut the beam free at support A and thus set free the support reactions A_V and M_A we are looking for. The horizontal bearing force A_H, which also occurs here, is not considered further due to the lack of load in the longitudinal direction of the beam; A_H results with the value zero. The balance of forces and moments with respect to the arbitrary intersection point x (Fig. 12.6, right) then results in:

$$V(x) = A_V - qx, \quad M(x) = A_V x + M_A - \frac{qx^2}{2}.$$

(12.46)

Thus, the internal forces $V(x)$ and $M(x)$ are known in dependence of the two unknown support reactions A_V and M_A. The inner complementary potential $\overline{\Pi}_i$ can then be written as:

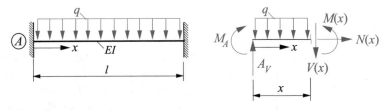

Fig. 12.6 Clamped beam

$$\overline{\Pi}_i = \frac{1}{2EI} \int_0^l M^2 dx = \frac{1}{2EI} \int_0^l \left(A_V x + M_A - \frac{qx^2}{2} \right)^2 dx$$

$$= \frac{1}{2EI} \left(\frac{1}{3} A_V^2 l^3 + A_V M_A l^2 - \frac{1}{4} A_V q l^4 + M_A^2 l - \frac{1}{3} M_A q l^3 + \frac{1}{20} q^2 l^5 \right). \quad (12.47)$$

The theorem of Menabrea is then:

$$\frac{\partial \overline{\Pi}_i}{\partial A_V} = w_A = 0, \quad \frac{\partial \overline{\Pi}_i}{\partial M_A} = \varphi_A = 0, \quad (12.48)$$

where w_A and φ_A are the deflection and rotation of the support A, respectively. These quantities must become zero due to the clamping, and the result is

$$\frac{\partial \overline{\Pi}_i}{\partial A_V} = \frac{2}{3} A_V l^3 + M_A l^2 - \frac{1}{4} q l^4 = 0,$$

$$\frac{\partial \overline{\Pi}_i}{\partial M_A} = A_V l^2 + 2 M_A l - \frac{1}{3} q l^3 = 0. \quad (12.49)$$

Thus, two equations are available for the determination of the two unknown support reactions A_V and M_A. It follows:

$$M_A = -\frac{1}{12} q l^2, \quad A_V = \frac{1}{2} q l. \quad (12.50)$$

12.8 The Force Method

The principle of virtual forces in conjunction with the unit load theorem is often summarized under the term force method. In this section, we want to formalize this method and work out further applications besides the determination of displacements on statically determinate systems.

12.8.1 Calculation of Deformations of Statically Determinate Systems

12.8.1.1 Truss Structures

We formalize the procedure for the determination of deformations in statically determinate systems as follows and start with the consideration of statically determinate plane truss structures. Consider a truss which consists of m rods. The load is such that only constant normal forces N_i ($i = 1.2, ..., m$) result in the m rods. The member i has the length l_i and the extensional stiffness EA_i. A deflection w occurring at any point of the truss can then be calculated as follows:

- At the given truss the forces N_i ($i = 1, 2, ..., m$) in the m members are calculated as a result of the given load.
- Furthermore, the forces $\delta N_i = \overline{N}_i$ of the m members are calculated as a result of a virtual single force $F = 1$ at the considered location in the direction of the desired displacement w.
- The displacement w can then be determined according to the following formula.

$$w = \sum_{i=1}^{m} \frac{N_i \overline{N}_i}{(EA)_i} l_i. \tag{12.51}$$

We discuss the procedure at the example of the truss shown in Fig. 12.7. We want to determine the indicated deflection w using the force method.

The member forces N_i ($i = 1.2, ..., 9$) resulting from the two acting point forces are obtained as:

$$N_1 = \frac{F}{2}, \quad N_2 = 2F, \quad N_3 = 0, \quad N_4 = -\frac{3F}{2}, \quad N_5 = \frac{F}{2},$$
$$N_6 = \frac{F}{2}, \quad N_7 = -\frac{F}{\sqrt{2}}, \quad N_8 = \frac{3F}{\sqrt{2}}, \quad N_9 = -F. \tag{12.52}$$

The member forces $\overline{N}_i$ due to the virtual unit force $F = 1$ can be determined as:

$$\overline{N}_1 = -\frac{1}{2}, \quad \overline{N}_2 = 0, \quad \overline{N}_3 = 0, \quad \overline{N}_4 = -\frac{1}{2}, \quad \overline{N}_5 = -\frac{1}{2},$$
$$\overline{N}_6 = -\frac{1}{2}, \quad \overline{N}_7 = \frac{1}{\sqrt{2}}, \quad \overline{N}_8 = \frac{1}{\sqrt{2}}, \quad \overline{N}_9 = 0. \tag{12.53}$$

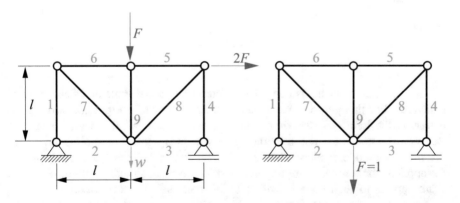

Fig. 12.7 Truss under point loads (left), truss under virtual load (right)

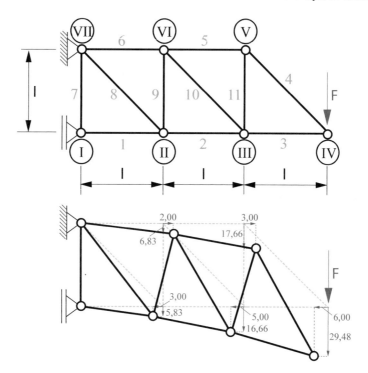

Fig. 12.8 Truss under point load F (top), deformed structure (bottom, displacements in the unit $\frac{Fl}{EA}$)

The deflection w then results as:

$$w = \sum_{i=1}^{9} \frac{N_i \overline{N}_i}{(EA)_i} l_i = \frac{1}{EA} \left[-\frac{F}{2}\frac{1}{2}l + \frac{3F}{2}\frac{1}{2}l - \frac{F}{2}\frac{1}{2}l - \frac{F}{2}\frac{1}{2}l - \frac{F}{\sqrt{2}}\frac{1}{\sqrt{2}}\sqrt{2}l + \frac{3F}{\sqrt{2}}\frac{1}{\sqrt{2}}\sqrt{2}l \right].$$
(12.54)

We obtain:

$$w = \sqrt{2}\frac{Fl}{EA}.$$
(12.55)

The force method can also be used to determine the deformation of entire truss structures. For this purpose we consider the truss shown in Fig. 12.8. In order to determine the deformed truss, the displacements in the individual nodes are required. Between the nodes the deformation figure will then consist of straight lines. To determine the node displacements the member forces N_i ($i = 1, 2, ..., 11$) due to the applied single force F are required first. Afterwards, virtual single forces $F = 1$ are applied to each node in horizontal as well as in vertical direction and the corresponding virtual member forces $\overline{N}_i$ are determined, which are marked with Roman numerals for the node number and with h or v (horizontal and vertical). In all cases the virtual forces were applied to the right respectively downwards. The following table shows an overview of the determined member forces (see Table 12.1).

Table 12.1 Member forces

Stab	1	2	3	4	5	6	7	8	9	10	11
l_i	l	l	l	$\sqrt{2}l$	l	l	l	$\sqrt{2}l$	l	$\sqrt{2}l$	l
N_i	$-3F$	$-2F$	$-F$	$\sqrt{2}F$	F	$2F$	0	$\sqrt{2}F$	$-F$	$\sqrt{2}F$	$-F$
$\overline{N}_{i,I,v}$	0	0	0	0	0	0	F	0	0	0	0
$\overline{N}_{i,II,h}$	1	0	0	0	0	0	0	0	0	0	0
$\overline{N}_{i,II,v}$	-1	0	0	0	0	0	0	$\sqrt{2}$	0	0	0
$\overline{N}_{i,III,h}$	1	1	0	0	0	0	0	0	0	0	0
$\overline{N}_{i,III,v}$	-2	-1	0	0	0	1	0	$\sqrt{2}$	-1	$\sqrt{2}$	0
$\overline{N}_{i,IV,h}$	1	1	1	0	0	0	0	0	0	0	0
$\overline{N}_{i,IV,v}$	-3	-2	-1	$\sqrt{2}$	1	2	0	$\sqrt{2}$	-1	$\sqrt{2}$	-1
$\overline{N}_{i,V,h}$	0	0	0	0	1	1	0	0	0	0	0
$\overline{N}_{i,V,v}$	-2	-1	0	0	0	1	0	$\sqrt{2}$	-1	$\sqrt{2}$	-1
$\overline{N}_{i,VI,h}$	0	0	0	0	0	1	0	0	0	0	0
$\overline{N}_{i,VI,v}$	-1	0	0	0	0	0	0	$\sqrt{2}$	-1	0	0

The displacements can now be determined from the following formula:

$$\delta = \sum_{i=1}^{m} \frac{N_i \overline{N}_i}{(EA)_i} l_i. \tag{12.56}$$

We have:

$$\delta_{I,h} = 0, \quad \delta_{I,v} = 0, \quad \delta_{II,h} = -3\frac{Fl}{EA},$$

$$\delta_{II,v} = \left(3 + 2\sqrt{2}\right)\frac{Fl}{EA} = 5.83\frac{Fl}{EA}, \quad \delta_{III,h} = -5\frac{Fl}{EA},$$

$$\delta_{III,v} = \left(11 + 4\sqrt{2}\right)\frac{Fl}{EA} = 16.66\frac{Fl}{EA}, \quad \delta_{IV,h} = -6\frac{Fl}{EA},$$

$$\delta_{IV,v} = \left(21 + 6\sqrt{2}\right)\frac{Fl}{EA} = 29.49\frac{Fl}{EA}, \quad \delta_{V,h} = 3\frac{Fl}{EA},$$

$$\delta_{V,v} = \left(12 + 4\sqrt{2}\right)\frac{Fl}{EA} = 17.66\frac{Fl}{EA}, \quad \delta_{VI,h} = 2\frac{Fl}{EA},$$

$$\delta_{VI,v} = \left(4 + 2\sqrt{2}\right)\frac{Fl}{EA} = 6.83\frac{Fl}{EA}.$$

$$\tag{12.57}$$

The deformed configuration of the truss can then be determined quite easily, it is shown in Fig. 12.8, bottom.

12.8.1.2 The Euler-Bernoulli Beam

We can proceed analogously when considering plane beam structures. For explanation we will look again at the example of Fig. 12.3, where we have determined the deflection by integrating the products of the moment lines $M_0(x_1)$ and $M_1(x_1)$ as well as $M_0(x_2)$ and $M_1(x_2)$:

$$w = \int_0^{\frac{l}{2}} \frac{M_0(x_1)M_1(x_1)}{EI} dx_1 + \int_0^{\frac{l}{2}} \frac{M_0(x_2)M_1(x_2)}{EI} dx_2. \qquad (12.58)$$

While the necessary integrations can easily be carried out on this simple example, this type of calculation can be much more complex and thus more error-prone on more complex systems. Therefore, it is useful to refer to the integral table of Fig. 5.8. At the current example we have:

$$\begin{aligned} w &= \int_0^{\frac{l}{2}} \frac{M_0(x_1)M_1(x_1)}{EI} dx_1 + \int_0^{\frac{l}{2}} \frac{M_0(x_2)M_1(x_2)}{EI} dx_2 \\ &= \frac{1}{EI}\left(\frac{1}{3}\cdot\frac{Fl}{4}\cdot\frac{l}{4}\cdot\frac{l}{2} + \frac{1}{3}\cdot\frac{Fl}{4}\cdot\frac{l}{4}\cdot\frac{l}{2}\right) = \frac{Fl^3}{48EI}. \end{aligned} \qquad (12.59)$$

Obviously this corresponds to the previously determined result (12.29).

Another example is the beam as shown in Fig. 12.9 with the length $2l$ and the constant bending stiffness EI, which is subjected to the constant line load q and the two point forces F. The deflection w at the end of the cantilever is to be determined.

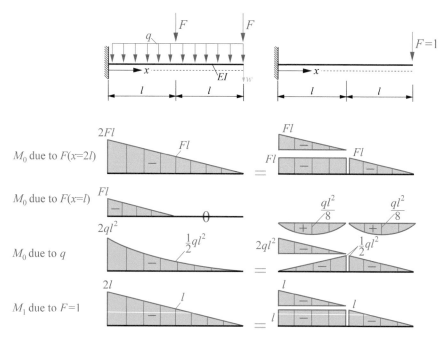

Fig. 12.9 Beam under line load q and point forces F (top left), beam under virtual load $F = 1$ (top right), associated moment lines (bottom)

First, we determine the moment M_0 due to the given load and the moment M_1 due to the virtual force $F = 1$ at the end of the cantilever. In order to use the integral table of Fig. 5.8, we divide the individual partial moment lines into corresponding segments which can then be evaluated using the integral table. In addition, we use the superposition principle here and determine the moment line M_0 separately for the respective loads. The integration is then carried out on the intervals $0 \leq x \leq l$ and $l \leq x \leq 2l$. The deflection w_1 due to the single load F at the position $x = 2l$ is then calculated as:

$$w_1 = \frac{1}{EI} \left(\frac{1}{3}Fl^3 + \frac{1}{2}Fl^3 + \frac{1}{2}Fl^3 + Fl^3 + \frac{1}{3}Fl^3 \right) = \frac{8}{3}\frac{Fl^3}{EI}. \qquad (12.60)$$

For the deflection w_2 due to the point load F at the position $x = l$ we get:

$$w_2 = \frac{1}{EI} \left(\frac{1}{3}Fl^3 + \frac{1}{2}Fl^3 \right) = \frac{5}{6}\frac{Fl^3}{EI}. \qquad (12.61)$$

The deflection w_3 due to the line load q follows as:

$$w_3 = \frac{1}{EI} \left(\frac{1}{12}ql^4 + \frac{2}{3}ql^4 - \frac{1}{24}ql^4 + \frac{1}{4}ql^4 + ql^4 - \frac{1}{12}ql^4 + \frac{1}{6}ql^4 - \frac{1}{24}ql^4 \right) = 2\frac{ql^4}{EI}. $$
$$\qquad (12.62)$$

The total deflection w then results as:

$$w = w_1 + w_2 + w_3 = \frac{8}{3}\frac{Fl^3}{EI} + \frac{5}{6}\frac{Fl^3}{EI} + 2\frac{ql^4}{EI} = \frac{1}{EI}\left(\frac{7}{2}Fl^3 + 2ql^4 \right). \quad (12.63)$$

For the frame shown in Fig. 12.10, top left, (constant bending stiffness EI) under the constant line load q we want to determine the mutual displacement w of the vertical beams as indicated. For this purpose, we apply opposing virtual single loads at the corresponding locations (Fig. 12.10, top right) and determine the moment lines for both the given loads and the virtual load case (Fig. 12.10, bottom). After superimposing the moment lines M_0 and M_1 and after performing the necessary integrations, we obtain the following result for the mutual displacement of the considered points:

$$w = \frac{ql^2h^2}{64EI}. \qquad (12.64)$$

Furthermore, we want to consider the beam shown in Fig. 12.11 and determine the rotation of the left support. The beam is under the line load q and has the length l and the constant bending stiffness EI. The moment lines M_0 und M_1 shown in Fig. 12.11 lead to the following rotation after superimposing M_0 and M_1:

$$\varphi = \frac{1}{EI} \cdot \frac{1}{3} \cdot \frac{ql^2}{8} \cdot 1 \cdot l = \frac{ql^3}{24EI}. \qquad (12.65)$$

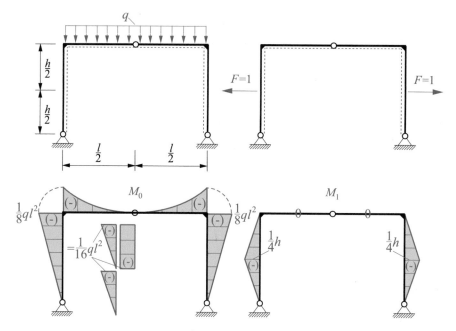

Fig. 12.10 Frame structure under line load q, determination of the mutual horizontal displacement at half height of the frame uprights

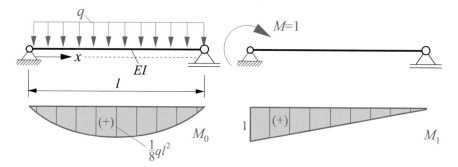

Fig. 12.11 Determination of the rotation of the left support of a simply supported beam under line load q

Finally we consider the static system (constant bending stiffness EI) of Fig. 12.12, and we want to determine the mutual rotation φ of the beam segments in the joint. For this purpose, we determine both the moment line M_0 for the given load case (line load q) and the moment curve M_1 as a result of the two virtual single moments $M = 1$. The superposition of these two moment lines then results in the desired mutual rotation φ as:

$$\varphi = \frac{q}{24EI}\left[l_2^3 - 2\left(3l_2 + 2l_1\right)l_1^2\right]. \tag{12.66}$$

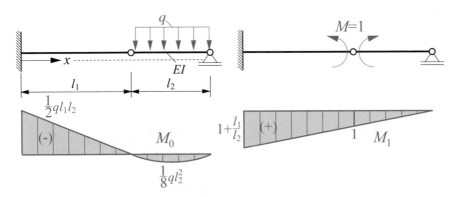

Fig. 12.12 Determination of the mutual rotation of the beam segments at the joint location

12.8.2 Analysis of Simply Statically Indeterminate Systems

12.8.2.1 The Euler-Bernoulli Beam

The force method is not only sui for determining deformations of statically determinate systems, but it can also be used to treat statically indeterminate systems. We will limit ourselves for the moment to the treatment of simply statically indeterminate systems and consider as an introductory example the beam shown in Fig. 12.13, top, as an introductory example. The bar is clamped at the left end and simply supported at the right end. The length l and the constant bending stiffness EI are assumed to be given. The load consists of the constant line load q, the moment line $M(x)$ is to be determined.

To determine the state variables at this statically indeterminate beam, any support reaction can be set free by removing the corresponding kinematic bond. Here we want to remove the support B, whereby the support reaction B is released and thus the beam becomes statically determinate. We want to call the resulting system the statically determinate basic system or the main system. The choice of the statically determinate basic system is arbitrary, but it is mandatory to make sure that the system does not become kinematically movable and thus unusable by removing the corresponding kinematic bond.

We now determine the moment line M_0 due to the applied line load q on the statically determinate system as shown in Fig. 12.13, middle. The deflection δ_0 at the support point B can then be determined with the means that have already been treated in this chapter. We apply a virtual single force $F = 1$ at the right end of the beam and calculate the moment line M_1 as shown. We have divided the moment line M_0 into its two components, namely a square parabola with the maximum value $\frac{ql^2}{8}$ and a triangular area with the maximum value $-\frac{ql^2}{2}$. The deflection δ_0 is then calculated as follows:

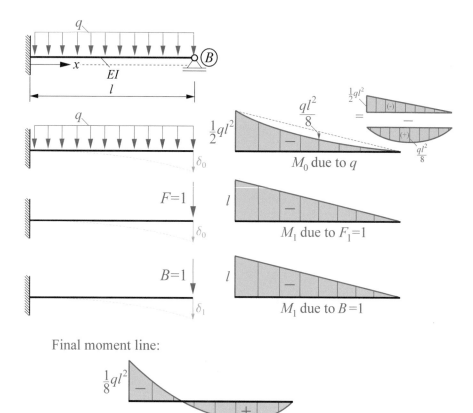

Fig. 12.13 Statically indeterminate beam under line load q, determination of the moment line

$$\delta_0 = \int_0^l \frac{M_0 M_1}{EI} \mathrm{d}x = \frac{1}{EI}\left(\frac{1}{3}\cdot l \cdot l \cdot \frac{1}{2}\cdot q \cdot l^2 - \frac{1}{3}\cdot l \cdot l \cdot \frac{1}{8}\cdot q \cdot l^2\right) = \frac{3ql^4}{24EI}.$$
(12.67)

The deflection δ_0 cannot occur in reality due to the support B. Therefore, we now calculate the deflection δ_1 that would result from the support force B. Since B is still unknown, we treat B as a unit force $B = 1$ and calculate the deflection δ_1 due to $B = 1$. We obtain:

$$\delta_1 = \int_0^l \frac{M_1 M_1}{EI} \mathrm{d}x = \frac{l^3}{3EI}.$$
(12.68)

The further procedure follows from the fact that the two deflections δ_0 and δ_1 cannot occur in reality due to the support B, so that we establish the following compatibility condition:

$$\delta_0 + B\delta_1 = 0.$$
(12.69)

This means that the deflection δ_0 and the deflection δ_1 multiplied by B must cancel each other out. From this the reaction force B can be determined directly as follows:

$$B = -\frac{\delta_0}{\delta_1} = -\frac{\dfrac{3ql^4}{24EI}}{\dfrac{l^3}{3EI}} = -\frac{3}{8}ql. \tag{12.70}$$

Thus the reaction force B is clearly defined. The negative sign indicates that B is in the opposite direction as shown in Fig. 12.13. With the reaction force B all further support reactions and all state variables can be determined. The final moment line $M(x)$ is shown in Fig. 12.13, bottom. Note the similarity of the calculation performed here with the determination of the shear flows in closed box sections due to transverse shear forces (see Chap. 6).

Support reactions and internal forces of the system (here symbolically denoted as S) can be determined from the superposition of the individual calculations as follows:

$$S = S_0 + BS_1. \tag{12.71}$$

Therein, S_0 are the internal forces on the statically determinate basic system due to the given external load. The internal forces S_1 are those internal forces at the statically determinate basic system due to the unit load $B = 1$ which must be multiplied in (12.71) by the calculated value for B. For the final moment line M on the statically indeterminate system, the following would be the result:

$$M = M_0 + BM_1. \tag{12.72}$$

Usually, the support reactions that are released by removal of the corresponding kinematic bond are symbolically called X, and the compatibility condition can be given in general form:

$$\delta_0 + X\delta_1 = 0. \tag{12.73}$$

The solution reads:

$$X = -\frac{\delta_0}{\delta_1}, \tag{12.74}$$

and the final internal forces and moments at the statically indeterminate system result as:

$$S = S_0 + XS_1. \tag{12.75}$$

12.8.2.2 Torsion

We again consider the example of Fig. 7.31, Chap. 7. As an alternative to the procedure shown there, a statically indeterminate analysis can also be carried out for this example using the force method (Fig. 12.14). For this purpose we remove the right

Fig. 12.14 Statically indeterminate analysis for a beam under torsion subjected to a single moment $M_{T,0}$

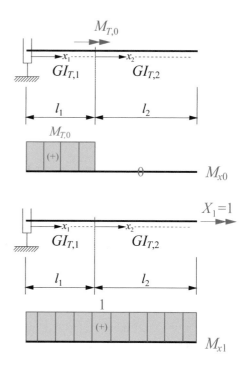

fork restraint in order to obtain the statically determinate basic system (Fig. 12.14, top). The corresponding moment line M_{x0} is obtained as shown. In addition, we analyze the system under a virtual single moment $X_1 = 1$ at the free end of the beam and determine the moment line M_{x1} as shown in Fig. 12.14, bottom.

To determine the moment X_1 we use the following compatibility equation:

$$\vartheta_0 + X_1\vartheta_1 = 0. \tag{12.76}$$

The rotations ϑ_0 and ϑ_1 can be determined as:

$$\vartheta_0 = \int_0^{l_1} \frac{M_{x0}M_{x,1}}{GI_{T,1}}\mathrm{d}x_1 + \int_0^{l_2} \frac{M_{x0}M_{x1}}{GI_{T,2}}\mathrm{d}x_2 = \frac{M_{T,0}l_1}{GI_{T,1}},$$

$$\vartheta_1 = \int_0^{l_1} \frac{M_{x1}M_{x,1}}{GI_{T,1}}\mathrm{d}x_1 + \int_0^{l_2} \frac{M_{x1}M_{x1}}{GI_{T,2}}\mathrm{d}x_2 = \frac{l_1}{GI_{T,1}} + \frac{l_2}{GI_{T,2}}. \tag{12.77}$$

We can now calculate the moment X_1 which at the same time corresponds to the torsional moment $M_{x,2}$ in the range $0 \leq x_2 \leq l_2$:

$$X_1 = -\frac{\vartheta_0}{\vartheta_1} = -\frac{M_{T0}l_1}{l_1 + l_2\dfrac{GI_{T1}}{GI_{T2}}} = M_{x,2}. \tag{12.78}$$

Hence, the moment line $M_{x,1}$ in the range $0 \leq x_1 \leq l_1$ can be determined:

$$M_{x,1} = M_{x,2} + M_{T,0} = M_{T0} \left(1 - \frac{l_1}{l_1 + l_2 \dfrac{GI_{T1}}{GI_{T2}}} \right). \tag{12.79}$$

These results are consistent with the results shown in Fig. 7.31.

12.9   Reciprocity Theorems

Reciprocity theorems are useful work and energy theorems which we will discuss in this section. We will show later on that they are important for the further application of the principle of virtual forces in the form of the force method. The reciprocity theorems to be discussed here are Betti's theorem and Maxwell's theorem.

12.9.1   Theorem of Betti

The theorem of Betti[3] can be adequately explained by the Fig. 12.15. A solid body is considered which is not specified in detail at this point and which is loaded by the two forces F_1 and F_2. The displacements at the force application points 1 and 2 in the direction of the forces F_1 and F_2 are u_1 and u_2, respectively.

We will now conduct a thought experiment and first consider the situation that only the force F_1 acts on the solid and the force F_2 is not yet present. In the case of linear elasticity, the external work performed can then be written as:

$$W_a = \frac{1}{2} F_1 u_1^{(1)}. \tag{12.80}$$

Fig. 12.15 On the theorem of Betti

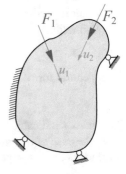

---

[3] Enrico Betti, 1823–1892, Italian mathematician.

The superscripted index (1) indicates that this displacement was caused by the force F_1.

We now also apply the force F_2 to the solid and assume that F_1 continues to act fully. The external work then reads:

$$W_a = \frac{1}{2}F_1 u_1^{(1)} + \frac{1}{2}F_2 u_2^{(2)} + F_1 u_1^{(2)}. \tag{12.81}$$

In addition to the portion $\frac{1}{2}F_1 u_1^{(1)}$ which has already been performed previously, now also the portion from the force F_2 along the displacement $u_2^{(2)}$ caused by F_2 is added. In addition, the proportion of the force F_1 must also be considered, which performs a further work along the displacement $u_1^{(2)}$ caused by F_2 (also known as passive work).

We now repeat the thought experiment in reverse order and first apply the force F_2. Then:

$$W_a = \frac{1}{2}F_2 u_2^{(2)}. \tag{12.82}$$

The force F_2 is fully effective and we now apply the force F_1 in addition. Then we have the following result for the entire external work:

$$W_a = \frac{1}{2}F_2 u_2^{(2)} + \frac{1}{2}F_1 u_1^{(1)} + F_2 u_2^{(1)}. \tag{12.83}$$

The performed external work must be identical in both cases, so that after equating (12.81) and (12.83) we obtain:

$$\frac{1}{2}F_1 u_1^{(1)} + \frac{1}{2}F_2 u_2^{(2)} + F_1 u_1^{(2)} = \frac{1}{2}F_2 u_2^{(2)} + \frac{1}{2}F_1 u_1^{(1)} + F_2 u_2^{(1)}. \tag{12.84}$$

The terms $\frac{1}{2}F_1 u_1^{(1)}$ and $\frac{1}{2}F_2 u_2^{(2)}$ cancel out so that:

$$F_1 u_1^{(2)} = F_2 u_2^{(1)}. \tag{12.85}$$

This is the theorem of Betti. It states that the work of a force F_1 along a displacement $u_1^{(2)}$ caused by a force F_2 acting at another point is identical to the work of the force F_2 which it performs along the displacement $u_2^{(1)}$ caused by F_1. Symbolically, we can also write:

$$W_{12} = W_{21}, \tag{12.86}$$

or in general form:

$$W_{ij} = W_{ji}, \quad i \neq j. \tag{12.87}$$

The index i indicates the location and the index j the cause of the performed work.

In the case of a continuum with the volume loads f_i, the surface loads t_i and the displacements u_i we get

$$\int_V f_i^{(1)} u_i^{(2)} \, dV + \int_{\partial V_t} t_i^{(1)} u_i^{(2)} \, dS = \int_V f_i^{(2)} u_i^{(1)} \, dV + \int_{\partial V_t} t_i^{(2)} u_i^{(1)} \, dS. \quad (12.88)$$

12.9.2 Theorem of Maxwell

The theorem of Maxwell[4] can be obtained directly from Betti's theorem by assuming the forces F_1 and F_2 as unit loads. It reads as follows:

$$u_1^{(2)} = u_2^{(1)}, \quad (12.89)$$

or in a notation that will be useful later on:

$$\delta_{ij} = \delta_{ji}, \quad (12.90)$$

with $i \neq j$.

Both theorems discussed in this section have many applications, a selection of which will be discussed below.

An example is the determination of shear flow due to transverse shear forces in multicellular cross-sections, as already discussed in detail in Chap. 6. In the compatibility equations (6.56) to be used there, the equality of the displacement discontinuities Δu_{ij} and Δu_{ji} is assumed. These are defined as (cf. Eq. (6.53)):

$$\Delta u_{ij} = -\oint_{i,j} \frac{ds}{Gt(s)}, \quad (12.91)$$

where the integration has to be performed using the walls common to both cells i and j. While Δu_{ij} can be understood as the displacement discontinuity due to the statically indeterminate unit shear flow T_{sj}, Δu_{ji} represents the displacement discontinuity due to the statically indeterminate unit shear flow T_{si} at the same position. According to Maxwell's theorem these two displacement discontinuities are identical.

12.10 Calculation of Multiple Statically Indeterminate Systems

The theorem of Betti or the theorem of Maxwell are used e.g. for the calculation of statically indeterminate systems using the force method. As an introductory example, we consider the beam of the Fig. 12.16, top left.

---

[4]James Clerk Maxwell, 1831–1879, Scottish physicist.

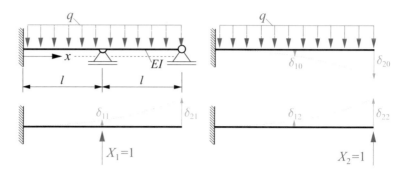

Fig. 12.16 Twofold statically indeterminate beam, determination of the support reactions

The beam has the total length $2l$ and the constant bending stiffness EI. It is clamped at its left end and is supported at midspan and at its right end. To apply the force method, we first make the given system statically determinate by removing the two supports (Fig. 12.16, top right). Since the supports have been removed, the displacements δ_{10} and δ_{20} occur at the two support locations, i.e. the displacements at points 1 and 2 as a result of the given loads. In the real, statically indeterminate system, however, these two displacements are not possible, so that we first apply the still unknown support force X_1 at the middle support point (Fig. 12.16 bottom left), which we initially assume to be a unit load. This still unknown reaction force X_1 then causes the two displacements δ_{11} and δ_{21}. The indexing indicates that these are the displacements at points 1 and 2 that were caused by the force X_1. In the same way, we proceed with the unknown reaction force at the right support, which we first assume as unit load $X_2 = 1$ and which causes the two displacements δ_{12} and δ_{22} (Fig. 12.16, bottom right).

All calculated displacements cannot occur in the real, statically indeterminate system, they must cancel each other out in combination with the unknown bearing forces. Thus, two compatibility conditions result in a twofold statically indeterminate system. Regarding δ_{10}, δ_{11} and δ_{12} at the support location at midspan, the following results:

$$\delta_{10} + X_1\delta_{11} + X_2\delta_{12} = 0. \tag{12.92}$$

At the bearing point at the right end of the beam we obtain:

$$\delta_{20} + X_1\delta_{21} + X_2\delta_{22} = 0. \tag{12.93}$$

From these two equations the two unknown reaction forces X_1 and X_2 can be determined.

The application of Maxwell's theorem is now such that we may assume the equality of the displacements δ_{12} and δ_{21}:

$$\delta_{12} = \delta_{21}, \tag{12.94}$$

or in a more general form:

$$\delta_{ij} = \delta_{ji}, \quad i \neq j. \tag{12.95}$$

For the two compatibility conditions we then obtain:

$$\delta_{10} + X_1\delta_{11} + X_2\delta_{12} = 0,$$
$$\delta_{20} + X_1\delta_{12} + X_2\delta_{22} = 0. \tag{12.96}$$

When considering twofold statically indeterminate systems using the force method, we thus obtain a total of 5 displacement quantities, which result as follows:

$$\delta_{ij} = \int \frac{N_i N_j}{EA} dx + \int \frac{M_i M_j}{EI} dx. \tag{12.97}$$

The compatibility conditions (12.96) can be solved as:

$$X_1 = \frac{-\delta_{10}\delta_{22} + \delta_{20}\delta_{12}}{\delta_{11}\delta_{22} - \delta_{12}^2}, \quad X_2 = \frac{-\delta_{20}\delta_{11} + \delta_{10}\delta_{12}}{\delta_{11}\delta_{22} - \delta_{12}^2}. \tag{12.98}$$

Once the unknown reaction forces are determined, all further reactions and internal forces of the statically indeterminate beam can be obtained.

If a static system is considered with a degree of static indeterminacy higher than two, then the force method can be applied analogously. For a triple statically indeterminate system, the compatibility conditions are as follows:

$$\delta_{10} + X_1\delta_{11} + X_2\delta_{12} + X_3\delta_{13} = 0,$$
$$\delta_{20} + X_1\delta_{21} + X_2\delta_{22} + X_3\delta_{23} = 0,$$
$$\delta_{30} + X_1\delta_{31} + X_2\delta_{32} + X_3\delta_{33} = 0. \tag{12.99}$$

Application of Maxwell's theorem yields:

$$\delta_{10} + X_1\delta_{11} + X_2\delta_{12} + X_3\delta_{13} = 0,$$
$$\delta_{20} + X_1\delta_{12} + X_2\delta_{22} + X_3\delta_{23} = 0,$$
$$\delta_{30} + X_1\delta_{13} + X_2\delta_{23} + X_3\delta_{33} = 0. \tag{12.100}$$

In a vector-matrix notation we have:

$$\begin{bmatrix} \delta_{11} & \delta_{12} & \delta_{13} \\ \delta_{12} & \delta_{22} & \delta_{23} \\ \delta_{13} & \delta_{23} & \delta_{33} \end{bmatrix} \begin{pmatrix} X_1 \\ X_2 \\ X_3 \end{pmatrix} = - \begin{pmatrix} \delta_{10} \\ \delta_{20} \\ \delta_{30} \end{pmatrix}. \tag{12.101}$$

These are three equations for the determination of the three unknown quantities X_1, X_2 and X_3. The coefficient matrix occurring here is symmetrical due to Maxwell's theorem.

Let us now assume a static system with the degree of static indeterminacy n. The n compatibility conditions are:

$$\delta_{10} + X_1\delta_{11} + X_2\delta_{12} + X_3\delta_{13} + \cdots + X_n\delta_{1n} = 0,$$
$$\delta_{20} + X_1\delta_{12} + X_2\delta_{22} + X_3\delta_{23} + \cdots + X_n\delta_{2n} = 0,$$
$$\delta_{30} + X_1\delta_{13} + X_2\delta_{23} + X_3\delta_{33} + \cdots + X_n\delta_{3n} = 0,$$
$$\cdots$$
$$\delta_{n0} + X_1\delta_{1n} + X_2\delta_{2n} + X_3\delta_{3n} + \cdots + X_n\delta_{nn} = 0, \qquad (12.102)$$

or in a vector-matrix notation:

$$
\begin{bmatrix}
\delta_{11} & \delta_{12} & \delta_{13} & \cdots & \delta_{1n} \\
\delta_{12} & \delta_{22} & \delta_{23} & \cdots & \delta_{2n} \\
\delta_{13} & \delta_{23} & \delta_{33} & \cdots & \delta_{3n} \\
\vdots & \vdots & \vdots & \ddots & \vdots \\
\delta_{1n} & \delta_{2n} & \delta_{3n} & \cdots & \delta_{nn}
\end{bmatrix}
\begin{pmatrix} X_1 \\ X_2 \\ X_3 \\ \vdots \\ X_n \end{pmatrix}
= -
\begin{pmatrix} \delta_{10} \\ \delta_{20} \\ \delta_{30} \\ \vdots \\ \delta_{n0} \end{pmatrix}. \qquad (12.103)
$$

In symbolic form we have:

$$\underline{\delta}\,\underline{X} = -\underline{\delta}_0. \qquad (12.104)$$

The matrix $\underline{\delta}$ is symmetrical in all cases due to Maxwell's theorem. Again, note the strong similarity of the procedure for determining the shear flows in closed cross-sections with several cells (see Chap. 6) under transverse shear forces.

The procedure is demonstrated at the example shown in Fig. 12.17, for which the support reactions and the moment line are to be determined. The given statically determinate basic system is to be used.

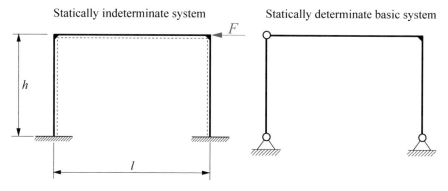

Fig. 12.17 Triple statically indeterminate system under single load (left), statically determinate basic system (right)

The compatibility conditions to be fulfilled here to determine the three quantities X_1, X_2 and X_3 are:

$$\delta_{10} + X_1\delta_{11} + X_2\delta_{12} + X_3\delta_{13} = 0,$$
$$\delta_{20} + X_1\delta_{12} + X_2\delta_{22} + X_3\delta_{23} = 0,$$
$$\delta_{30} + X_1\delta_{13} + X_2\delta_{23} + X_3\delta_{33} = 0. \tag{12.105}$$

The displacements δ_{ij} can be determined as (see Fig. 12.18):

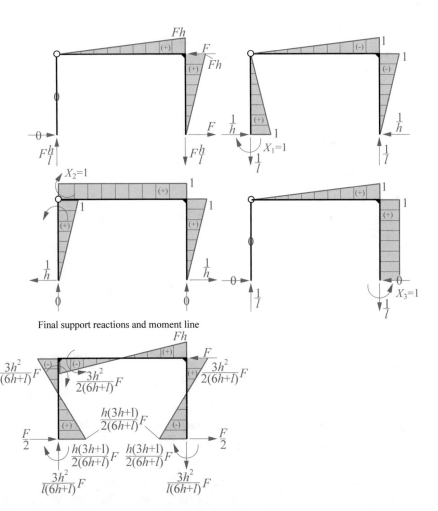

Fig. 12.18 Moment diagrams at the statically indeterminate system

$$\delta_{10} = -\frac{1}{3EI}Fh\,(l+h), \quad \delta_{11} = \frac{1}{3EI}\,(l+2h), \quad \delta_{12} = -\frac{1}{EI}\left(\frac{1}{2}l+\frac{1}{6}h\right),$$

$$\delta_{13} = -\frac{1}{EI}\left(\frac{1}{3}l+\frac{1}{2}h\right), \quad \delta_{20} = \frac{1}{EI}Fh\left(\frac{1}{2}l+\frac{1}{3}h\right), \quad \delta_{22} = \frac{1}{EI}\left(l+\frac{2}{3}h\right),$$

$$\delta_{23} = \frac{1}{2EI}\,(l+h), \quad \delta_{30} = \frac{1}{EI}Fh\left(\frac{1}{3}l+\frac{1}{2}h\right), \quad \delta_{33} = \frac{1}{EI}\left(\frac{1}{3}l+h\right). \quad (12.106)$$

Then we obtain from (12.105):

$$\begin{bmatrix} \frac{1}{3}(l+2h) & -\left(\frac{1}{2}l+\frac{1}{6}h\right) & -\left(\frac{1}{3}l+\frac{1}{2}h\right) \\[2mm] -\left(\frac{1}{2}l+\frac{1}{6}h\right) & \left(l+\frac{2}{3}h\right) & \frac{1}{2}(l+h) \\[2mm] -\left(\frac{1}{3}l+\frac{1}{2}h\right) & \frac{1}{2}(l+h) & \left(\frac{1}{3}l+h\right) \end{bmatrix} \begin{pmatrix} X_1 \\ X_2 \\ X_3 \end{pmatrix} = \begin{pmatrix} \frac{1}{3}Fh\,(l+h) \\[2mm] -Fh\left(\frac{1}{2}l+\frac{1}{3}h\right) \\[2mm] -Fh\left(\frac{1}{3}l+\frac{1}{2}h\right) \end{pmatrix}. \quad (12.107)$$

The solution reads:

$$X_1 = \frac{h\,(3h+l)}{2\,(6h+l)}F, \quad X_2 = -\frac{3h^2}{2\,(6h+l)}F, \quad X_3 = -\frac{h\,(3h+l)}{2\,(6h+l)}F. \quad (12.108)$$

The support reactions result as:

$$A_V = A_{V0} + X_1 A_{V1} + X_2 A_{V2} + X_3 A_{V3} = \frac{3h^2}{l\,(6h+l)}F,$$

$$A_H = A_{h0} + X_1 A_{H1} + X_2 A_{H2} + X_3 A_{H3} = \frac{F}{2},$$

$$B_V = B_{V0} + X_1 B_{V1} + X_2 B_{V2} + X_3 B_{V3} = -\frac{3h^2}{l\,(6h+l)}F,$$

$$B_H = B_{H0} + X_1 B_{H1} + X_2 B_{H2} + X_3 B_{H3} = -\frac{F}{2}. \quad (12.109)$$

The final moment line can be determined analogously, it is shown in Fig. 12.18, bottom.

12.11 Influence Lines for Deformations of Statically Determinate Systems

The theorems of Betti or Maxwell can also be applied when determining the influence lines of deformations on statically determinate structures. For illustration consider the beam shown in Fig. 12.19 which is loaded by the two single forces F_1 and F_2. The force F_1 causes the displacement δ_{11} at its point of application. In the same way the force F_2 causes the displacement δ_{22} at its point of application. In addition, F_1

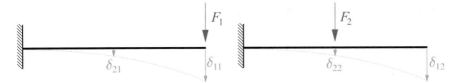

Fig. 12.19 Relation between influence line and bending line of a beam under moving load

causes the displacement δ_{21} at the point of application of F_2. Similarly, the force F_2 causes the displacement δ_{12} at the point of application of the force F_1. Betti's theorem then yields the following conclusion:

$$F_1 \delta_{12} = F_2 \delta_{21}. \tag{12.110}$$

If the loads F_1 and F_2 are unit loads, then the theorem of Maxwell yields:

$$\delta_{12} = \delta_{21}. \tag{12.111}$$

The following important conclusion can be drawn from this. The displacement δ_{21}, which is caused at location 2 by force 1 and thus represents the bending line of the beam $w(x)$ under the unit load $F_1 = 1$, is identical to the deflection δ_{12} at location 1, which is caused by the arbitrarily positioned force F_2. Thus, δ_{12} can be interpreted as the influence line of the deflection at the end of the cantilever beam.

Consider the cantilever beam (length l, constant bending stiffness EI) of Fig. 12.20. We want to determine the influence line IL δ as indicated.

We start from the constitutive law of the beam in the form

$$EI w'''' = q. \tag{12.112}$$

The bending line due to a unit load F at the cantilever end $x = l$ can easily be determined with the means of Chap. 5 and follows as:

$$w(x) = \frac{1}{EI}\left(-\frac{1}{6}x^3 + \frac{1}{2}x^2 l\right). \tag{12.113}$$

Fig. 12.20 Cantilever beam under moving load

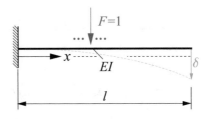

This corresponds exactly to the influence line IL δ:

$$\text{IL}\delta = \frac{1}{EI}\left(-\frac{1}{6}x^3 + \frac{1}{2}x^2 l\right). \tag{12.114}$$

Thus, if the deflection δ is sought for a certain position of the moving load, then the coordinate x where the moving load is located must be inserted in (12.114). The Eq. (12.114) then yields the deflection at the free end of the cantilever.

12.12 The Reduction Theorem of Statics

A further important theorem that comes into effect in connection with the force method is the so-called reduction theorem of statics, which is important e.g. when considering deformations of statically indeterminate systems. We consider the beam shown in Fig. 12.21 which is clamped at both ends and is under the arbitrarily positioned single force F. We are looking for the deflection of the beam at the point of force application and use the force method.

We first establish the statically determinate basic system by replacing the clampings with simple supports. The two clamping moments are denoted as X_1 and X_2. The corresponding moment lines are shown in Fig. 12.21. The compatibility conditions are then:

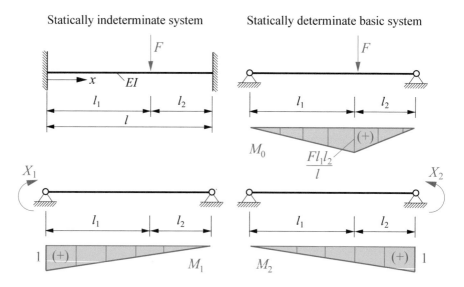

Fig. 12.21 Clamped beam under single load; statically determinate basic system and respective moment lines

$$\delta_{10} + X_1\delta_{11} + X_2\delta_{12} = 0,$$
$$\delta_{20} + X_1\delta_{12} + X_2\delta_{22} = 0, \tag{12.115}$$

with

$$\delta_{10} = \int_0^l \frac{M_0 M_1}{EI}\mathrm{d}x = \frac{Fl_1l_2(l_1 + 2l_2)}{6EIl},$$

$$\delta_{20} = \int_0^l \frac{M_0 M_2}{EI}\mathrm{d}x = \frac{Fl_1l_2(2l_1 + l_2)}{6EIl},$$

$$\delta_{11} = \int_0^l \frac{M_1 M_1}{EI}\mathrm{d}x = \frac{l}{3EI}, \quad \delta_{12} = \int_0^l \frac{M_1 M_2}{EI}\mathrm{d}x = \frac{l}{6EI},$$

$$\delta_{22} = \int_0^l \frac{M_2 M_2}{EI}\mathrm{d}x = \frac{l}{3EI}. \tag{12.116}$$

The two clamping moments result as:

$$X_1 = -\frac{Fl_1l_2^2}{l^2}, \quad X_2 = -\frac{Fl_1^2 l_2}{l^2}. \tag{12.117}$$

The moment line at the statically indeterminate beam is shown in Fig. 12.22.

To determine the beam deflection at the point of force application, we apply the virtual single force $F = 1$ at the point of force application and determine the corresponding moment line M'. It is shown in Fig. 12.23 and can be determined simply from the previously determined moment line M (Fig. 12.22) by replacing the force F with the unit load $F = 1$. In general, however, the determination of the moment line M' requires a further statically indeterminate calculation, which, depending on the complexity of the system and the degree of static indeterminacy, can involve a considerable amount of computational effort.

The deflection to be determined then follows as:

$$w = \int_0^l \frac{M M'}{EI}\mathrm{d}x = \frac{Fl_1^3 l_2^3}{3EIl^3}. \tag{12.118}$$

Fig. 12.22 Moment line of the clamped beam

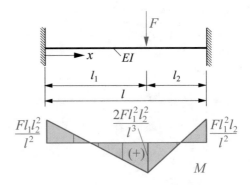

M

Fig. 12.23 Moment line M'
due to a virtual single force

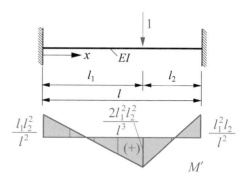

At this point the reduction theorem of statics is applied. It states that for the determination of the desired deflection w by

$$w = \int_0^l \frac{M M'}{E I} \, \mathrm{d}x \tag{12.119}$$

only one of the two moment lines M or M' needs to be determined at the statically indeterminate system. The remaining moment line can then be determined at any permissible statically determinate basic system. Application to the present example yields the following result for the deflection if the moment area M'_0 was determined at the statically determinate basic system (Fig. 12.24):

$$w = \int_0^l \frac{M M'_0}{E I} \, \mathrm{d}x = = \frac{1}{6} \frac{l_1}{E I} \frac{l_1 l_2}{l} \left(-\frac{F l_1 l_2^2}{l^2} + 2F \frac{2 l_1^2 l_2^2}{l^3} \right)$$
$$+ \frac{1}{6} \frac{l_2}{E I} \frac{l_1 l_2}{l} \left(2F \frac{2 l_1^2 l_2^2}{l^3} - \frac{F l_1^2 l_2}{l^2} \right) = \frac{F l_1^3 l_2^3}{3 E I l^3}. \tag{12.120}$$

This is apparently identical with the result which is obtained when both moment lines are determined at the statically indeterminate system.

Fig. 12.24 Moment line M'_0
at the statically determinate
basic system due to a virtual
single force

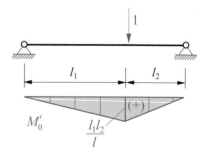

We can prove the reduction theorem as follows. We assume here that both moment lines to be superimposed were calculated according to (12.119) on the statically indeterminate system. However, we already know from the previous remarks on the force method that the two moment lines can be composed of partial moment lines as:

$$M = M_0 + \sum_{i=1}^{m} X_i M_i = M_0 + X_1 M_1 + X_2 M_2 + ... + X_n M_n,$$

$$M' = M_0' + \sum_{j=1}^{n} X_j' M_j' = M_0' + X_1' M_1' + X_2' M_2' + ... + X_n' M_n'. \quad (12.121)$$

This results in the desired deflection as follows:

$$w = \int_0^l \frac{(M_0 + X_1 M_1 + X_2 M_2 + ... + X_n M_n)(M_0' + X_1' M_1' + X_2' M_2' + ... + X_n' M_n')}{EI} dx,$$

$$(12.122)$$

or:

$$w = \int_0^l \frac{M M_0'}{EI} dx$$
$$+ \int_0^l \frac{1}{EI} \Big[X_1' M_1' (M_0 + X_1 M_1 + X_2 M_2 + \cdots + X_n M_n)$$
$$+ \quad X_2' M_2' (M_0 + X_1 M_1 + X_2 M_2 + \cdots + X_n M_n)$$
$$\vdots$$
$$+ \quad X_n' M_n' (M_0 + X_1 M_1 + X_2 M_2 + \cdots + X_n M_n) \Big] dx, \quad (12.123)$$

respectively:

$$w = \int_0^l \frac{M M_0'}{EI} dx$$
$$+ \int_0^l \frac{1}{EI} \Big[X_1' \left(M_0 M_1' + X_1 M_1 M_1' + X_2 M_2 M_1' + \cdots + X_n M_n M_1' \right)$$
$$+ \quad X_2' \left(M_0 M_2' + X_1 M_1 M_2' + X_2 M_2 M_2' + \cdots + X_n M_n M_2' \right)$$
$$\vdots$$
$$+ \quad X_n' \left(M_0 M_n' + X_1 M_1 M_n' + X_2 M_2 M_n' + \cdots + X_n M_n M_n' \right) \Big] dx. \quad (12.124)$$

The integrals over the products of two moment lines represent exactly the displacements δ_{ij}:

$$\int_0^l \frac{M_i M_j'}{EI} dx = \delta_{ij}, \quad (12.125)$$

so that:

$$w = \int_0^l \frac{M M_0'}{E I} dx$$

$$+ \int_0^l \frac{1}{E I} \big[X_1' \left(\delta_{10} + X_1 \delta_{11} + X_2 \delta_{12} + \cdots + X_n \delta_{1n} \right)$$

$$+ \quad X_2' \left(\delta_{20} + X_1 \delta_{12} + X_2 \delta_{22} + \cdots + X_n \delta_{2n} \right)$$

$$\vdots$$

$$+ \quad X_n' \left(\delta_{n0} + X_1 \delta_{1n} + X_2 \delta_{2n} + \cdots + X_n \delta_{nn} \right) \big] dx. \qquad (12.126)$$

The parenthesized expressions of the second integral correspond exactly to the compatibility conditions (12.102) which must become zero. Therefore, the whole second integral term disappears, and we obtain:

$$w = \int_0^l \frac{M M_0'}{E I} dx. \qquad (12.127)$$

This is the proof of the reduction theorem. The formulation worked out here states that it is sufficient to calculate the moment line due to the given loads at the statically indeterminate system and to determine the line of moments due to a virtual single force / virtual single moment for the calculation of a system deformation at any permissible statically determinate basic system.

Finally, it should be noted that the reduction theorem can also be derived for the opposite case, i.e.:

$$w = \int_0^l \frac{M_0 M'}{E I} dx. \qquad (12.128)$$

12.13 Analysis of Continuous Beams

Consider the beam shown in Fig. 12.25 which is divided by its supports into $n + 1$ fields. Each field i ($i = 1, 2, ..., n + 1$) has the length l_i and the constant bending stiffness $E I_i$ and is loaded by the constant line load q_i. The supports are numbered from left to right, whereby the first support is assigned the number 0. The support on the far right is thus assigned the number $n + 1$. We now want to determine the system of equations for the determination of the statically indeterminate quantities for the cases $n = 1$, $n = 2$, $n = 3$ as well as for the case of any number n. In all cases we will create the statically determinate basic system by inserting a full joint at each support i ($i = 1, 2, 3, ..., n$). The basic procedure is shown in Fig. 12.26. The moment area M_0 results in each subfield i due to the constant line load q_i acting here as a square parabola with the respective maximum value $\frac{q_i l_i^2}{8}$. Due to the selection of the statically determinate basic system, zero values result for each partial moment

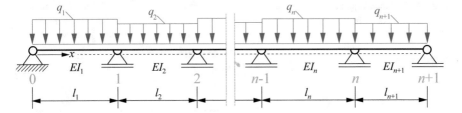

Fig. 12.25 Continuous beam with $n + 1$ fields

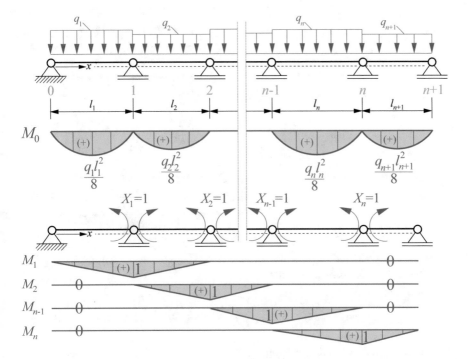

Fig. 12.26 Moment line M_0 and moment lines M_i at the statically determinate basic system

area at the two adjacent supports $i - 1$ and i. The moment lines $M_1, M_2, \ldots M_{n-1}, M_n$ show that the influence of the supporting moments $X_1, X_2, \ldots X_{n-1}, X_n$, which we initially regard as unit moments, is limited exclusively to the two directly adjacent fields. The moment area M_i thus consists of two sub-areas with linear distributions in the fields i and $i + 1$.

We now take a closer look at the case $n = 1$ (see Fig. 12.27, top). The moment areas M_0 and M_1 are shown in Fig. 12.27, middle and bottom. The compatibility equation results as:

$$X_1 = -\frac{\delta_0}{\delta_1}, \tag{12.129}$$

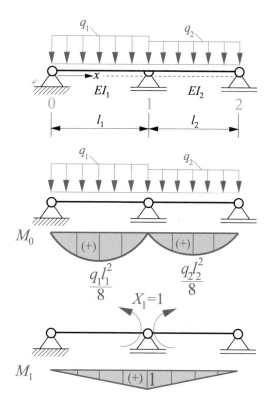

Fig. 12.27 Static system (top); moment line M_0 (middle) and moment line M_1 (bottom) at the statically determinate system for $n = 1$

with the two rotations δ_0 and δ_1 as:

$$\delta_0 = \frac{q_1 l_1^3}{24 E I_1} + \frac{q_2 l_2^3}{24 E I_2}, \quad \delta_1 = \frac{l_1}{3 E I_1} + \frac{l_2}{3 E I_2}. \tag{12.130}$$

The statically indeterminate supporting moment $X_1 = M_1$ then follows as:

$$M_1 = -\frac{\dfrac{q_1 l_1^3}{8 E I_1} + \dfrac{q_2 l_2^3}{8 E I_2}}{\dfrac{l_1}{E I_1} + \dfrac{l_2}{E I_2}}. \tag{12.131}$$

In the special case of identical field lengths, line loads and bending stiffness we obtain

$$M_1 = -\frac{q l^2}{8}. \tag{12.132}$$

Furthermore, the case $n = 2$ is examined (Fig. 12.28, top). This is a twofold statically indeterminate system. The moment area M_0 and the two moment areas M_1 and

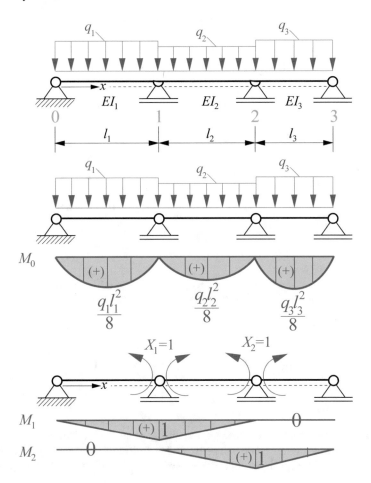

Fig. 12.28 Static system (top); moment line M_0 (middle) and moment lines M_1 and M_2 (bottom) at the statically determinate system for $n = 2$

M_2 due to the two moments $X_1 = 1$ and $X_2 = 1$ are shown in Fig. 12.28, middle and bottom.

The compatibility equations to be applied here read:

$$\delta_{10} + X_1\delta_{11} + X_2\delta_{12} = 0,$$
$$\delta_{20} + X_1\delta_{12} + X_2\delta_{22} = 0, \qquad (12.133)$$

with:

$$\delta_{10} = \int \frac{M_0 M_1}{EI}\,\mathrm{d}x = \frac{q_1 l_1^3}{24 E I_1} + \frac{q_2 l_2^3}{24 E I_2},$$

$$\delta_{11} = \int \frac{M_1 M_1}{EI} dx = \frac{l_1}{3EI_1} + \frac{l_2}{3EI_2},$$

$$\delta_{12} = \int \frac{M_1 M_2}{EI} dx = \frac{l_2}{6EI_2},$$

$$\delta_{20} = \int \frac{M_0 M_2}{EI} dx = \frac{q_2 l_2^3}{24EI_2} + \frac{q_3 l_3^3}{24EI_3},$$

$$\delta_{22} = \int \frac{M_2 M_2}{EI} dx = \frac{l_2}{3EI_2} + \frac{l_3}{3EI_3}. \tag{12.134}$$

In a vector-matrix notation and with $X_1 = M_1$ and $X_2 = M_2$ we get:

$$\begin{bmatrix} \dfrac{l_1}{3EI_1} + \dfrac{l_2}{3EI_2} & \dfrac{l_2}{6EI_2} \\[2ex] \dfrac{l_2}{6EI_2} & \dfrac{l_2}{3EI_2} + \dfrac{l_3}{3EI_3} \end{bmatrix} \begin{pmatrix} M_1 \\ M_2 \end{pmatrix} = - \begin{pmatrix} \dfrac{q_1 l_1^3}{24EI_1} + \dfrac{q_2 l_2^3}{24EI_2} \\[2ex] \dfrac{q_2 l_2^3}{24EI_2} + \dfrac{q_3 l_3^3}{24EI_3} \end{pmatrix}. \tag{12.135}$$

For the case $n = 3$, which we want to deal with here without a detailed representation, the following resulting system of equations results:

$$\begin{bmatrix} \dfrac{l_1}{3EI_1} + \dfrac{l_2}{3EI_2} & \dfrac{l_2}{6EI_2} & 0 \\[2ex] \dfrac{l_2}{6EI_2} & \dfrac{l_2}{3EI_2} + \dfrac{l_3}{3EI_3} & \dfrac{l_3}{6EI_3} \\[2ex] 0 & \dfrac{l_3}{6EI_3} & \dfrac{l_3}{3EI_3} + \dfrac{l_4}{3EI_4} \end{bmatrix} \begin{pmatrix} M_1 \\ M_2 \\ M_3 \end{pmatrix} = - \begin{pmatrix} \dfrac{q_1 l_1^3}{24EI_1} + \dfrac{q_2 l_2^3}{24EI_2} \\[2ex] \dfrac{q_2 l_2^3}{24EI_2} + \dfrac{q_3 l_3^3}{24EI_3} \\[2ex] \dfrac{q_3 l_3^3}{24EI_2} + \dfrac{q_4 l_4^3}{24EI_4} \end{pmatrix}.$$

Based on the cases considered so far, the general case of an n-fold statically indeterminate continuous beam can be treated as follows. In the case of a continuous beam with n supporting moments, the system of equations for their determination can be given as:

$$\begin{bmatrix} \dfrac{l_1}{3EI_1} + \dfrac{l_2}{3EI_2} & \dfrac{l_2}{6EI_2} & 0 & \cdots & 0 & 0 & 0 \\[2ex] \dfrac{l_2}{6EI_2} & \dfrac{l_2}{3EI_2} + \dfrac{l_3}{3EI_3} & \dfrac{l_3}{6EI_3} & \cdots & 0 & 0 & 0 \\[2ex] 0 & \dfrac{l_3}{6EI_3} & \dfrac{l_3}{3EI_3} + \dfrac{l_4}{3EI_4} & \cdots & 0 & 0 & 0 \\[2ex] \vdots & \vdots & \vdots & \ddots & \vdots & \vdots \\[2ex] 0 & 0 & 0 & \cdots & \dfrac{l_{n-1}}{6EI_{n-1}} & \dfrac{l_{n-1}}{3EI_{n-1}} + \dfrac{l_n}{3EI_n} & \dfrac{l_n}{6EI_n} \\[2ex] 0 & 0 & 0 & \cdots & 0 & \dfrac{l_n}{3EI_n} + \dfrac{l_n}{6EI_n} & \dfrac{l_{n+1}}{6EI_{n+1}} \end{bmatrix}$$

$$
\times
\begin{pmatrix}
M_1 \\
M_2 \\
M_3 \\
\vdots \\
M_{n-1} \\
M_n
\end{pmatrix}
= -
\begin{pmatrix}
\dfrac{q_1 l_1^3}{24 E I_1} + \dfrac{q_2 l_2^3}{24 E I_2} \\[2mm]
\dfrac{q_2 l_2^3}{24 E I_2} + \dfrac{q_3 l_3^3}{24 E I_3} \\[2mm]
\dfrac{q_3 l_3^3}{24 E I_2} + \dfrac{q_4 l_4^3}{24 E I_4} \\[2mm]
\vdots \\[1mm]
\dfrac{q_{n-1} l_{n-1}^3}{24 E I_{n-1}} + \dfrac{q_n l_n^3}{24 E I_n} \\[2mm]
\dfrac{q_n l_n^3}{24 E I_n} + \dfrac{q_{n+1} l_{n+1}^3}{24 E I_{n+1}}
\end{pmatrix}.
$$

$$\tag{12.136}$$

It can be seen that all equations in the above system of equations are of a similar form. Equation i is therefore in a general notation:

$$
\frac{l_i}{6 E I_i} M_{i-1} + \left(\frac{l_i}{3 E I_i} + \frac{l_{i+1}}{3 E I_{i+1}} \right) M_i + \frac{l_{i+1}}{6 E I_{i+1}} M_{i+1} = \frac{q_i l_i^3}{24 E I_i} + \frac{q_{i+1} l_{i+1}^3}{24 E I_{i+1}},
$$

$$\tag{12.137}$$

where i can take all values between $i = 1$ and $i = n$. The moments M_0 and M_{n+1} at the two outer supports of the continuous beam are to be set to zero.

If the beam is loaded by other arbitrary load types in its fields, we can generalize the above representation as follows:

$$
\frac{l_i}{6 E I_i} M_{i-1} + \left(\frac{l_i}{3 E I_i} + \frac{l_{i+1}}{3 E I_{i+1}} \right) M_i + \frac{l_{i+1}}{6 E I_{i+1}} M_{i+1} = L_i + L_{i+1}, \tag{12.138}
$$

where the terms L_i and L_{i+1} are called load terms which result from the displacement quantities δ_{i0}. This equation is usually referred to as the so-called three-moment equation.

Bibliography

Becker W, Gross D (2002) Mechanik elastischer Körper und Strukturen. Springer, Berlin, Germany

Gross D, Hauger W, Wriggers P (2014) Technische Mechanik 4: Hydromechanik, Elemente der Höheren Mechanik, Numerische Methoden, 9th edn. Springer, Berlin, Germany

Hirschfeld K (2006) Baustatik, 5th edn. Springer, Berlin, Germany

Kossira H (1996) Grundlagen des Leichtbaus. Springer, Berlin, Germany

Lanczos C (1986) The variational principles of mechanics, 4th edn. Dover, New York, USA

Langhaar HL (2016) Energy methods in applied mechanics. Dover, New York, USA

Linke M, Nast E (2015) Festigkeitslehre für den Leichtbau. Springer, Berlin, Germany

Meskouris K, Hake E (2009) Statik der Stabtragwerke: Einführung in die Tragwerkslehre, 2nd edn. Springer, Berlin, Germany

Mittelstedt C (2017) Energiemethoden der Elastostatik, Studienbereich Mechanik, Technische Universität Darmstadt, Germany

Reddy JN (2017) Energy principles and variational methods in applied mechanics, 3rd edn. Wiley, New York, USA

Tauchert TR (1974) Energy principles in structural mechanics. McGraw-Hill, New York, USA

Chapter 13
Energy-Based Approximation Methods

13.1 Introduction

The work and energy principles discussed so far offer the possibility to solve static problems in an exact way. In many practical applications, however, one will be confronted with the fact that a given problem can no longer be treated in the form of exact closed-analytical solutions or only with disproportionate computational effort. At this point, a number of energy-based approximate analysis methods have been established, a selection of which we will discuss in this chapter. In addition to classical methods such as the Ritz method and the Galerkin method which are the subject matter of the present chapter, we want to discuss in detail a modern and firmly established numerical method, namely the so-called Finite Element Method (short: FEM). All mentioned methods have in common that they are based on suitable approaches for the derivation of the state variables of the system under consideration, here mostly in the form of approximations for the displacements, from which statements about strain and stress states can be obtained. Based on these approximations, solutions for a given static problem are then determined from energy principles. A distinction is usually made between continuous methods and discretizing methods. In the continuous methods (here the Ritz method and the Galerkin method) approximations are used on the entire structure under consideration, whereas when using discretizing methods the structure under consideration is divided into sub-areas in which corresponding approximations are used. These sub-areas, the so-called elements, are then assembled at a later point by means of continuity conditions to form the entire structure under consideration. The FEM nowadays is the most important representative of discretizing methods in engineering practice. This chapter deals with the continuous methods, the finite element method is the subject of the following Chap. 14.

© The Author(s), under exclusive license to Springer Nature Switzerland AG 2021   469
C. Mittelstedt, *Structural Mechanics in Lightweight Engineering*,
https://doi.org/10.1007/978-3-030-75193-7_13

13.2 The Ritz Method

The Ritz method[1] is based on the principle of the minimum of the elastic total potential and requires that the total potential Π of the structure under consideration is given as a function of the displacements u, v, w, so we can formally write $\Pi = \Pi(u, v, w)$. For the displacements u, v, w of the considered solid the following approaches are made:

$$u(x, y, z) \simeq U(x, y, z) = \sum_{i=1}^{i=n_u} A_i U_i(x, y, z),$$

$$v(x, y, z) \simeq V(x, y, z) = \sum_{i=1}^{i=n_v} B_i V_i(x, y, z),$$

$$w(x, y, z) \simeq W(x, y, z) = \sum_{i=1}^{i=n_w} C_i W_i(x, y, z). \tag{13.1}$$

The functions $U(x, y, z)$, $V(x, y, z)$, $W(x, y, z)$ are the so-called shape functions which are composed of the n_u, n_v and n_w functions $U_i(x, y, z)$, $V_i(x, y, z)$, $W_i(x, y, z)$. The quantities A_i, B_i and C_i are constants yet to be determined and are denoted as the so-called Ritz constants. The shape functions are to be specified by the user. They are required to fulfill at least the geometric boundary conditions of the given system. Both the shape functions $U_i(x, y, z)$, $V_i(x, y, z)$, $W_i(x, y, z)$ and the degrees n_u, n_v, n_w of the series expansions in (13.1) are to be specified by the user. Thus, the Ritz constants A_i, B_i, C_i are the actual goal of the calculation.

The approach given with (13.1) is now inserted into the total elastic potential Π of the considered solid. It is important to note that the principle of the stationary value of the elastic total potential is a variation problem, but the approach (13.1) already completely defines the displacement field (and consequently also the strain and stress field) in its mathematical form. The only quantities which are accessible to a variation at all are the Ritz constants A_i, B_i, C_i, and as a consequence, the first variation $\delta\Pi$ of the total elastic potential changes into the following form:

$$\begin{aligned}
\delta\Pi &= \frac{\partial\Pi}{\partial A_1}\delta A_1 + \frac{\partial\Pi}{\partial A_2}\delta A_2 + \frac{\partial\Pi}{\partial A_3}\delta A_3 + ... + \frac{\partial\Pi}{\partial A_{n_u}}\delta A_{n_u} \\
&+ \frac{\partial\Pi}{\partial B_1}\delta B_1 + \frac{\partial\Pi}{\partial B_2}\delta B_2 + \frac{\partial\Pi}{\partial B_3}\delta B_3 + ... + \frac{\partial\Pi}{\partial B_{n_v}}\delta B_{n_v} \\
&+ \frac{\partial\Pi}{\partial C_1}\delta C_1 + \frac{\partial\Pi}{\partial C_2}\delta C_2 + \frac{\partial\Pi}{\partial C_3}\delta C_3 + ... + \frac{\partial\Pi}{\partial C_{n_w}}\delta C_{n_w} = 0. \tag{13.2}
\end{aligned}$$

The variations of the constants are arbitrary and also independent of each other, so that (13.2) can only be fulfilled if the following applies:

---

[1] Walter Ritz, 1878–1909, Swiss mathematician and physicist.

$$\frac{\partial \Pi}{\partial A_1} = 0, \quad \frac{\partial \Pi}{\partial A_2} = 0, \quad \frac{\partial \Pi}{\partial A_3} = 0, \quad \dots \quad \frac{\partial \Pi}{\partial A_{n_u}} = 0,$$

$$\frac{\partial \Pi}{\partial B_1} = 0, \quad \frac{\partial \Pi}{\partial B_2} = 0, \quad \frac{\partial \Pi}{\partial B_3} = 0, \quad \dots \quad \frac{\partial \Pi}{\partial B_{n_v}} = 0,$$

$$\frac{\partial \Pi}{\partial C_1} = 0, \quad \frac{\partial \Pi}{\partial C_2} = 0, \quad \frac{\partial \Pi}{\partial C_3} = 0, \quad \dots \quad \frac{\partial \Pi}{\partial C_{n_w}} = 0. \tag{13.3}$$

These are the so-called Ritz equations, from which the constants A_i, B_i, C_i can be determined. In the context of geometrical linear problems this leads to a linear system of equations with $n_u + n_v + n_w$ equations for the $n_u + n_v + n_w$ unknown constants A_i, B_i, C_i. Thus, the approximate displacement field is known and the approximate strain field can be determined from the kinematic equations of the given problem. Using the constitutive law, the approximate stress field can then be calculated. In this chapter, we limit ourselves exclusively to linear-elastic materials and geometrically linear problems.

13.2.1 The Euler-Bernoulli Beam

13.2.1.1 A Simple Example

The application of the Ritz method is demonstrated at the simple example of the cantilever beam as shown in Fig. 13.1. The beam has the length l and the constant bending stiffness EI. The cantilever is loaded by a force F at its free end and we want to approximate the bending line $w(x)$ using the Ritz method. The Euler-Bernoulli beam theory is assumed.

As a reference solution we first determine the exact solution of the given problem and determine the bending line $w(x)$ by fourfold integration of the beam differential equation $EIw'''' = q$ (cf. Chap. 5, Eq. (5.101)):

$$w(x) = \frac{1}{EI} \left(-\frac{F}{6} x^3 + \frac{F}{2} l x^2 \right). \tag{13.4}$$

The maximum deflection w_{max} results at the free end of the cantilever at the position $x = l$ as:

Fig. 13.1 Cantilever beam under single load F at the free end

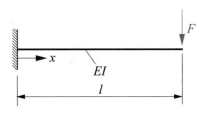

$$w_{max} = \frac{1}{EI}\left(-\frac{F}{6}l^3 + \frac{F}{2}l^3\right) = \frac{Fl^3}{3EI}. \tag{13.5}$$

To solve the problem using the Ritz method, the total elastic potential Π of the given beam situation is required. It reads:

$$\Pi = \frac{1}{2}EI\int_0^l w''^2 dx - Fw(x = l). \tag{13.6}$$

First we use an approach $W(x)$ of the form

$$w(x) \simeq W(x) = C_1W_1(x) + C_2W_2(x), \tag{13.7}$$

with the shape functions $W_1(x)$ and $W_2(x)$ as follows:

$$W_1(x) = x^2, \quad W_2(x) = x^3. \tag{13.8}$$

This approach fulfills the geometrical boundary conditions $w = 0$ and $w' = 0$ at the clamping point identically, it is admissible within the scope of the Ritz method. Thus, the approximate approach for the bending line of the beam is

$$w(x) \simeq W(x) = C_1x^2 + C_2x^3. \tag{13.9}$$

For the total elastic potential of the beam we then obtain:

$$\Pi = \frac{1}{2}EI\int_0^l w''^2 dx - Fw(x = l) = \frac{1}{2}EI\int_0^l (2C_1 + 6C_2x)^2 dx - F\left(C_1l^2 + C_2l^3\right)$$
$$= 2EIl\left(C_1^2 + 3C_1C_2l + 3C_2^2l^2\right) - F\left(C_1l^2 + C_2l^3\right) = \Pi(C_1, C_2). \tag{13.10}$$

We now determine the constants C_1 and C_2 with the help of the Ritz equations which currently read as follows:

$$\frac{\partial \Pi}{\partial C_1} = 2EIl(2C_1 + 3C_2l) - Fl^2 = 0,$$

$$\frac{\partial \Pi}{\partial C_2} = 2EIl\left(3C_1l + 6C_2l^2\right) - Fl^3 = 0. \tag{13.11}$$

This is a linear system of equations for determining the constants C_1 and C_2, which after a short rearrangement reads as follows:

$$2C_1 + 3C_2l = \frac{Fl}{2EI}, \quad 3C_1 + 6C_2l = \frac{Fl}{2EI}. \tag{13.12}$$

It has the following solution:

$$C_1 = \frac{Fl}{2EI}, \quad C_2 = -\frac{F}{6EI}. \tag{13.13}$$

The approximate solution (13.9) of the bending line $w(x)$ is therefore as follows:

$$w(x) = \frac{1}{EI}\left(-\frac{F}{6}x^3 + \frac{F}{2}lx^2\right). \tag{13.14}$$

This solution by the Ritz method leads exactly to the previously determined exact solution of the given bending problem. This is due to the fact that the selected approach (13.9) already contains the exact solution of the given problem. Since the variational principle used here is not only a consequence of the equilibrium conditions but is also completely equivalent to them, the use of an approach of the form (13.9) inevitably results in the exact solution of the problem under consideration.

Next we use an approach $W(x)$ of the form

$$w(x) \simeq W(x) = C_1 W_1(x) = C_1 \left[1 - \cos\left(\frac{\pi x}{2l}\right)\right]. \tag{13.15}$$

This approach also fulfils the geometric boundary conditions identically and is therefore permissible. Since this is a trigonometric function, the exact solution is not included here. We then obtain the approximate total potential Π as:

$$\Pi = \frac{1}{2}EI \int_0^l C_1^2 \frac{\pi^4}{16l^4} \cos^2\left(\frac{\pi x}{2l}\right) dx - FC_1 = C_1^2 \frac{EI\pi^4}{64l^3} - FC_1 = \Pi(C_1). \tag{13.16}$$

The Ritz equation then results as:

$$\frac{d\Pi}{dC_1} = \frac{EI\pi^4}{32l^3}C_1 - F = 0, \tag{13.17}$$

from which we can determine the following Ritz constant C_1:

$$C_1 = \frac{32Fl^3}{EI\pi^4}. \tag{13.18}$$

The approximate solution for the bending line $w(x)$ is therefore:

$$w(x) = \frac{32Fl^3}{EI\pi^4}\left[1 - \cos\left(\frac{\pi x}{2l}\right)\right]. \tag{13.19}$$

We calculate the approximate maximum deflection as:

$$w_{max} = \frac{32Fl^3}{EI\pi^4} = 0.3285\frac{Fl^3}{EI}. \tag{13.20}$$

This result is already quite close to the exact solution although an approach with only one shape function was used here.

13.2.1.2 General Formulation

We now want to derive a more general description of the Ritz method for beam bending problems and therefore consider the simply supported beam (length l) shown in Fig. 13.2. It has the variable but continuous bending stiffness $EI = EI(x)$ and is loaded by the arbitrary but continuously distributed line load $q = q(x)$. An approximate solution is sought for the bending line $w(x)$. The total elastic potential of the beam is calculated for the present case as:

$$\Pi = \Pi_i + \Pi_a = \frac{1}{2} \int_0^l EI(x) w(x)''^2 dx - \int_0^l q(x) w(x) dx. \tag{13.21}$$

Now an approximation for the bending line $w(x)$ of the beam consisting of n shape functions of the form

$$w(x) \simeq W(x) = \sum_{i=1}^{i=n} C_i W_i(x) = C_1 W_1(x) + C_2 W_2(x) + \dots + C_n W_n(x) \tag{13.22}$$

is used, where we assume that the n shape functions $W_i(x)$ identically fulfill the given geometric boundary conditions $w(x = 0) = w(x = l) = 0$. Inserting the approach (13.22) into the elastic total potential (13.21) yields

$$\Pi = \frac{1}{2} \int_0^l EI(x) \left(\sum_{i=1}^{i=n} C_i W_i''(x) \right)^2 dx - \int_0^l q(x) \sum_{i=1}^{i=n} C_i W_i(x) dx, \tag{13.23}$$

respectively after rearrangement:

$$\Pi = \frac{1}{2} \sum_{i=1}^{i=n} \sum_{j=1}^{j=n} \int_0^l EI(x) C_i C_j W_i''(x) W_j''(x) dx - \int_0^l q(x) \sum_{i=1}^{i=n} C_i W_i(x) dx. \tag{13.24}$$

Fig. 13.2 Beam under line load $q(x)$

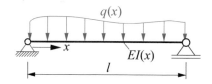

For $n = 3$ we obtain:

$$
\begin{aligned}
\Pi = {} & \frac{1}{2} \int_0^l EI(x) C_1 C_1 W_1''(x) W_1''(x) \mathrm{d}x + \int_0^l EI(x) C_1 C_2 W_1''(x) W_2''(x) \mathrm{d}x \\
& + \int_0^l EI(x) C_1 C_3 W_1''(x) W_3''(x) \mathrm{d}x + \frac{1}{2} \int_0^l EI(x) C_2 C_2 W_2''(x) W_2''(x) \mathrm{d}x \\
& + \int_0^l EI(x) C_2 C_3 W_2''(x) W_3''(x) \mathrm{d}x + \frac{1}{2} \int_0^l EI(x) C_3 C_3 W_3''(x) W_3''(x) \mathrm{d}x \\
& - \int_0^l q(x) C_1 W_1(x) \mathrm{d}x - \int_0^l q(x) C_2 W_2(x) \mathrm{d}x - \int_0^l q(x) C_3 W_3(x) \mathrm{d}x (13.25)
\end{aligned}
$$

Herein, systematically recurring integral terms appear which we want to abbreviate as:

$$
K_{ij} = \int_0^l EI(x) W_i''(x) W_j''(x) \mathrm{d}x, \quad F_i = \int_0^l q(x) W_i(x) \mathrm{d}x. \quad (13.26)
$$

For (13.25) we then have:

$$
\begin{aligned}
\Pi = {} & \frac{1}{2} C_1 C_1 K_{11} + C_1 C_2 K_{12} + C_1 C_3 K_{13} + \frac{1}{2} C_2 C_2 K_{22} + C_2 C_3 K_{23} + \frac{1}{2} C_3 C_3 K_{33} \\
& - C_1 F_1 - C_2 F_2 - C_3 F_3. \quad (13.27)
\end{aligned}
$$

Thus the Ritz equations are:

$$
\begin{aligned}
\frac{\partial \Pi}{\partial C_1} &= C_1 K_{11} + C_2 K_{12} + C_3 K_{13} - F_1 = 0, \\
\frac{\partial \Pi}{\partial C_2} &= C_1 K_{12} + C_2 K_{22} + C_3 K_{23} - F_2 = 0, \\
\frac{\partial \Pi}{\partial C_3} &= C_1 K_{13} + C_2 K_{23} + C_3 K_{33} - F_3 = 0. \quad (13.28)
\end{aligned}
$$

In a vector-matrix notation this can be represented as:

$$
\begin{bmatrix} K_{11} & K_{12} & K_{13} \\ K_{12} & K_{22} & K_{23} \\ K_{13} & K_{23} & K_{33} \end{bmatrix} \begin{pmatrix} C_1 \\ C_2 \\ C_3 \end{pmatrix} = \begin{pmatrix} F_1 \\ F_2 \\ F_3 \end{pmatrix}. \quad (13.29)
$$

In the case of n shape functions we obtain:

$$\begin{bmatrix} K_{11} & K_{12} & K_{13} & \dots & K_{1n} \\ K_{12} & K_{22} & K_{23} & \dots & K_{2n} \\ K_{13} & K_{23} & K_{33} & \dots & K_{3n} \\ \vdots & \vdots & \vdots & \ddots & \vdots \\ K_{n1} & K_{n2} & K_{n3} & \dots & K_{nn} \end{bmatrix} \begin{pmatrix} C_1 \\ C_2 \\ C_3 \\ \vdots \\ C_n \end{pmatrix} = \begin{pmatrix} F_1 \\ F_2 \\ F_3 \\ \vdots \\ F_n \end{pmatrix}. \tag{13.30}$$

In symbolic notation this equation system reads:

$$\underline{\underline{K}}\,\underline{C} = \underline{F}. \tag{13.31}$$

The matrix $\underline{\underline{K}}$ is referred to as the stiffness matrix. It is always symmetric and has $n \times n$ rows and columns when n shape functions are considered. The vectors $\underline{C}$ and $\underline{F}$ contain the constants C_i and the force resultants F_i. This provides the equations necessary to determine the constants C_i, and the approximate bending line $w(x)$ can be completely specified once the constants C_i are known.

13.2.1.3 Quality of the Solution, Convergence Behavior

When using approximation methods such as the Ritz method, it is of fundamental importance to obtain clarity regarding the quality of the approximate solution. This can concern both the choice of the shape functions and the degree of the approaches (in the case of beam bending, the degree n) and must always be considered in detail. As an example, the simply supported beam shown in Fig. 13.3 with the length l and the constant bending stiffness EI under the constant line load q is considered. Again, the reference solution is the exact solution of the problem which we can specify in the form of the bending line $w(x)$ and the moment and shear force lines $M(x)$ and $V(x)$ as follows:

$$EIw''' = -V = qx - \frac{1}{2}ql,$$

$$EIw'' = -M = \frac{1}{2}qx^2 - \frac{1}{2}qlx,$$

$$EIw = \frac{1}{24}qx^4 - \frac{1}{12}qlx^3 + \frac{1}{24}ql^3x. \tag{13.32}$$

Fig. 13.3 Simply supported beam under line load q

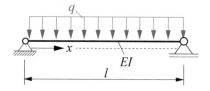

To provide reference values for checking the quality of the approximate solution, we arbitrarily use the values $E = 1$, $I = 1$. $l = 1$ and $q = 24$ and obtain:

$$w(x) = x^4 - 2x^3 + x, \quad M(x) = 12(x - x^2), \quad V(x) = 12(1 - 2x). \quad (13.33)$$

To set up an approximate solution using the Ritz method, we now use trigonometric functions of the type

$$W_i(x) = \sin\left(\frac{i\pi x}{l}\right), \quad (13.34)$$

so that:

$$w(x) \simeq W(x) = \sum_{i=1}^{i=n} C_i W_i(x) = C_1 \sin\left(\frac{\pi x}{l}\right) + C_2 \sin\left(\frac{2\pi x}{l}\right) + \ldots + C_n \sin\left(\frac{n\pi x}{l}\right).$$
$$(13.35)$$

The components K_{ij} of the stiffness matrix $\underline{\underline{K}}$ and the components F_i of the vector $\underline{F}$ then result as:

$$K_{ii} = \int_0^l EI(x)W_i''(x)W_i''(x)dx = EI\left(\frac{i\pi}{l}\right)^4 \int_0^l \sin\left(\frac{i\pi x}{l}\right)\sin\left(\frac{i\pi x}{l}\right)dx$$

$$= \frac{EIl}{2}\left(\frac{i\pi}{l}\right)^4 = \frac{(i\pi)^4}{2},$$

$$F_i = \int_0^l q(x)W_i(x)dx = q\int_0^l \sin\left(\frac{i\pi x}{l}\right)dx$$

$$= 0 \quad \text{wenn} \quad i = \text{gerade}$$

$$= \frac{2ql}{i\pi} \quad \text{wenn} \quad i = \text{ungerade}. \quad (13.36)$$

It is remarkable that the components K_{ij} for $i \neq j$ all become zero. Shape functions with the property

$$\int_0^l EI(x)W_i''(x)W_j''(x)dx = 0 \quad (13.37)$$

(with $i \neq j$) are called orthogonal functions. If the series development (13.35) is carried out exemplarily up to the degree $n = 7$, then the following system of equations results for the Ritz constants $C_1, C_2, \ldots, C_7$:

$$
\begin{bmatrix}
\dfrac{1}{2}\pi^4 & 0 & 0 & 0 & 0 & 0 & 0 \\
0 & 8\pi^4 & 0 & 0 & 0 & 0 & 0 \\
0 & 0 & \dfrac{81}{2}\pi^4 & 0 & 0 & 0 & 0 \\
0 & 0 & 0 & 128\pi^4 & 0 & 0 & 0 \\
0 & 0 & 0 & 0 & \dfrac{625}{2}\pi^4 & 0 & 0 \\
0 & 0 & 0 & 0 & 0 & 648\pi^4 & 0 \\
0 & 0 & 0 & 0 & 0 & 0 & \dfrac{2401}{2}\pi^4
\end{bmatrix}
\begin{pmatrix}
C_1 \\ C_2 \\ C_3 \\ C_4 \\ C_5 \\ C_6 \\ C_7
\end{pmatrix}
=
\begin{pmatrix}
\dfrac{48}{\pi} \\ 0 \\ \dfrac{16}{\pi} \\ 0 \\ \dfrac{48}{5\pi} \\ 0 \\ \dfrac{48}{7\pi}
\end{pmatrix}. \qquad (13.38)
$$

The vector $\underline{C}$ can be calculated immediately due to the advantageous exclusive assignment of the stiffness matrix on the main diagonal as:

$$
\underline{C} = \left(\frac{96}{\pi^5}\ 0\ \frac{32}{81\pi^5}\ 0\ \frac{96}{3125\pi^5}\ 0\ \frac{96}{16807\pi^5} \right)^T. \qquad (13.39)
$$

It turns out that in the series development (13.35) only the odd functions remain, which can easily be explained by the symmetry of the considered situation as well as the type of shape functions chosen here. The comparison of the exact and the approximated state variables $w(x)$, $M(x)$, and $V(x)$ is shown in Fig. 13.4. From this figure some important conclusions can be drawn about the typical behavior of approximate solutions in the framework of the Ritz method. First of all, it can be seen that it is apparently sufficient to use an approach with the degree $n = 1$ of the form

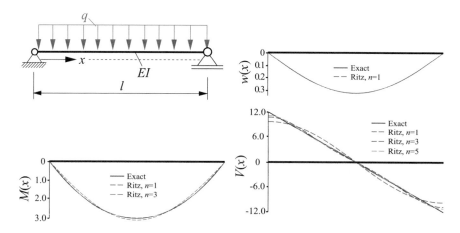

Fig. 13.4 Approximation results for the simply supported beam under uniform load q, determined with trigonometric shape functions

$$w(x) \simeq \frac{96}{\pi^5} \sin(\pi x) \tag{13.40}$$

in order to appropriately approximate the exact solution for the deflection $w(x)$. On the other hand, it is equally obvious that the approximation of the bending moment $M(x)$ and the transverse shear force $V(x)$ requires a higher number of shape functions. For $M(x)$, an approach of degree $n = 3$ is sufficient to generate an acceptable approximation, whereas for the transverse shear force $V(x)$, functions up to degree $n = 5$ should be considered. Thus, it is an important conclusion that the quality of an approximation using the Ritz method depends not only on the quality of the directly approximated quantity (here the deflection $w(x)$), but also on the quality of the approximation of quantities derived from it (here the bending moment $M(x)$ and the transverse shear force $V(x)$). The choice of the appropriate degree of approximation as well as the verification of the used approach functions therefore makes convergence analyses necessary without which an approximation remains incomplete. A wide variety of convergence indicators can be used for this purpose, but this is not described here.

Another illustrative example is the beam (length l, constant bending stiffness EI) under a the point force F as shown in Fig. 13.5. The exact solution for all state variables can easily be determined with the means of the Chap. 5, their determination is not described here in detail. With the arbitrary values $E = I = l = 1$ and $F = 6$, in the range $0 \leq x \leq \frac{l}{2}$ the following results:

$$w(x) = \frac{1}{8}x\left(3 - 4x^2\right), \quad M(x) = 3x, \quad V(x) = 3. \tag{13.41}$$

For $\frac{l}{2} \leq x \leq l$ we obtain:

$$w(x) = \frac{1}{8}\left(4x^3 - 12x^2 + 9x - 1\right), \quad M(x) = 3 - 3x, \quad V(x) = -3. \tag{13.42}$$

An approximation using the Ritz method is given with sin −functions as follows:

$$W_i(x) = \sin\left(\frac{i\pi x}{l}\right). \tag{13.43}$$

Fig. 13.5 Simply supported beam under single force F

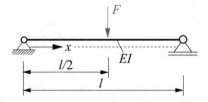

The total elastic potential Π is for the present case:

$$\Pi = \frac{1}{2}EI \int_0^l w''^2 dx - Fw\left(x = \frac{l}{2}\right). \tag{13.44}$$

For the approximation degree $n = 9$ the equation system for determining the Ritz constants C_i is as follows:

$$
\begin{bmatrix}
\frac{1}{2}\pi^4 & 0 & 0 & 0 & 0 & 0 & 0 & 0 & 0 \\
0 & 8\pi^4 & 0 & 0 & 0 & 0 & 0 & 0 & 0 \\
0 & 0 & \frac{81}{2}\pi^4 & 0 & 0 & 0 & 0 & 0 & 0 \\
0 & 0 & 0 & 128\pi^4 & 0 & 0 & 0 & 0 & 0 \\
0 & 0 & 0 & 0 & \frac{625}{2}\pi^4 & 0 & 0 & 0 & 0 \\
0 & 0 & 0 & 0 & 0 & 648\pi^4 & 0 & 0 & 0 \\
0 & 0 & 0 & 0 & 0 & 0 & \frac{2401}{2}\pi^4 & 0 & 0 \\
0 & 0 & 0 & 0 & 0 & 0 & 0 & 2048\pi^4 & 0 \\
0 & 0 & 0 & 0 & 0 & 0 & 0 & 0 & \frac{6561}{2}\pi^4
\end{bmatrix}
\times
\begin{pmatrix}
C_1 \\ C_2 \\ C_3 \\ C_4 \\ C_5 \\ C_6 \\ C_7 \\ C_8 \\ C_9
\end{pmatrix}
=
\begin{pmatrix}
6 \\ 0 \\ -6 \\ 0 \\ 6 \\ 0 \\ -6 \\ 0 \\ 6
\end{pmatrix}. \tag{13.45}
$$

This yields:

$$\underline{C} = \left(\frac{12}{\pi^4}\ 0\ -\frac{4}{27\pi^4}\ 0\ \frac{12}{625\pi^4}\ 0\ -\frac{12}{2\,401\pi^4}\ 0\ \frac{4}{2\,187\pi^4}\right)^T. \tag{13.46}$$

Again, due to the symmetry of the given situation and the nature of the shape functions, only the odd elements remain. Figure 13.6 shows a comparison of the determined approximate solution with the exact solution. It can be seen that the approximation of the bending line $w(x)$ with the shape functions selected here is already possible with a very low degree of the employed series expansion, whereas the satisfactory representation of the moment area $M(x)$ requires a higher computational effort. On the other hand, it becomes obvious that the approximation of the transverse shear force line $V(x)$ with the selected maximum degree $n = 9$ is not satisfactory at all. This underlines once again the necessity of thoroughly performed convergence analyses when using approximation methods such as the Ritz method.

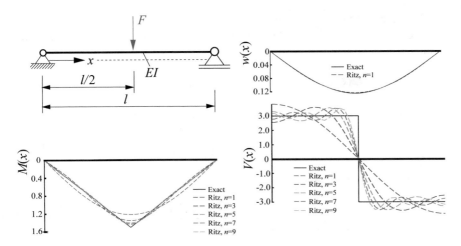

Fig. 13.6 Approximation results for the simply supported beam under point force F, determined with trigonometric shape functions

13.2.1.4 Shape Functions

For boundary conditions other than simple supports, the available literature provides a number of suitable shape functions a selection of which is briefly outlined below.

If the beam under consideration is clamped at both ends, a polynomial formulation can be used as follows:

$$W_i(x) = \left(\frac{x}{l}\right)^{i+1}\left(1 - \frac{x}{l}\right)^2.$$

The so-called beam eigenfunctions can also be used:

$$W_i(x) = \sin(\lambda_i x) - \sinh(\lambda_i x) + \alpha_i(\cosh(\lambda_i x) - \cos(\lambda_i x)). \qquad (13.47)$$

Here the quantities λ_i are the roots of the following characteristic equation:

$$\cos(\lambda_i l)\cosh(\lambda_i l) - 1 = 0. \qquad (13.48)$$

The quantities α_i can be determined as follows:

$$\alpha_i = \frac{\sinh(\lambda_i l) - \sin(\lambda_i l)}{\cosh(\lambda_i l) - \cos(\lambda_i l)} = \frac{\cosh(\lambda_i l) - \cos(\lambda_i l)}{\sinh(\lambda_i l) + \sin(\lambda_i l)}. \qquad (13.49)$$

A series expansion using $\cos$ − functions can also be employed:

$$W_i(x) = \cos\left(\frac{(i-2)\pi x}{l}\right) - \cos\left(\frac{i\pi x}{l}\right), \qquad (13.50)$$

where here the summation has to be performed from $i = 2$ to $i = n + 1$.

A beam which is clamped at $x = 0$ and is free at $x = l$ can be handled with the following shape functions:

$$W_i(x) = \sin(\lambda_i x) - \sinh(\lambda_i x) + \alpha_i(\cosh(\lambda_i x) - \cos(\lambda_i x)). \qquad (13.51)$$

The quantities λ_i are the roots of the following characteristic equation:

$$\cos(\lambda_i l)\cosh(\lambda_i l) + 1 = 0. \qquad (13.52)$$

The quantities α_i result as follows:

$$\alpha_i = \frac{\sinh(\lambda_i l) + \sin(\lambda_i l)}{\cosh(\lambda_i l) + \cos(\lambda_i l)}. \qquad (13.53)$$

In the case of a simple support at $x = 0$ and a clamping at $x = l$ the following shape functions can be used:

$$W_i(x) = \sinh(\lambda_i l)\sin(\lambda_i x) + \sin(\lambda_i l)\sinh(\lambda_i x), \qquad (13.54)$$

where the quantities λ_m result from the following characteristic equation:

$$\tan(\lambda_i l) - \tanh(\lambda_i l) = 0. \qquad (13.55)$$

13.2.1.5   Beams with Variable Bending Stiffness

The treatment of an example of a beam with variable bending stiffness $EI(x)$ is briefly discussed below. The cantilever beam of Fig. 13.7 is considered. The beam is subjected to a constant line load q and has a rectangular cross section with a variable moment of inertia $I(x)$ along the longitudinal axis x. The approximation of the bending line to be determined by the Ritz method is carried out with the help of a shape function in the form of a square parabola. The moment of inertia $I(x)$

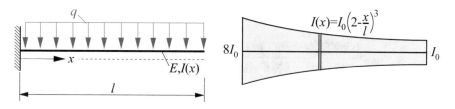

Fig. 13.7 Cantilever beam with variable moment of inertia $I(x)$ under uniform load q

is given as $I(x) = I_0 \left(2 - \frac{x}{l}\right)^3$, where $I_0 = \frac{b_0 h_0^3}{12}$ is the constant section width and h_0 is the section height which is variable over x. At $x = 0$ the moment of inertia $I(x = 0) = 8I_0$ is given. At the cantilever end at $x = l$ the value $I(x = l) = I_0$ is given.

The following shape function is used:

$$w(x) \simeq W(x) = C_1 W_1(x), \tag{13.56}$$

with

$$W_1(x) = x^2. \tag{13.57}$$

The total elastic potential Π reads:

$$\Pi = \frac{1}{2} \int_0^l E I(x) w(x)''^2 \mathrm{d}x - q \int_0^l w(x) \mathrm{d}x. \tag{13.58}$$

Inserting (13.57) yields:

$$\Pi = 2E I_0 C_1^2 \int_0^l \left(2 - \frac{x}{l}\right)^3 \mathrm{d}x - q C_1 \int_0^l x^2 \mathrm{d}x. \tag{13.59}$$

Performing the necessary integrations leads to:

$$\Pi = \frac{15}{2} E I_0 l C_1^2 - \frac{q l^3}{3} C_1. \tag{13.60}$$

Evaluating the Ritz equation $\frac{\mathrm{d}\Pi}{\mathrm{d}C_1} = 0$ leads to the following expression:

$$15 E I_0 l C_1 - \frac{q l^3}{3} = 0. \tag{13.61}$$

Solving for the Ritz constant C_1 yields:

$$C_1 = \frac{q l^2}{45 E I_0}. \tag{13.62}$$

The approximation $w(x) \simeq W(x)$ is therefore:

$$w(x) \simeq W(x) = \frac{q l^2 x^2}{45 E I_0}. \tag{13.63}$$

The maximum deflection occurs at the free end of the cantilever:

$$W(x = l) = \frac{q l^4}{45 E I_0} = 0.0222 \frac{q l^4}{E I_0}. \tag{13.64}$$

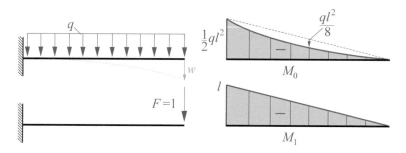

Fig. 13.8 Moment lines M_0 and M_1

This example allows for an exact-analytical solution which is briefly explained below for comparison. For this purpose, we calculate the displacement of the beam at the free end $x = l$ using the force method (Chap. 12) by first determining the two moment lines M_0 and M_1 due to the given load q and the virtual unit force $F = 1$ at the end of the cantilever. They are shown in Fig. 13.8. We have

$$M_0(x) = -\frac{q}{2}(x - l)^2, \quad M_1(x) = x - l. \tag{13.65}$$

The deflection w of the free end of the cantilever beam follows as:

$$
\begin{aligned}
w &= \int_0^l \frac{M_0 M_1}{EI}\,\mathrm{d}x \\
&= -\frac{q}{2}\int_0^l \frac{(x-l)^3}{EI_0\left(2 - \frac{x}{l}\right)^3}\,\mathrm{d}x = -\frac{q}{2EI_0}\int_0^l \left(\frac{x-l}{2-\frac{x}{l}}\right)^3 \mathrm{d}x = \frac{ql^3}{2EI_0}\int_0^l \left(\frac{x-l}{x-2l}\right)^3 \mathrm{d}x.
\end{aligned}
\tag{13.66}
$$

The integration to be performed here is handled by the substitution $z = x - l$:

$$\int_0^l \left(\frac{x-l}{x-2l}\right)^3 \mathrm{d}x = \int_{-l}^0 \left(\frac{z}{z-l}\right)^3 \mathrm{d}z. \tag{13.67}$$

We have:

$$\int_{-l}^0 \left(\frac{z}{z-l}\right)^3 \mathrm{d}z = \left[z - l + 3l\ln(z-l) - \frac{3l^2}{z-l} - \frac{l^3}{2(z-l)^2}\right]_{-l}^0 = l\left(\frac{17}{8} - 3\ln(2)\right). \tag{13.68}$$

The maximum deflection at the end of the cantilever arm can then be given as:

$$w = \frac{ql^4}{2EI_0}\left(\frac{17}{8} - 3\ln(2)\right) = 0.0228\frac{ql^4}{EI_0}.$$

The comparison shows that in this case the approximation using the Ritz method provides a very good result quality.

13.2.2   The Rod

The Ritz method for rod structures can be derived in a completely analogous way. Consider the situation as shown in Fig. 13.9. A rod of length l is given which is loaded by the line load $n(x)$ that is distributed arbitrarily but continuously over x. The member has the arbitrarily but continuously distributed extensional stiffness $EA(x)$.

The displacement $u(x)$ is approximated as follows:

$$u(x) \simeq U(x) = \sum_{i=1}^{m} C_i U_i(x), \tag{13.69}$$

where the upper limit of the summation is denoted as m to avoid confusion with the applied line load $n(x)$. The quantities $U_i(x)$ are shape functions which have to be conform to the underlying geometric boundary conditions.

For the given situation the total elastic potential $\Pi = \Pi_i + \Pi_a$ results as follows:

$$\Pi_i = \frac{1}{2}\int_0^l EA(x)\left(u(x)'\right)^2 dx, \quad \Pi_a = -\int_0^l n(x)u(x)dx. \tag{13.70}$$

Inserting (13.69) results in:

$$\Pi_i = \frac{1}{2}\sum_{i=1}^{m}\sum_{j=1}^{m}\int_0^l EA(x)C_iC_jU_i'(x)U_j'(x)dx,$$

$$\Pi_a = -\int_0^l n(x)\sum_{i=1}^{m}C_iU_i'(x)dx. \tag{13.71}$$

Fig. 13.9 Rod under line load $n(x)$

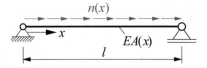

The system of equations to be solved for the Ritz constants is obtained from the Ritz equations

$$\frac{\partial \Pi}{\partial C_i} = 0 \tag{13.72}$$

analogously to (13.30) as:

$$\begin{bmatrix} K_{11} & K_{12} & K_{13} & \dots & K_{1m} \\ K_{12} & K_{22} & K_{23} & \dots & K_{2m} \\ K_{13} & K_{23} & K_{33} & \dots & K_{3m} \\ \vdots & \vdots & \vdots & \ddots & \vdots \\ K_{m1} & K_{m2} & K_{m3} & \dots & K_{mm} \end{bmatrix} \begin{pmatrix} C_1 \\ C_2 \\ C_3 \\ \vdots \\ C_n \end{pmatrix} = \begin{pmatrix} F_1 \\ F_2 \\ F_3 \\ \vdots \\ F_n \end{pmatrix}, \tag{13.73}$$

or in a symbolic notation:

$$\underline{\underline{K}}\, \underline{C} = \underline{F}. \tag{13.74}$$

Here the entries K_{ij} of the stiffness matrix $\underline{\underline{K}}$ are defined as:

$$K_{ij} = \int_0^l EA(x)U_i'(x)U_j'(x)\mathrm{d}x. \tag{13.75}$$

The force resultants F_i read:

$$F_i = \int_0^l n(x)U_i(x)\mathrm{d}x. \tag{13.76}$$

13.2.3   Torsion

A Ritz formulation for beams under torsion is established here in the context of St. Venant torsion (Chap. 7). We consider a straight beam (Fig. 13.10) with fork restraints on both sides which exhibits the torsional stiffness $GI_T(x)$ which is distributed

Fig. 13.10 Beam subjected to the distributed torsional moment $m_T(x)$

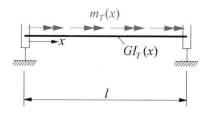

arbitrarily but continuously over x. The load consists of the torsional moment m_T which is arbitrarily but continuously distributed over x. The total elastic potential $\Pi = \Pi_i + \Pi_a$ is given as:

$$\Pi_i = \frac{1}{2} \int_0^l GI_T(x) \left(\vartheta(x)'\right)^2 dx, \quad \Pi_a = -\int_0^l m_T(x)\vartheta(x)dx. \quad (13.77)$$

It can easily be shown that the system of equations resulting here is again of the form (13.73), although the following entries of the stiffness matrix $\underline{K}$ are to be employed in the present case:

$$K_{ij} = \int_0^l GI_T(x)\theta_i'(x)\theta_j'(x)dx. \quad (13.78)$$

The force resultants F_i read:

$$F_i = \int_0^l m_T(x)\theta_i(x)dx. \quad (13.79)$$

In this context, the quantities $\theta_i(x)$ are arbitrary allowable shape functions for the approximation of $\vartheta(x)$.

13.3 The Galerkin Method

A further classical approximation method is the Galerkin method,[2] which we will explain in the following at the example of a beam of length l with a constant bending stiffness EI. We consider a simply supported beam (see Fig. 13.3) under a constant line load q. The Galerkin method is based on the variational statement

$$\int_0^l \left(EIw'''' - q\right) \delta w dx - V\delta w(x = 0) + M\delta w'(x = 0)$$
$$-M\delta w'(x = l) + V\delta w(x = l) = 0, \quad (13.80)$$

which can be obtained from the principle of virtual displacements. Analogous to the Ritz method, an approximation of the bending line $w(x)$ of the beam of the form

$$w(x) \simeq W(x) = \sum_{i=1}^{i=n} C_i W_i(x) = C_1 W_1(x) + C_2 W_2(x) + ... + C_n W_n(x) \quad (13.81)$$

---

[2] Boris Grigorjewitsch Galerkin, 1871–1945, Russian mathematician.

is employed where now the shape functions are required to fulfill all boundary conditions of the considered problem. In this case all boundary terms disappear in (13.80) after the insertion of (13.81), and the following expression remains:

$$\int_0^l \left(EIW'''' - q \right) \delta W \mathrm{d}x = 0, \tag{13.82}$$

or:

$$\int_0^l \left[\left(EIC_1 W_1'''' + EIC_2 W_2'''' + ... + EIC_n W_n'''' - q \right) \right.$$
$$\left. (\delta C_1 W_1 + \delta C_2 W_2 + ... + \delta C_n W_n) \right] \mathrm{d}x = 0. \tag{13.83}$$

It has already been considered here that the only variables in this approach that are accessible to a variation are the constants C_1, C_2,..., C_n. The variations of the constants C_i are independent of one another so that (13.83) can only be fulfilled meaningfully if the following holds:

$$\int_0^l \left(EIC_1 W_1'''' + EIC_2 W_2'''' + ... + EIC_n W_n'''' - q \right) W_1 \mathrm{d}x = 0,$$

$$\int_0^l \left(EIC_1 W_1'''' + EIC_2 W_2'''' + ... + EIC_n W_n'''' - q \right) W_2 \mathrm{d}x = 0,$$

$$\vdots$$

$$\int_0^l \left(EIC_1 W_1'''' + EIC_2 W_2'''' + ... + EIC_n W_n'''' - q \right) W_n \mathrm{d}x = 0. \tag{13.84}$$

These are the so-called Galerkin equations. After performing the prescribed integrations they represent a linear system of n equations for the n constants $C_1, C_2, ..., C_n$.

As already outlined for the Ritz method, also when using the Galerkin method convergence analyses are necessary to determine a sufficient degree of approximation. The practical application of this method is generally made more difficult by the fact that, depending on the problem, it can prove very difficult to find suitable shape functions that fulfill all given boundary conditions. Therefore, we do not describe the Galerkin method for the rod or the case of torsion. However, the following is a brief illustration of an elementarily simple example using a simply supported beam under a uniformly distributed load q with constant bending stiffness EI (cf. Fig. 13.3). The Galerkin method is used to find an approximate solution for the bending line $w(x)$ of the beam.

First an approximation of the form

$$w(x) \simeq W(x) = C_1 W_1(x) = C_1 \sin\left(\frac{\pi x}{l}\right) \tag{13.85}$$

is employed. The Galerkin equations are then reduced to the following equation:

$$\int_0^l \left(EIC_1 W_1'''' - q \right) W_1 dx = 0. \tag{13.86}$$

The approach (13.85) results in:

$$\int_0^l \left[EIC_1 \left(\frac{\pi}{l} \right)^4 \sin \left(\frac{\pi x}{l} \right) - q \right] \sin \left(\frac{\pi x}{l} \right) dx = 0. \tag{13.87}$$

With the integrals

$$\int_0^l \sin^2 \left(\frac{\pi x}{l} \right) dx = \frac{l}{2}, \quad \int_0^l \sin \left(\frac{\pi x}{l} \right) dx = \frac{2l}{\pi} \tag{13.88}$$

the constant C_1 can be determined as:

$$C_1 = \frac{4q}{EIl} \left(\frac{l}{\pi} \right)^5. \tag{13.89}$$

The approximate solution according to the Galerkin method can then be written down as:

$$w(x) \simeq W(x) = \frac{4q}{EIl} \left(\frac{l}{\pi} \right)^5 \sin \left(\frac{\pi x}{l} \right). \tag{13.90}$$

The maximum deflection occurs in the center of the beam at $x = \frac{l}{2}$ and is according to the approximate solution:

$$w(x = \frac{l}{2}) \simeq \frac{4q}{EIl} \left(\frac{l}{\pi} \right)^5 = \frac{4ql^4}{EI\pi^5} = 0.01307 \frac{ql^4}{EI}. \tag{13.91}$$

The exact solution for the bending line $w(x)$ is for the given case:

$$EIw = \frac{1}{24} qx^4 - \frac{1}{12} qlx^3 + \frac{1}{24} ql^3 x. \tag{13.92}$$

The maximum deflection at the center of the beam at $x = \frac{l}{2}$ results as:

$$w \left(x = \frac{l}{2} \right) = \frac{5ql^4}{384EI} = 0.01302 \frac{ql^4}{EI}. \tag{13.93}$$

It is shown that the Galerkin method leads to a very plausible result even with one single suitable shape function.

The maximum bending moment in the center of the beam is also considered. The exact solution is:

$$M\left(x = \frac{l}{2}\right) = \frac{ql^2}{8} = 0.125ql^2. \tag{13.94}$$

The approximate solution leads to the following moment $M(x)$:

$$M(x) \simeq= \frac{4q}{l}\left(\frac{l}{\pi}\right)^3 \sin\left(\frac{\pi x}{l}\right). \tag{13.95}$$

This results in the maximum value at $x = \frac{l}{2}$ as:

$$M(x = \frac{l}{2}) \simeq= \frac{4q}{l}\left(\frac{l}{\pi}\right)^3 = 0.12901ql^2. \tag{13.96}$$

The estimation for the maximum bending moment is therefore also satisfactory.

We repeat the above calculation and now use the following approximation $W(x)$ for the bending line of the beam:

$$w(x) \simeq W(x) = C_1 W_1(x) + C_2 W_2(x) + C_3 W_3(x)$$
$$= C_1 \sin\left(\frac{\pi x}{l}\right) + C_2 \sin\left(\frac{2\pi x}{l}\right) + C_3 \sin\left(\frac{3\pi x}{l}\right). \tag{13.97}$$

The Galerkin equations then result as:

$$\int_0^l \left(EIC_1 W_1'''' + EIC_2 W_2'''' + EIC_3 W_3'''' - q\right) W_1 dx = 0,$$

$$\int_0^l \left(EIC_1 W_1'''' + EIC_2 W_2'''' + EIC_3 W_3'''' - q\right) W_2 dx = 0,$$

$$\int_0^l \left(EIC_1 W_1'''' + EIC_2 W_2'''' + EIC_3 W_3'''' - q\right) W_3 dx = 0. \tag{13.98}$$

As the shape functions chosen here are orthogonal, the integrals

$$\int_0^l W_i''''(x) W_j(x) dx \tag{13.99}$$

disappear for $i \neq j$. The Galerkin equations are therefore significantly simplified:

$$\int_0^l \left(EIC_1 W_1'''' W_1 - q W_1\right) dx = 0,$$

$$\int_0^l \left(EIC_2 W_2'''' W_2 - q W_2\right) dx = 0,$$

$$\int_0^l \left(EIC_3 W_3'''' W_3 - q W_3\right) dx = 0. \tag{13.100}$$

The integrals appearing here result in:

$$\int_0^l EIC_1 W_1'''' W_1 dx = EIC_1 \left(\frac{\pi}{l}\right)^4 \int_0^l \sin^2\left(\frac{\pi x}{l}\right) dx = \frac{1}{2}EIC_1 l \left(\frac{\pi}{l}\right)^4,$$

$$\int_0^l EIC_2 W_2'''' W_2 dx = EIC_2 \left(\frac{2\pi}{l}\right)^4 \int_0^l \sin^2\left(\frac{2\pi x}{l}\right) dx = 8EIC_2 l \left(\frac{\pi}{l}\right)^4,$$

$$\int_0^l EIC_3 W_3'''' W_3 dx = EIC_3 \left(\frac{3\pi}{l}\right)^4 \int_0^l \sin^2\left(\frac{3\pi x}{l}\right) dx = \frac{81}{2}EIC_3 l \left(\frac{\pi}{l}\right)^4,$$

$$\tag{13.101}$$

and:

$$\int_0^l q W_1 dx = q \int_0^l \sin\left(\frac{\pi x}{l}\right) dx = \frac{2ql}{\pi},$$

$$\int_0^l q W_2 dx = q \int_0^l \sin\left(\frac{2\pi x}{l}\right) dx = 0,$$

$$\int_0^l q W_3 dx = q \int_0^l \sin\left(\frac{3\pi x}{l}\right) dx = \frac{2ql}{3\pi}. \tag{13.102}$$

We therefore obtain the Galerkin equations in the following form:

$$\frac{1}{2}EIC_1 l \left(\frac{\pi}{l}\right)^4 = \frac{2ql}{\pi}, \quad 8EIC_2 l \left(\frac{\pi}{l}\right)^4 = 0, \quad \frac{81}{2}EIC_3 l \left(\frac{\pi}{l}\right)^4 = \frac{2ql}{3\pi}. \tag{13.103}$$

These can be solved directly for the constants:

$$C_1 = \frac{4q}{EIl}\left(\frac{l}{\pi}\right)^5, \quad C_2 = 0, \quad C_3 = \frac{4q}{243EIl}\left(\frac{l}{\pi}\right)^5. \tag{13.104}$$

The approximation for the bending line $w(x)$ can then be formulated as:

$$w(x) \simeq W(x) = C_1 W_1(x) + C_2 W_2(x) + C_3 W_3(x)$$

$$= \frac{4q}{EIl}\left(\frac{l}{\pi}\right)^5 \sin\left(\frac{\pi x}{l}\right) + \frac{4q}{243EIl}\left(\frac{l}{\pi}\right)^5 \sin\left(\frac{3\pi x}{l}\right). \tag{13.105}$$

The resulting maximum value for the deflection w at $x = \frac{l}{2}$ is

$$w\left(x = \frac{l}{2}\right) = \frac{4q}{EIl}\left(\frac{l}{\pi}\right)^5 \sin\left(\frac{\pi}{2}\right) + \frac{4q}{243EIl}\left(\frac{l}{\pi}\right)^5 \sin\left(\frac{3\pi}{2}\right)$$

$$= 0.01302\frac{ql^4}{EI}, \tag{13.106}$$

which corresponds to the exact result with a very high accuracy.

The approximate maximum value of the bending moment at the center of the beam is

$$M\left(x = \frac{l}{2}\right) \simeq \frac{4q}{l}\left(\frac{l}{\pi}\right)^3 \sin\left(\frac{\pi}{2}\right) + \frac{4q}{27l}\left(\frac{l}{\pi}\right)^3 \sin\left(\frac{3\pi}{2}\right) = 0.12423ql^2. \qquad (13.107)$$

This result also shows an improved quality compared to the approximation with a single shape function.

Bibliography

Becker W, Gross D (2002) Mechanik elastischer Körper und Strukturen, Springer. Berlin, Germany

Betten J (2004) Finite Elemente für Ingenieure Teil 2: Variationsrechnung, Energiemethoden, Näherungsverfahren, Nichtlinearitäten, numerische Integrationen, 2nd edn. Springer, Berlin, Germany

Gross D, Hauger W, Wriggers P (2014) Technische Mechanik 4: Hydromechanik, Elemente der Höheren Mechanik, Numerische Methoden, Ninth edition, Springer. Berlin, Germany

Kossira H (1996) Grundlagen des Leichtbaus, Springer. Berlin, Germany

Lanczos C (1986) The variational principles of mechanics, 4th edn. Dover, New York, USA

Langhaar HL (2016) Energy methods in applied mechanics. Dover, New York, USA

Mittelstedt C (2017) Energiemethoden der Elastostatik. Technische Universität Darmstadt, Germany, Studienbereich Mechanik

Reddy JN (2017) Energy principles and variational methods in applied mechanics, 3rd edn. John Wiley and Sons, New York, USA

Tauchert TR (1974) Energy principles in structural mechanics. McGraw-Hill, New York, USA

Chapter 14
The Finite Element Method

14.1 Introduction

The finite element method (short: FEM) is an energy-based approximation method that has found its firm place in lightweight engineering applications. It has largely replaced classical methods such as the previously discussed methods according to Ritz and Galerkin in many fields of application, and practical lightweight engineering work is unthinkable without the finite element method. In this chapter we will discuss this method in detail at the example of rod and beam structures.

The FEM is based on the idea to formulate approximations not, as e.g. in the context of the Ritz method, on the entire structure under consideration, but rather to divide the structure into sub-areas, the so-called elements (i.e. to perform a so-called discretization) and to make approximations for the state variables within these sub-areas. By means of suitable continuity conditions, the structure is then computationally assembled at a later stage. This division into subareas or elements is also called discretization or meshing. A typical example from lightweight engineering application is shown in Fig. 14.1.

The finite elements are coupled at discrete points, the so-called nodes, and the approaches used in FEM work with the idea that the behavior of the structure can be described by discrete numerical values, namely exactly the values of the sought-after state variables in the element nodes. The FEM can be used for a variety of physical tasks. If it is a problem of lightweight engineering, usually in the framework of the theory of elasticity, then appropriate approximations are often formulated in the form of nodal displacements, and the displacement field within an element is represented by corresponding shape functions. In the following we will show that this approach is equivalent to using the Ritz method within each finite element. Thus, once the displacement field in the form of the nodal displacements as well as the shape functions is given inside each element, strain and stress fields can be determined in an approximate manner. Depending on the nature of the given problem, a large number of element types is available which can be distinguished according

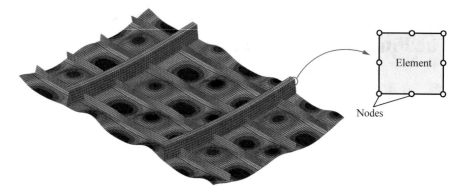

Fig. 14.1 Finite element model of a stiffened fuselage shell of an aircraft, showing the first buckling mode under longitudinal compressive load

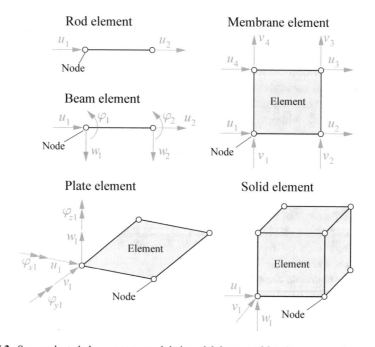

Fig. 14.2 Some selected element types and their nodal degrees of freedom

to their dimension (one- or multi-dimensional elements), their number of nodes and, consequently, their shape functions. A selection of common element types for problems in elasticity theory is shown in Fig. 14.2.

The approximation used within each finite element must ensure that compatibility conditions between adjacent elements are not violated. Continuous distributions of state variables within each element as well as on the element boundaries must be ensured by the employed approaches, and the element formulations must be able to represent at least constant stress and strain states. Rigid body displacements must not cause deformations in the elements. Due to the element-wise formulation of shape functions, usually quite simple types of functions are sufficient. In contrast to the preceding Chap. 13, we will denote the element shape functions as N in this chapter.

In this chapter, we limit ourselves to displacement based elements for linear-elastic and geometrically linear problems in statics. For advanced topics, the reader is referred to the widely available technical literature, a selection of which can be found in the bibliography at the end of this chapter.

A structural analysis using the FEM typically contains the following essential steps:

- Creation of a structural model (loads, material behavior, geometry) and adaptation of a suitable discretization or meshing,
- Derivation of a suitable element formulation and setting up the necessary element equations, or use already existing element libraries which are available in commercial finite element program systems,
- Combining the element equations into a system of equations for the entire structure and taking all boundary conditions into account,
- Solving the system equations and calculation of the state variables of the structure in a so-called follow-up calculation or post-processing at all relevant points.

14.2 Finite Elements for Plane Trusses

14.2.1 Element Formulation and Calculation Steps

We start the detailed presentation of the FEM at the example of plane truss structures. The considered truss structures are exclusively loaded by point forces in their nodes. An introductory example is shown in Fig. 14.3. The displacements u and w of node 2 as well as the member forces of the members 1 and 2 are to be determined using the FEM. Identical linear-elastic material (modulus of elasticity E) is present in both members, and both members have the same cross-sectional area A. For the calculation we introduce a global coordinate system x, y as indicated.

To verify the FEM results we determine the exact solution of the given truss problem.

The member forces can be determined as:

$$N_1 = F, \quad N_2 = -\sqrt{2}F. \tag{14.1}$$

The node displacements can be calculated e.g. by means of the force method, and we obtain:

$$u = \frac{Fl}{EA}, \quad w = -(1 + 2\sqrt{2})\frac{Fl}{EA}. \tag{14.2}$$

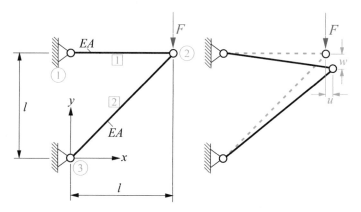

Fig. 14.3 Truss under point load F (left), deformed structure (right)

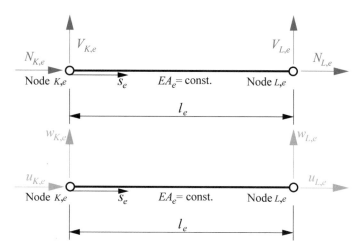

Fig. 14.4 Two-noded rod element

The negative sign for w results from the choice of the global reference system x, y.

We want to perform the calculation of the given truss in the framework of the FEM with a very elementary element type, namely the rod element shown in Fig. 14.4 which has two nodes. The two element nodes are designated as K and L. For the sake of a generally valid formulation we consider the rod element e. The force quantities in the nodes K, e and L, e are denoted by $N_{K,e}$ and $N_{L,e}$ for the normal forces and by $V_{K,e}$ and $V_{L,e}$ for the transverse shear forces. We refer to the displacements of the nodes K, e and L, e as $u_{K,e}$ and $u_{L,e}$ for the nodal displacements in the direction of the local reference axis s_e. Analogously, the nodal displacements perpendicular to the beam element axis are denoted as $w_{K,e}$ and $w_{L,e}$. The element has the length l_e and the constant extensional stiffness EA_e. In the present case, we want to represent or discretize each member of the truss with a single finite element. Thus, e currently assumes the values $e = 1$ and $e = 2$.

The FEM will lead to the following general form of an equation system. Since we want to apply a finite rod element with two nodes, we will have to determine statements about the displacements u_1, w_1, $u_2 = u$, $w_2 = w$, u_3 and w_3. This is contrasted by the forces H_1, V_1, H_2, V_2, H_3 and V_3 in the nodes. All these quantities are related to the global coordinate system x, y. The system of equations to be solved finally takes on the following form:

$$\underline{\underline{K}} \, \underline{v} = \underline{F}, \tag{14.3}$$

with:

$$\begin{bmatrix} k_{11} & k_{12} & k_{13} & k_{14} & k_{15} & k_{16} \\ k_{21} & k_{22} & k_{23} & k_{24} & k_{25} & k_{26} \\ k_{31} & k_{32} & k_{33} & k_{34} & k_{35} & k_{36} \\ k_{41} & k_{42} & k_{43} & k_{44} & k_{45} & k_{46} \\ k_{51} & k_{52} & k_{53} & k_{54} & k_{55} & k_{56} \\ k_{61} & k_{62} & k_{63} & k_{64} & k_{65} & k_{66} \end{bmatrix} \begin{pmatrix} u_1 \\ w_1 \\ u_2 \\ w_2 \\ u_3 \\ w_3 \end{pmatrix} = \begin{pmatrix} H_1 \\ V_1 \\ H_2 \\ V_2 \\ H_3 \\ V_3 \end{pmatrix}. \tag{14.4}$$

The matrix $\underline{\underline{K}}$ with the components k_{ij} is called the stiffness matrix of the system. The vector $\underline{v}$ contains all finite element nodal degrees of freedom, whereas the vector $\underline{F}$ contains the nodal forces. The nodal displacements and nodal forces in the framework of the FEM are shown in Fig. 14.5.

The support conditions of the truss under consideration allow the following statements about some of the displacement quantities:

$$u_1 = w_1 = u_3 = w_3 = 0. \tag{14.5}$$

The nodal displacements u_2 and w_2 are unknown and are the actual aim of the calculation. For the forces H_2 and V_2 the following results are obvious:

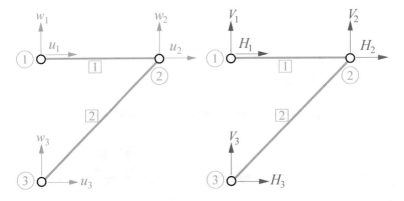

Fig. 14.5 Nodal displacements and forces

$$H_2 = 0, \quad V_2 = -F. \tag{14.6}$$

The remaining nodal forces, here interpretable as support reactions, will result at a later point from the FEM calculation. Obviously we first have to get clarity about the stiffness matrix $\underline{\underline{K}}$. The goal of the calculations is then to determine the two node displacements u_2 and w_2 as well as the forces H_1, V_1, H_3 and V_3 in a follow-up calculation.

For the rod element shown in Fig. 14.4 we now formulate the following approximation for the longitudinal displacement $u_e(s_e)$:

$$u_e(s_e) = u_{K,e}N_{1,e}(s_e) + u_{L,e}N_{2,e}(s_e). \tag{14.7}$$

The functions $N_{1,e}$ and $N_{2,e}$ which are shown in Fig. 14.6 are called element shape functions. The designation with the capital letter N is quite common in the literature and should not lead to confusion with the normal forces of the members. The shape functions must meet the following requirements:

$$N_{1,e}(s_e = 0) = 1, \quad N_{1,e}(s_e = l_e) = 0, \quad N_{2,e}(s_e = 0) = 0, \quad N_{2,e}(s_e = l_e) = 1. \tag{14.8}$$

They can therefore be formulated as:

$$N_{1,e}(s_e) = 1 - \frac{s_e}{l_e}, \quad N_{2,e}(s_e) = \frac{s_e}{l_e}. \tag{14.9}$$

This kind of approach means that between the nodal displacements $u_{K,e}$ and $u_{L,e}$ a linear interpolation is performed using the shape functions $N_{1,e}$ and $N_{2,e}$. This also means that this type of FEM modeling using rod elements with linear shape func-

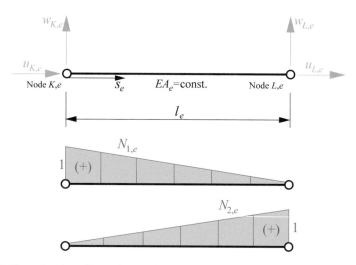

Fig. 14.6 Shape functions $N_{1,e}$ and $N_{2,e}$ of the two-noded rod element.

tions is a so-called natural discretization, since this kind of analysis leads to the exact results as they would also be obtained with the help of the calculation methods of elastostatics already discussed, provided that the structures in question are exclusively truss structures under discrete nodal loads. From Fig. 14.3 it can be concluded that in a truss under discrete loads in the nodes the members always remain straight and the deformations can be expressed exclusively by the node displacements, which is also expressed by the functions $N_{1,e}$ and $N_{2,e}$ (see also Fig. 12.8 and the explanations there in Chap. 12).

With (14.7) and the shape functions (14.9) we now formulate the total elastic potential Π of the rod element. The inner potential Π_i can be written as:

$$\Pi_i = \frac{1}{2}EA_e \int_0^{l_e} (u'_e)^2 \, ds_e = \frac{1}{2}EA_e \int_0^{l_e} \left(-\frac{u_{K,e}}{l_e} + \frac{u_{L,e}}{l_e}\right)^2 \, ds_e$$
$$= \frac{EA_e}{2l_e}\left(u^2_{K,e} - 2u_{K,e}u_{L,e} + u^2_{L,e}\right). \tag{14.10}$$

The external potential Π_a follows from the negative work of the nodal forces $N_{K,e}$, $V_{K,e}$, $N_{L,e}$ and $V_{L,e}$ performed along the nodal displacements $u_{K,e}$, $u_{L,e}$, $w_{K,e}$ and $w_{L,e}$. In an ideal truss which is loaded exclusively by nodal forces the transverse displacements $w_{K,e}$ and $w_{L,e}$ of the element will take place free of forces. The work of the forces $V_{K,e}$ and $V_{L,e}$ is therefore not considered in Π_a. Hence:

$$\Pi_a = -N_{K,e}u_{K,e} - N_{L,e}u_{L,e}. \tag{14.11}$$

Application of the principle of minimum total elastic potential $\delta\Pi = \delta(\Pi_i + \Pi_a) = 0$ yields:

$$\delta\Pi = \frac{1}{2EA_e l_e}\left(2u_{K,e}\delta u_{K,e} - 2\delta u_{K,e}u_{L,e} - 2u_{K,e}\delta u_{L,e} + 2u_{L,e}\delta u_{L,e}\right)$$
$$- N_{K,e}\delta u_{K,e} - N_{L,e}\delta u_{L,e} = 0. \tag{14.12}$$

Sorting according to the variations $\delta u_{K,e}$ and $\delta u_{L,e}$ results in:

$$\delta\Pi = \left[\frac{EA_e}{l_e}\left(u_{K,e} - u_{L,e}\right) - N_{K,e}\right]\delta u_{K,e}$$
$$+ \left[\frac{EA_e}{l_e}\left(-u_{K,e} + u_{L,e}\right) - N_{L,e}\right]\delta u_{L,e} = 0. \tag{14.13}$$

The two variations occurring here are arbitrary within the given boundary conditions and are also completely independent of each other. Therefore the variational statement can be fulfilled only if the two bracketed terms become zero:

$$\frac{EA_e}{l_e}\left(u_{K,e} - u_{L,e}\right) = N_{K,e},$$

$$\frac{EA_e}{l_e}\left(-u_{K,e} + u_{L,e}\right) = N_{L,e}. \qquad (14.14)$$

In a vector-matrix notation we get:

$$\frac{EA_e}{l_e}\begin{bmatrix} 1 & -1 \\ -1 & 1 \end{bmatrix}\begin{pmatrix} u_{K,e} \\ u_{L,e} \end{pmatrix} = \begin{pmatrix} N_{K,e} \\ N_{L,e} \end{pmatrix}. \qquad (14.15)$$

The same result would be obtained if the partial derivatives of Π with respect to the nodal displacements $u_{K,e}$ and $u_{L,e}$ were evaluated which would be equivalent to the Ritz equations:

$$\frac{\partial \Pi}{\partial u_{K,e}} = \frac{EA_e}{l_e}\left(u_{K,e} - u_{L,e}\right) - N_{K,e} = 0,$$

$$\frac{\partial \Pi}{\partial u_{L,e}} = \frac{EA_e}{l_e}\left(-u_{K,e} + u_{L,e}\right) - N_{L,e} = 0. \qquad (14.16)$$

Obviously, the finite element method can be interpreted as a special form of the Ritz method in such a way that the FEM consists of a Ritz formulation within each element. In contrast to the Ritz method, however, the constants have a clear physical meaning in the context of the FEM since they represent exactly the displacements in the element nodes of the finite element under consideration. On the other hand, the constants in the context of the Ritz method are usually not easily accessible to a clear interpretation.

The element formulation (14.15) for the two-node finite rod element can also be determined using the 1st theorem of Castigliano, as briefly shown below. It reads (cf. Eq. (11.102), Chap. 11):

$$\frac{\partial \Pi_i}{\partial u_i} = F_i. \qquad (14.17)$$

The inner potential Π_i of the element e is already available with (14.10) for the finite two-node rod element considered here. The normal force $N_{K,e}$ acts on node K, e, and the normal force $N_{L,e}$ is present on node L, e. Therefore the partial derivatives of the inner potential Π_i with respect to the two displacements $u_{K,e}$ and $u_{L,e}$ must correspond exactly to these forces. It follows:

$$\frac{\partial \Pi_{i,e}}{\partial u_{K,e}} = N_{K,e} = \frac{EA_e}{l_e}(u_{K,e} - u_{L,e}),$$

$$\frac{\partial \Pi_{i,e}}{\partial u_{L,e}} = N_{L,e} = \frac{EA_e}{l_e}(-u_{K,e} + u_{L,e}). \qquad (14.18)$$

This corresponds exactly to the already derived formulation (14.15).

In a truss, the transverse displacements $w_{K,e}$ and $w_{L,e}$ usually also occur at the nodes. In order to take these into account, the system of Eqs. (14.15) is extended by corresponding contributions:

$$\frac{EA_e}{l_e}\begin{bmatrix} 1 & 0 & -1 & 0 \\ 0 & 0 & 0 & 0 \\ -1 & 0 & 1 & 0 \\ 0 & 0 & 0 & 0 \end{bmatrix}\begin{pmatrix} u_{K,e} \\ w_{K,e} \\ u_{L,e} \\ w_{L,e} \end{pmatrix} = \begin{pmatrix} N_{K,e} \\ V_{K,e} \\ N_{L,e} \\ V_{L,e} \end{pmatrix}, \tag{14.19}$$

or in symbolic form:

$$\underline{\underline{\bar{K}}}_e\, \underline{\bar{v}}_e = \underline{\bar{F}}_e. \tag{14.20}$$

The matrix $\underline{\underline{\bar{K}}}_e$ is the so-called element stiffness matrix, the vector $\underline{\bar{v}}_e$ contains the nodal displacements of the element. The element force vector $\underline{\bar{F}}_e$ contains the nodal forces. The superscript indicates that these are element quantities which are related to the local element coordinate system s_e and still have to be transformed to the global coordinates x and y. This transformation to the global coordinate system x, y is shown in Fig. 14.7. We assume that element e is rotated by the angle α_e to the global coordinate system. From Fig. 14.7 we can then conclude the following relationship between the nodal displacements $u_{K,e}$ and $w_{K,e}$ of node K and the displacements u_K and w_K of node K in global coordinates:

$$u_{K,e} = u_K \cos(\alpha_e) + w_K \sin(\alpha_e), \quad w_{K,e} = -u_K \sin(\alpha_e) + w_K \cos(\alpha_e), \tag{14.21}$$

or:

$$\begin{bmatrix} \cos(\alpha_e) & \sin(\alpha_e) \\ -\sin(\alpha_e) & \cos(\alpha_e) \end{bmatrix}\begin{pmatrix} u_K \\ w_K \end{pmatrix} = \begin{pmatrix} u_{K,e} \\ w_{K,e} \end{pmatrix}. \tag{14.22}$$

We then have for (14.19):

$$\frac{EA_e}{l_e}\begin{bmatrix} 1 & 0 & -1 & 0 \\ 0 & 0 & 0 & 0 \\ -1 & 0 & 1 & 0 \\ 0 & 0 & 0 & 0 \end{bmatrix}\begin{bmatrix} \cos(\alpha_e) & \sin(\alpha_e) & 0 & 0 \\ -\sin(\alpha_e) & \cos(\alpha_e) & 0 & 0 \\ 0 & 0 & \cos(\alpha_e) & \sin(\alpha_e) \\ 0 & 0 & -\sin(\alpha_e) & \cos(\alpha_e) \end{bmatrix}\begin{pmatrix} u_K \\ w_K \\ u_L \\ w_L \end{pmatrix}$$

$$= \begin{pmatrix} N_{K,e} \\ V_{K,e} \\ N_{L,e} \\ V_{L,e} \end{pmatrix}, \tag{14.23}$$

Fig. 14.7 Transformation of the nodal displacements and forces

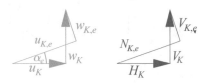

or:

$$\underline{\underline{\bar{K}}}_e \underline{\underline{T}}_e \underline{v}_e = \underline{\bar{F}}_e. \tag{14.24}$$

The matrix $\underline{\underline{T}}_e$ is called the transformation matrix of element e.

The transformation of the nodal forces $N_{K,e}$, $V_{K,e}$, $N_{L,e}$ and $V_{L,e}$ is performed analogously:

$$\begin{pmatrix} N_{K,e} \\ V_{K,e} \\ N_{L,e} \\ V_{L,e} \end{pmatrix} = \begin{bmatrix} \cos(\alpha_e) & \sin(\alpha_e) & 0 & 0 \\ -\sin(\alpha_e) & \cos(\alpha_e) & 0 & 0 \\ 0 & 0 & \cos(\alpha_e) & \sin(\alpha_e) \\ 0 & 0 & -\sin(\alpha_e) & \cos(\alpha_e) \end{bmatrix} \begin{pmatrix} N_K \\ Q_K \\ N_L \\ Q_L \end{pmatrix}, \tag{14.25}$$

or:

$$\underline{\bar{F}}_e = \underline{\underline{T}}_e \underline{F}_e. \tag{14.26}$$

We thus obtain for (14.24):

$$\underline{\underline{T}}_e^{-1} \underline{\underline{\bar{K}}}_e \underline{\underline{T}}_e \underline{v}_e = \underline{F}_e. \tag{14.27}$$

A transformation of the type $\underline{\underline{T}}_e^{-1} \underline{\underline{\bar{K}}}_e \underline{\underline{T}}_e$ is called a similarity transformation. If an orthogonal coordinate system is subjected to a pure rotation, then the transformation matrices in question result as so-called orthogonal matrices. Then the inverted matrix is identical with the transposed transformation matrix:

$$\underline{\underline{T}}_e^{-1} = \underline{\underline{T}}_e^{T}. \tag{14.28}$$

Thus:

$$\underline{\underline{T}}_e^{T} \underline{\underline{\bar{K}}}_e \underline{\underline{T}}_e \underline{v}_e = \underline{F}_e. \tag{14.29}$$

With

$$\underline{\underline{K}}_e = \underline{\underline{T}}_e^{T} \underline{\underline{\bar{K}}}_e \underline{\underline{T}}_e \tag{14.30}$$

we obtain:

$$\underline{\underline{K}}_e \underline{v}_e = \underline{F}_e. \tag{14.31}$$

Thus, the matrix $\underline{\underline{K}}_e$ is the stiffness matrix of the plane two-noded rod element related to the global coordinate system x, y. The vectors $\underline{v}_e$ and $\underline{F}_e$ are also related to the global coordinate system x, y. In full the stiffness matrix $\underline{\underline{K}}_e$ reads:

$$\underline{\underline{K}}_e = \frac{EA_e}{l_e} \begin{bmatrix} \cos^2(\alpha_e) & \cos(\alpha_e)\sin(\alpha_e) & -\cos^2(\alpha_e) & -\cos(\alpha_e)\sin(\alpha_e) \\ \cos(\alpha_e)\sin(\alpha_e) & \sin^2(\alpha_e) & -\cos(\alpha_e)\sin(\alpha_e) & -\sin^2(\alpha_e) \\ -\cos^2(\alpha_e) & -\cos(\alpha_e)\sin(\alpha_e) & \cos^2(\alpha_e) & \cos(\alpha_e)\sin(\alpha_e) \\ -\cos(\alpha_e)\sin(\alpha_e) & -\sin^2(\alpha_e) & \cos(\alpha_e)\sin(\alpha_e) & \sin^2(\alpha_e) \end{bmatrix}. \tag{14.32}$$

The stiffness matrix is always symmetrical.

The stiffness matrices of the elements at hand are later assembled to the stiffness matrix $\underline{\underline{K}}$ of the complete system (see also Eqs. (14.3) and (14.4)). Here, the forces occurring in a node are added together by summing the contributions of all elements connected to the node under consideration. Currently, each element e has two nodes K and L whose displacements determine the forces occurring in the element and thus the nodal contributions of the element. The assembly of the total stiffness matrix is thus done by adding up the contributions of all elements of the present FEM model at appropriate locations of the total stiffness matrix.

In the present example, two element matrices are to be formulated. In the case of rod 1, the angle $\alpha_1 = 0°$ applies so that the stiffness matrix $\underline{\underline{K}}_1$ is obtained as:

$$\underline{\underline{K}}_1 = \begin{bmatrix} k_{11}^1 & k_{12}^1 & k_{13}^1 & k_{14}^1 \\ k_{21}^1 & k_{22}^1 & k_{23}^1 & k_{24}^1 \\ k_{31}^1 & k_{32}^1 & k_{33}^1 & k_{34}^1 \\ k_{41}^1 & k_{42}^1 & k_{43}^1 & k_{44}^1 \end{bmatrix} = \frac{EA}{l} \begin{bmatrix} 1 & 0 & -1 & 0 \\ 0 & 0 & 0 & 0 \\ -1 & 0 & 1 & 0 \\ 0 & 0 & 0 & 0 \end{bmatrix}. \tag{14.33}$$

The superscript 1 is not an exponent, but rather indicates the number of the element under consideration. Element 1 is connected to nodes 1 and 2, node K, 1 is identical to node 1, and element node L, 1 corresponds to node 2, see also Fig. 14.8. Thus, the element stiffness matrix $\underline{\underline{K}}_1$ can be inserted into the system of Eqs. (14.4) as follows:

$$\begin{bmatrix} k_{11}^1 & k_{12}^1 & k_{13}^1 & k_{14}^1 & 0 & 0 \\ k_{21}^1 & k_{22}^1 & k_{23}^1 & k_{24}^1 & 0 & 0 \\ k_{31}^1 & k_{32}^1 & k_{33}^1 & k_{34}^1 & 0 & 0 \\ k_{41}^1 & k_{42}^1 & k_{43}^1 & k_{44}^1 & 0 & 0 \\ 0 & 0 & 0 & 0 & 0 & 0 \\ 0 & 0 & 0 & 0 & 0 & 0 \end{bmatrix} \begin{pmatrix} u_1 \\ w_1 \\ u_2 \\ w_2 \\ u_3 \\ w_3 \end{pmatrix} = \begin{pmatrix} H_1 \\ V_1 \\ H_2 \\ V_2 \\ H_3 \\ V_3 \end{pmatrix}. \tag{14.34}$$

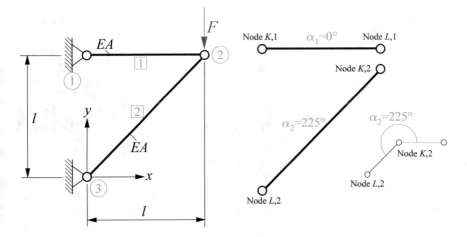

Fig. 14.8 Assignment of element nodes to system nodes

For element 2 with $\alpha_2 = 225°$ we have:

$$\underline{\underline{K}}_2 = \begin{bmatrix} k_{11}^2 & k_{12}^2 & k_{13}^2 & k_{14}^2 \\ k_{21}^2 & k_{22}^2 & k_{23}^2 & k_{24}^2 \\ k_{31}^2 & k_{32}^2 & k_{33}^2 & k_{34}^2 \\ k_{41}^2 & k_{42}^2 & k_{43}^2 & k_{44}^2 \end{bmatrix} = \frac{EA}{2\sqrt{2}l} \begin{bmatrix} 1 & 1 & -1 & -1 \\ 1 & 1 & -1 & -1 \\ -1 & -1 & 1 & 1 \\ -1 & -1 & 1 & 1 \end{bmatrix}. \tag{14.35}$$

At this element, element node $K, 2$ corresponds to node 2, and element node $L, 2$ coincides with node 3 of the system. Thus, the matrix $\underline{\underline{K}}_2$ can be integrated into the system of Eqs. (14.34) as follows:

$$\begin{bmatrix} k_{11}^1 & k_{12}^1 & k_{13}^1 & k_{14}^1 & 0 & 0 \\ k_{21}^1 & k_{22}^1 & k_{23}^1 & k_{24}^1 & 0 & 0 \\ k_{31}^1 & k_{32}^1 & k_{33}^1 + k_{11}^2 & k_{34}^1 + k_{12}^2 & k_{13}^2 & k_{14}^2 \\ k_{41}^1 & k_{42}^1 & k_{43}^1 + k_{21}^2 & k_{44}^1 + k_{22}^2 & k_{23}^2 & k_{24}^2 \\ 0 & 0 & k_{31}^2 & k_{32}^2 & k_{33}^2 & k_{34}^2 \\ 0 & 0 & k_{41}^2 & k_{42}^2 & k_{43}^2 & k_{44}^2 \end{bmatrix} \begin{pmatrix} u_1 \\ w_2 \\ u_2 \\ w_2 \\ u_3 \\ w_3 \end{pmatrix} = \begin{pmatrix} H_1 \\ V_1 \\ H_2 \\ V_2 \\ H_3 \\ V_3 \end{pmatrix}. \tag{14.36}$$

The two element matrices $\underline{\underline{K}}_1$ and $\underline{\underline{K}}_2$ are thus superimposed by appropriate placement at the corresponding points in the system of Eqs. (14.36) (see Fig. 14.9).

The system of Eqs. (14.36) has the form (14.3), thus:

$$\underline{\underline{K}}\upsilon = \underline{F}. \tag{14.37}$$

The matrix present in (14.36) is thus the total stiffness matrix $\underline{\underline{K}}$ of the FEM model composed of two members. The system of Eqs. (14.36) then allows the calculation of the nodal displacements, and it is occupied as follows:

Fig. 14.9 Assembly of the total stiffness matrix

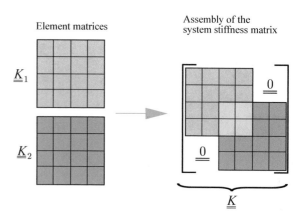

Element matrices

Assembly of the system stiffness matrix

$\underline{\underline{K}}_1$

$\underline{\underline{K}}_2$

$\underline{\underline{K}}$

$$
\underline{\underline{K}} = \frac{EA}{l}
\begin{bmatrix}
1 & 0 & -1 & 0 & 0 & 0 \\
0 & 0 & 0 & 0 & 0 & 0 \\
-1 & 0 & 1+\dfrac{1}{2\sqrt{2}} & \dfrac{1}{2\sqrt{2}} & -\dfrac{1}{2\sqrt{2}} & -\dfrac{1}{2\sqrt{2}} \\
0 & 0 & \dfrac{1}{2\sqrt{2}} & \dfrac{1}{2\sqrt{2}} & -\dfrac{1}{2\sqrt{2}} & -\dfrac{1}{2\sqrt{2}} \\
0 & 0 & -\dfrac{1}{2\sqrt{2}} & -\dfrac{1}{2\sqrt{2}} & \dfrac{1}{2\sqrt{2}} & \dfrac{1}{2\sqrt{2}} \\
0 & 0 & -\dfrac{1}{2\sqrt{2}} & -\dfrac{1}{2\sqrt{2}} & \dfrac{1}{2\sqrt{2}} & \dfrac{1}{2\sqrt{2}}
\end{bmatrix}
\begin{pmatrix}
u_1 \\ v_1 \\ u_2 \\ w_2 \\ u_3 \\ w_3
\end{pmatrix}
=
\begin{pmatrix}
H_1 \\ V_1 \\ 0 \\ -F \\ H_3 \\ V_3
\end{pmatrix},
$$

$$(14.38)$$

where we have already considered the conditions $H_2 = 0$ and $V_2 = -F$.

The total stiffness matrix $\underline{\underline{K}}$ always has the following properties:

- The stiffness matrix $\underline{\underline{K}}$ is always a symmetric matrix.
- Its main diagonal is always occupied by entries that have values greater than zero.
- It is always singular. This means that its determinant is identically zero. Thus, its inverse cannot be formed. As a consequence, the system of Eqs. (14.38) cannot be solved. The reason is that up to this point no boundary conditions of the system have been considered. Thus, the equation system (14.38) describes a total of three rigid body degrees of freedom, namely two translations as well as one rotation in the $x, y-$plane.
- In the second column and row all entries are identically zero. Thus, it would be possible that the vertical displacement w_1 of node 1 could occur force-free and, in addition, could assume arbitrary values, namely in the form of a rotation of the element 1 around the node 2. If all entries disappear in a row and column, this indicates that the degree of freedom concerned is a rigid-body degree of freedom. This means that the degree of freedom in question must be set to zero in the associated node so that the system does not become kinematically displaceable. However, this is ensured here by the support conditions of node 1 ($w_1 = 0$ due to the support at this location).

The conditions $u_1 = w_1 = u_3 = w_3 = 0$ are valid for the nodes 1 and 3 of the system. For this reason, columns 1, 2, 5 and 6 of the system of Eqs. (14.38) can be disregarded in its solution since the corresponding matrix elements would always be multiplied by zero. Moreover, the forces on the right side of Eqs. 1, 2, 5 and 6 are unknown, they represent the still unknown reaction forces of the system. Therefore, for the calculation of the nodal displacements, the lines 1, 2, 5 and 6 can also be neglected, they do not contribute to the solution of the system of equations. However, lines 1, 2, 5 and 6 can be used in a follow-up calculation to determine the support reactions and from them also the member forces of the truss, given that the nodal displacements u_2 and w_2 are known as will be shown later.

The remaining system of equations after deleting rows and columns 1, 2, 5, and 6 (see also Fig. 14.10) is as follows:

Fig. 14.10 Deleting rows and columns of the total stiffness matrix

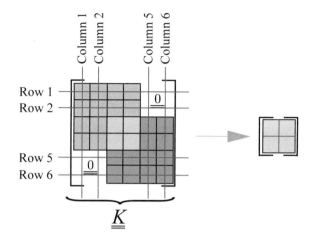

$$\frac{EA}{l}\begin{bmatrix} 1 + \dfrac{1}{2\sqrt{2}} & \dfrac{1}{2\sqrt{2}} \\[2mm] \dfrac{1}{2\sqrt{2}} & \dfrac{1}{2\sqrt{2}} \end{bmatrix}\begin{pmatrix} u_2 \\ w_2 \end{pmatrix} = \begin{pmatrix} 0 \\ -F \end{pmatrix}. \tag{14.39}$$

The system of equations to be finally solved therefore always has exactly as many rows and columns as there are nodal degrees of freedom to be determined. Accordingly, there are two equations for the two unknown nodal displacements u_2 and w_2. We obtain:

$$u_2 = \frac{Fl}{EA}, \quad w_2 = -(1 + 2\sqrt{2})\frac{Fl}{EA}. \tag{14.40}$$

Obviously, these are exactly the results that have already been obtained at the beginning of this section using the force method.

The still unknown nodal forces H_1, V_1, H_3 and V_3 can be calculated in a post-calculation from those equations that were deleted from the system of equations. The horizontal force H_1 follows from the first equation in (14.38) considering $u_1 = 0$:

$$H_1 = -\frac{EA}{l}u_2 = -\frac{EA}{l}\frac{Fl}{EA} = -F. \tag{14.41}$$

From the second equation in (14.38) we determine the force V_1 as:

$$V_1 = 0. \tag{14.42}$$

We proceed analogously for the two forces H_3 and V_3 and obtain:

$$H_3 = F, \quad V_3 = F. \tag{14.43}$$

Fig. 14.11 Determination of member forces

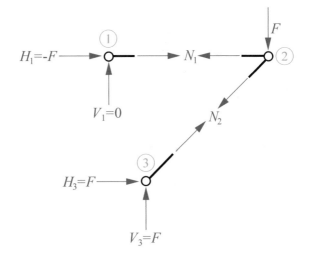

Thus, once the nodal forces are determined we can also determine the member forces of the truss from elementary equilibrium considerations at the nodes (see Fig. 14.11).

We obtain the following result:

$$N_1 = F, \quad N_2 = -\sqrt{2}F. \tag{14.44}$$

Alternatively, the nodal forces can also be transformed from the global coordinate system into the respective element reference systems using (14.25), and the member forces can then be determined. However, this remains without further explanation at this point.

14.2.2 Statically Indeterminate Trusses

The previously discussed truss is a statically determinate structure so that all support forces as well as all member forces can be determined from simple equilibrium considerations. However, the FEM procedure does not change even for statically indeterminate systems as we can easily see by extending the previous example. A statically indeterminate system can be generated from the system under consideration, for example, by inserting an additional bar which, starting from node 2, runs horizontally to the right and ends in an additional node 4 (Fig. 14.12). The general procedure within the framework of the FEM does not change in this case. We only have to take into account the additional degrees of freedom u_4 and w_4 as well as the element stiffness matrix of the newly added element 3. The resulting system of equations with the element stiffness matrix $\underline{\underline{K}}_3$

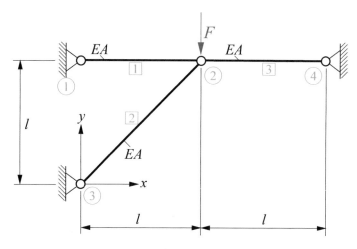

Fig. 14.12 Statically indeterminate truss

$$\underline{\underline{K}}_3 = \begin{bmatrix} k_{11}^3 & k_{12}^3 & k_{13}^3 & k_{14}^3 \\ k_{21}^3 & k_{22}^3 & k_{23}^3 & k_{24}^3 \\ k_{31}^3 & k_{32}^3 & k_{33}^3 & k_{34}^3 \\ k_{41}^3 & k_{42}^3 & k_{43}^3 & k_{44}^3 \end{bmatrix} = \frac{EA}{l} \begin{bmatrix} 1 & 0 & -1 & 0 \\ 0 & 0 & 0 & 0 \\ -1 & 0 & 1 & 0 \\ 0 & 0 & 0 & 0 \end{bmatrix} \tag{14.45}$$

is then occupied as follows:

$$\frac{EA}{l} \begin{bmatrix} 1 & 0 & -1 & 0 & 0 & 0 & 0 & 0 \\ 0 & 0 & 0 & 0 & 0 & 0 & 0 & 0 \\ -1 & 0 & 2+\dfrac{1}{2\sqrt{2}} & \dfrac{1}{2\sqrt{2}} & \dfrac{1}{2\sqrt{2}} & -\dfrac{1}{2\sqrt{2}} & -1 & 0 \\ 0 & 0 & \dfrac{1}{2\sqrt{2}} & \dfrac{1}{2\sqrt{2}} & -\dfrac{1}{2\sqrt{2}} & -\dfrac{1}{2\sqrt{2}} & 0 & 0 \\ 0 & 0 & -\dfrac{1}{2\sqrt{2}} & -\dfrac{1}{2\sqrt{2}} & \dfrac{1}{2\sqrt{2}} & \dfrac{1}{2\sqrt{2}} & 0 & 0 \\ 0 & 0 & -\dfrac{1}{2\sqrt{2}} & -\dfrac{1}{2\sqrt{2}} & \dfrac{1}{2\sqrt{2}} & \dfrac{1}{2\sqrt{2}} & 0 & 0 \\ 0 & 0 & -1 & 0 & 0 & 0 & 1 & 0 \\ 0 & 0 & 0 & 0 & 0 & 0 & 0 & 0 \end{bmatrix} \begin{pmatrix} u_1 \\ v_1 \\ u_2 \\ w_2 \\ u_3 \\ w_3 \\ u_4 \\ w_4 \end{pmatrix} = \begin{pmatrix} H_1 \\ V_1 \\ 0 \\ -F \\ H_3 \\ V_3 \\ H_4 \\ V_4 \end{pmatrix}.$$

$$\tag{14.46}$$

The additional entries due to the stiffness matrix $\underline{\underline{K}}_3$ occur in rows and columns 3, 4, 7, and 8 (see Fig. 14.13). The degrees of freedom u_1, w_1, u_2, w_2, u_4 and w_4 are zero with the given support conditions, and the associated forces are unknown. Therefore, rows and columns 1, 2, 5, 6, 7, and 8 can be discarded, and ultimately we obtain the following linear system of equations:

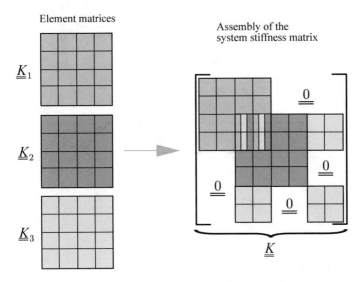

Fig. 14.13 Assembly of the stiffness matrix for the statically indeterminate truss

$$\frac{EA}{l}\begin{bmatrix} 2+\dfrac{1}{2\sqrt{2}} & \dfrac{1}{2\sqrt{2}} \\ \dfrac{1}{2\sqrt{2}} & \dfrac{1}{2\sqrt{2}} \end{bmatrix}\begin{pmatrix} u_2 \\ w_2 \end{pmatrix} = \begin{pmatrix} 0 \\ -F \end{pmatrix}. \tag{14.47}$$

The solution reads:

$$u_2 = \frac{Fl}{2EA}, \quad w_2 = -\frac{Fl}{2EA}\left(4\sqrt{2}+1\right). \tag{14.48}$$

The still unknown node and member forces can then be determined in a post-calculation which, however, will not be discussed further at this point.

14.2.3 Examples

For further illustration of the procedure consider the statically determinate truss shown in Fig. 14.14, consisting of four members of identical extensional stiffnesses EA, where each member is discretized with one finite element. The system of equations $\underline{K}\,\underline{v} = \underline{F}$ that ultimately results and must be solved will be of the following form:

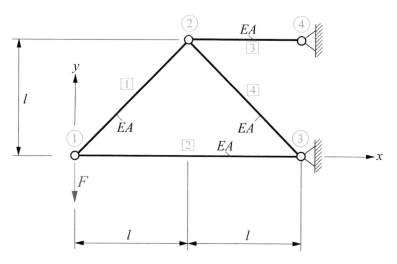

Fig. 14.14 Statically determinate truss, consisting of four members

$$
\begin{bmatrix} k_{11} \, k_{12} \, k_{13} \, k_{14} \, k_{15} \, k_{16} \, k_{17} \, k_{18} \\ k_{21} \, k_{22} \, k_{23} \, k_{24} \, k_{25} \, k_{26} \, k_{27} \, k_{28} \\ k_{31} \, k_{32} \, k_{33} \, k_{34} \, k_{35} \, k_{36} \, k_{37} \, k_{38} \\ k_{41} \, k_{42} \, k_{43} \, k_{44} \, k_{45} \, k_{46} \, k_{47} \, k_{48} \\ k_{51} \, k_{52} \, k_{53} \, k_{54} \, k_{55} \, k_{56} \, k_{57} \, k_{58} \\ k_{61} \, k_{62} \, k_{63} \, k_{64} \, k_{65} \, k_{66} \, k_{67} \, k_{68} \\ k_{71} \, k_{72} \, k_{73} \, k_{74} \, k_{75} \, k_{76} \, k_{77} \, k_{78} \\ k_{81} \, k_{82} \, k_{83} \, k_{84} \, k_{85} \, k_{86} \, k_{87} \, k_{88} \end{bmatrix}
\begin{pmatrix} u_1 \\ w_1 \\ u_2 \\ w_2 \\ u_3 \\ w_3 \\ u_4 \\ w_4 \end{pmatrix}
=
\begin{pmatrix} H_1 \\ V_1 \\ H_2 \\ V_2 \\ H_3 \\ V_3 \\ H_4 \\ V_4 \end{pmatrix} .
\tag{14.49}
$$

The following nodal conditions are given:

$$
u_3 = w_3 = u_4 = w_4 = 0,
\tag{14.50}
$$

and

$$
H_1 = 0, \quad V_1 = -F, \quad H_2 = 0, \quad V_2 = 0.
\tag{14.51}
$$

We first consider rod 1 whose element stiffness matrix $\underline{\underline{K}}_1$ is obtained with (14.32) and $\alpha_1 = 45°$ as follows:

$$
\underline{\underline{K}}_1 = \frac{EA}{2\sqrt{2}l}
\begin{bmatrix} 1 & 1 & -1 & -1 \\ 1 & 1 & -1 & -1 \\ -1 & -1 & 1 & 1 \\ -1 & -1 & 1 & 1 \end{bmatrix} .
\tag{14.52}
$$

Element 1 is connected to the nodes 1 and 2 so that the element stiffness matrix is arranged in rows and columns 1, 2, 3 and 4 of the total stiffness matrix $\underline{\underline{K}}$ in (14.49):

$$\underline{\underline{K}} = \frac{EA}{2\sqrt{2}l} \begin{bmatrix} 1 & 1 & -1 & -1 & 0 & 0 & 0 & 0 \\ 1 & 1 & -1 & -1 & 0 & 0 & 0 & 0 \\ -1 & -1 & 1 & 1 & 0 & 0 & 0 & 0 \\ -1 & -1 & 1 & 1 & 0 & 0 & 0 & 0 \\ 0 & 0 & 0 & 0 & 0 & 0 & 0 & 0 \\ 0 & 0 & 0 & 0 & 0 & 0 & 0 & 0 \\ 0 & 0 & 0 & 0 & 0 & 0 & 0 & 0 \\ 0 & 0 & 0 & 0 & 0 & 0 & 0 & 0 \end{bmatrix}. \tag{14.53}$$

The stiffness matrix $\underline{\underline{K}}_2$ of element 2 follows with $\alpha_2 = 0°$ as:

$$\underline{\underline{K}}_2 = \frac{EA}{2l} \begin{bmatrix} 1 & 0 & -1 & 0 \\ 0 & 0 & 0 & 0 \\ -1 & 0 & 1 & 0 \\ 0 & 0 & 0 & 0 \end{bmatrix}. \tag{14.54}$$

Element 2 is connected to nodes 1 and 3. Its entries are therefore found in rows and columns 1, 2, 5 and 6 of the total stiffness matrix $\underline{\underline{K}}$:

$$\underline{\underline{K}} = \frac{EA}{2\sqrt{2}l} \begin{bmatrix} 1+\sqrt{2} & 1 & -1 & -1 & -\sqrt{2} & 0 & 0 & 0 \\ 1 & 1 & -1 & -1 & 0 & 0 & 0 & 0 \\ -1 & -1 & 1 & 1 & 0 & 0 & 0 & 0 \\ -1 & -1 & 1 & 1 & 0 & 0 & 0 & 0 \\ -\sqrt{2} & 0 & 0 & 0 & \sqrt{2} & 0 & 0 & 0 \\ 0 & 0 & 0 & 0 & 0 & 0 & 0 & 0 \\ 0 & 0 & 0 & 0 & 0 & 0 & 0 & 0 \\ 0 & 0 & 0 & 0 & 0 & 0 & 0 & 0 \end{bmatrix}. \tag{14.55}$$

For element 3 we obtain the following stiffness matrix with $\alpha_3 = 0°$:

$$\underline{\underline{K}}_3 = \frac{EA}{l} \begin{bmatrix} 1 & 0 & -1 & 0 \\ 0 & 0 & 0 & 0 \\ -1 & 0 & 1 & 0 \\ 0 & 0 & 0 & 0 \end{bmatrix}. \tag{14.56}$$

Element 3 is connected to nodes 2 and 4. The entries of $\underline{\underline{K}}_3$ are therefore found in rows and columns 3, 4, 7 and 8 of the total stiffness matrix:

$$\underline{\underline{K}} = \frac{EA}{2\sqrt{2}l} \begin{bmatrix} 1+\sqrt{2} & 1 & -1 & -1 & -\sqrt{2} & 0 & 0 & 0 \\ 1 & 1 & -1 & -1 & 0 & 0 & 0 & 0 \\ -1 & -1 & 1+2\sqrt{2} & 1 & 0 & 0 & -2\sqrt{2} & 0 \\ -1 & -1 & 1 & 1 & 0 & 0 & 0 & 0 \\ -\sqrt{2} & 0 & 0 & 0 & \sqrt{2} & 0 & 0 & 0 \\ 0 & 0 & 0 & 0 & 0 & 0 & 0 & 0 \\ 0 & 0 & -2\sqrt{2} & 0 & 0 & 0 & 2\sqrt{2} & 0 \\ 0 & 0 & 0 & 0 & 0 & 0 & 0 & 0 \end{bmatrix}. \tag{14.57}$$

For element 4, the element angle $\alpha_4 = -45°$ is given. With the element length $\sqrt{2}l$ we obtain the following element stiffness matrix:

$$\underline{\underline{K}}_4 = \frac{EA}{2\sqrt{2}l} \begin{bmatrix} 1 & -1 & -1 & 1 \\ -1 & 1 & 1 & -1 \\ -1 & 1 & 1 & -1 \\ 1 & -1 & -1 & 1 \end{bmatrix}. \tag{14.58}$$

Element 4 is connected to system nodes 2 and 3, the entries of $\underline{\underline{K}}_4$ must therefore be added to rows and columns 3, 4, 5 and 6 of the total stiffness matrix $\underline{\underline{K}}$:

$$\underline{\underline{K}} = \frac{EA}{2\sqrt{2}l} \begin{bmatrix} 1+\sqrt{2} & 1 & -1 & -1 & -\sqrt{2} & 0 & 0 & 0 \\ 1 & 1 & -1 & -1 & 0 & 0 & 0 & 0 \\ -1 & -1 & 2+2\sqrt{2} & 0 & -1 & 1 & -2\sqrt{2} & 0 \\ -1 & -1 & 0 & 2 & 1 & -1 & 0 & 0 \\ -\sqrt{2} & 0 & -1 & 1 & 1+\sqrt{2} & -1 & 0 & 0 \\ 0 & 0 & 1 & -1 & -1 & 1 & 0 & 0 \\ 0 & 0 & -2\sqrt{2} & 0 & 0 & 0 & 2\sqrt{2} & 0 \\ 0 & 0 & 0 & 0 & 0 & 0 & 0 & 0 \end{bmatrix}. \tag{14.59}$$

To solve the system of Eqs. (14.49) with (14.59), the rows and columns 5, 6, 7 and 8 can be discarded due to the nodal conditions (14.50). Thus, the following system of equations remains to be solved:

$$\frac{EA}{2\sqrt{2}l} \begin{bmatrix} 1+\sqrt{2} & 1 & -1 & -1 \\ 1 & 1 & -1 & -1 \\ -1 & -1 & 2+2\sqrt{2} & 0 \\ -1 & -1 & 0 & 2 \end{bmatrix} \begin{pmatrix} u_1 \\ w_1 \\ u_2 \\ w_2 \end{pmatrix} = \begin{pmatrix} 0 \\ -F \\ 0 \\ 0 \end{pmatrix}. \tag{14.60}$$

Herein (14.51) was considered. The nodal displacements are:

$$\begin{pmatrix} u_1 \\ w_1 \\ u_2 \\ w_2 \end{pmatrix} = \frac{Fl}{EA} \begin{pmatrix} 2 \\ -(6+4\sqrt{2}) \\ -2 \\ -2(1+\sqrt{2}) \end{pmatrix}. \tag{14.61}$$

The reaction forces H_3, V_3, H_4, and V_4 can be obtained from lines 5, 6, 7, and 8 of the original system of Eqs. (14.49):

$$H_3 = \frac{EA}{2\sqrt{2}l} \left(-\sqrt{2}\frac{2Fl}{EA} + \frac{2Fl}{EA} - 2(1+\sqrt{2})\frac{Fl}{EA} \right) = -2F,$$

$$V_3 = \frac{EA}{2\sqrt{2}l} \left(-\frac{2Fl}{EA} + 2(1+\sqrt{2})\frac{Fl}{EA} \right) = F,$$

Fig. 14.15 Determination of member forces

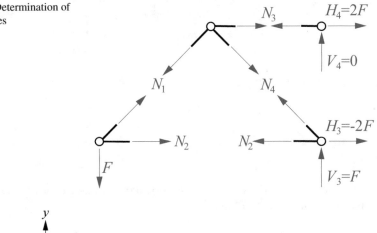

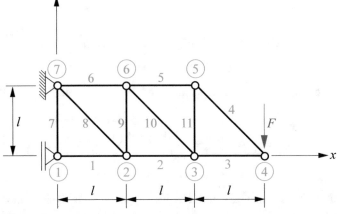

Fig. 14.16 Statically determinate truss

$$H_4 = \frac{EA}{2\sqrt{2}l} 2\sqrt{2} \frac{2Fl}{EA} = 2F. \tag{14.62}$$

The vertical force V_4 is zero: $V_4 = 0$.

The member forces can be determined with the now known reaction forces from observation of equilibrium at nodal intersections (Fig. 14.15). We obtain:

$$N_1 = \sqrt{2}F, \quad N_2 = -F, \quad N_3 = 2F, \quad N_4 = -\sqrt{2}F. \tag{14.63}$$

As a final illustrative example, the truss of Fig. 12.8, Chap. 12, is considered. It is shown in Fig. 14.16 again for clarity. Again, one two-noded rod element is used per truss member.

The stiffness matrices of the individual elements can be set up using (14.32). The elements 1, 2 and 3 have the orientation angle $\alpha = 0°$, so that:

$$\underline{K}_1 = \underline{K}_2 = \underline{K}_3 = \frac{EA}{l} \begin{bmatrix} 1 & 0 & -1 & 0 \\ 0 & 0 & 0 & 0 \\ -1 & 0 & 1 & 0 \\ 0 & 0 & 0 & 0 \end{bmatrix}. \tag{14.64}$$

Elements 5 and 6 are oriented under $\alpha = 180°$:

$$\underline{K}_5 = \underline{K}_6 = \frac{EA}{l} \begin{bmatrix} 1 & 0 & -1 & 0 \\ 0 & 0 & 0 & 0 \\ -1 & 0 & 1 & 0 \\ 0 & 0 & 0 & 0 \end{bmatrix}. \tag{14.65}$$

For elements 7, 9 and 11 with the orientation angles $\alpha = 90°$ we obtain the following element stiffness matrices:

$$\underline{K}_7 = \underline{K}_9 = \underline{K}_{11} = \frac{EA}{l} \begin{bmatrix} 0 & 0 & 0 & 0 \\ 0 & 1 & 0 & -1 \\ 0 & 0 & 0 & 0 \\ 0 & -1 & 0 & 1 \end{bmatrix}. \tag{14.66}$$

The stiffness matrices of elements 4, 8, and 10 with the angle $\alpha = 135°$ follow as:

$$\underline{K}_4 = \underline{K}_8 = \underline{K}_{10} = \frac{EA}{2\sqrt{2}l} \begin{bmatrix} 1 & -1 & -1 & 1 \\ -1 & 1 & 1 & -1 \\ -1 & 1 & 1 & -1 \\ 1 & -1 & -1 & 1 \end{bmatrix}. \tag{14.67}$$

The total stiffness matrix $\underline{K}$ is then obtained by assembling the individual element matrices to the total stiffness matrix. A detailed explanation of this process is omitted here. The total stiffness matrix can be given with the abbreviation $\beta = \frac{1}{2\sqrt{2}}$ as:

$$\underline{K} = \frac{EA}{l}$$

$$\times \begin{bmatrix}
1 & 0 & -1 & 0 & 0 & 0 & 0 & 0 & 0 & 0 & 0 & 0 & 0 & 0 \\
0 & 1 & 0 & 0 & 0 & 0 & 0 & 0 & 0 & 0 & 0 & 0 & 0 & -1 \\
-1 & 0 & 2+\beta & -\beta & -1 & 0 & 0 & 0 & 0 & 0 & 0 & 0 & -\beta & \beta \\
0 & 0 & -\beta & 1+\beta & 0 & 0 & 0 & 0 & 0 & 0 & 0 & -1 & \beta & -\beta \\
0 & 0 & -1 & 0 & 2+\beta & -\beta & -1 & 0 & 0 & 0 & -\beta & \beta & 0 & 0 \\
0 & 0 & 0 & 0 & -\beta & 1+\beta & 0 & 0 & 0 & -1 & \beta & -\beta & 0 & 0 \\
0 & 0 & 0 & 0 & -1 & 0 & 1+\beta & -\beta & -\beta & \beta & 0 & 0 & 0 & 0 \\
0 & 0 & 0 & 0 & 0 & 0 & -\beta & \beta & \beta & -\beta & 0 & 0 & 0 & 0 \\
0 & 0 & 0 & 0 & 0 & 0 & -\beta & \beta & 1+\beta & -\beta & -1 & 0 & 0 & 0 \\
0 & 0 & 0 & 0 & 0 & -1 & \beta & -\beta & -\beta & 1+\beta & 0 & 0 & 0 & 0 \\
0 & 0 & 0 & 0 & 0 & 0 & 0 & 0 & -1 & 0 & 2+\beta & -\beta & -1 & 0 \\
0 & 0 & 0 & -1 & \beta & -\beta & 0 & 0 & 0 & 0 & -\beta & 1+\beta & 0 & 0 \\
0 & 0 & -\beta & \beta & 0 & 0 & 0 & 0 & 0 & 0 & -1 & 0 & 1+\beta & -\beta \\
0 & -1 & \beta & -\beta & 0 & 0 & 0 & 0 & 0 & 0 & 0 & 0 & -\beta & 1+\beta
\end{bmatrix}.$$

With the nodal conditions $u_1 = u_7 = w_7 = 0$, rows and columns 1, 13 and 14 can be omitted in the global equation system $\underline{K}\underline{v} = \underline{F}$. With the force vector

$$\underline{F} = (0, 0, 0, 0, 0, 0, -F, 0, 0, 0, 0)^T \tag{14.68}$$

we obtain the following solution vector:

$$
\begin{pmatrix} w_1 \\ u_2 \\ w_2 \\ u_3 \\ w_3 \\ u_4 \\ w_4 \\ u_5 \\ w_5 \\ u_6 \\ w_6 \end{pmatrix} = \frac{Fl}{EA}
\begin{pmatrix} 0 \\ -3 \\ -\left(3 + 2\sqrt{2}\right) \\ -5 \\ -\left(11 + 4\sqrt{2}\right) \\ -6 \\ -\left(21 + 6\sqrt{2}\right) \\ 3 \\ -\left(12 + 4\sqrt{2}\right) \\ 2 \\ -\left(4 + 2\sqrt{2}\right) \end{pmatrix}. \tag{14.69}
$$

Obviously these results agree with the values already determined in Chap. 12. The differences of the signs result from the different assumptions of the positive directions.

14.3   Finite Elements for Plane Systems of Straight Rods

14.3.1   The Two-Noded Rod Element

We now want to extend the explanations to systems of plane straight members which are not necessarily ideally hinged at intersection points and which can also be subjected to loads other than ideal nodal loads. As an example, we consider a bar of length l and of arbitrarily variable, but continuously distributed extensional stiffness $EA(x)$, which is loaded by the arbitrarily, but continuously distributed axial line load $n(x)$ (Fig. 14.17) as well as the point load F at the free end. The procedure within the framework of the FEM is then such that the considered member is subdivided over its length into a certain number of rod elements. In each of these elements, an approach for the element displacements is then formulated as already shown in the case of truss structures. Again, we will initially restrict the considerations to rod elements that have two nodes, so that a linear interpolation is again performed between the two occurring nodal displacements $u_{K,e}$ and $u_{L,e}$ using the shape functions $N_{1,e}(s_e)$ and $N_{2,e}(s_e)$. The procedure is shown schematically in Fig. 14.18. By this procedure, the displacement field u occurring in the straight member can be approximated polygonally. In the following element formulations, we assume that the line load n and the

Fig. 14.17 Rod under axial line load and point load

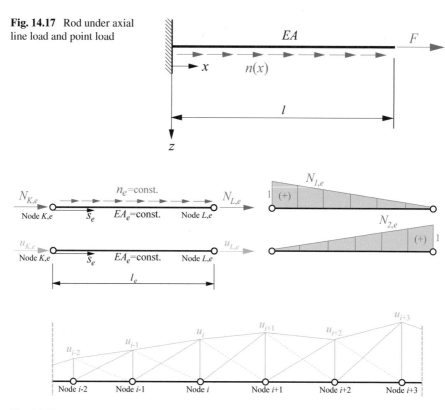

Fig. 14.18 Two-noded finite rod element under the axial line load n (top), approximation of the axial member displacement u (bottom).

extensional stiffness EA may well be variable over x, but are assumed to be constant within a single finite element.

The displacement approach to be used here is identical to the approach already formulated in the case of rod elements without line load (Eq. 14.7). It reads:

$$u_e(s_e) = u_{K,e} N_{1,e}(s_e) + u_{L,e} N_{2,e}(s_e), \tag{14.70}$$

where the shape functions $N_{1,e}$ and $N_{2,e}$ are defined as follows:

$$N_{1,e}(s_e) = 1 - \frac{s_e}{l_e}, \quad N_{2,e}(s_e) = \frac{s_e}{l_e}. \tag{14.71}$$

The starting point of the considerations is again the principle of the minimum of the elastic total potential. The inner potential of the rod element e is identical with (14.10) and reads:

$$\Pi_i = \frac{EA_e}{2l_e}\left(u_{K,e}^2 - 2u_{K,e}u_{L,e} + u_{L,e}^2\right). \tag{14.72}$$

The external potential Π_a now includes not only the negative work of the nodal forces $N_{K,e}$ and $N_{L,e}$ along the nodal displacements $u_{K,e}$ and $u_{L,e}$, but the line load n_e, assumed to be constant, must also be included here:

$$\Pi_a = -N_{K,e}u_{K,e} - N_{L,e}u_{L,e} - \int_0^{l_e} n_e u_e(s_e)\mathrm{d}s_e. \tag{14.73}$$

Performing the integration leads to:

$$\Pi_a = -N_{K,e}u_{K,e} - N_{L,e}u_{L,e} - \frac{1}{2}n_e l_e(u_{K,e} + u_{L,e}). \tag{14.74}$$

The principle of the minimum of the total elastic potential $\delta\Pi = \delta(\Pi_i + \Pi_a) = 0$ then results in:

$$\delta\Pi = \frac{EA_e}{2l_e}\left(2u_{K,e}\delta u_{K,e} - 2\delta u_{K,e}u_{L,e} - 2u_{K,e}\delta u_{L,e} + 2u_{L,e}\delta u_{L,e}\right)$$

$$- N_{K,e}\delta u_{K,e} - N_{L,e}\delta u_{L,e} - \frac{1}{2}n_e l_e\delta u_{K,e} - \frac{1}{2}n_e l_e\delta u_{L,e} = 0. \tag{14.75}$$

Sorting by the variations $\delta u_{K,e}$ and $\delta u_{L,e}$ yields:

$$\left[\frac{EA_e}{l_e}\left(u_{K,e} - u_{L,e}\right) - N_{K,e} - \frac{1}{2}n_e l_e\right]\delta u_{K,e}$$

$$+ \left[\frac{EA_e}{l_e}\left(-u_{K,e} + u_{L,e}\right) - N_{L,e} - \frac{1}{2}n_e l_e\right]\delta u_{L,e} = 0. \tag{14.76}$$

Equating the bracket terms to zero gives the element equations as:

$$\frac{EA_e}{l_e}\left(u_{K,e} - u_{L,e}\right) = N_{K,e} + \frac{1}{2}n_e l_e,$$

$$\frac{EA_e}{l_e}\left(-u_{K,e} + u_{L,e}\right) = N_{L,e} + \frac{1}{2}n_e l_e. \tag{14.77}$$

In vector matrix form we get:

$$\frac{EA_e}{l_e}\begin{bmatrix} 1 & -1 \\ -1 & 1 \end{bmatrix}\begin{pmatrix} u_{K,e} \\ u_{L,e} \end{pmatrix} = \begin{pmatrix} N_{K,e} + \frac{1}{2}n_e l_e \\ N_{L,e} + \frac{1}{2}n_e l_e \end{pmatrix}. \tag{14.78}$$

The system of Eqs. (14.78) can be interpreted as converting the line load n_e, assumed as constant, into statically equivalent nodal forces with $\frac{1}{2}n_e l_e$ applied in the

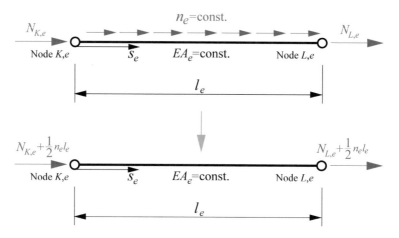

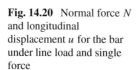

Fig. 14.19 Converting the line load n_e into statically equivalent nodal forces

Fig. 14.20 Normal force N and longitudinal displacement u for the bar under line load and single force

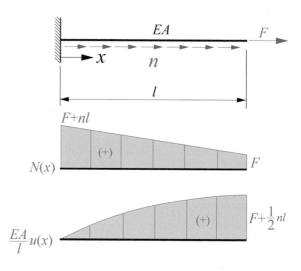

element nodes (see Fig. 14.19) and added to the nodal forces $N_{K,e}$ und $N_{L,e}$. The procedure is illustrated using the example of Fig. 14.20. Let the line load n and the extensional stiffness EA be constant over the rod length l. Based on the differential equation $EAu'' = -n$, an exact solution of the given rod problem can be derived as:

$$N = n(l - x) + F, \quad u = \frac{1}{EA}\left(-\frac{1}{2}nx^2 + (F + nl)x\right). \tag{14.79}$$

Obviously, the distribution of the normal force N results to be linearly dependent on x, whereas the longitudinal displacement is parabolic over x. The displacement at the rod end amounts to

$$u(x = l) = \frac{l}{EA}\left(F + \frac{1}{2}nl\right).$$ (14.80)

The distribution of the state variables is shown in Fig. 14.20.

We now solve the problem using the finite element method with the help of the element formulation derived above and start the considerations with a single finite rod element over the entire length l (Fig. 14.21). The system of equations to be solved then corresponds to the element equation system (14.78), where the labels 1 and 2 were applied for the nodes:

$$\frac{EA}{l}\begin{bmatrix} 1 & -1 \\ -1 & 1 \end{bmatrix}\begin{pmatrix} u_1 \\ u_2 \end{pmatrix} = \begin{pmatrix} N_1 + \frac{1}{2}nl \\ F + \frac{1}{2}nl \end{pmatrix}.$$ (14.81)

Here $n_1 = n$, $EA_1 = EA$, $l_1 = l$ were employed. The force at node 2 must then correspond to the external point force F plus the fraction $\frac{1}{2}nl$ due to the line load n. Node 1 is completely blocked, therefore the nodal displacement u_1 must be set to zero. Moreover, the still unknown reaction force N_1 occurs at this node, and therefore to solve the system of equations the first row and column may be disregarded. Accordingly, we have:

$$\frac{EA}{l}u_2 = F + \frac{1}{2}nl.$$ (14.82)

Fig. 14.21 Discretization with one rod element

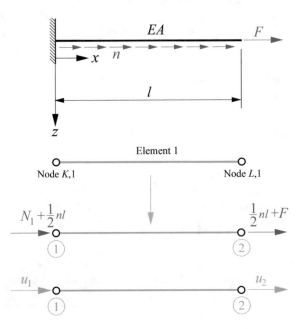

This can be solved immediately for the displacement u_2:

$$u_2 = \frac{l}{EA}\left(F + \frac{1}{2}nl\right). \qquad (14.83)$$

From the first line in (14.81), the unknown nodal force N_1 can then be determined in a post calculation:

$$N_1 = -F - nl. \qquad (14.84)$$

The results of the FEM analysis can be interpreted as follows. The approach (14.70) with (14.71) allows a linearly varying representation of the displacement u over the rod length, but the exact solution (14.79) results in a parabolic distribution. It is remarkable, however, that the nodal values of the displacement u always correspond to the exact solution. The course of the displacements is thus approximated over the rod length with a linear distribution, which can be understood as a linear interpolation between the exact nodal values (Fig. 14.22). The FEM thus underestimates the actual displacement distribution and overestimates the stiffness of the system.

Furthermore, it is important to investigate the normal force distribution $N(x)$ according to the FEM. The exact solution leads to a linear distribution of $N(x)$, but the FEM analysis assumes constant normal forces due to the linear displacement approach. This is shown in Fig. 14.23. The nodal force N_1 which results as negative and thus as tensile force is added to the portion $\frac{1}{2}nl$ from the line load n. The resulting tensile force then takes the value $F + \frac{1}{2}nl$ at the left end of the rod, which also corresponds exactly to the resulting force at the right end. For reasons of equilibrium, this is a reasonable result. Apparently, then, the FEM assumes a normal force constant across the length of the rod with magnitude $F + \frac{1}{2}nl$, which is shown in Fig. 14.23.

Fig. 14.22 Comparison of displacement u after exact solution and FEM analysis for the rod under line load and single force

Fig. 14.23 Comparison of the normal force N after exact solution and FEM analysis for the rod under line load and single force

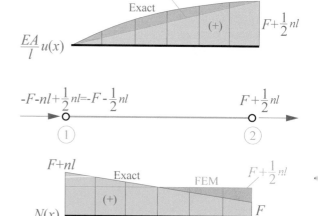

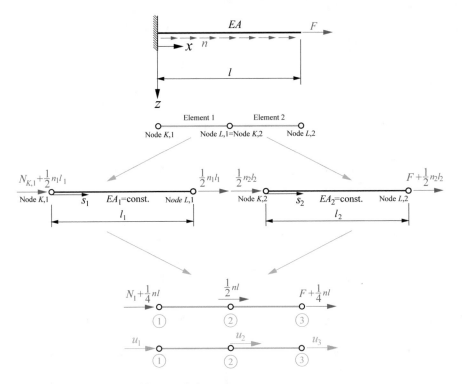

Fig. 14.24 Discretization with two rod elements

We now extend the considerations to the use of two elements of equal length. The discretization is given in Fig. 14.24. The element equations are then:

$$\frac{EA_1}{l_1}\begin{bmatrix} 1 & -1 \\ -1 & 1 \end{bmatrix}\begin{pmatrix} u_{K,1} \\ u_{L,1} \end{pmatrix} = \begin{pmatrix} N_{K,1} + \frac{1}{2}n_1 l_1 \\ N_{L,1} + \frac{1}{2}n_1 l_1 \end{pmatrix},$$

$$\frac{EA_2}{l_2}\begin{bmatrix} 1 & -1 \\ -1 & 1 \end{bmatrix}\begin{pmatrix} u_{K,2} \\ u_{L,2} \end{pmatrix} = \begin{pmatrix} N_{K,2} + \frac{1}{2}n_2 l_2 \\ N_{L,2} + \frac{1}{2}n_2 l_2 \end{pmatrix}. \qquad (14.85)$$

The equation system for the entire rod reads with $EA_1 = EA_2 = EA$, $l_1 = l_2 = \frac{l}{2}$, $u_{K,1} = u_1$, $u_{L,1} = u_{K,2} = u_2$, $u_{L,2} = u_3$ and $N_{K,1} = N_1 + \frac{1}{4}nl$ and $N_{L,1} = \frac{1}{4}nl$, $N_{K,2} = \frac{1}{4}nl$, $N_{L,2} = F + \frac{1}{4}nl$:

$$\frac{2EA}{l}\begin{bmatrix} 1 & -1 & 0 \\ -1 & 2 & -1 \\ 0 & -1 & 1 \end{bmatrix}\begin{pmatrix} u_1 \\ u_2 \\ u_3 \end{pmatrix} = \begin{pmatrix} N_1 + \frac{1}{4}nl \\ \frac{1}{2}nl \\ F + \frac{1}{4}nl \end{pmatrix}. \tag{14.86}$$

Discarding the first row and column provides:

$$\frac{2EA}{l}\begin{bmatrix} 2 & -1 \\ -1 & 1 \end{bmatrix}\begin{pmatrix} u_2 \\ u_3 \end{pmatrix} = \begin{pmatrix} \frac{1}{2}nl \\ F + \frac{1}{4}nl \end{pmatrix}. \tag{14.87}$$

We then obtain the nodal displacements as:

$$u_2 = \frac{l}{EA}\left(\frac{F}{2} + \frac{3nl}{8}\right), \quad u_3 = \frac{l}{EA}\left(F + \frac{nl}{2}\right). \tag{14.88}$$

Again, the nodal displacements correspond to the displacements after the exact solution.

The unknown nodal force $N_{K,1} = N_1$ can be obtained from the first line in (14.86) in a post-calculation:

$$N_{K,1} = -F - nl. \tag{14.89}$$

The distributions of the state variables are shown in Fig. 14.25. Obviously, the approximation of the displacements u is much better when using two rod elements than when using a single element. The normal force line is again approximated by

Fig. 14.25 Comparison of the state variables u and N after exact solution and FEM approximation using two rod elements

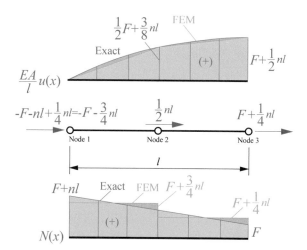

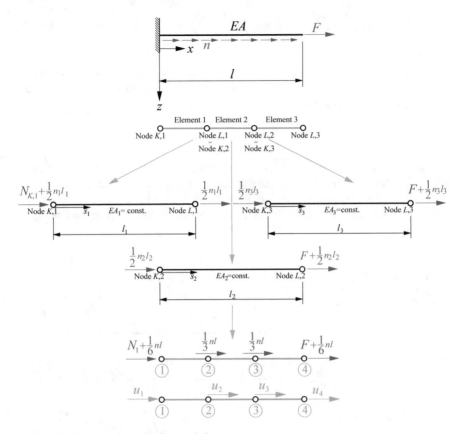

Fig. 14.26 Discretization with three rod elements

section-wise constant values, so that in the case of the normal force a less performant convergence of the results must be assumed than for the displacements.

We now briefly discuss the results obtained when three rod elements of equal length are used. The corresponding discretization is shown in Fig. 14.26. The system of equations to be solved here results as:

$$\frac{2EA}{l} \begin{bmatrix} 1 & -1 & 0 & 0 \\ -1 & 2 & -1 & 0 \\ 0 & -1 & 1 & -1 \\ 0 & 0 & -1 & 1 \end{bmatrix} \begin{pmatrix} u_1 \\ u_2 \\ u_3 \\ u_4 \end{pmatrix} = \begin{pmatrix} N_1 + \dfrac{1}{6}nl \\ \dfrac{1}{3}nl \\ \dfrac{1}{3}nl \\ F + \dfrac{1}{6}nl \end{pmatrix}. \tag{14.90}$$

Deleting the first row and column then ultimately leads to the following solution:

$$u_2 = \frac{l}{EA}\left(\frac{F}{3} + \frac{5nl}{18}\right), \quad u_3 = \frac{l}{EA}\left(\frac{2F}{3} + \frac{4nl}{9}\right), \quad u_4 = \frac{l}{EA}\left(F + \frac{nl}{2}\right). \quad (14.91)$$

Again, these values correspond to the exact values of the displacements at the locations $x = \frac{l}{3}$, $x = \frac{2l}{3}$ and $x = l$. A post-calculation gives the following nodal force $N_{K,1} = N_1$ from the first line in (14.90):

$$N_{K,1} = -F - nl. \quad (14.92)$$

The displacements and normal forces according to the FEM are shown in Fig. 14.27. As another example, consider the situation of Fig. 5.35, Chap. 5. We consider a rod that is firmly clamped on both sides and which is loaded by a single force F at the point $x = l_1$ (Fig. 14.28, top). The exact solution for the rod displace-

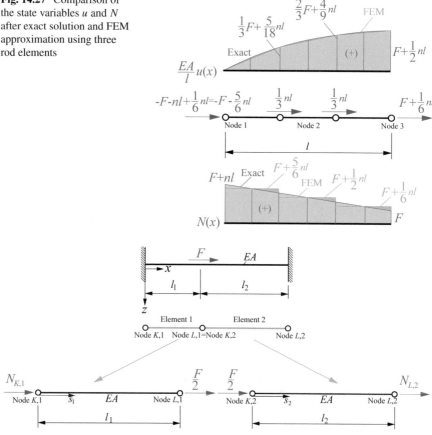

Fig. 14.27 Comparison of the state variables u and N after exact solution and FEM approximation using three rod elements

Fig. 14.28 Given structure (top), discretization with two rod elements (bottom)

Fig. 14.29 Exact solution
for $N(x)$ and $u(x)$

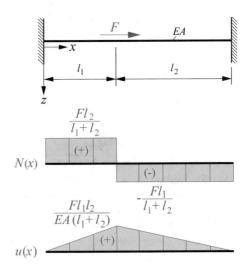

ment u and the normal force N, obtained in Chap. 5 by integrating the constitutive
relation $EAu''(x) = -n$, is shown again for clarity in Fig. 14.29.

We now solve the given problem using the finite element method and use two rod
elements as shown in Fig. 14.28, bottom. The system equations here are given by:

$$
EA \begin{bmatrix} \dfrac{1}{l_1} & -\dfrac{1}{l_1} & 0 \\[2ex] -\dfrac{1}{l_1} & \dfrac{1}{l_1} + \dfrac{1}{l_2} & -\dfrac{1}{l_2} \\[2ex] 0 & -\dfrac{1}{l_2} & \dfrac{1}{l_2} \end{bmatrix} \begin{pmatrix} u_1 \\ u_2 \\ u_3 \end{pmatrix} = \begin{pmatrix} N_1 \\ F \\ N_3 \end{pmatrix}. \tag{14.93}
$$

The displacements u_1 and u_3 must become zero because of the clamped ends.
Thus, the first and the third row and column can be discarded, which leads directly
to the displacement u_2 as follows:

$$
u_2 = \frac{Fl_1 l_2}{EA\,(l_1 + l_2)}. \tag{14.94}
$$

From the first and third lines of the system of Eqs. (14.93), the two forces $N_1 =
N_{K,1}$ and $N_3 = N_{L,2}$ can then be calculated:

$$
N_1 = -\frac{Fl_2}{l_1 + l_2}, \qquad N_3 = -\frac{Fl_1}{l_1 + l_2}. \tag{14.95}
$$

It is easy to see that these are, in contrast to the previous example, the exact results.
This can be justified by the linear displacement approach in the element formulation

chosen here as well as by the resulting constant distribution of the normal force in each element, which corresponds exactly to the conditions present in the given example.

14.3.2 The Three-Noded Rod Element

To achieve a higher accuracy of a finite element calculation, elements with an arbitrary number of nodes can be developed. As the number of nodes increases, higher order shape functions become necessary. It can be expected that the quality of the calculation results also improves by the use of higher-order shape functions. However, it must be accepted as a disadvantage that this goes hand in hand with an increased computational effort.

In the following, we want to derive the element equations for a three-noded rod element, where we assume a constant line load n_e as element load. The element is shown in Fig. 14.30. We assume the element nodes to be equidistant, thus dividing the element of length l_e into two halves of lengths $\frac{l_e}{2}$. The stiffness EA_e of the element is also assumed to be constant. The three-node rod element has three nodes K, L and M in which the nodal degrees of freedom $u_{K,e}$, $u_{L,e}$ and $u_{M,e}$ occur. Let the associated nodal forces be denoted as $N_{K,e}$, $N_{L,e}$ and $N_{M,e}$.

Let the displacement approach used here be of the following form:

$$u_e(s_e) = u_{K,e}N_{1,e}(s_e) + u_{L,e}N_{2,e}(s_e) + u_{M,e}N_{3,e}(s_e). \qquad (14.96)$$

The shape functions $N_{1,e}(s_e)$, $N_{2,e}(s_e)$, and $N_{3,e}(s_e)$ must then satisfy the following requirements:

$$N_{1,e}\left(s_e = 0\right) = 1, \quad N_{1,e}\left(s_e = \frac{l_e}{2}\right) = 0, \quad N_{1,e}\left(s_e = l_e\right) = 0,$$

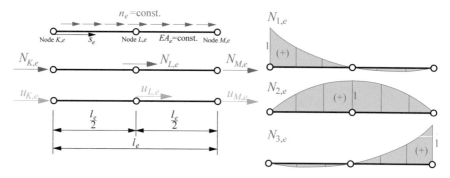

Fig. 14.30 Three-noded rod element

$$N_{2,e}\left(s_e = 0\right) = 0, \quad N_{2,e}\left(s_e = \frac{l_e}{2}\right) = 1, \quad N_{2,e}\left(s_e = l_e\right) = 0,$$

$$N_{3,e}\left(s_e = 0\right) = 0, \quad N_{3,e}\left(s_e = \frac{l_e}{2}\right) = 0, \quad N_{3,e}\left(s_e = l_e\right) = 1. \quad (14.97)$$

A parabolic formulation of the type $f\left(s_e\right) = as_e^2 + bs_e + c$ yields the following shape functions $N_{1,e}(s_e)$, $N_{2,e}(s_e)$ and $N_{3,e}(s_e)$ after fitting to the above conditions (14.97):

$$N_{1,e}\left(s_e\right) = 1 - 3\frac{s_e}{l_e} + 2\frac{s_e^2}{l_e^2}, \quad N_{2,e}\left(s_e\right) = 4\left(\frac{s_e}{l_e} - \frac{s_e^2}{l_e^2}\right), \quad N_{3,e}\left(s_e\right) = -\frac{s_e}{l_e} + 2\frac{s_e^2}{l_e^2}.$$

$$(14.98)$$

The shape functions $N_{1,e}(s_e)$, $N_{2,e}(s_e)$, and $N_{3,e}(s_e)$ are shown in Fig. 14.30.

The element equations can be derived from the principle of the minimum of the total elastic potential $\delta\Pi = \delta\Pi_i + \delta\Pi_a = 0$:

$$\delta\Pi_i = \frac{1}{2}EA_e\delta\int_0^{l_e}\left(\left(u_e(s_e)\right)'\right)^2 ds_e = \frac{EA_e}{l_e}\left[\left(\frac{7}{3}u_{K,e} - \frac{8}{3}u_{L,e} + \frac{1}{3}u_{M,e}\right)\delta u_{K,e}\right.$$

$$\left. + \left(-\frac{8}{3}u_{K,e} + \frac{16}{3}u_{L,e} - \frac{8}{3}u_{M,e}\right)\delta u_{L,e} + \left(\frac{1}{3}u_{K,e} - \frac{8}{3}u_{L,e} + \frac{7}{3}u_{M,e}\right)\delta u_{M,e}\right],$$

$$\delta\Pi_a = -N_{K,e}\delta u_{K,e} - N_{L,e}\delta u_{L,e} - N_{M,e}\delta u_{M,e} - \frac{n_e l_e}{6}\left(\delta u_{K,e} + 4\delta u_{L,e} + \delta u_{M,e}\right).$$

$$(14.99)$$

Imposing $\delta\Pi = 0$ leads to:

$$\left[\frac{EA_e}{3l_e}\left(7u_{K,e} - 8u_{L,e} + u_{M,e}\right) - N_{K,e} - \frac{n_e l_e}{6}\right]\delta u_{K,e}$$

$$+ \left[\frac{EA_e}{3l_e}\left(-8u_{K,e} + 16u_{L,e} - 8u_{M,e}\right) - N_{L,e} - \frac{2n_e l_e}{3}\right]\delta u_{L,e}$$

$$+ \left[\frac{EA_e}{3l_e}\left(u_{K,e} - 8u_{L,e} + 7u_{M,e}\right) - N_{M,e} - \frac{n_e l_e}{6}\right]\delta u_{M,e} = 0. \quad (14.100)$$

Equating the bracketed terms to zero leads to:

$$\frac{EA_e}{3l_e}\left(7u_{K,e} - 8u_{L,e} + u_{M,e}\right) = N_{K,e} + \frac{n_e l_e}{6},$$

$$\frac{EA_e}{3l_e}\left(-8u_{K,e} + 16u_{L,e} - 8u_{M,e}\right) = N_{L,e} + \frac{2n_e l_e}{3},$$

$$\frac{EA_e}{3l_e}\left(u_{K,e} - 8u_{L,e} + 7u_{M,e}\right) = N_{M,e} + \frac{n_e l_e}{6}, \quad (14.101)$$

or in a vector-matrix-notation:

$$\frac{EA_e}{3l_e}\begin{bmatrix} 7 & -8 & 1 \\ -8 & 16 & -8 \\ 1 & -8 & 7 \end{bmatrix}\begin{pmatrix} u_{K,e} \\ u_{L,e} \\ u_{M,e} \end{pmatrix} = \begin{pmatrix} N_{K,e} + \dfrac{n_e l_e}{6} \\ N_{L,e} + \dfrac{2n_e l_e}{3} \\ N_{M,e} + \dfrac{n_e l_e}{6} \end{pmatrix}.$$ (14.102)

These are the element equations for the three-noded rod element.

14.4   Finite Elements for Plane Systems of Straight Beams

14.4.1   The Two-Noded Beam Element

We now consider finite elements as they are used in the FEM analysis of systems of plane straight beams. For the plane beam structures now to be discussed, the loads are applied exclusively in the structural plane, and it is assumed that one of the two cross-sectional principal axes is also in the structural plane so that we may assume uniaxial bending. We will now consider the formulation for the two-noded finite beam element shown in Fig. 14.31. Let the finite beam element have the two

Fig. 14.31 Finite beam element with two nodes

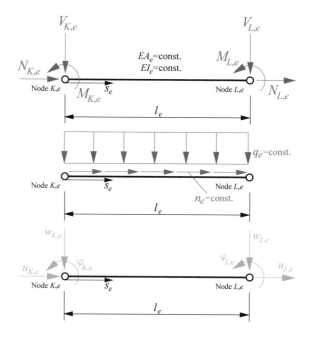

nodes K and L with nodal forces and moments $N_{K,e}$, $V_{K,e}$, $M_{K,e}$ and $N_{L,e}$, $V_{L,e}$, $M_{L,e}$ and the associated nodal degrees of freedom $u_{K,e}$, $w_{K,e}$, $\varphi_{K,e}$ and $u_{L,e}$, $w_{L,e}$, $\varphi_{L,e}$, respectively. Let the length of the element be l_e. The extensional stiffness is assumed to be constant with the value EA_e, as well as the flexural stiffness which is assumed as constant with the value EI_e. Let the load consist of the constant line loads n_e and q_e.

With regard to the element equations concerning the two longitudinal displacements $u_{K,e}$ and $u_{L,e}$, the formulations (14.78) of the plane beam element can be adopted directly which we write down again for the sake of clarity:

$$\frac{EA_e}{l_e}\begin{bmatrix} 1 & -1 \\ -1 & 1 \end{bmatrix}\begin{pmatrix} u_{K,e} \\ u_{L,e} \end{pmatrix} = \begin{pmatrix} N_{K,e} + \frac{1}{2}n_e l_e \\ N_{L,e} + \frac{1}{2}n_e l_e \end{pmatrix}. \tag{14.103}$$

The formulations for the transverse displacements $w_{K,e}$ and $w_{L,e}$ and the nodal rotations $\varphi_{K,e}$ and $\varphi_{L,e}$ are explained below. For the element deflection $w_e(s_e)$, we use a formulation as follows:

$$w_e(s_e) = N_{1,e}(s_e)\,w_{K,e} + N_{2,e}(s_e)\,\varphi_{K,e} + N_{3,e}(s_e)\,w_{L,e} + N_{4,e}(s_e)\,\varphi_{L,e}. \tag{14.104}$$

The shape functions $N_{1,e}(s_e)$, $N_{2,e}(s_e)$, $N_{3,e}(s_e)$, $N_{4,e}(s_e)$ used here are the so-called Hermitian[1] polynomials. They are defined as follows:

$$N_{1,e}(s_e) = 1 - 3\frac{s_e^2}{l_e^2} + 2\frac{s_e^3}{l_e^3}, \quad N_{2,e}(s_e) = s_e - 2\frac{s_e^2}{l_e} + \frac{s_e^3}{l_e^2},$$

$$N_{3,e}(s_e) = 3\frac{s_e^2}{l_e^2} - 2\frac{s_e^3}{l_e^3}, \quad N_{4,e}(s_e) = -\frac{s_e^2}{l_e} + \frac{s_e^3}{l_e^2}. \tag{14.105}$$

They are shown in Fig. 14.32 and have the following properties:

$$
\begin{array}{llll}
N_{1,e}(s_e = 0) = 1, & N'_{1,e}(s_e = 0) = 0, & N_{1,e}(s_e = l_e) = 0, & N'_{1,e}(s_e = l_e) = 0, \\
N_{2,e}(s_e = 0) = 0, & N'_{2,e}(s_e = 0) = 1, & N_{2,e}(s_e = l_e) = 0, & N'_{2,e}(s_e = l_e) = 0, \\
N_{3,e}(s_e = 0) = 0, & N'_{3,e}(s_e = 0) = 0, & N_{3,e}(s_e = l_e) = 1, & N'_{3,e}(s_e = l_e) = 0, \\
N_{4,e}(s_e = 0) = 0, & N'_{4,e}(s_e = 0) = 0, & N_{4,e}(s_e = l_e) = 0, & N'_{4,e}(s_e = l_e) = 1.
\end{array}
$$

$$\tag{14.106}$$

It is noteworthy here that in the approach form (14.104) for the element deflection w_e both the nodal displacements $w_{K,e}$ and $w_{L,e}$ and the nodal rotations $\varphi_{K,e}$ and $\varphi_{L,e}$ occur in combination.

The inner potential Π_i of the beam element can now be given as follows:

---

[1] Charles Hermite, 1822–1901, French mathematician.

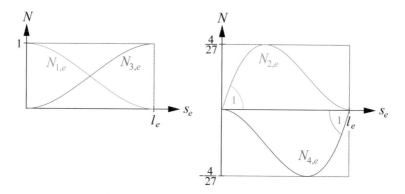

Fig. 14.32 Hermitian polynomials for a two-noded beam element

$$\Pi_i = \frac{1}{2} E I_e \int_0^{l_e} \left(w_e''(s_e) \right)^2 \mathrm{d}s_e$$

$$= \frac{1}{2} E I_e \int\limits_0^{l_e} \left(N_{1,e}'' w_{K,e} + N_{2,e}'' \varphi_{K,ei} + N_{3,e}'' w_{L,e} + N_{4,e}'' \varphi_{L,e} \right)^2 \mathrm{d}s_e. \quad (14.107)$$

Inserting the shape functions (14.105) and performing the integrations of the twofold derivatives of the shape functions prescribed in (14.107) and finally forming the first variation $\delta \Pi_i$ results in:

$$\delta \Pi_i = \frac{2E I_e}{l_e^3} \left(6 w_{K,e} + 3 \varphi_{K,e} l_e - 6 w_{L,e} + 3 \varphi_{L,e} l_e \right) \delta w_{K,e}$$

$$+ \frac{2E I_e}{l_e^3} \left(3 w_{K,e} l_e + 2 \varphi_{K,e} l_e^2 - 3 w_{L,e} l_e + \varphi_{L,e} l_e^2 \right) \delta \varphi_{K,e}$$

$$+ \frac{2E I_e}{l_e^3} \left(-6 w_{K,e} - 3 \varphi_{K,e} l_e + 6 w_{L,e} - 3 \varphi_{L,e} l_e \right) \delta w_{L,e}$$

$$+ \frac{2E I_e}{l_i e^3} \left(3 w_{K,e} l_e + \varphi_{K,e} l_e^2 - 3 w_{L,e} l_e + 2 \varphi_{L,e} l_e^2 \right) \delta \varphi_{L,e}. \quad (14.108)$$

The external potential Π_a of the beam element reads:

$$\Pi_a = -V_{K,e} w_{K,e} - V_{L,e} w_{L,e} - M_{K,e} \varphi_{K,e} - M_{L,e} \varphi_{L,e} - \int_0^{l_e} q_e w_e(s_e) \mathrm{d}s_e. \quad (14.109)$$

The first variation $\delta \Pi_a$ is then obtained as:

$$\delta \Pi_a = \left(-V_{K,e} - \frac{1}{2} q_e l_e \right) \delta w_{K,e} + \left(-M_{K,e} - \frac{1}{12} q_e l_e^2 \right) \delta \varphi_{K,e}$$

$$+ \left(-V_{L,e} - \frac{1}{2} q_e l_e \right) \delta w_{L,e} + \left(-M_{L,e} + \frac{1}{12} q_e l_e^2 \right) \delta \varphi_{L,e}. \quad (14.110)$$

The variations $\delta w_{K,e}$, $\delta w_{L,e}$, $\delta \varphi_{K,e}$ and $\delta \varphi_{L,e}$ are independent of each other and arbitrary. Therefore, the requirement $\delta \Pi = \delta \Pi_i + \delta \Pi_a = 0$ can be satisfied only if the bracketed terms vanish. With (14.108) and (14.110) this results in:

$$\frac{2E I_e}{l_e^3} \left(6 w_{K,e} + 3 \varphi_{K,e} l_e - 6 w_{L,e} + 3 \varphi_{L,e} l_e \right) = V_{K,e} + \frac{1}{2} q_e l_e,$$

$$\frac{2E I_e}{l_e^3} \left(3 w_{K,e} l_e + 2 \varphi_{K,e} l_e^2 - 3 w_{L,e} l_e + \varphi_{L,e} l_e^2 \right) = M_{K,e} + \frac{1}{12} q_e l_e^2,$$

$$\frac{2E I_e}{l_e^3} \left(-6 w_{K,e} - 3 \varphi_{K,e} l_e + 6 w_{L,e} - 3 \varphi_{L,e} l_e \right) = V_{L,e} + \frac{1}{2} q_e l_e,$$

$$\frac{2E I_e}{l_i e^3} \left(3 w_{K,e} l_e + \varphi_{K,e} l_e^2 - 3 w_{L,e} l_e + 2 \varphi_{L,e} l_e^2 \right) = M_{L,e} - \frac{1}{12} q_e l_e^2. \quad (14.111)$$

In a vector-matrix notation we have:

$$\frac{E I_e}{l_e^3} \begin{bmatrix} 12 & 6 l_e & -12 & 6 l_e \\ 6 l_e & 4 l_e^2 & -6 l_e & 2 l_e^2 \\ -12 & -6 l_e & 12 & -6 l_e \\ 6 l_e & 2 l_e^2 & -6 l_e & 4 l_e^2 \end{bmatrix} \begin{pmatrix} w_{K,e} \\ \varphi_{K,e} \\ w_{L,e} \\ \varphi_{L,e} \end{pmatrix} = \begin{pmatrix} V_{K,e} + \frac{1}{2} q_e l_e \\ M_{K,e} + \frac{1}{12} q_e l_e^2 \\ V_{L,e} + \frac{1}{2} q_e l_e \\ M_{L,e} - \frac{1}{12} q_e l_e^2 \end{pmatrix}. \quad (14.112)$$

Again, this result is of the form $\underline{K}_e \underline{v}_e = \underline{F}_e$. Apparently, the line load q_e has been converted into statically equivalent nodal forces and nodal moments, as shown in Fig. 14.33.

Fig. 14.33 Conversion of the line load q_e into statically equivalent nodal forces and moments

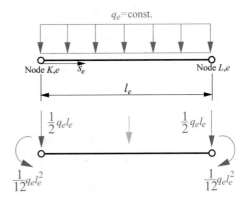

The equations derived with (14.112) are valid for the case of pure uniaxial bending and do not consider the influence of longitudinal displacements u so far. If, however, a beam is present which is loaded not only by the transverse load q but also by an axial line load n, i.e. longitudinal displacements u can occur, then the rod element Eqs. (14.103) can be superimposed on the derived element Eqs. (14.112) so that:

$$
\begin{bmatrix}
\dfrac{EA_e}{l_e} & 0 & 0 & -\dfrac{EA_e}{l_e} & 0 & 0 \\[2mm]
0 & 12\dfrac{EI_e}{l_e^3} & 6\dfrac{EI_e}{l_e^2} & 0 & -12\dfrac{EI_e}{l_e^3} & 6\dfrac{EI_e}{l_e^2} \\[2mm]
0 & 6\dfrac{EI_e}{l_e^2} & 4\dfrac{EI_e}{l_e} & 0 & -6\dfrac{EI_e}{l_e^2} & 2\dfrac{EI_e}{l_e} \\[2mm]
-\dfrac{EA_e}{l_e} & 0 & 0 & \dfrac{EA_e}{l_e} & 0 & 0 \\[2mm]
0 & -12\dfrac{EI_e}{l_e^3} & -6\dfrac{EI_e}{l_e^2} & 0 & 12\dfrac{EI_e}{l_e^3} & -6\dfrac{EI_e}{l_e^2} \\[2mm]
0 & 6\dfrac{EI_e}{l_e^2} & 2\dfrac{EI_e}{l_e} & 0 & -6\dfrac{EI_e}{l_e^2} & 4\dfrac{EI_e}{l_e}
\end{bmatrix}
\begin{pmatrix}
u_{K,e} \\
w_{K,e} \\
\varphi_{K,e} \\
u_{L,e} \\
w_{L,e} \\
\varphi_{L,e}
\end{pmatrix}
$$

$$
=
\begin{pmatrix}
N_{K,e} + \dfrac{1}{2}n_e l_e \\[2mm]
V_{K,e} + \dfrac{1}{2}q_e l_e \\[2mm]
M_{K,e} + \dfrac{1}{12}q_e l_e^2 \\[2mm]
N_{L,e} + \dfrac{1}{2}n_e l_e \\[2mm]
V_{L,e} + \dfrac{1}{2}q_e l_e \\[2mm]
M_{K,e} - \dfrac{1}{12}q_e l_e^2
\end{pmatrix}
. \quad (14.113)
$$

The transformation of the element equations from the element reference system into an arbitrarily oriented global reference system can be performed as already shown for rod elements, where the transformation equations have to be extended by the contribution of the nodal moments. Since both moments and rotations are invariant with respect to a pure rotation of a Cartesian reference system, this contribution of the nodal moments can be easily accounted for by appropriate unit values, and the transformation matrix $\underline{\underline{T}}_e$ is obtained as follows:

$$
\underline{\underline{T}}_e =
\begin{bmatrix}
\cos(\alpha_e) & \sin(\alpha_e) & 0 & 0 & 0 & 0 \\
-\sin(\alpha_e) & \cos(\alpha_e) & 0 & 0 & 0 & 0 \\
0 & 0 & 1 & 0 & 0 & 0 \\
0 & 0 & 0 & \cos(\alpha_e) & \sin(\alpha_e) & 0 \\
0 & 0 & 0 & -\sin(\alpha_e) & \cos(\alpha_e) & 0 \\
0 & 0 & 0 & 0 & 0 & 1
\end{bmatrix}
. \quad (14.114)
$$

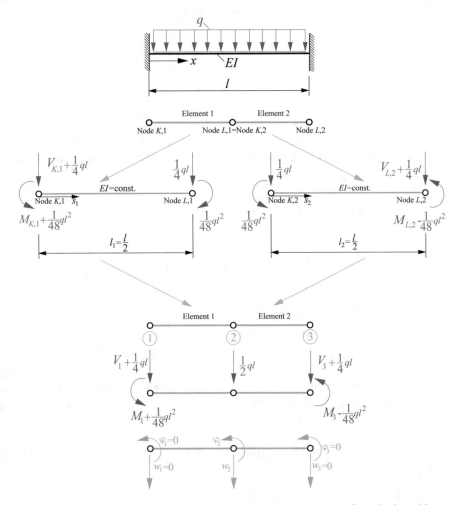

Fig. 14.34 Beam clamped on both sides under constant line load q (top), discretization with two beam elements (bottom)

The assembly of the individual finite elements to the complete system can then be carried out as already shown for trusses and rod structures.

We illustrate the procedure for FEM analysis of plane beam structures using the example of Fig. 14.34, top. Let a beam of length l and constant bending stiffness EI be given, clamped on both sides and loaded by the constant line load q. Obviously, this is a statically indeterminate system. We start with a discretization consisting of two beam elements (Fig. 14.34, bottom). With $l_1 = l_2 = \frac{l}{2}$ and $EI_1 = EI_2 = EI$, we obtain the nodal forces and moments as shown, and the element matrices result as:

$$\underline{\underline{K}}_1 = \underline{\underline{K}}_2 = \frac{8EI}{l^3} \begin{bmatrix} 12 & 3l & -12 & 3l \\ 3l & l^2 & -3l & \frac{1}{2}l^2 \\ -12 & -3l & 12 & -3l \\ 3l & \frac{1}{2}l^2 & -3l & l^2 \end{bmatrix}. \tag{14.115}$$

The nodal displacements and rotations in the assembled system are w_1, w_2, w_3 and φ_1, φ_2, φ_3, respectively, and we denote the nodal forces and moments as V_1, $V_2 = \frac{1}{2}ql$, V_3 and M_1, $M_2 = 0$, M_3. We then obtain the following system of equations to be solved:

$$\frac{8EI}{l^3} \begin{bmatrix} 12 & 3l & -12 & 3l & 0 & 0 \\ 3l & l^2 & -3l & \frac{1}{2}l^2 & 0 & 0 \\ -12 & -3l & 24 & 0 & -12 & 3l \\ 3l & \frac{1}{2}l^2 & 0 & 2l^2 & -3l & \frac{1}{2}l^2 \\ 0 & 0 & -12 & -3l & 12 & -3l \\ 0 & 0 & 3l & \frac{1}{2}l^2 & -3l & l^2 \end{bmatrix} \begin{pmatrix} w_1 \\ \varphi_1 \\ w_2 \\ \varphi_2 \\ w_3 \\ \varphi_3 \end{pmatrix} = \begin{pmatrix} Q_1 + \frac{1}{4}ql \\ M_1 + \frac{1}{48}ql^2 \\ \frac{1}{2}ql \\ 0 \\ Q_3 + \frac{1}{4}ql \\ M_3 - \frac{1}{48}ql^2 \end{pmatrix}. \tag{14.116}$$

Due to the nodal conditions $w_1 = 0$, $\varphi_1 = 0$, $w_3 = 0$, $\varphi_3 = 0$, the rows and columns 1, 2, 5 and 6 can be disregarded, so that:

$$\frac{8EI}{l^3} \begin{bmatrix} 24 & 0 \\ 0 & 2l^2 \end{bmatrix} \begin{pmatrix} w_2 \\ \varphi_2 \end{pmatrix} = \begin{pmatrix} \frac{1}{2}ql \\ 0 \end{pmatrix}. \tag{14.117}$$

The solution to this system of equations results as:

$$w_2 = \frac{ql^4}{384EI}, \quad \varphi_2 = 0. \tag{14.118}$$

The disappearance of the rotation φ_2 is plausible due to the symmetry of the system.

The initially neglected rows and columns 1, 2, 5, and 6 of the system of Eqs. (14.116) are now used to determine the support reactions of the system in a follow-up calculation. We obtain:

$$V_1 = -\frac{1}{2}ql, \quad M_1 = -\frac{1}{12}ql^2, \quad V_3 = -\frac{1}{2}ql, \quad M_3 = \frac{1}{12}ql^2. \tag{14.119}$$

14.4.2 Quality of the Solution, Convergence Behavior

The use of beam elements also requires a thorough convergence analysis to get clarity on the quality of the solution when using different discretizations. We will illustrate this with the simple example of Fig. 14.35. Let us consider a simply supported beam of length l loaded by the constant line load q. It is now of fundamental interest to consider the quality of the results when different numbers of two-noded elements are used for the discretization of the beam. We want to use the symmetry of the given system and consider only the left half of the beam. We now consider the results obtained when using a single element, for two elements and also for three elements (Fig. 14.36).

We will not give the resulting system equations at this point and turn directly to the results. They are shown in Fig. 14.37 for $EI = 1, l = 1$ and $q = 24$. It is shown that already with one element for the considered half of the system (which is equivalent to the use of two elements over the entire beam length l) a very good approximation is obtained for the beam deflection $w(x)$. However, it is also shown that more elements are required for a satisfactory approximation of the moment $M(x)$ and, in particular, the transverse shear force $V(x)$. The reason for this can be found in the type of beam element formulation used. The Hermitian polynomials used here are third degree polynomials, which leads to a linear approximation of the moment line $M(x)$ and a constant approximation of the transverse shear force line $V(x)$. Thus, in particular for quantities derived from displacements such as $M(x)$ and $V(x)$, more elements have to be used than would be necessary for a satisfactory approximation of the deflection $w(x)$. In any case, a convergence study must clarify which number of elements leads to which quality of results.

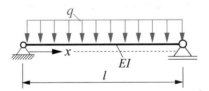

Fig. 14.35 Simply supported beam under line load q

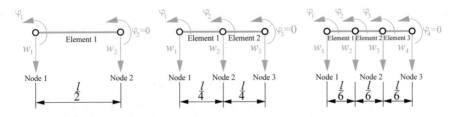

Fig. 14.36 Employed discretizations, use of symmetry

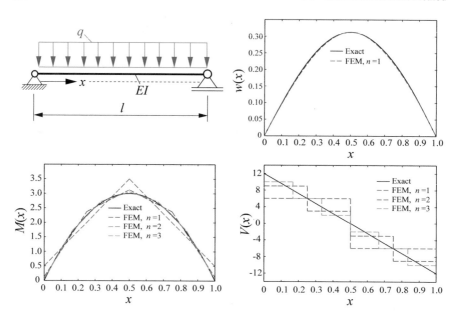

Fig. 14.37 Finite element results for the simply supported beam under constant line load q, comparison with the exact solution

14.4.3   The Three-Noded Beam Element

We want to briefly outline below the way to determine the stiffness matrix $\underline{\underline{K}}$ for a three-noded beam element. The considered element with equidistant nodes is shown in Fig. 14.38. We only consider the case of uniaxial bending, i.e. normal forces are not present. The three-noded element exhibits the deflections $w_{K,e}$, $w_{L,e}$ and $w_{M,e}$ and the rotations $\varphi_{K,e}$, $\varphi_{L,e}$ and $\varphi_{M,e}$ as nodal degrees of freedom. Let the associated forces

Fig. 14.38 Three-noded beam element

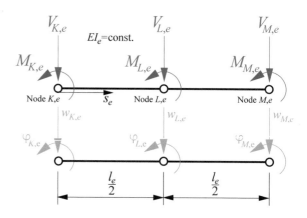

and moments be the transverse shear forces $V_{K,e}$, $V_{L,e}$ and $V_{M,e}$ and the bending moments $M_{K,e}$, $M_{L,e}$ and $M_{M,e}$. We choose the following approach for the element deflection $w_e(s_e)$:

$$w_e(s_e) = w_{K,e}N_{1,e} + \varphi_{K,e}N_{2,e} + w_{L,e}N_{3,e} + \varphi_{L,e}N_{4,e} + w_{M,e}N_{5,e} + \varphi_{M,e}N_{6,e}.$$
(14.120)

The shape functions $N_{1,e}$, $N_{2,e}$, $N_{3,e}$, $N_{4,e}$, $N_{5,e}$ and $N_{6,e}$ are chosen here as fifth degree polynomials:

$$w_e(s_e) = as_e^5 + bs_e^4 + cs_e^3 + ds_e^2 + es_e + f.$$
(14.121)

They have to fulfill the following conditions:

$$N_{1,e}(s_e = 0) = 1, \quad N_{1,e}\left(s_e = \frac{l_e}{2}\right) = 0, \quad N_{1,e}(s_e = l_e) = 0,$$

$$N'_{1,e}(s_e = 0) = 0, \quad N'_{1,e}\left(s_e = \frac{l_e}{2}\right) = 0, \quad N'_{1,e}(s_e = l_e) = 0,$$

$$N_{2,e}(s_e = 0) = 0, \quad N_{2,e}\left(s_e = \frac{l_e}{2}\right) = 0, \quad N_{2,e}(s_e = l_e) = 0,$$

$$N'_{2,e}(s_e = 0) = 1, \quad N'_{2,e}\left(s_e = \frac{l_e}{2}\right) = 0, \quad N'_{2,e}(s_e = l_e) = 0,$$

$$N_{3,e}(s_e = 0) = 0, \quad N_{3,e}\left(s_e = \frac{l_e}{2}\right) = 1, \quad N_{3,e}(s_e = l_e) = 0,$$

$$N'_{3,e}(s_e = 0) = 0, \quad N'_{3,e}\left(s_e = \frac{l_e}{2}\right) = 0, \quad N'_{3,e}(s_e = l_e) = 0,$$

$$N_{4,e}(s_e = 0) = 0, \quad N_{4,e}\left(s_e = \frac{l_e}{2}\right) = 0, \quad N_{4,e}(s_e = l_e) = 0,$$

$$N'_{4,e}(s_e = 0) = 0, \quad N'_{4,e}\left(s_e = \frac{l_e}{2}\right) = 1, \quad N'_{4,e}(s_e = l_e) = 0,$$

$$N_{5,e}(s_e = 0) = 0, \quad N_{5,e}\left(s_e = \frac{l_e}{2}\right) = 0, \quad N_{5,e}(s_e = l_e) = 1,$$

$$N'_{5,e}(s_e = 0) = 0, \quad N'_{5,e}\left(s_e = \frac{l_e}{2}\right) = 0, \quad N'_{5,e}(s_e = l_e) = 0,$$

$$N_{6,e}(s_e = 0) = 0, \quad N_{6,e}\left(s_e = \frac{l_e}{2}\right) = 0, \quad N_{6,e}(s_e = l_e) = 0,$$

$$N'_{6,e}(s_e = 0) = 0, \quad N'_{6,e}\left(s_e = \frac{l_e}{2}\right) = 0, \quad N'_{6,e}(s_e = l_e) = 1. \quad (14.122)$$

This results in the shape functions as shown below:

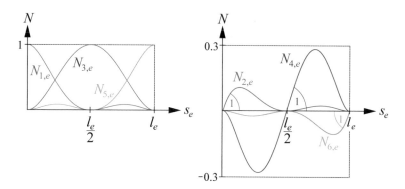

Fig. 14.39 Shape functions for the three-noded beam element

$$N_{1,e}(s_e) = 1 - 23\left(\frac{s_e}{l_e}\right)^2 + 66\left(\frac{s_e}{l_e}\right)^3 - 68\left(\frac{s_e}{l_e}\right)^4 + 24\left(\frac{s_e}{l_e}\right)^5,$$

$$N_{2,e}(s_e) = s_e\left[1 - \frac{s_e}{l_e} + 13\left(\frac{s_e}{l_e}\right)^2 - 12\left(\frac{s_e}{l_e}\right)^3 + 4\left(\frac{s_e}{l_e}\right)^4\right],$$

$$N_{3,e}(s_e) = \left(\frac{s_e}{l_e}\right)^2\left[16 - 32\frac{s_e}{l_e} + 16\left(\frac{s_e}{l_e}\right)^2\right],$$

$$N_{4,e}(s_e) = \frac{s_e^2}{l_e}\left[-8 + 32\frac{s_e}{l_e} - 40\left(\frac{s_e}{l_e}\right)^2 + 16\left(\frac{s_e}{l_e}\right)^3\right],$$

$$N_{5,e}(s_e) = \left(\frac{s_e}{l_e}\right)^2\left[7 - 34\frac{s_e}{l_e} + 52\left(\frac{s_e}{l_e}\right)^2 - 24\left(\frac{s_e}{l_e}\right)^3\right],$$

$$N_{6,e}(s_e) = \frac{s_e^2}{l_e}\left[-1 + 5\frac{s_e}{l_e} - 8\left(\frac{s_e}{l_e}\right)^2 + 4\left(\frac{s_e}{l_e}\right)^3\right]. \tag{14.123}$$

A graphical representation can be found in Fig. 14.39. The internal potential Π_i of the element is:

$$\Pi_i = \frac{1}{2}EI_e\int_0^{l_e}\left(w_e''(s_e)\right)^2\,ds_e, \tag{14.124}$$

and evaluation of the Ritz equations

$$\frac{\partial\Pi_i}{\partial w_{K,e}} = 0, \quad \frac{\partial\Pi_i}{\partial\varphi_{K,e}} = 0, \quad \frac{\partial\Pi_i}{\partial w_{L,e}} = 0,$$

$$\frac{\partial\Pi_i}{\partial\varphi_{L,e}} = 0, \quad \frac{\partial\Pi_i}{\partial w_{M,e}} = 0, \quad \frac{\partial\Pi_i}{\partial\varphi_{M,e}} = 0$$

$$\tag{14.125}$$

finally yields, after some arithmetic operations, the following stiffness matrix $\underline{\underline{K}}_e$ of the three-noded beam element considered here:

$$
\underline{\underline{K}}_e = \frac{2EI}{35l_e}
\begin{bmatrix}
2546 & 569l_e & -1792 & 960l_e & -754 & 121l_e \\
569l_e & 166l_e^2 & -448l_e & 160l_e^2 & -121l_e & 19l_e^2 \\
-1792 & -448l_e & 3584 & 0 & -1792 & 448l_e \\
960l_e & 160l_e^2 & 0 & 640l_e^2 & -960l_e & 160l_e^2 \\
-754 & -121l_e & -1792 & -960l_e & 2546 & -569l_e \\
121l_e & 19l_e^2 & 448l_e & 160l_e^2 & -569l_e & 166l_e^2
\end{bmatrix}.
\tag{14.126}
$$

In a very similar way, the force vector $\underline{F}$ can also be derived for this element class which, however, remains without further representation at this point.

14.4.4   Comparison Ritz / FEM

This section shows a comparison of the results obtained by the Ritz method and the finite element method for the beam problem of Fig. 14.40. Consider a beam on two supports for which the bending line $w(x)$, the moment line $M(x)$ and the transverse shear force line $V(x)$ are sought. The given beam has the total length l and is loaded by the constant line load q. The left half of the beam with length αl has a constant bending stiffness EI_1, whereas the right half of the beam with the length $l(1 - \alpha)$ has a constant bending stiffness EI_2. The value α may be arbitrary in the interval $0 \le \alpha \le 1$.

For this beam problem, an exact solution can be obtained by the means of Chap. 5 which we will use here as a reference for later comparison with the results of the Ritz method and the FEM. In the beam section on the left, we obtain by integrating the differential equation of the beam bending line:

$$
E I_1 w_1''''(x_1) = q,
$$
$$
E I_1 w_1'''(x_1) = q x_1 + C_{1,1},
$$

Fig. 14.40 Static system

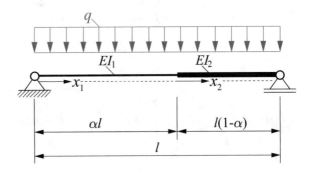

$$EI_1w_1''(x_1) = \frac{1}{2}qx_1^2 + C_{1,1}x_1 + C_{1,2},$$

$$EI_1w_1'(x_1) = \frac{1}{6}qx_1^3 + \frac{1}{2}C_{1,1}x_1^2 + C_{1,2}x_1 + C_{1,3},$$

$$EI_1w_1(x_1) = \frac{1}{24}qx_1^4 + \frac{1}{6}C_{1,1}x_1^3 + \frac{1}{2}C_{1,2}x_1^2 + C_{1,3}x_1 + C_{1,4}. \quad (14.127)$$

The following boundary conditions must be observed here:

$$w_1''(x_1 = 0) = 0, \quad w_1(x_1 = 0) = 0. \quad (14.128)$$

In the right beam section results:

$$EI_2w_2''''(x_2) = q,$$
$$EI_2w_2'''(x_2) = qx_2 + C_{2,1},$$
$$EI_2w_2''(x_2) = \frac{1}{2}qx_2^2 + C_{2,1}x_2 + C_{2,2},$$
$$EI_2w_2'(x_2) = \frac{1}{6}qx_2^3 + \frac{1}{2}C_{2,1}x_2^2 + C_{2,2}x_2 + C_{2,3},$$
$$EI_2w_2(x_2) = \frac{1}{24}qx_2^4 + \frac{1}{6}C_{2,1}x_2^3 + \frac{1}{2}C_{2,2}x_2^2 + C_{2,3}x_2 + C_{2,4}. \quad (14.129)$$

The boundary conditions for this beam section read:

$$w_2''(x_2 = l(1 - \alpha)) = 0, \quad w_2(x_2 = l(1 - \alpha)) = 0. \quad (14.130)$$

The following transition conditions must be met between the two beam sections:

$$EI_1w_1'''(x_1 = \alpha l) = EI_2w_2'''(x_2 = 0),$$
$$EI_1w_1''(x_1 = \alpha l) = EI_2w_2''(x_2 = 0),$$
$$w_1'(x_1 = \alpha l) = w_2'(x_2 = 0),$$
$$w_1(x_1 = \alpha l) = w_2(x_2 = 0). \quad (14.131)$$

From (14.128), (14.130), and (14.131), the following system of linear equations can be obtained to determine the integration constants:

$$C_{1,2} = 0,$$
$$C_{1,4} = 0,$$
$$\frac{1}{2}q(l(1 - \alpha))^2 + C_{2,1}(l(1 - \alpha)) + C_{2,2} = 0,$$
$$\frac{1}{24}q(l(1 - \alpha))^4 + \frac{1}{6}C_{2,1}(l(1 - \alpha))^3 + \frac{1}{2}C_{2,2}(l(1 - \alpha))^2 + C_{2,3}(l(1 - \alpha)) + C_{2,4} = 0,$$
$$q_0\alpha l + C_{1,1} - C_{2,1} = 0,$$

$$\frac{1}{2}q(\alpha l)^2 + C_{1,1}\alpha l + C_{1,2} - C_{2,2} = 0,$$

$$\frac{1}{EI_1}\left[\frac{1}{6}q(\alpha l)^3 + \frac{1}{2}C_{1,1}(\alpha l)^2 + C_{1,2}\alpha l + C_{1,3}\right] - \frac{1}{EI_2}C_{2,3} = 0,$$

$$\frac{1}{EI_1}\left[\frac{1}{24}q(\alpha l)^4 + \frac{1}{6}C_{1,1}(\alpha l)^3 + \frac{1}{2}C_{1,2}(\alpha l)^2 + C_{1,3}\alpha l + C_{1,4}\right] - \frac{1}{EI_2}C_{2,3} = 0.$$

$$(14.132)$$

The explicit specification of the resulting solution is omitted here. The exact solution is evaluated below for the following ratios of the bending stiffnesses EI_1 and EI_2 and provided as a reference:

$$EI_2 = 1.0EI_1, \quad EI_2 = 2.5EI_1, \quad EI_2 = 5.0EI_1, \quad EI_2 = 10.0EI_1, \quad EI_2 = 100.0EI_1.$$

$$(14.133)$$

As concrete numerical values for the bending stiffness EI_1, the length l and the line load q the following values are provided:

$$EI_1 = 1, \quad l = 1, \quad q = 24, \quad \alpha = \frac{1}{2}. \qquad (14.134)$$

An evaluation of the exact solution is given in Fig. 14.41, where for the sake of clarity the continuous longitudinal axis x with $0 \le x \le l$ was used. Obviously,

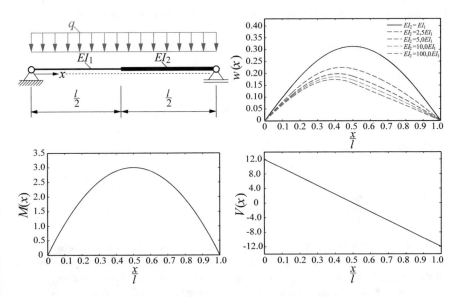

Fig. 14.41 Exact solution for beam deflection, bending moment line and transverse shear force line for different ratios of bending stiffnesses EI_1 and EI_2

the maximum deflection w decreases with increasing bending stiffness EI_2. The location of the maximum deflection moves further and further into the left beam section with increasing EI_2, which is also an obvious result due to the stiffness ratios occurring here. On the other hand, it also shows that both the moment line $M(x)$ and the transverse shear force line $V(x)$ are independent of the ratio of the stiffnesses EI_1 and EI_2, which can be explained by the fact that the given system is statically determinate.

We now want to apply the Ritz method to the problem at hand, noting here that the inner potential Π_i splits into two partial integrals due to the discontinuous bending stiffness distribution as follows:

$$\Pi = \Pi_i + \Pi_a = \frac{1}{2} \int_0^{\frac{l}{2}} EI_1(x)w(x)''^2 dx + \frac{1}{2} \int_{\frac{l}{2}}^l EI_2(x)w(x)''^2 dx - \int_0^l q(x)w(x)dx.$$

(14.135)

We formulate the Ritz approach despite the discontinuous distribution of the bending stiffnesses over the entire beam length l. Results are obtained here for the case $EI_2 = 2.5EI_1$. As shape functions $\sin$ −functions of the following kind are used:

$$w(x) \simeq W(x) = \sum_{i=1}^{i=5} C_i W_i(x) = C_1 \sin\left(\frac{\pi x}{l}\right) + C_2 \sin\left(\frac{2\pi x}{l}\right) + ... + C_5 \sin\left(\frac{5\pi x}{l}\right).$$

(14.136)

We restrict ourselves at this point to a series expansion with the degree $n = 5$. Evaluation of the elastic total potential Π according to (14.135) as well as the Ritz equations $\frac{\partial \Pi}{\partial C_i} = 0$ (for $i = 1, 2, ..., 5$) finally results in the following linear system of equations for the determination of the constants $C_1, C_2, ..., C_5$:

$$
\begin{bmatrix}
\dfrac{7\pi^4}{8} & -4\pi^3 & 0 & \dfrac{32\pi^3}{5} & 0 \\
-4\pi^3 & 14\pi^4 & -\dfrac{108\pi^3}{5} & 0 & \dfrac{100\pi^3}{7} \\
0 & -\dfrac{108\pi^3}{5} & \dfrac{567\pi^4}{8} & -\dfrac{864\pi^3}{7} & 0 \\
\dfrac{32\pi^3}{5} & 0 & -\dfrac{864\pi^3}{7} & 224\pi^4 & -\dfrac{800\pi^3}{3} \\
0 & \dfrac{100\pi^3}{7} & 0 & -\dfrac{800\pi^3}{3} & \dfrac{4375\pi^4}{8}
\end{bmatrix}
\begin{pmatrix} C_1 \\ C_2 \\ C_3 \\ C_4 \\ C_5 \end{pmatrix}
=
\begin{pmatrix} \dfrac{48}{\pi} \\ 0 \\ \dfrac{16}{\pi} \\ 0 \\ \dfrac{48}{5\pi} \end{pmatrix}.
$$

(14.137)

The advantages of the orthogonality properties of the sinusoidal functions used here, which have already been discussed elsewhere, are lost due to the discontinuous stiffness distribution of the given beam. In the stiffness matrix $\underline{K}$ presented here, terms outside the main diagonals also appear. The explicit specification of the constants $C_1, C_2, ..., C_5$ is omitted here. The results of the Ritz method are shown in Fig. 14.42

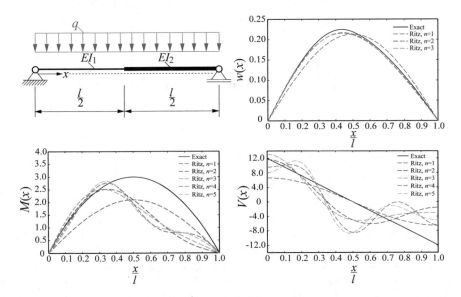

Fig. 14.42 Approximation of beam deflection, bending moment and transverse shear force employing the Ritz method using a sinusoidal series expansion for $EI_2 = 2.5EI_1$

and show that the approximation of the bending line $w(x)$ apparently succeeds quite well. On the other hand, it also shows that the approximation of the bending moment and transverse shear force is not very satisfactory even with a higher degree of approximation.

For comparison, let us now examine the approximation quality provided by the FEM when using two-noded elements for the example at hand. If two beam elements with length $l_1 = l_2 = \frac{l}{2} = \frac{1}{2}$ are used, then the two element stiffness matrices $\underline{\underline{K}}_1$ and $\underline{\underline{K}}_2$ result as:

$$
\underline{\underline{K}}_1 = \begin{bmatrix} 96 & 24 & -96 & 24 \\ 24 & 8 & -24 & 4 \\ -96 & -24 & 96 & -24 \\ 24 & 4 & -24 & 8 \end{bmatrix}, \quad \underline{\underline{K}}_2 = \begin{bmatrix} 240 & 60 & -240 & 60 \\ 60 & 20 & -60 & 10 \\ -240 & -60 & 240 & -60 \\ 60 & 10 & -60 & 20 \end{bmatrix} = \frac{5}{2}\underline{\underline{K}}_1 .
$$

(14.138)

The element load vectors in this case are:

$$
\underline{F}_1 = \underline{F}_2 = \left(6 \; \frac{1}{2} \; 6 \; -\frac{1}{2} \right)^T .
$$

(14.139)

Assembling to the total system of equations $\underline{K}\underline{v} = \underline{F}$ considering the nodal conditions $w_1 = w_3 = 0$ finally yields:

$$
\begin{bmatrix}
8 & -24 & 4 & 0 \\
-24 & 336 & 36 & 60 \\
4 & 36 & 28 & 10 \\
0 & 60 & 10 & 20
\end{bmatrix}
\begin{pmatrix}
\varphi_1 \\
w_2 \\
\varphi_2 \\
\varphi_3
\end{pmatrix}
=
\begin{pmatrix}
1 \\
\dfrac{1}{2} \\
12 \\
0 \\
-\dfrac{1}{2}
\end{pmatrix}.
\tag{14.140}
$$

The solution is:

$$
\begin{pmatrix}
\varphi_1 \\
w_2 \\
\varphi_2 \\
\varphi_3
\end{pmatrix}
=
\left(\frac{87}{12} \quad \frac{45}{224} \quad -\frac{25}{112} \quad -\frac{57}{112} \right)^T.
\tag{14.141}
$$

The procedure is identical for higher element numbers, but will not be elaborated here. The FEM results for different element numbers are shown in Fig. 14.43. It can be stated that with a higher element number a better approximation of $w(x)$, $M(x)$ and $V(x)$ results. However, for reasons already discussed elsewhere, the greater the degree of derivatives of the state variables under consideration, the more the quality of the approximate solution decreases. Consequently, the bending line $w(x)$ (exact solution fourth order polynomial, FEM third order polynomial) is best represented, whereas the transverse shear force line $V(x)$ (exact solution a linear function) is approximated piecewise by constant values.

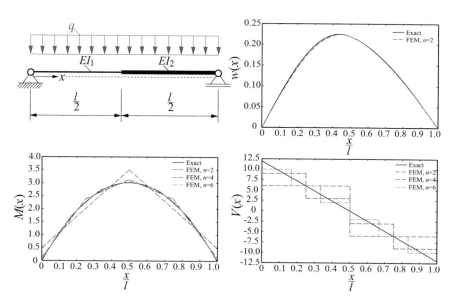

Fig. 14.43 Approximation of beam deflection, bending moment and transverse shear force using FEM employing two-noded beam elements for $EI_2 = 2.5EI_1$

Finally, we consider what quality of results we obtain when we three-noded beam elements are used. If we use two elements for discretization, then we obtain the element stiffness matrices as:

$$
\underline{\underline{K}}_1 = \frac{35}{4}
\begin{bmatrix}
10184 & 1138 & -7168 & 1920 & -3016 & 242 \\
1138 & 166 & -896 & 160 & -242 & 19 \\
-7168 & -896 & 14336 & 0 & -7168 & 896 \\
1920 & 160 & 0 & 640 & -1920 & 160 \\
-3016 & -242 & -7168 & -1920 & 10184 & -1138 \\
242 & 19 & 896 & 160 & -1138 & 166
\end{bmatrix},
$$

$$
\underline{\underline{K}}_2 = \frac{7}{2}
\begin{bmatrix}
10184 & 1138 & -7168 & 1920 & -3016 & 242 \\
1138 & 166 & -896 & 160 & -242 & 19 \\
-7168 & -896 & 14336 & 0 & -7168 & 896 \\
1920 & 160 & 0 & 640 & -1920 & 160 \\
-3016 & -242 & -7168 & -1920 & 10184 & -1138 \\
242 & 19 & 896 & 160 & -1138 & 166
\end{bmatrix}
= \frac{2}{5}\underline{\underline{K}}_1. \qquad (14.142)
$$

The element load vectors then result as:

$$
\underline{F}_1 = \underline{F}_2 = \left(\frac{14}{5}\ \frac{1}{10}\ \frac{32}{5}\ 0\ \frac{14}{5}\ -\frac{1}{10} \right)^T. \qquad (14.143)
$$

Assembling to the total stiffness matrix considering the nodal conditions $w_1 = 0$ and $w_5 = 0$ yields the total stiffness matrix $\underline{\underline{K}}$ as:

$$
\underline{\underline{K}} = \frac{35}{2}
\begin{bmatrix}
332 & -1792 & 320 & -484 & 38 & 0 & 0 & 0 \\
-1792 & 28672 & 0 & -14336 & 1792 & 0 & 0 & 0 \\
320 & 0 & 1280 & -3840 & 320 & 0 & 0 & 0 \\
-484 & -14336 & -3840 & 71288 & 3414 & -35840 & 9600 & 1210 \\
38 & 1792 & 320 & 3414 & 1162 & -4480 & 800 & 95 \\
0 & 0 & 0 & -35840 & -4480 & 71680 & 0 & 4480 \\
0 & 0 & 0 & 9600 & 800 & 0 & 3200 & 800 \\
0 & 0 & 0 & 1210 & 95 & 4480 & 800 & 830
\end{bmatrix}.
$$

$$(14.144)$$

The force vector $\underline{F}$ reads:

$$
\underline{F} = \left(\frac{1}{10}\ \frac{32}{5}\ 0\ \frac{28}{5}\ 0\ \frac{32}{5}\ 0\ -\frac{1}{10} \right)^T. \qquad (14.145)
$$

The solution vector then results in:

$$\underline{v} = \begin{pmatrix} \varphi_1 \\ w_2 \\ \varphi_2 \\ w_3 \\ \varphi_3 \\ w_4 \\ \varphi_4 \\ \varphi_5 \end{pmatrix} = \begin{pmatrix} \dfrac{13}{16} \\ \dfrac{45}{256} \\ \dfrac{1}{2} \\ \dfrac{2}{7} \\ \dfrac{32}{3} \\ -\dfrac{3}{16} \\ \dfrac{87}{640} \\ -\dfrac{37}{80} \\ -\dfrac{47}{80} \end{pmatrix}. \tag{14.146}$$

In contrast to the two-noded beam element, the three-noded beam element used here yields the exact distributions of all state variables due to the higher-order shape functions. A graphical representation of the results can therefore be omitted at this point.

14.5   Finite Elements for Torsion

In this section we consider finite elements for straight beams in the framework of St. Venant's torsion theory (see Chap. 7). We will show that the derivation of the corresponding element formulations is largely the same as that for straight rods, so we can keep the present section quite short.

Consider a two-noded element for St. Venant torsion, as shown in Fig. 14.44. The element has the length l_e and the constant torsional stiffness $GI_{T,e}$ and is loaded by the constant distributed torsional moment $m_{T,e}$ as well as the two torsional moments $M_{K,e}$ and $M_{L,e}$ in the two nodes K, e and L, e, whereby we omit the index x in the former for the sake of a clearer notation. Let the corresponding nodal degrees of freedom be the two nodal rotations $\vartheta_{K,e}$ and $\vartheta_{L,e}$ about the local reference axis s_e.

We formulate the rotation $\vartheta(s_e)$ as follows:

$$\vartheta_e(s_e) = \vartheta_{K,e} N_{1,e}(s_e) + \vartheta_{L,e} N_{2,e}(s_e), \tag{14.147}$$

with the shape functions $N_{1,e}$ and $N_{2,e}$ as:

$$N_{1,e}(s_e) = 1 - \frac{s_e}{l_e}, \quad N_{2,e}(s_e) = \frac{s_e}{l_e}. \tag{14.148}$$

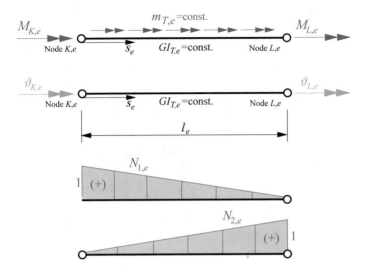

Fig. 14.44 The two-noded element for torsion

The comparison with the rod formulation (14.70) and (14.71) shows a formal agreement.

We again use the principle of the minimum of the total elastic potential to derive the element equations. The internal potential Π_i of the beam element e with respect to St. Venant torsion is:

$$\Pi_i = \frac{1}{2} G I_{T,e} \int_0^{l_e} (\vartheta_e')^2 \, \mathrm{d}s_e = \frac{1}{2} G I_{T,e} \int_0^{l_e} \left(-\frac{\vartheta_{K,e}}{l_e} + \frac{\vartheta_{L,e}}{l_e} \right)^2 \mathrm{d}s_e$$

$$= \frac{G I_{T,e}}{2 l_e} \left(\vartheta_{K,e}^2 - 2\vartheta_{K,e}\vartheta_{L,e} + \vartheta_{L,e}^2 \right). \tag{14.149}$$

The external potential Π_a includes the contributions of the nodal moments $M_{K,e}$ and $M_{L,e}$ along the nodal rotations $\vartheta_{K,e}$ and $\vartheta_{L,e}$, as well as the torsional moment $m_{T,e}$ assumed to be constant:

$$\Pi_a = -M_{K,e}\vartheta_{K,e} - M_{L,e}\vartheta_{L,e} - \int_0^{l_e} m_{T,e}\vartheta_e(s_e)\mathrm{d}s_e, \tag{14.150}$$

which, after integration, leads to the following expression:

$$\Pi_a = -M_{K,e}\vartheta_{K,e} - M_{L,e}\vartheta_{L,e} - \frac{1}{2}m_{T,e}l_e(\vartheta_{K,e} + \vartheta_{L,e}). \tag{14.151}$$

The principle of the minimum of the total elastic potential $\delta\Pi = \delta(\Pi_i + \Pi_a) = 0$ then provides the following statement:

$$
\left[\frac{GI_{T,e}}{l_e} \left(\vartheta_{K,e} - \vartheta_{L,e} \right) - M_{K,e} - \frac{1}{2} m_{T,e} l_e \right] \delta\vartheta_{K,e}
$$

$$
+ \left[\frac{GI_{T,e}}{l_e} \left(-\vartheta_{K,e} + \vartheta_{L,e} \right) - M_{L,e} - \frac{1}{2} m_{T,e} l_e \right] \delta\vartheta_{L,e} = 0. \quad (14.152)
$$

Equating the bracketed terms to zero then yields the element equations we are looking for, which we can write in vector-matrix form as:

$$
\frac{GI_{T,e}}{l_e}
\begin{bmatrix} 1 & -1 \\ -1 & 1 \end{bmatrix}
\begin{pmatrix} \vartheta_{K,e} \\ \vartheta_{L,e} \end{pmatrix}
=
\begin{pmatrix} M_{K,e} + \frac{1}{2} m_{T,e} l_e \\ M_{L,e} + \frac{1}{2} m_{T,e} l_e \end{pmatrix}. \quad (14.153)
$$

It turns out that the element equations formally agree with those of the two-noded rod element (see Eq. 14.78), so that a further elaboration on this class of finite elements can be omitted here.

Bibliography

Bathe KJ (1996) Finite element procedures. Prentice Hall, New Jersey, USA

Betten J (2003) Finite Elemente für Ingenieure Teil 1: Grundlagen, Matrixmethoden, Elastisches Kontinuum, 2nd edn. Springer, Berlin, Germany

Betten J (2004) Finite Elemente für Ingenieure Teil 2: Variationsrechnung, Energiemethoden, Näherungsverfahren, Nichtlinearitäten, numerische Integrationen, 2nd edn. Springer, Berlin, Germany

Crisfield MA (1991) Non-linear finite element analysis of solids and structures, vol 1: essentials. John Wiley and sons, Chichester, UK

Crisfield MA (1997) Non-linear finite element analysis of solids and structures, vol 2: advanced topics. John Wiley and sons, Chichester, UK

Gross D, Hauger W, Wriggers P (2014) Technische Mechanik 4: Hydromechanik, Elemente der Höheren Mechanik, Numerische Methoden, 9th edn. Springer. Berlin, Germany

Langhaar HL (2016) Energy methods in applied mechanics. Dover, New York, USA

Mittelstedt C (2017) Energiemethoden der Elastostatik. Technische Universität Darmstadt, Germany, Studienbereich Mechanik

Reddy JN (2014) An introduction to nonlinear finite element analysis with applications to heat transfer, fluid mechanics, and solid mechanics, 2nd edn. Oxford University Press. New York, USA

Reddy JN (2005) An introduction to the finite element method, 3rd edn. Mcgraw Hill. New York, USA

Reddy JN (2017) Energy principles and variational methods in applied mechanics, 3rd edn. John Wiley and Sons, New York, USA

Tauchert TR (1974) Energy principles in structural mechanics. McGraw-Hill, New York, USA

Wriggers P (2001) Nichtlineare finite-element-methoden. Springer, Berlin, Germany

Zienkiewicz OC, Taylor RL, Zhu JZ (2013a) The finite element method: its basis and fundamentals, 7th edn. Butterworth Heinemann, Oxford, USA

Zienkiewicz OC, Taylor RL (2013b) The finite element method for solid and structural mechanics, 7th edn. Butterworth Heinemann, Oxford, USA

Part IV
Advanced Beam Models

Chapter 15
Shear Field Beams

15.1 Introduction

The modeling concept of shear field beams (Fig. 15.1) is an idealization very often used in lightweight engineering and which takes into account some very important principles of structural lightweight design. Shear field beams are flat planar structures that are loaded exclusively in their plane. The structure is therefore behaving similar to a disk. Since these are typically very thin structures spread over a large area, not only the strength but also the stability (here in the form of buckling out of the structural plane) must be considered in the verification. The load-bearing capacity of such a planar structure can be significantly increased by reinforcing the structure with stiffeners. Not only are the stiffeners significantly involved in the load transfer, but they also stiffen the structure against buckling, thus contributing to the safety of such thin-walled structures. The stiffener arrangement is then usually along the expected main load paths within the structure, with rectangular bays or other elementary simple bay shapes being used in many cases in lightweight construction practice. In principle, however, any stiffening pattern can be arranged, depending on the intended application. In addition, stiffeners can be used to introduce suitable local loads into a thin-walled structure.

In the following, we want to assume that the stiffeners have large extensional stiffnesses compared to the disk and that they act as rods, i.e. that they only carry normal forces. The stiffeners are connected to each other by hinges at their intersection points, and external loads are applied exclusively in the form of point forces at the hinge points. The stiffeners are assumed to be continuously connected with the skin fields. Thus, a shear field beam behaves similar to a truss structure, and we will refer to the sections located between the stiffeners as the so-called skin fields, a terminology that is very common in lightweight engineering. Since the stiffeners are thus essentially involved in the transfer of normal forces, let us assume that the skin fields have the task of transferring the shear stresses occurring in such structures in the form of shear flows (which explains the term shear field beam). The skin then performs the task in a shear field beam that is usually performed by diagonal

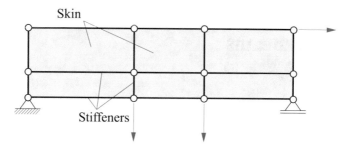

Fig. 15.1 Exemplary plane shear field beam with rectangular skin fields

members in a truss structure. The explanations of this chapter are following e.g. the works of Czerwenka and Schnell (1970), Linke and Nast (2015), Göldner et al. (1979, 1985), Dieker and Reimerdes (1992), from which further details can be obtained.

Since we have already made some rather strict assumptions for the calculation of shear field beams at this point, it becomes clear that this is a rather simple model for the analysis of stiffened structures which is mainly used for predesign, but of course does not do justice to the real conditions in such structures in every respect.

15.2 Rectangular Skin Fields

15.2.1 Determination of Stiffener Forces and Shear Flows

We start the explanations on the basis of the simplest shear field beam structures, namely statically determinate shear field beams with rectangular skin fields. Consider the structure shown in Fig. 15.2 which consists of a single rectangular skin field reinforced by stiffeners at all four edges. Let the shear field beam have the dimensions a and b. Furthermore, it is supported in a statically determinate manner as shown, and is loaded by a single force F as indicated. The stiffeners and nodes are numbered as shown.

To explain the general analysis approach, we cut the stiffeners free from the skin field and consider the resulting free-body diagram according to Fig. 15.3. Herein, $N_{i,j}$ is the normal force of the stiffener i at node j ($i, j = 1, 2, 3, 4$). The distinction according to the node is necessary because the free-body image of Fig. 15.3 clearly shows that the stiffener forces will be variable along the stiffener length due to the acting shear flow T_i. If the shear flow in the skin field is constant, this will result in linearly varying stiffener forces N_i (see also Chap. 5).

We begin the considerations by examining the skin field and establish the equilibrium of forces in the horizontal direction. We obtain:

$$T_1 a - T_3 a = 0, \tag{15.1}$$

which immediately leads to the equality of the two shear flows T_1 and T_3:

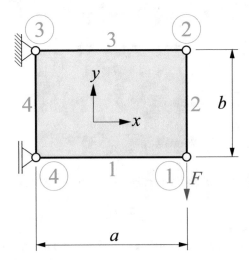

Fig. 15.2 Plane shear field beam under point load with a rectangular skin field

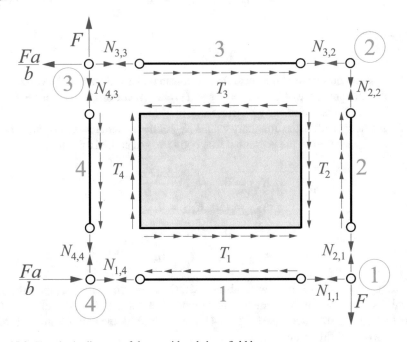

Fig. 15.3 Free-body diagram of the considered shear field beam

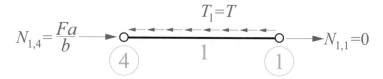

$$N_{1,4}=\frac{Fa}{b} \qquad\qquad\qquad N_{1,1}=0$$

Fig. 15.4 Free-body diagram of stiffener 1

$$T_1 = T_3. \tag{15.2}$$

Analogously, the vertical force balance leads to the equality of the two shear flows T_2 and T_4:

$$T_2 = T_4. \tag{15.3}$$

If, in addition, we set up the moment balance around the lower left corner of the skin field, we get:

$$T_2ba - T_3ab = 0, \tag{15.4}$$

which leads to

$$T_2 = T_3. \tag{15.5}$$

Thus, it can be seen that the shear flow is constant at all edges and thus also in the entire skin field because of $T_1 = T_2 = T_3 = T_4 = T$. Accordingly, the normal forces N_i in each stiffener are linearly distributed.

We further calculate the boundary values $N_{i,j}$ of the stiffener forces N_i and obtain from the observation of the presented free body images at the individual nodes:

$$
\begin{aligned}
N_{1,1} &= 0, & N_{1,4} &= -\frac{Fa}{b}, \\
N_{2,1} &= F, & N_{2,2} &= 0, \\
N_{3,2} &= 0, & N_{3,3} &= \frac{Fa}{b}, \\
N_{4,3} &= F, & N_{4,4} &= 0.
\end{aligned}
\tag{15.6}
$$

From the observation of the equilibrium at any stiffener, here for example stiffener 1, we can then determine both the magnitude of the shear flow T and its actual direction. Using stiffener 1 (see Fig. 15.4), we obtain from the equilibrium of forces in the horizontal direction:

$$Ta = \frac{Fa}{b}, \tag{15.7}$$

and thus:

$$T = \frac{F}{b}. \tag{15.8}$$

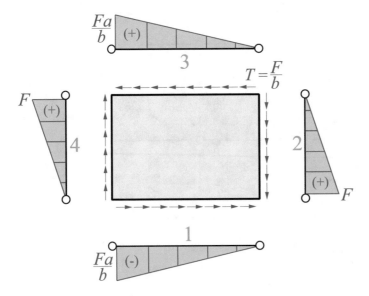

Fig. 15.5 Distributions of stiffener forces

The distributions of the stiffener forces are shown in Fig. 15.5.

For comparison, Fig. 15.6 shows the distribution of the member forces for a similar truss, where the skin has been replaced by a diagonal strut between nodes 3 and 1. It can be clearly seen here that the two members 2 and 3 exhibit zero normal forces and the force transfer is essentially provided by the tensile diagonal as well as by the

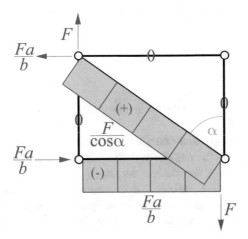

Fig. 15.6 Bar forces in a similar ideal truss

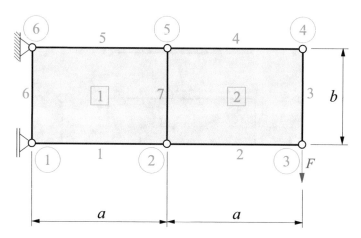

Fig. 15.7 Plane two-span shear field beam under point force with two rectangular skin fields stiffened on all sides

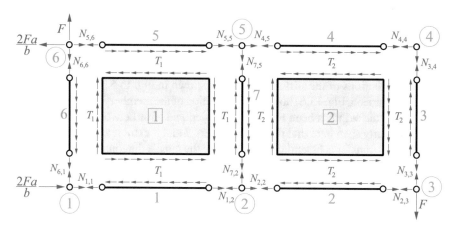

Fig. 15.8 Free-body diagram of the two-span shear field beam

compression member 1. However, the analogy of the skin field to the diagonal bar is clearly evident from this figure.

The procedure shown can be extended quite easily to shear field beams consisting of several skin fields. Let us consider the two-span shear field beam of Fig. 15.7 under the single force F. Again, the stiffener forces $N_1, \ldots, N_7$ and the shear flows T_1, T_2 are to be determined. The corresponding free-body diagram is shown in Fig. 15.8.

We start the considerations with the investigation of stiffener 3. The boundary values of the stiffener force N_3 can be read directly from the free-body diagram and are given as:

$$N_{3,3} = F, \quad N_{3,4} = 0. \tag{15.9}$$

In this stiffener, the normal force distribution is linear as shown before due to the shear flow T_2 assumed to be constant. We can determine the magnitude and direction of the shear flow T_2 from the vertical force balance at stiffener 3:

$$T_2 b = F, \tag{15.10}$$

so that:

$$T_2 = \frac{F}{b}. \tag{15.11}$$

The direction of action of the shear flow T_2 thus corresponds to the direction assumed in Fig. 15.8.

If we now turn to stiffener 7, we obtain the two boundary values $N_{7,2}$ and $N_{7,5}$ to be zero, that is:

$$N_{7,2} = N_{7,5} = 0. \tag{15.12}$$

Stiffener 7 is therefore not involved in the load transfer under this specific load case. Based on the free-body diagram of stiffener 7, we can then directly conclude the magnitude and direction of the shear flow T_1. It follows immediately that $T_1 = T_2$ must hold, thus:

$$T_1 = \frac{F}{b}. \tag{15.13}$$

The direction of action of T_1 is identical to that assumed in Fig. 15.8.

We further consider stiffener 2 which has the boundary value $N_{2,3} = 0$. We cannot determine the boundary value $N_{2,2}$ directly from the observation of the nodal free-body image at node 2. Therefore, we form the horizontal force equilibrium on the free-body diagram of stiffener 2 and obtain:

$$N_{2,2} + T_2 a = 0, \tag{15.14}$$

from which we obtain:

$$N_{2,2} = -\frac{Fa}{b}. \tag{15.15}$$

Thus, stiffener 2 is a compression member with the boundary values 0 at node 3 and $-\frac{Fa}{b}$ at node 2. From the free-body diagram of node 2 we can also observe that $N_{1,2} = N_{2,2} = -\frac{Fa}{b}$ must hold, so that apparently the value of the stiffener force N_2 at this node transfers directly into stiffener 1. At the free-body diagram of node 1 we also note that the boundary value $N_{1,1}$ of the stiffener force N_1 must take the value $N_{1,1} = -\frac{2Fa}{b}$. Obviously, in stiffener 1 the force N_1 is linear with the boundary values $N_{1,1} = -\frac{2Fa}{b}$ and $N_{1,2} = -\frac{Fa}{b}$.

The calculation of all other stiffener forces is carried out in an analogous way and will not be explained further at this point. The resulting member forces and shear flows are shown in Fig. 15.9.

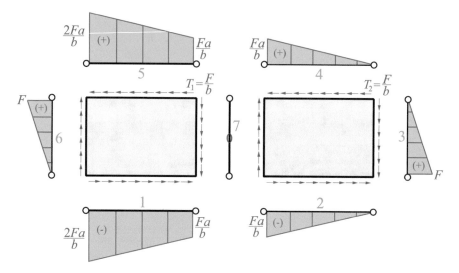

Fig. 15.9 Stiffener forces and shear flows

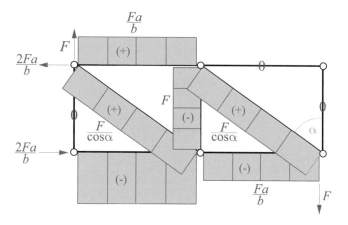

Fig. 15.10 Bar forces in a similar ideal truss

A comparison with a similar truss with diagonal bars instead of skin fields is given in Fig. 15.10. Again, the similarity of the load transfer behavior can be seen.

15.2.2 Determination of Deformations

Once the stiffener forces and shear flows in the skin panels are determined, dimensioning of the structure can be carried out and the static strength can be verified. However, the determination of deformations of shear field beams is also of interest which we will consider in the following. For this purpose, we again consider the shear field beam of Fig. 15.2 and apply the principle of work and energy $W_i = W_a$ to determine the vertical displacement of the force application point. We want to assume linear-elastic material behavior, and for simplicity all stiffeners have identical and constant extensional stiffnesses EA.

We first consider the internal work. For a stiffener with constant extensional stiffness EA, we obtain the internal work W_i as follows:

$$W_i = \frac{1}{2EA} \int_0^l N^2 dx. \tag{15.16}$$

Since there are a total of four stiffeners, the contributions of the individual stiffeners are added up:

$$W_i = \frac{1}{2EA} \sum_{j=1}^4 \int_0^{l_j} N_j^2 dx, \tag{15.17}$$

where $l_1 = l_3 = a$ and $l_2 = l_4 = b$. The integration of the linear stiffener forces is again performed with the help of the integral table of Fig. 5.8. After a short calculation we obtain:

$$W_i = \frac{F^2 b}{3EA} \left[1 + \left(\frac{a}{b}\right)^3 \right]. \tag{15.18}$$

We furthermore introduce the global reference system x, y as shown in Fig. 15.2 and consider the internal work of the skin field, where only the shear stress τ_{xy} requires consideration:

$$dW_i = \int_0^{\gamma_{xy}} \tau_{xy} d\gamma_{xy} dV = U_0 dV. \tag{15.19}$$

Assuming linear elasticity with $\tau_{xy} = G\gamma_{xy}$, we obtain:

$$dW_i = \frac{1}{2} \tau_{xy} \gamma_{xy} dV, \tag{15.20}$$

and after integration over the volume:

$$W_i = \frac{1}{2} \int_V \tau_{xy} \gamma_{xy} dV. \tag{15.21}$$

With $dV = dx dy dz$ we obtain:

$$W_i = \frac{1}{2} \int_{-\frac{t}{2}}^{\frac{t}{2}} \int_0^b \int_0^a \tau_{xy} \gamma_{xy} dx dy dz. \tag{15.22}$$

The integration with respect to z can be easily performed with the skin field thickness t and the relation $T = \tau_{xy} t$:

$$W_i = \frac{1}{2} \int_0^b \int_0^a T \gamma_{xy} \mathrm{d}x \mathrm{d}y. \tag{15.23}$$

With $\gamma_{xy} = \frac{T}{Gt}$, (15.23) can be easily solved. Accordingly, we obtain for the strain energy of the rectangular skin field:

$$W_i = \frac{T^2 ab}{2Gt}. \tag{15.24}$$

In the case of a shear field beam with n skin fields, the individual contributions are to be summed up:

$$W_i = \sum_{j=1}^{n} \frac{T_j^2 A_j}{2G_j t_j}. \tag{15.25}$$

For the present case, we obtain the total internal work W_i as:

$$W_i = \frac{F^2 b}{3EA} \left[1 + \left(\frac{a}{b} \right)^3 \right] + \frac{T^2 ab}{2Gt}. \tag{15.26}$$

The external work is:

$$W_a = \frac{1}{2} Fw. \tag{15.27}$$

Application of $W_i = W_a$ then results with $T = \frac{F}{b}$:

$$\frac{F^2 b}{3EA} \left[1 + \left(\frac{a}{b} \right)^3 \right] + \frac{F^2 a}{2Gbt} = \frac{1}{2} Fw. \tag{15.28}$$

Solving for w yields:

$$w = \frac{2Fb}{3EA} \left[1 + \left(\frac{a}{b} \right)^3 \right] + \frac{Fa}{Gbt}. \tag{15.29}$$

To determine deformations of shear field beams, we can also use the force method. If there is a shear field beam with m stiffeners and n skin fields, a displacement w can be calculated as:

$$w = \sum_{i=1}^{m} \int_{l_i} \frac{N_i \bar{N}_i}{E_i A_i} \mathrm{d}x + \sum_{i=1}^{n} \frac{T_i \bar{T}_i A_i}{G_i t_i}, \tag{15.30}$$

where $\bar{N}_i$ and $\bar{T}_i$ are the stiffener forces and shear flows, respectively, due to a virtual single force at the point where the deformation is to be determined, pointing in the direction of this deformation. The quantities A_i, G_i and t_i are the area, the shear modulus and the thickness of the skin field i, respectively.

15.3 Parallelogram Skin Fields

As the next basic form of skin fields we want to treat skin fields with parallelogram planforms. Such a skin field is shown in Fig. 15.11. Let the considered skin field lie in the $xy-$plane and have the side lengths a and b as well as the inclination angle α as shown. Again, let us assume that the shear flows T_1, T_2, T_3 and T_4 are constant along the edges of the skin field.

We first consider the vertical equilibrium of forces on the parallelogram field under consideration and obtain:

$$T_1 \frac{b}{\cos \alpha} \cos \alpha - T_3 \frac{b}{\cos \alpha} \cos \alpha = 0, \tag{15.31}$$

from which immediately follows the equality of the two shear flows T_1 and T_3:

$$T_1 = T_3. \tag{15.32}$$

Analogously, we form the equilibrium of forces in the horizontal direction and obtain:

$$T_2 a - T_4 a + T_1 b \tan \alpha - T_3 b \tan \alpha = 0. \tag{15.33}$$

With $T_1 = T_3$ this results in the identity of T_2 and T_4:

$$T_2 = T_4. \tag{15.34}$$

Finally, we consider the moment equilibrium with respect to the bottom left corner point of the parallelogram and obtain:

Fig. 15.11 Parallelogram skin field

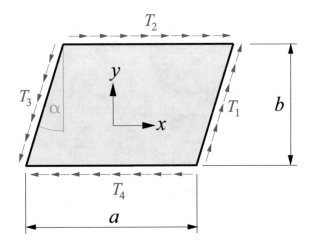

$$T_2ab - T_1\frac{b}{\cos\alpha}a\cos\alpha = 0, \tag{15.35}$$

from which we can conclude the equality of T_1 and T_2:

$$T_1 = T_2. \tag{15.36}$$

This proves that all shear flows shown in Fig. 15.11 assumed to be constant are identical with the value T:

$$T_1 = T_2 = T_3 = T_4 = T. \tag{15.37}$$

Thus, linear stiffener forces will again result in shear field beams with skin fields of parallelogram-shaped planforms.

Regarding the use of energy-based methods for the determination of deformations some remarks have to be made. It turns out that in a parallelogram-shaped skin field under constant shear flow T, a normal force flow n_{xx} in the $x-$direction must also occur as can easily be seen from the sectional view of Fig. 15.12. The horizontal sum of forces then results in:

$$n_{xx}b - T\frac{b}{\cos\alpha}\sin\alpha - Tb\tan\alpha = 0, \tag{15.38}$$

which leads to

$$n_{xx} = 2T\tan\alpha \tag{15.39}$$

and thus the normal stress

$$\sigma_{xx} = \frac{n_{xx}}{t} = 2\frac{T}{t}\tan\alpha. \tag{15.40}$$

The contribution of the normal force flow must be taken into account when determining the internal work or strain energy W_i. Assuming linear elasticity, we obtain:

Fig. 15.12 Section through the parallelogram field, resulting normal force flow n_{xx}

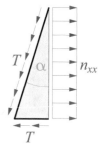

$$W_i = \frac{1}{2} \int_V \left(\sigma_{xx} \varepsilon_{xx} + \tau_{xy} \gamma_{xy} \right) dV. \tag{15.41}$$

The integration with respect to the thickness direction z can be performed in an elementary simple way and leads to an expression in the two force flows n_{xx} and T:

$$W_i = \frac{1}{2} \int_A \left(n_{xx} \varepsilon_{xx} + T \gamma_{xy} \right) dA. \tag{15.42}$$

With $\varepsilon = \frac{\sigma_{xx}}{E} = \frac{n_{xx}}{Et}$, $\gamma_{xy} = \frac{\tau_{xy}}{G} = \frac{T}{GT}$ and $n_{xx} = 2T \tan \alpha$ we obtain:

$$W_i = \frac{1}{2} \int_A \left(\frac{4T^2 \tan^2 \alpha}{Et} + \frac{T^2}{Gt} \right) dA. \tag{15.43}$$

All quantities appearing in the integrand are constant over the planform of the parallelogram field, so we can also write $\int_A dA = A$ and $E = 2G(1 + \nu)$:

$$W_i = \frac{T^2}{2Gt} \left(1 + \frac{2 \tan^2 \alpha}{1 + \nu} \right) A. \tag{15.44}$$

This expression for W_i can then be used, for example, to calculate deformations of parallelogram-shaped fields using the work and energy theorem $W_i = W_a$, whereby the proportion of the stiffener forces are to be added here.

15.4 Trapezoidal Skin Fields

We consider a trapezoidal skin field (see Göldner et al. (1985) or Linke and Nast (2015) for further elaborations), as shown in Fig. 15.13. The trapezoidal field is defined by two straight lines which intersect at one point. This point is at the same time the origin of the x−axis. Let the two angles α_0 and α_1 be enclosed between the x−axis and the two straight lines bounding the trapezoidal field. The distance from the origin to the left edge of the trapezoidal field is a_0, the width of the field with respect to the x−direction is a_1. The lengths of the two vertical field edges are b_0 and b_1. The trapezoidal field is bounded by the stiffeners 1, 2, 3 and 4. Since in a trapezoidal field constructed in this way we can no longer assume a priori that the shear flows along the field edges are constant, we have instead sketched the mean values $\bar{T}_1$, $\bar{T}_2$, $\bar{T}_3$ and $\bar{T}_4$ which can be determined from the actual edge shear flows T_1, T_2, T_3 and T_4 as:

$$\bar{T}_i = \frac{1}{l_i} \int_0^{l_i} T_i(x_i) dx_i. \tag{15.45}$$

Here l_i is the length of the edge under consideration which is distinguished by the edge-parallel local axis x_i (not shown in Fig. 15.13). The quantity T_i is the occurring edge shear flow which will generally be a function of the coordinate x_i.

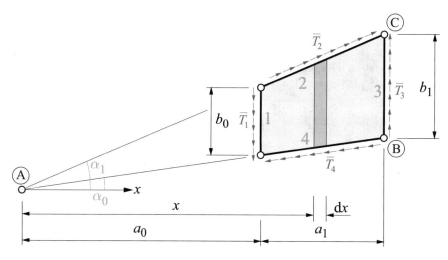

Fig. 15.13 Trapezoidal skin field

We now use the relation $\frac{b_0}{b_1} = \frac{a_0}{a_0+a_1}$ and consider the three moment equilibria around the points A, B and C marked in Fig. 15.13, respectively. It follows:

$$\bar{T}_3 b_1 (a_0 + a_1) - \bar{T}_1 b_0 a_0 = 0,$$

$$\bar{T}_2 \frac{a_1}{\cos \alpha_1} b_1 \cos \alpha_1 - \bar{T}_1 b_0 a_1 = 0,$$

$$\bar{T}_4 \frac{a_1}{\cos \alpha_0} b_1 \cos \alpha_0 - \bar{T}_1 b_0 a_1 = 0. \tag{15.46}$$

Hereby it is possible to express the mean shear flows $\bar{T}_2$, $\bar{T}_3$ and $\bar{T}_4$ by the mean shear flow $\bar{T}_1$:

$$\bar{T}_3 = \bar{T}_1 \left(\frac{b_0}{b_1}\right)^2, \quad \bar{T}_2 = \bar{T}_4 = \bar{T}_1 \frac{b_0}{b_1}. \tag{15.47}$$

This shows that the mean shear flows at the individual field edges of a trapezoidal field are not identical. However, since the shear flows meeting at the field corners must be identical for reasons of equilibrium, it follows that the shear flows along the edges cannot be constant.

We now consider a sectional element of the trapezoidal field of Fig. 15.14, as already indicated in Fig. 15.13. The section element has the width dx and is bounded by the two vertical edges with the lengths $b(x)$ and $b(x) + db(x)$. The shear flows $T(x)$ and $T(x) + dT(x)$ act on these edges, respectively. We want to consider here the actually occurring shear flows T_2 and T_4 at the top and bottom. The equilibrium of forces in the x-direction then results in:

$$T_4 \frac{dx}{\cos \alpha_0} \cos \alpha_0 - T_2 \frac{dx}{\cos \alpha_1} \cos \alpha_1 = 0, \tag{15.48}$$

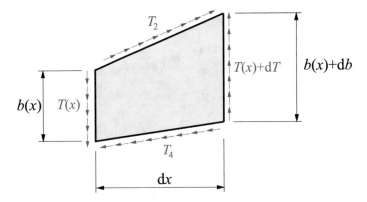

Fig. 15.14 Sectional element of the trapezoidal skin field

which leads directly to the equality of the two shear flows T_2 and T_4:

$$T_2 = T_4. \tag{15.49}$$

Thus, not only the two mean values $\bar{T}_2$ and $\bar{T}_4$ are identical, but also the shear flows T_2 and T_4 acting locally at these edges. It is also easy to prove that no normal force flow n_{xx} occurs in such a trapezoidal field.

It also follows from Eq. (15.48) and the resulting equality of the two shear flows T_2 and T_4 that the shear flows at the two field edges in question do not change over x, so that the mean shear flows coincide with the actual shear flows. It can also be concluded that the two shear flows T_1 and T_3 are constant along their respective edges, so that the use of the mean values can be dispensed with here:

$$\bar{T}_1 = T_1, \quad \bar{T}_3 = T_3. \tag{15.50}$$

Accordingly, it holds for Eq. (15.47):

$$T_3 = T_1 \left(\frac{b_0}{b_1}\right)^2, \quad \bar{T}_2 = \bar{T}_4 = T_1 \frac{b_0}{b_1}. \tag{15.51}$$

We now consider the vertical equilibrium of forces at the section element of Fig. 15.14:

$$- Tb + (T + dT)(b + db) - T_4 \frac{dx}{\cos \alpha_0} \sin \alpha_0 + T_2 \frac{dx}{\cos \alpha_1} \sin \alpha_1 = 0, \tag{15.52}$$

where we can set $T_2 = T_4 = T$. With $db = dx(\tan \alpha_1 - \tan \alpha_0)$ we get:

$$\frac{dT}{T(x)} = -2 \frac{db}{b(x)}. \tag{15.53}$$

We integrate this equation over x and use the limits T_1 and $T(x)$ respectively b_0 and $b(x)$ and get:

$$\ln\left(\frac{T(x)}{T_1}\right) = \ln\left(\frac{b_0}{b(x)}\right)^2,$$ (15.54)

which can be solved directly for $T(x)$:

$$T(x) = T_1\left(\frac{b_0}{b(x)}\right)^2 = T_1\left(\frac{a_0}{a_0+x}\right)^2.$$ (15.55)

As mentioned, the shear flows along the edges of stiffeners 2 and 4 do not show constant distributions. The corresponding stiffener forces $N_2(x)$ and $N_4(x)$ can be determined from the following equation:

$$N_i(x) = \int T(x)\mathrm{d}x + C,$$ (15.56)

where the integration constant follows from the initial value $N_{i,0}$ of the stiffener force at the position $x_i = 0$. Inserting (15.55) then results in:

$$N_2(x) = N_{2,0} + \frac{T_1}{\cos\alpha_1} \frac{x}{1 + \dfrac{x}{a_1}\left(\dfrac{b_1}{b_0} - 1\right)},$$

$$N_4(x) = N_{4,0} + \frac{T_1}{\cos\alpha_0} \frac{x}{1 + \dfrac{x}{a_1}\left(\dfrac{b_1}{b_0} - 1\right)}.$$ (15.57)

The two initial values $N_{2,0}$ and $N_{4,0}$ are the values of the corresponding stiffener force at the location $x = 0$, i.e. according to Fig. 15.13 at the respective left end of the stiffener. Equation (15.51) shows that the product of the shear flows of two opposite field edges is identical, i.e. $T_1 T_3 = \bar{T}_2 \bar{T}_4$. We can use this fact to calculate an average shear flow T_m for the trapezoidal field which we want to approximately assume to act at every location of the entire trapezoidal field. Thus:

$$T_m = \sqrt{T_1 T_3} = \sqrt{\bar{T}_2 \bar{T}_4}.$$ (15.58)

We will use this mean shear flow T_m to determine deformations of a trapezoidal skin field. The shear flow $T(x)$ can then be determined as a function of the mean shear flow T_m at each location x of the trapezoidal field:

$$T(x) = T_m \frac{b_1}{b_0} \frac{a_0^2}{(a_0 + x)^2}.$$ (15.59)

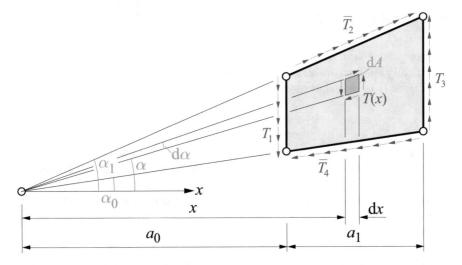

Fig. 15.15 Infinitesimal element of the trapezoidal field

To specify the strain energy of the trapezoidal field we assume that the infinitesimal element dA of Fig. 15.15 may be taken approximately as a parallelogram. This allows us to employ the solution already provided for the parallelogram skin field. The area element dA then results as:

$$dA = \frac{(a_0 + x)d\alpha}{\cos \alpha} \frac{dx}{\cos \alpha}. \tag{15.60}$$

For the incremental strain energy dW_i we then obtain:

$$dW_i = \frac{T^2(x)}{2Gt} \left(1 + \frac{2}{1+\nu} \tan^2 \alpha\right) dA. \tag{15.61}$$

With (15.59) and (15.61) we achieve:

$$W_i = \int_0^{a_1} \int_{\alpha_0}^{\alpha_1} \frac{T_m^2}{2Gt} \left(\frac{b_1}{b_0}\right)^2 a_0^4 \frac{1 + \frac{2}{1+\nu} \tan^2 \alpha}{(a_0 + x)^3 \cos^2 \alpha} d\alpha dx. \tag{15.62}$$

Performing the integrations prescribed here finally yields:

$$W_i = \frac{T_m^2 A}{2Gt} \left[1 + \frac{2\left(\tan^2 \alpha_0 + \tan \alpha_0 \tan \alpha_1 + \tan^2 \alpha_1\right)}{3(1+\nu)}\right]. \tag{15.63}$$

15.5 Statically Indeterminate Shear Field Beams

The analysis of shear field beams is to be distinguished according to the calculation of statically determinate structures and statically indeterminate shear field beams. The degree U of static indeterminacy is defined as:

$$U = r + s + m - 2k. \tag{15.64}$$

Herein r is the number of support reactions, s the number of stiffeners, m the number of shear fields, and k the number of nodes of the considered shear field beam. If $U = 0$ holds, then the beam is statically determinate. For negative values of U, the structure is statically underdetermined. However, if there are positive values for U, then the shear field beam is statically indeterminate, and appropriate calculation means must be used.

We illustrate the procedure using the example of Fig. 15.16 in which the shear field beam of Fig. 15.7 is shown again, but here it consists of four skin panels of identical dimensions and properties. All of the stiffeners also have identical properties. The load consists of a single force F as shown. It can be easily shown that here the degree of indeterminacy U takes the value $U = 1$. Since the support situation of the shown shear field beam is statically determinate, an internal indeterminacy is present here. We want to use the force method for the analysis of the statically indeterminate shear field beam.

To determine the stiffener forces N_i and the shear flows T_i of the skin fields, we make the system statically determinate by removing skin field 2. The resulting stiffener forces and shear flows are shown in Fig. 15.17.

Next, we apply the unit shear flow $\bar{T}_2 = 1$ and calculate the resulting stiffener forces $\bar{N}_i$ and shear flows $\bar{T}_i$. The resulting internal forces are shown in Fig. 15.18.

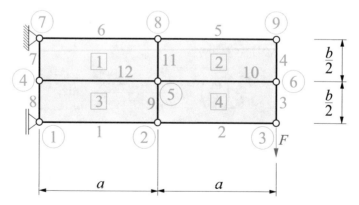

Fig. 15.16 Plane shear field beam under single force with four rectangular skin fields stiffened on all sides

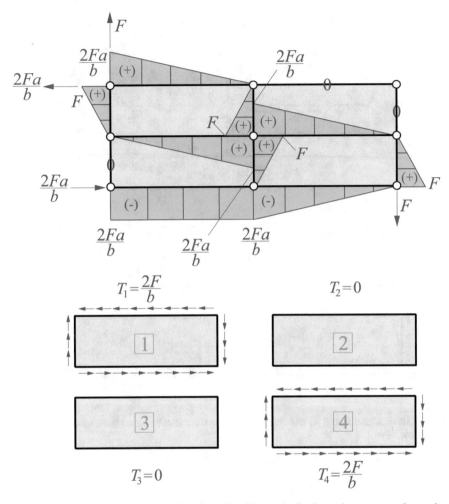

Fig. 15.17 Stiffener forces N_i and shear flows T_i of the statically determinate system due to the single force F

The compatibility condition from which the statically indeterminate quantity can be determined reads:

$$\delta_{10} + X\delta_{11} = 0, \qquad (15.65)$$

with:

$$\delta_{10} = \sum_{i=1}^{12} \int_0^{l_i} \frac{N_i \bar{N}_i}{E_i A_i} dx_i + \sum_{j=1}^{4} \frac{T_i \bar{T}_i A_i}{G_i t_i},$$

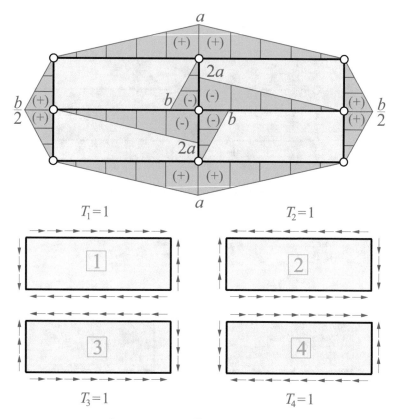

Fig. 15.18 Stiffener forces $\bar{N}_i$ and shear flows $\bar{T}_i$ on the statically determinate system due to unit shear flow $\bar{T}_2 = 1$

$$\delta_{11} = \sum_{i=1}^{12} \int_0^{l_i} \frac{\bar{N}_i \bar{N}_i}{E_i A_i} dx_i + \sum_{j=1}^{4} \frac{\bar{T}_i \bar{T}_i A_i}{G_i t_i}. \tag{15.66}$$

Evaluation yields:

$$\delta_{10} = -\frac{F}{EA}\left(\frac{4a^3}{b} + \frac{b^2}{4}\right) - \frac{2Fa}{Gt},$$

$$\delta_{11} = \frac{1}{EA}\left(4a^3 + \frac{b^3}{2}\right) + \frac{2ab}{Gt}, \tag{15.67}$$

so that:

$$X = -\frac{\delta_{10}}{\delta_{11}} = \frac{\dfrac{F}{EA}\left(\dfrac{4a^3}{b} + \dfrac{b^2}{4}\right) + \dfrac{2Fa}{Gt}}{\dfrac{1}{EA}\left(4a^3 + \dfrac{b^3}{2}\right) + \dfrac{2ab}{Gt}}. \tag{15.68}$$

Thus, the statically indeterminate shear flow is known, and all further internal forces can be calculated. However, this remains without representation at this point.

15.6 Applications of the Shear Field Beam Model

Shear field beam idealizations are used in a wide variety of applications in lightweight construction and design. A selection is presented in the following.

15.6.1 Flexurally Rigid Beam Connections

The shear field beam model is suitable for the analysis of flexurally stiff beam connections such as frame corners or beam sections with variable cross-sectional properties. The present section is adapted from explanations in Petersen (1997) and Hanswille (2013). For motivation, consider the frame of Fig. 12.10 (Chap. 12) (see also Fig. 15.19). Let the given frame (height h, width l) be supported at the base points of the members and loaded by the constant line load q on the horizontal beam. The

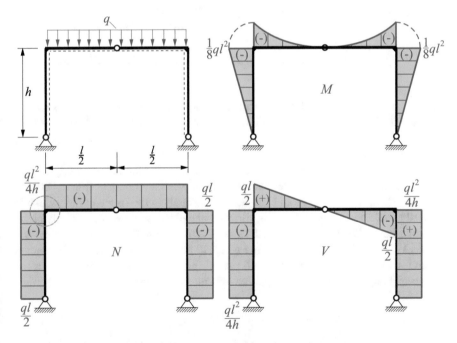

Fig. 15.19 Frame structure under line load q and associated internal forces and moments M, N, V

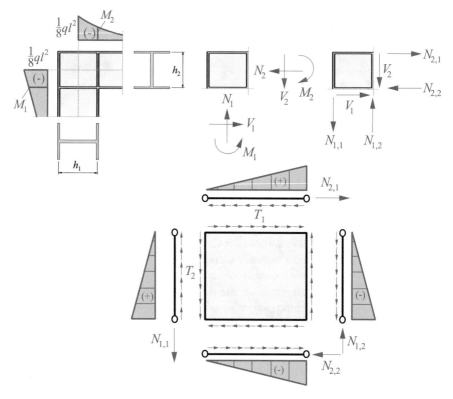

Fig. 15.20 Frame corner with internal forces and moments; associated shear field beam model

beam is partitioned by an ideal moment joint. Both the uprights and the horizontal beam are designed as I-sections (web heights h_1 and h_2, Fig. 15.20). The structural design of a frame corner, i.e. the flexurally rigid connection between the uprights and the horizontal beam, depends on the task at hand, but in order to avoid a local risk of buckling it will in all cases be designed in such a way that the flanges are guided in a cross pattern up to the opposite flanges of the adjacent component (Fig. 15.20, top left). The shear field beam model can be used to analyze the frame corner which is primarily subjected to shear. For this purpose, one places two sections exactly at the bounding flange lines as indicated and calculates the internal forces and moment N_1, N_2, V_1, V_2, M_1 and M_2 acting there (in Fig. 15.20 indicated for the bending moments M_1 and M_2), which are converted into partial internal forces $N_{2,1}$ and $N_{2,2}$ in the horizontal beam and $N_{1,1}$ and $N_{1,2}$ in the upright as follows:

$$N_{2,1} = -\frac{N_2}{2} + \frac{M_2}{h_2}, \quad N_{2,2} = \frac{N_2}{2} + \frac{M_2}{h_2},$$

$$N_{1,1} = -\frac{N_1}{2} + \frac{M_1}{h_1}, \quad N_{1,2} = \frac{N_1}{2} + \frac{M_1}{h_1}. \tag{15.69}$$

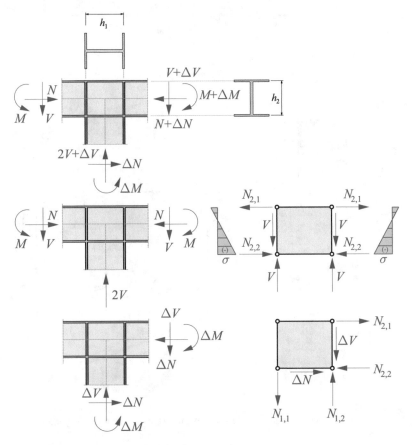

Fig. 15.21 Shear field beam model of a continuous beam

The resulting shear field beam model is shown in Fig. 15.20, bottom. From this, the linear stiffener forces and the shear flows $T_1 = T_2$ can be determined as:

$$T_1 = \frac{N_{2,1}}{h_1}, \quad T_2 = \frac{N_{1,1}}{h_2}. \tag{15.70}$$

If we consider a flexurally stiff connection between a continuous beam and a column perpendicular to it (Fig. 15.21, top), then the internal forces can be conveniently divided into a portion from the continuous action of the beam (Fig. 15.21, middle) and a portion from the redirecting action between the column and the beam (Fig. 15.21, bottom).

A similar approach can be employed in the case of a beam with variable cross-sectional properties, for instance at the transition between two beam sections of different heights (Fig. 15.22, top). The applied bending moments M_L, M_R are converted into partial internal forces $N_{L,1}$, $N_{L,2}$ and $N_{R,1}$, $N_{R,2}$ of the flanges (Fig. 15.22,

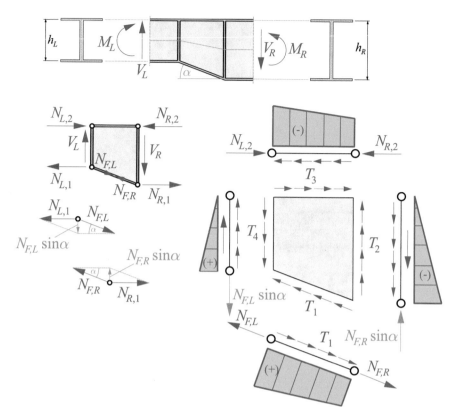

Fig. 15.22 Shear field beam model for a beam with variable cross-sectional properties

bottom left). It should be noted that due to the inclination α of the lower flange, a division of the forces $N_{L,1}$ and $N_{R,1}$ into chord-parallel and vertical components has to be performed. The resulting shear field beam model is shown in Fig. 15.22, bottom right.

15.6.2 Large Area Stiffened Structures

Shear field beam models are very simple to use and, due to the closed-analytical approach, are particularly suitable for purposes such as the preliminary analysis or optimization of stiffened plane structures, i.e. especially when the calculation time is an important factor and when the accuracy of a structural calculation does not have to capture each detail. Shear field beam models allow the fast and uncomplicated analysis of even large-scale planar structures with a large number of stiffeners. However, when interpreting the results of such a calculation, the assumed division of tasks between stiffeners and skin fields must always be taken into account, i.e. the

Fig. 15.23 Shear field beam model for the introduction of a single force into a stiffened disc with three stiffeners, qualitative force distributions, after Wiedemann (2007)

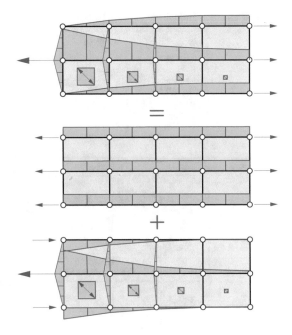

strict separation of the normal force and strain effect on the one hand, and the shear effect on the other, which necessitates a high ratio between the extensional stiffness of the stiffeners and the shear stiffness of the skin fields.

Application examples that emphasize the value of the shear field beam model include load introductions as shown in Fig. 15.23 (Wiedemann (2007)). A separate section will be devoted to the topic of load introductions later on, but it should be noted at this point that a shear field beam model can be used to provide a fairly simple analysis approach to the interplay of stiffener and skin forces in such structures. The division of the situation (15.23, top) into a homogeneous part (15.23, middle) and an edge disturbance part (15.23, bottom) can be clearly seen here. The shear flows in the skin fields are shown as rectangles with inscribed tension diagonals.

Another application example is a disk with a rectangular opening as shown in Fig. 15.24 (after Wiedemann (2007)). For clarity, shear flows and vertical stiffener forces are limited to one side each, horizontal stiffener forces are not shown. The results clearly show the expected boundary perturbation due to the hole as well as its characteristic decay behavior, highlighting the benefit of practicable shear field beam models.

15.6.3 Load Introductions

The introduction of concentrated loads into thin-walled structures, whether they are bars, beams or thin-walled disks, plates or shells, is one of the main tasks of

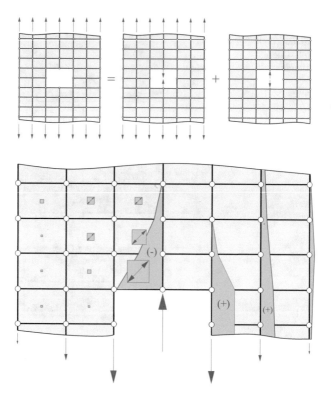

Fig. 15.24 Shear field beam model for a disk with a rectangular hole, qualitative representation of the perturbance forces under tensile load (horizontal forces not shown), after Wiedemann (2007)

lightweight construction and design. In this section, we want to explore to what extent shear field beam models can help to solve this complex and extremely important engineering task with manageable effort under certain idealizations. This section is based on explanations given by e.g. Klein (1994), Dieker and Reimerdes (1992) and Wiedemann (2007). We first turn to the shear field beam model as given by (Fig. 15.25, top). We consider an orthogonally stiffened disk (stiffnesses A_1, A_2, A_V as indicated) loaded by the single force F at its left end and supported at its right end as shown. This situation could be exemplary for the load introduction into a thin-walled beam. The field length is a, and the field height is given with h. The wall thickness is constant with the value t. The given fourfold statically indeterminate problem is solved by a calculation using the force method, and the statically determinate system is shown in Fig. 15.25, middle. It is obtained by removing the lower support and by separating the three marked lower longitudinal stiffeners. The force distribution in the statically determinate system is particularly simple in this case and consists of a constant stiffener force with magnitude F in the middle longitudinal stiffener. All vertical stiffeners as well as skin fields remain unloaded in the statically determinate system under the single force F. The corresponding statically indeterminate stress

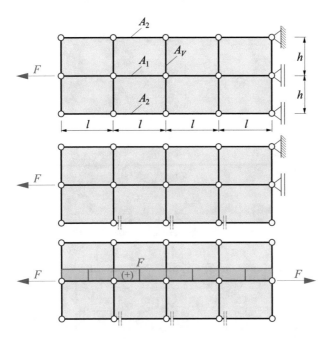

Fig. 15.25 Introduction of a single force into a disk stiffened by three stiffeners, after Dieker and Reimerdes (1992)

states are shown in Fig. 15.26. The compatibility requirements belonging to this system are thus (with $i = 1, 2, 3, 4$):

$$\delta_{i0} + X_1\delta_{1i} + X_2\delta_{2i} + X_3\delta_{3i} + X_4\delta_{4i} = 0, \tag{15.71}$$

where the displacements δ_{ij} result as follows (Dieker and Reimerdes (1992)):

$$\delta_{10} = -\frac{Fl}{EA_1}, \quad \delta_{20} = \delta_{30} = \delta_{40} = 2\delta_{10},$$

$$\delta_{11} = 2\left(\frac{l}{3EA_2} + \frac{2h}{3EA_V} + \frac{h}{Glt}\right) + \frac{4l}{3EA_1},$$

$$\delta_{22} = \delta_{33} = \delta_{44} = 4\left(\frac{l}{3EA_2} + \frac{h}{3EA_V} + \frac{h}{Glt}\right) + 2\left(\frac{4l}{3EA_1} + \frac{4h}{3EA_V}\right),$$

$$\delta_{12} = 2\left(\frac{l}{6EA_2} - \frac{h}{EA_V} - \frac{h}{Glt}\right) + \frac{2l}{3EA_1},$$

$$\delta_{13} = \delta_{24} = \frac{2h}{3EA_V},$$

$$\delta_{23} = \delta_{34} = 2\left(\frac{l}{6EA_2} - \frac{4h}{3EA_V} - \frac{h}{Glt}\right) + \frac{2l}{3EA_1}, \quad \delta_{14} = 0. \tag{15.72}$$

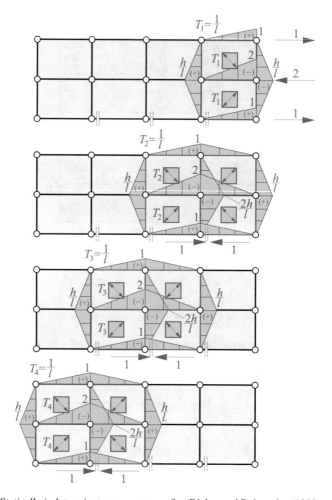

Fig. 15.26 Statically indeterminate stress states, after Dieker and Reimerdes (1992)

With $l = h = 300t$, $A_1 = A_2 = A_V = 200t^2$, and $\frac{G}{A} = \frac{5}{13}$, (15.71) gives the following values for the statically indeterminate quantities:

$$X_1 = 0.333F, \quad X_2 = 0.330F, \quad X_3 = 0.309F, \quad X_4 = 0.225F. \quad (15.73)$$

The force distribution in the stiffened beam has already been qualitatively shown in Fig. 15.23. The local character of the force concentrations in the area of load introduction becomes very obvious here; the upper and lower stiffeners are still completely unloaded at the free edge, and their forces build up gradually with increasing distance from the loaded edge by force transfer from the skin fields. At a sufficient distance from the loaded edge, there is an almost complete uniformity of the stiffener forces;

each stiffener carries one third of the applied force. At the same time, it can be seen that the shear flows in the skin fields decay almost completely with increasing distance from the loaded edge.

An improved and yet quite easily manageable computational model is obtained (so-called semi-continuous model, see Göldner et al. (1985), Klein (1994), Wiedemann (2007), Dieker and Reimerdes (1992)), if it is assumed that in one direction of the model the stiffnesses of both the longitudinal stiffeners and the skin fields are constant and the transverse stiffeners are not placed too far apart. Then this situation can be idealized in such a way that the transverse stiffeners are computationally combined with the skin fields to form a homogeneous orthotropic disk with corresponding effective properties. These ideal skin panels are then able to absorb both shear flows and normal force flows in the transverse direction, whereas longitudinal forces are still applied exclusively to the longitudinal stiffeners. As a result, even with this type of idealization, the shear flow in the skin panels remains constant in the direction perpendicular to the stiffeners and variable in the longitudinal direction, whereas the transverse normal force flow now to be considered here is linear. This type of calculation approach leads to a differential equation with respect to the shear flow of the stiffener-disk model, which explains the term semi-continuous model. In the following, only a certain special case of this model approach is shown, which can also be found in this way, for example, in Dieker and Reimerdes (1992) and Klein (1994).

Again, we consider a disk stiffened by three stiffeners according to Fig. 15.27, top, where we want to assume for simplicity that the transverse stiffeners are not only regularly arranged but also have a sufficiently high extensional stiffness so that we can neglect transverse displacements perpendicular to the stiffeners in the subsequent modeling. Figure 15.27, center left, shows the free-body diagram of an infinitesimal sectional element of length dx, from which the interaction of forces between skin field and stiffener can be concluded. For reasons of symmetry, it is sufficient to consider only one half of the model. The force equilibria at the stiffeners result in:

$$N_2' = T, \quad N_1' = -T, \tag{15.74}$$

where N_1 and N_2 are the stiffener forces, T is the shear flow of the skin field.

The kinematics of the skin field can be seen in Fig. 15.27, middle right. The difference of the two strains $\varepsilon_1 = u_1'$ and $\varepsilon_2 = u_2'$, related to the skin field height h, must correspond to the change of the shear strain γ of the skin field, i.e.:

$$\frac{d\gamma}{dx} = \frac{u_2' - u_1'}{h}. \tag{15.75}$$

With the constitutive law $u_1' = \frac{N_1}{EA_1}, u_2' = \frac{N_2}{EA_2}$ we obtain:

$$\frac{d\gamma}{dx} = \frac{1}{h}\left(\frac{N_2}{EA_2} - \frac{N_1}{EA_1}\right). \tag{15.76}$$

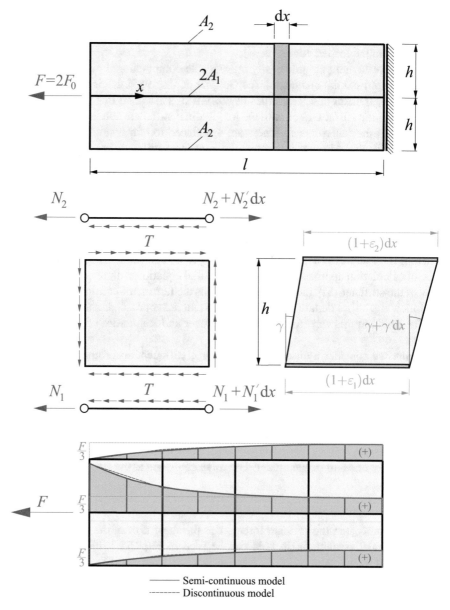

Fig. 15.27 Semi-continuous model of the disk with three stiffeners (top), free-body diagram of an infinitesimal element of the upper half (middle left) as well as associated kinematics (middle right), qualitative distribution of the stiffener forces (bottom), after Dieker and Reimerdes (1992)

Differentiating once with respect to x and using the constitutive law $\gamma = \frac{\tau}{G} = \frac{T}{Gt}$ with (15.74) results in:

$$\frac{d^2 T}{dx^2} - T \frac{Gt}{Eh} \left(\frac{1}{A_1} + \frac{1}{A_2} \right) = 0. \tag{15.77}$$

Introduction of the factor

$$\alpha^2 = \frac{Gt}{Eh} \left(\frac{1}{A_1} + \frac{1}{A_2} \right) \tag{15.78}$$

leads to the following expression:

$$\frac{d^2 T}{dx^2} - \alpha^2 T = 0. \tag{15.79}$$

This is an ordinary linear homogeneous differential equation of second order with constant coefficients for the shear flow T. Once the shear flow has been determined, then the stiffener forces can also be determined using (15.74).

The differential equation (15.79) has the following general solution:

$$T(x) = C_1 \cosh \alpha x + C_2 \sinh \alpha x. \tag{15.80}$$

The two constants C_1 and C_2 are to be adapted to given boundary conditions. For this purpose, it is assumed that the shear flow at the location $x = l$ has completely decayed to zero, i.e. only the stiffeners are involved in the load transfer at a sufficient distance from the edge, i.e. $T(x = l) = 0$. Moreover, at the edge $x = 0$, the outer stiffeners do not carry any load, while the center stiffener carries the full applied load which means $N_1(x = 0) = F_0$ and $N_2(x = 0) = 0$. We now evaluate the equation (15.76) with $\gamma = \frac{T}{Gt}$ at the point $x = 0$ and obtain:

$$\left. \frac{dT}{dx} \right|_{x=0} = -\frac{Gt F_0}{hE A_1}. \tag{15.81}$$

Substituting (15.80) then leads to the constant C_2 as follows:

$$C_2 = -\frac{Gt F_0}{\alpha h E A_1}. \tag{15.82}$$

From the condition $T(x = l) = 0$ the constant C_1 can then be determined with (15.80) as:

$$C_1 = \frac{Gt F_0}{\alpha h E A_1} \frac{\sinh \alpha l}{\cosh \alpha l}. \tag{15.83}$$

Thus, the distribution of the shear flow over x is completely determined and can be formulated as:

$$T(x) = \alpha F_0 \frac{A_2}{A_1 + A_2} \frac{\sinh \alpha(x - l)}{\cosh \alpha l}. \tag{15.84}$$

We obtain the normal force $N_2(x)$ from (15.74) by integrating $T(x)$ as:

$$N_2(x) = -F_0 \frac{A_2}{A_1 + A_2} \frac{\cosh \alpha(x - l)}{\cosh \alpha l} + C_3, \tag{15.85}$$

where the constant C_3 can be determined from the condition $N_2(x = 0) = 0$ as:

$$C_3 = F_0 \frac{A_2}{A_1 + A_2}. \tag{15.86}$$

Finally, for the stiffener force $N_2(x)$ we have:

$$N_2(x) = F_0 \frac{A_2}{A_1 + A_2} \left(1 - \frac{\cosh \alpha(x - l)}{\cosh \alpha l} \right). \tag{15.87}$$

The stiffener force N_1 can be determined in a similar way and follows as:

$$N_1(x) = F_0 \frac{A_1}{A_1 + A_2} \left(1 + \frac{A_2}{A_1} \frac{\cosh \alpha(x - l)}{\cosh \alpha l} \right). \tag{15.88}$$

The qualitative distribution of the stiffener forces N_1 and N_2 is compared with the results of the discontinuous model in Fig. 15.27, bottom. Here, the present semi-continuous method leads to steady and continuous distributions for both the stiffener forces and the shear flow T. Both models obviously yield values of the same order of magnitude. A renewed discussion of the results of Fig. 15.27 can thus be omitted at this point.

15.6.4 Adhesive Overlap Joints

Adhesive overlap joints (Fig. 15.28) are a way of joining two or more thin-walled structural elements and are a typical kind of joint in lightweight engineering. Such overlap joints typically require quite large bonding areas over which the acting forces can be transmitted, bond strengths are typically quite low. Advantages of bonded joints include the ability to join different materials together and the avoidance of notch effects, in contrast to bolted joints. Disadvantages include non-uniform stress states as well as typically low transverse tensile strengths, which may lead to fail-

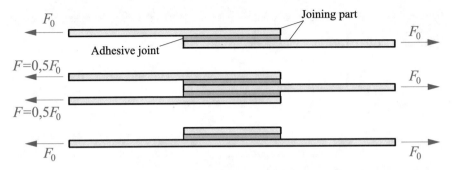

Fig. 15.28 Single lap joint (top), double lap joint (middle), local reinforcement by a doubler (bottom)

ure by peeling, and an increase in structural weight due to the need for overlap surfaces. Adhesive bonding in lightweight construction and the according analysis models will be discussed in detail in a subsequent volume of this book series. At this point, it should suffice to point out that common analysis models can often be understood as idealizations within the framework of shear field beam models. One of the most important representatives of this type of modeling is the so-called Volkersen model (see Volkersen 1938, 1953), which uses a semi-continuous shear field beam model similar to the previous section. Very detailed descriptions of bonded joints in lightweight construction can be found, for example, in Wiedemann (2007) and Schürmann (2007), but a detailed discussion of this specific application of shear field beam models will be omitted here.

References

Czerwenka G, Schnell W (1970) Einführung in die Rechenmethoden des Leichtbaus, Band 1, Bibliographisches Institut, Mannheim, Germany

Dieker S, Reimerdes HG (1992) Elementare Festigkeitslehre im Leichtbau. Donat Verlag, Bremen, Germany

Göldner H, Altenbach J, Eschke K, Garz KF, Sähn S (1979) Lehrbuch Höhere Festigkeitslehre, vol 1. Physik Verlag, Weinheim, Germany

Göldner H, Altenbach J, Eschke K, Garz KF, Sähn S (1985) Lehrbuch Höhere Festigkeitslehre, vol 2. Physik Verlag, Weinheim, Germany

Hanswille G (2013) Vorlesung Stahlbau. Bergische Universität Wuppertal, Germany, Department of civil engineering

Klein B (1994) Leichtbau-Konstruktion. Vieweg, Braunschweig, Germany

Linke M, Nast E (2015) Festigkeitslehre für den Leichtbau. Springer, Berlin, Germany

Petersen C (1997) Stahlbau, 3rd edn. Vieweg, Braunschweig, Germany

Schürmann H (2007) Konstruieren mit Faser-Kunststoff-Verbunden, 2nd edn. Springer, Berlin, Germany

Volkersen O (1938) Die Nietkraftverteilung in zugbeanspruchten Nietverbindungen mit konstanten Laschenquerschnitten. Luftfahrtforschung 1, 41–47

Volkersen O (1953) Die Schubkraftverteilung in Leim-. Niet- und Bolzenverbindungen, Energie und Technik 5, 68–71

Wiedemann J (2007) Leichtbau 1: Elemente. Springer, Berlin, Germany

Chapter 16
The Timoshenko Beam

16.1 Introduction

The Euler-Bernoulli beam theory is based on the fundamental hypothesis that the cross sections remain plane and that the normal hypothesis is valid, i.e. a beam is assumed where shear strains of the cross section are explicitly excluded. While this theory provides sufficiently reliable results for many practical purposes, there are also many applications where these assumptions are no longer acceptable and lead to incorrect results. A qualitative example is shown in Fig. 16.1. If a beam structure consisting of a shear deformable material is considered, or if we treat a beam whose length l is no longer significantly larger than, for example, its cross-sectional height h, then the shear deformations (Fig. 16.1, middle right) occurring in addition to the bending deformation (Fig. 16.1, top right) can no longer be neglected, and an overall deformation pattern occurs as shown qualitatively in Fig. 16.1, bottom right. In such cases, the Euler-Bernoulli beam theory (see Chap. 5) with its strict specifications in the form of the hypothesis of plane cross-sections and the normal hypothesis reaches its limits, and improved theories have to be found. Therefore, in this chapter we want to deal with a beam theory which relaxes the rather strict restrictions of the Euler-Bernoulli beam theory to a degree and which can provide an improvement of the calculation results especially for structures where shear strains play a role. This is the beam theory according to Timoshenko,[1] and in this context one also speaks of the so-called Timoshenko beam or simply of a shear deformable beam.

---

[1]Stephen Timoshenko, 1878–1972, Ukrainian scientist with important contributions to modern structural mechanics

C. Mittelstedt, *Structural Mechanics in Lightweight Engineering*,
https://doi.org/10.1007/978-3-030-75193-7_16

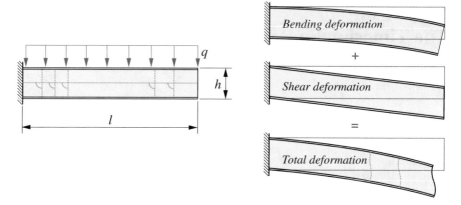

Fig. 16.1 Deformation of a shear deformable beam

16.2   Kinematics and Constitutive Law

The beam theory according to Timoshenko is based on the following assumptions:

- Similar to the Euler-Bernoulli beam theory, we assume that cross-sections remain plane and no warping occurs during deformation. However, the normal hypothesis is discarded, meaning that cross-sections still do not warp when the beam deflects, but that the cross-sections in the deformed state do not necessarily have to be orthogonal to the longitudinal axis of the beam (Fig. 16.1). Shear strains in the beam are thus explicitly allowed, but we assume that they are constant over the cross-section thickness.
- We assume linear-elastic material behavior and assume that Hooke's law is valid.
- It is assumed that the cross-sectional shape does not change during deformation as a result of a given load.
- We assume geometric linearity and assume that the deformations are small with respect to the length of the beam, but also with respect to the dimensions of its cross-section.
- In this chapter, straight beam structures are considered exclusively.

The resulting kinematics is shown in Fig. 16.2. Figure 16.2 shows that here the normal hypothesis has been discarded, but that it is assumed that cross sections remain plane during deformation. As a consequence, the angle φ_y occurs, which here can no longer be related to the inclination of the bending line $\frac{dw}{dx} = w'$ as in Chap. 5 due to the missing requirement that cross sections remain perpendicular to the longitudinal axis x also in the deformed state. Thus, the angle φ_y must be included in the calculations as another unknown quantity and treated accordingly as an independent degree of freedom. In this chapter, we will restrict ourselves exclusively to the case of bending in the $xz-$plane, and let the reference system x, y, z be a principal axis system. The displacement field in the framework of Timoshenko's beam theory is then given as:

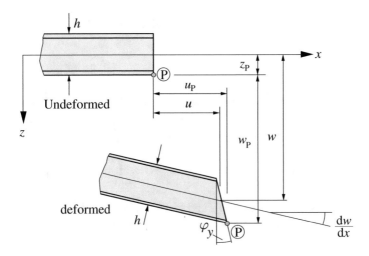

Fig. 16.2 Kinematics of the Timoshenko beam theory

$$u_P = u + z_P \varphi_y,$$
$$w_P = w. \tag{16.1}$$

From this, the strain field of the Timoshenko beam can be calculated, where it should be noted here that due to the kinematics shown in Fig. 16.2, in addition to the longitudinal strain ε_{xx} the shear strain γ_{xz} also occurs. With (2.70) we obtain with respect to the point P:

$$\varepsilon_{xxP} = \frac{du_P}{dx} = \frac{du}{dx} + z_P \frac{d\varphi_y}{dx} = u' + z_P \varphi'_y,$$
$$\gamma_{xzP} = \frac{du_P}{dz} + \frac{dw_P}{dx} = \varphi_y + \frac{dw}{dx} = \varphi_y + w'. \tag{16.2}$$

The considerations made so far apply to any point P on the cross-section of the beam, so we will drop the index P in all that follows.

Using Hooke's law $\sigma_{xx} = E\varepsilon_{xx}$ as well as $\tau_{xz} = G\gamma_{xz}$, the two stress components occurring in the Timoshenko beam can be determined as follows:

$$\sigma_{xx} = E\varepsilon_{xx} = E\left(u' + z\varphi'_y\right),$$
$$\tau_{xz} = G\gamma_{xz} = G\left(\varphi_y + w'\right). \tag{16.3}$$

Once the stress components are known, the internal forces and moments of the Timoshenko beam can be calculated (cf. Chap. 5, Eq. (5.11)). Since we restrict ourselves here to uniaxial bending, the normal stress σ_{xx} yields only the normal force N and the bending moment M_y. In addition, however, we also consider the transverse shear force V_z which is the resultant of the shear stress τ_{xz}. We have:

$$N = \int_A \sigma_{xx} dA, \quad M_y = \int_A \sigma_{xx} z dA, \quad V_z = \int_A \tau_{xz} dA. \tag{16.4}$$

Evaluation with (16.3) results in:

$$N = \int_A \sigma_{xx} dA = E \int_A \left(u' + z\varphi'_y \right) dA = Eu' \int_A dA + E\varphi'_y \int_A z dA,$$

$$M_y = \int_A \sigma_{xx} z dA = E \int_A \left(u' + z\varphi'_y \right) z dA = Eu' \int_A z dA + E\varphi'_y \int_A z^2 dA,$$

$$V_z = \int_A \tau_{xz} dA = G \int_A \left(\varphi_y + w' \right) dA = G\varphi_y \int_A dA + Gw' \int_A dA. \tag{16.5}$$

Using the integrals (5.13), (5.14), and (5.15) and noting that the static moment $S_y = \int_A z dA$ vanishes in a principal axis system yields:

$$N = EAu',$$
$$M_y = EI_{yy}\varphi'_y,$$
$$V_z = GA\left(\varphi_y + w' \right). \tag{16.6}$$

These are the constitutive equations of the Timoshenko beam. Note that in contrast to the Euler-Bernoulli beam theory a constitutive equation for the transverse shear force V_z is available. Since the differential equation concerning the longitudinal displacement remains unchanged compared to the explanations in Chap. 5, we do not consider the associated rod problem further in this chapter.

16.3 Displacements and Stresses

Consider the deflection of a prismatic beam with invariant elastic and cross-sectional properties along its length. Again, we assume bending in the $xz-$plane and the presence of a principal axis system. At this point we make use of the equilibrium conditions (5.95) which apply also to the Timoshenko beam. From $V'_z = -q_z$ we obtain:

$$GA\left(\varphi'_y + w'' \right) = -q_z. \tag{16.7}$$

Using $M'_y = V_z$ yields:

$$EI_{yy}\varphi''_y = GA\left(\varphi_y + w' \right). \tag{16.8}$$

We now integrate (16.7) once with respect to x and obtain:

$$GA\left(\varphi_y + w' \right) = -\int q_z dx + C_1. \tag{16.9}$$

Herein C_1 is an integration constant. We substitute the expression (16.9) into (16.8):

$$EI_{yy}\varphi_y'' = -\int q_z dx + C_1.$$ (16.10)

From this we can derive an expression for φ_y by twofold integration:

$$EI_{yy}\varphi_y' = -\int\int q_z dx dx + C_1 x + C_2,$$

$$EI_{yy}\varphi_y = -\int\int\int q_z dx dx dx + \frac{1}{2}C_1 x^2 + C_2 x + C_3.$$ (16.11)

From Eq. (16.9) we can obtain the following expression for w':

$$w' = \frac{1}{GA}\left(-\int q_z dx + C_1\right) - \varphi_y.$$ (16.12)

Substituting (16.11) then yields:

$$w' = \frac{1}{GA}\left(-\int q_z dx + C_1\right)$$
$$- \frac{1}{EI_{yy}}\left(-\int\int\int q_z dx dx dx + \frac{1}{2}C_1 x^2 + C_2 x + C_3\right).$$ (16.13)

Integrating with respect to x then yields the following expression from which the deflection of the beam can be determined:

$$w = \frac{1}{GA}\left(-\int\int q_z dx dx + C_1 x\right)$$
$$- \frac{1}{EI_{yy}}\left(-\int\int\int\int q_z dx dx dx dx + \frac{1}{6}C_1 x^3 + \frac{1}{2}C_2 x^2 + C_3 x + C_4\right).$$ (16.14)

The integration constants C_1, C_2, C_3, C_4 can be adapted to the given boundary conditions of the considered beam as already explained in Chap. 5. If the beam deflection $w(x)$ is available, the angle $\varphi_y(x)$ can be calculated from (16.11). It should be noted, however, that compared to the Euler-Bernoulli beam theory, in the framework of the Timoshenko theory different boundary conditions have to be applied. This will be discussed later.

At this point, some remarks should be made on the nature of the beam theory presented in this chapter. Timoshenko's beam theory, in contrast to the Euler-Bernoulli beam theory, explicitly allows for shear strains γ_{xz} which, however, will be constant over the section height due to the angle φ_y being considered invariant over h. Accordingly, this is also true for the shear stress τ_{xz}. However, Chap. 6 has clearly shown that the shear flow resulting from the transverse shear force V_z is parabolic over those segments of the cross-section that run in the direction of action of V_z. This discrepancy is accounted for, at least approximately, within the Timoshenko beam theory by multiplying the shear force V_z according to (16.6) by the so-called shear correction factor K:

$$V_z = K \int_A \tau_{xz} \mathrm{d}A = KGA \left(\varphi_y + w'\right).$$ (16.15)

The calculation of the deflection $w(x)$ of the Timoshenko beam can then be performed as follows:

$$w = \frac{1}{KGA} \left(-\int\int q_z \mathrm{d}x\mathrm{d}x + C_1 x \right)$$

$$- \frac{1}{EI_{yy}} \left(-\int\int\int\int q_z \mathrm{d}x\mathrm{d}x\mathrm{d}x\mathrm{d}x + \frac{1}{6}C_1 x^3 + \frac{1}{2}C_2 x^2 + C_3 x + C_4 \right).$$ (16.16)

The introduction of the shear correction factor K can be interpreted as a manipulation of the shear stiffness GA of the beam. Although this does not eliminate the ultimately incorrect result of a constant shear stress, it does provide a correction of the shear stiffness in such a way that this effect is accounted for in an approximate sense. We will discuss the determination of the shear correction factor in more detail later on.

The stresses acting in the shear deformable beam can be calculated by Hooke's law (16.3) in conjunction with (16.6) and are obtained as:

$$\sigma_{xx} = \frac{N_x}{A} + \frac{M_y}{I_{yy}}z, \quad \tau_{xz} = \frac{V_z}{KA}.$$ (16.17)

It can be seen that the expression for the calculation of the normal stress σ_{xx} is the same as that for the shear-rigid beam (Chap. 5, Eq. (5.65)) if uniaxial bending with $M_z = 0$ is assumed. The second equation in (16.17) indicates that in the framework of the Timoshenko beam theory, the shear stress τ_{xz} is found to be constant across the cross section.

It makes sense at this point to go into more detail on the Eq. (16.16) for the calculation of the beam deflection $w(x)$. Indeed, it turns out that the second term in this equation corresponds exactly to the Eq. (5.101) in Chap. 5 for the Euler-Bernoulli beam. Let us denote this term as $w_{EB}(x)$ (the subscript EB stands for the names Euler and Bernoulli). Obviously, the assumption of the shear deformable beam adds another term, which we will denote as $w_S(x)$ (the index S stands for shear). Thus:

$$w(x) = w_{EB}(x) + w_S(x),$$ (16.18)

where:

$$w_{EB}(x) = -\frac{1}{EI_{yy}} \left(-\int\int\int\int q_z \mathrm{d}x\mathrm{d}x\mathrm{d}x\mathrm{d}x + \frac{1}{6}C_1 x^3 + \frac{1}{2}C_2 x^2 + C_3 x + C_4 \right),$$

$$w_S(x) = \frac{1}{KGA} \left(-\int\int q_z \mathrm{d}x\mathrm{d}x + C_1 x \right). \tag{16.19}$$

In the case that the shear stiffness KGA assumes very high values, the expression $w_S(x)$ in (16.18) and (16.19), respectively, loses influence, and the Timoshenko beam theory changes to the Euler-Bernoulli beam theory.

16.4 Elementary Examples

We consider again the example of Chap. 5, Fig. 5.28 (see also Fig. 16.3). It is easy to show for this elementary example that the deflection $w(x)$ can be calculated with the boundary conditions

$$w(x = 0) = 0, \quad \varphi_y(x = 0) = 0, \quad M_y(x = l) = 0, \quad V_z(x = l) = F \tag{16.20}$$

and the resulting integration constants C_1, C_2, C_3 and C_4

$$C_1 = F, \quad C_2 = -Fl, \quad C_3 = 0, \quad C_4 = 0 \tag{16.21}$$

as

$$w(x) = \frac{Fl^3}{6EI_{yy}} \left[3 \left(\frac{x}{l} \right)^2 - \left(\frac{x}{l} \right)^3 \right] + \frac{Fl}{KGA} \left(\frac{x}{l} \right). \tag{16.22}$$

The maximum deflection w_{max} at the end of the beam is then given as:

$$w_{max} = \frac{Fl^3}{3EI_{yy}} + \frac{Fl}{KGA}. \tag{16.23}$$

It can be seen that this solution agrees with the solution already determined in (5.105) and (5.106) for the bending line and the maximum value of the deflection of

Fig. 16.3 Cantilever beam under point load

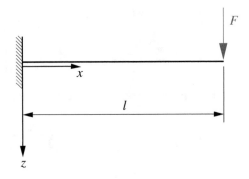

Fig. 16.4 Simply supported
beam under constant line
load

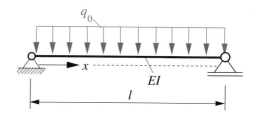

the shear rigid beam. In addition, however, the component due to transverse shear
comes into play here.

As another elementary example we consider the beam simply supported at both
ends under the constant line load q_0 (Fig. 16.4). The boundary conditions to be
considered here are:

$$w(x = 0) = 0, \quad M_y(x = 0) = 0, \quad w(x = l) = 0, \quad M_y(x = l) = 0. \quad (16.24)$$

The Eq. (16.16) takes the following form with $q = q_0$:

$$w(x) = \frac{1}{KGA}\left(-\frac{1}{2}q_0x^2 + C_1x\right)$$
$$-\frac{1}{EI_{yy}}\left(-\frac{1}{24}q_0x^4 + \frac{1}{6}C_1x^3 + \frac{1}{2}C_2x^2 + C_3x + C_4\right). \quad (16.25)$$

The first equation in (16.11) represents the bending moment M_y and can presently
be given as:

$$M_y = -\frac{1}{2}q_0x^2 + C_1x + C_2. \quad (16.26)$$

Evaluating the boundary conditions (16.24) leads to the following constants C_1,
C_2, C_3, C_4:

$$C_1 = \frac{1}{2}q_0l, \quad C_2 = 0, \quad C_3 = -\frac{1}{24}q_0l^3, \quad C_4 = 0. \quad (16.27)$$

The expression (16.25) for the bending line $w(x)$ then takes the following form:

$$w(x) = \frac{q_0xl^3}{24EI_{yy}}\left[\left(\frac{x}{l}\right)^3 - 2\left(\frac{x}{l}\right)^2 + 1\right] + \frac{q_0l^2}{2KGA}\left[\frac{x}{l} - \left(\frac{x}{l}\right)^2\right]. \quad (16.28)$$

The maximum deflection w_{max} is obtained at the center of the beam at $x = \frac{l}{2}$ and
results in:

$$w_{max} = \frac{5q_0l^4}{384EI_{yy}} + \frac{q_0l^2}{8KGA}. \quad (16.29)$$

16.5   Shear Correction Factor K

The shear correction factor K depends on the shape of the considered cross-section, and we want to determine K here for a rectangular cross-section with width b and height h. It is known from Chap. 6 that for the shear stress τ_{xz} due to V_z on a simple rectangular cross-section a parabolic distribution over z occurs (Fig. 16.5). The mathematical representation of $\tau_{xz}(z)$ in this case reads:

$$\tau_{xz}(z) = \frac{3V_z}{2A}\left(1 - \frac{4z^2}{h^2}\right), \tag{16.30}$$

and the maximum value of $\tau_{xz}(z)$ at the level of the center of gravity S is $\tau_{xz} = \frac{3V_z}{2A}$. In contrast, as shown by (16.17), the Timoshenko beam theory leads to a constant shear stress τ_{xz} over the entire cross section. A common way to determine the shear correction factor K is to calculate the strain energy introduced into the system due to shear for both the actual parabolic case (16.30) and the constant case (16.17) and equate them from which the shear correction K can be obtained. The general expression for the strain energy of a three-dimensional solid under pure shear is:

$$\Pi_i = \frac{1}{2}\int_V \tau_{xz}\gamma_{xz}dV. \tag{16.31}$$

Since at this point we are interested exclusively in the strain energy Π_i at a given location x, we transform the expression (16.31) into an area integral:

$$\Pi_i = \frac{1}{2}\int_A \tau_{xz}\gamma_{xz}dA = \frac{1}{2}\int_{-\frac{h}{2}}^{\frac{h}{2}}\int_{-\frac{b}{2}}^{\frac{b}{2}} \tau_{xz}\gamma_{xz}dydz. \tag{16.32}$$

Fig. 16.5 Shear stress distribution $\tau_{xz}(z)$ on a rectangular cross-section

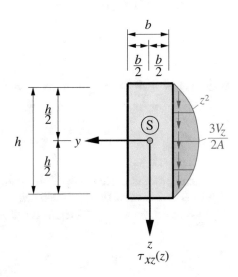

Using Hooke's law $\gamma_{xz} = \frac{\tau_{xz}}{G}$ and the expression (16.30) we obtain after some elementary arithmetic operations:

$$\Pi_i = \frac{3V_z^2}{5GA}. \tag{16.33}$$

On the other hand, for the case (16.17) according to the present Timoshenko beam theory it follows:

$$\Pi_i = \frac{1}{2} \int_{-\frac{h}{2}}^{\frac{h}{2}} \int_{-\frac{b}{2}}^{\frac{b}{2}} \tau_{xz}\gamma_{xz}\,\mathrm{d}y\mathrm{d}z = \frac{V_z^2}{2KGA}. \tag{16.34}$$

We now require the two expressions (16.33) and (16.34) to be identical, and we obtain the value

$$K = \frac{5}{6} \tag{16.35}$$

for a rectangular cross-section. Obviously, the shear correction factor $K = \frac{5}{6}$ determined in this way does not depend on the concrete dimensions b and h of the rectangular section, but results as a constant value.

A more general way to represent the influence of transverse shear deformations is based on the calculation of the so-called reference or effective cross section A_{eff}. The reference cross-section or the effective area A_{eff} represents the area of the cross-section which is effectively involved in the shear transfer. Again, we obtain statements about A_{eff} via energetic considerations and first consider the related shear strain energy provided by the shear force V_z along the shear strain $\gamma_s(s)$. For the strain energy Π_i we get:

$$\Pi_i = \frac{1}{2} \int_A \tau_s(s)\gamma_s(s)\mathrm{d}A, \tag{16.36}$$

or with Hooke's law $\gamma_s(s) = \frac{\tau_s(s)}{G}$:

$$\Pi_i = \frac{1}{2G} \int_A \tau_s^2(s)\mathrm{d}A. \tag{16.37}$$

With $\tau(s) = -\frac{V_z S_y(s)}{I_{yy}t(s)}$ and $\mathrm{d}A = t\mathrm{d}s$ we obtain:

$$\Pi_i = \frac{V_z^2}{2GI_{yy}^2} \int_s \frac{S_y^2(s)}{t(s)}\mathrm{d}s. \tag{16.38}$$

We can proceed analogously if we assume that the acting transverse shear force V_z results in shear stresses τ_{xz} distributed constantly over the effective area A_{eff}, i.e. if we assume $\tau_s(s) = \frac{V_z}{A_{eff}}$. Then the result for the shear strain energy is:

$$\Pi_i = \frac{V_z^2}{2GA_{eff}}. \tag{16.39}$$

Equating the expressions (16.38) and (16.39) then yields an equation for the effective area A_{eff} as follows:

$$A_{eff} = \frac{I_{yy}^2}{\int_s \frac{S_y^2(s)}{t(s)} ds}. \tag{16.40}$$

Again, this can be expressed by a shear correction factor K as follows:

$$K = \frac{A_{eff}}{A} = \frac{I_{yy}^2}{A \int_s \frac{S_y^2(s)}{t(s)} ds}. \tag{16.41}$$

Let us illustrate the calculation of the effective area A_{eff} or the shear correction factor K using the I-section of Fig. 16.6. The cross-section has the constant thickness t_W in the web and constant thickness t_F in the flanges. The distribution of the static moment $S_y(s)$ is shown in Fig. 16.6, right, where we have introduced here the web and flange areas A_W and A_F as auxiliary quantities:

$$A_W = t_W h, \quad A_F = 2 t_F b. \tag{16.42}$$

The moment of inertia I_{yy} is calculated as:

$$I_{yy} = A_W \frac{h^2}{12} + A_F \frac{h^2}{2}, \tag{16.43}$$

and

$$I_{yy}^2 = \frac{h^4}{12} \left(3 A_F^2 + A_F A_W + \frac{A_W^2}{12} \right). \tag{16.44}$$

Evaluating the integral expression $\int_s \frac{S_y^2(s)}{t(s)} ds$ yields based on Fig. 16.6:

$$\int_s \frac{S_y^2(s)}{t(s)} ds = \frac{A_W^2 h^2}{4 t_W} \left(\frac{A_F^2 b t_W}{3 A_W^2 t_F} + \frac{A_F^2 h}{A_W^2} + \frac{h}{30} \right). \tag{16.45}$$

The effective area A_{eff} is then calculated according to (16.40) as follows:

$$A_{eff} = \frac{t_W h^2}{3 A_W^2} \frac{3 A_F^2 + A_F A_W + \frac{A_W^2}{12}}{\frac{A_F^2 b t_W}{3 A_W^2 t_F} + \frac{A_F^2 h}{A_W^2} + \frac{h}{30}}. \tag{16.46}$$

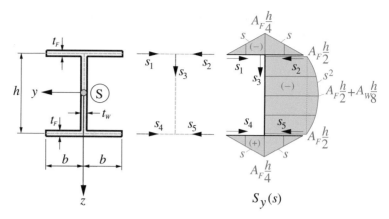

Fig. 16.6 Static moment $S_y(s)$ for the I-section

For the special case of a rectangular cross section with height h and vanishing flange cross sections ($A_F = 0$), A_{eff} simplifies quite substantially to:

$$A_{eff} = \frac{5}{6} t_w h, \tag{16.47}$$

which with $A_{eff} = K A$ is equivalent to $K = \frac{5}{6}$ which we have already deduced for a simple rectangular cross-section.

Transverse shear forces can occur both in $z-$direction and in $y-$direction, so that generally two shear correction factors are required, namely K_z (shear correction factor with respect to the transverse shear force V_z) and K_y (shear correction factor with respect to the transverse shear force V_y). Figure 16.7 contains an overview of some selected cross-sections, the corresponding values for K_z and K_y are given below. The rectangular cross-section shown in Fig. 16.7 has shear correction factors K_z and K_y which are not only identical in both directions, but also completely independent of the concrete dimensions of the cross-section:

$$K_z = K_y = \frac{5}{6}. \tag{16.48}$$

Likewise, identical shear correction factors K_z and K_y are obtained for the circular cross-section shown in Fig. 16.7, independent of the radius of the circle:

$$K_z = K_y = 0,90. \tag{16.49}$$

The thin-walled I-section of Fig. 16.7 has the following values for K_z and K_y:

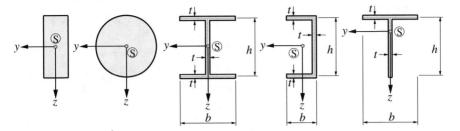

Fig. 16.7 Selected cross-sections for the calculation of shear correction factors K_z and K_y (Schürmann 2015)

$$K_z = \frac{\left[6\left(\dfrac{h}{b}\right) + \left(\dfrac{h}{b}\right)^2\right]^2}{\left[\left(\dfrac{h}{b}\right) + 2\right]\left[6 + 36\left(\dfrac{h}{b}\right) + 12\left(\dfrac{h}{b}\right)^2 + \dfrac{6}{5}\left(\dfrac{h}{b}\right)^3\right]},$$

$$K_y = \frac{5}{3\left[\left(\dfrac{h}{b}\right) + 2\right]}. \tag{16.50}$$

For the thin-walled C cross-section of Fig. 16.7 results:

$$K_z = \frac{\left[6\left(\dfrac{h}{b}\right) + \left(\dfrac{h}{b}\right)^2\right]^2}{\left[\left(\dfrac{h}{b}\right) + 2\right]\left[24 + 36\left(\dfrac{h}{b}\right) + 12\left(\dfrac{h}{b}\right)^2 + \dfrac{6}{5}\left(\dfrac{h}{b}\right)^3\right]},$$

$$K_y = \frac{\left[2 + 5\left(\dfrac{h}{b}\right) + 2\left(\dfrac{h}{b}\right)^2\right]^2}{3\left[\left(\dfrac{h}{b}\right) + 2\right]^3\left[\dfrac{1}{5} + \dfrac{7}{10}\left(\dfrac{h}{b}\right) + \dfrac{4}{5}\left(\dfrac{h}{b}\right)^2 + \dfrac{1}{4}\left(\dfrac{h}{b}\right)^3\right]}. \tag{16.51}$$

The T-section shown in Fig. 16.7 has the following values for K_z and K_y:

$$K_z = \frac{\left(\dfrac{h}{b}\right)^2 \left[4 + 5\left(\dfrac{h}{b}\right) + \left(\dfrac{h}{b}\right)^2\right]^2}{12\left[\left(\dfrac{h}{b}\right) + 1\right]^3 \left[\dfrac{1}{4} + \dfrac{8}{5}\left(\dfrac{h}{b}\right) + \dfrac{7}{10}\left(\dfrac{h}{b}\right)^2 + \dfrac{1}{10}\left(\dfrac{h}{b}\right)^3\right]},$$

$$K_y = \frac{5}{6\left[\left(\dfrac{h}{b}\right) + 1\right]}. \tag{16.52}$$

16.6   Energetic Consideration

In this section we use the principle of the minimum of the total elastic potential (see Chap. 11) to derive the governing equations as well as all potential boundary conditions for the shear-deformable beam in the framework of the Timoshenko beam theory. The inner potential Π_i is given by the two stress components σ_{xx} and τ_{xz}:

$$\Pi_i = \frac{1}{2}\int_V (\sigma_{xx}\varepsilon_{xx} + \tau_{xz}\gamma_{xz})\,\mathrm{d}V. \tag{16.53}$$

With (16.3) this results in:

$$\Pi_i = \frac{1}{2}\int_V \left[E\left(u' + z\varphi_y'\right)^2 + G\left(\varphi_y + w'\right)^2\right]\mathrm{d}V. \tag{16.54}$$

Splitting the volume integral into an area integral and an integral with respect to the longitudinal axis x, after integration over the area A and assuming a principal axis system this leads to:

$$\Pi_i = \frac{1}{2}\int_0^l \left[EA\left(u'\right)^2 + EI_{yy}\left(\varphi_y'\right)^2 + KGA\varphi_y^2 + 2KGAw'\varphi_y + KGA\left(w'\right)^2\right]\mathrm{d}x. \tag{16.55}$$

The transverse line load $q_z = q_z(x)$ and the longitudinal line load $n = n(x)$ are present as loads over the beam length l, so that the external potential Π_a results as follows:

$$\Pi_a = -\int_0^l nu\,\mathrm{d}x - \int_0^l q_z w\,\mathrm{d}x. \tag{16.56}$$

The principle of the minimum of the total elastic potential Π requires the disappearance of the first variation $\delta\Pi = \delta\Pi_i + \delta\Pi_a$, which after some calculation steps leads to the following expression:

$$- \int_0^l \left(EAu'' + n \right) \delta u \, dx$$

$$- \int_0^l \left[KGA \left(\varphi_y' + w'' \right) + q_z \right] \delta w \, dx$$

$$+ \int_0^l \left[KGA \left(\varphi_y + w' \right) - EI_{yy}\varphi_y'' \right] \delta\varphi_y \, dx$$

$$+ EAu' \delta u \Big|_0^l + KGA \left(\varphi_y + w' \right) \delta w \Big|_0^l + EI_{yy}\varphi_y' \delta\varphi_y \Big|_0^l = 0. \quad (16.57)$$

From the three integral expressions in (16.57), the underlying differential equations can be deduced:

$$EAu'' = -n, \quad KGA \left(\varphi_y' + w'' \right) = -q_z, \quad KGA \left(\varphi_y + w' \right) = EI_{yy}\varphi_y''. \quad (16.58)$$

These can be expressed in terms of the internal forces and moments N, V_z and M_y as follows:

$$N' = -n, \quad V_z' = -q_z, \quad M_y' = V_z. \quad (16.59)$$

While the first equation in (16.58) and (16.59), respectively, represents the relation between normal force N and line load n already known from rod structures (cf. Chap. 5), the two remaining equations are identical to (16.7) and (16.8).

The boundary terms in (16.57) represent the potential boundary conditions for the Timoshenko beam. The first boundary term gives at the locations $x = 0$ and $x = l$:

$$\text{Either} \quad EAu' = N = 0 \quad \text{or} \quad \delta u = 0. \quad (16.60)$$

This means that at the locations $x = 0$ and $x = l$ either the normal force N becomes zero, or the variation δu of the displacement u vanishes, which is true if the displacement at these locations is prescribed with a fixed value.

Analogous conclusions hold for the two remaining boundary terms in (16.57):

$$\text{Either} \quad KGA \left(\varphi_y + w' \right) = V_z = 0 \quad \text{or} \quad \delta w = 0,$$
$$\text{Either} \quad EI_{yy}\varphi_y' = M_y = 0 \quad \text{or} \quad \delta\varphi_y = 0. \quad (16.61)$$

A selection of elementary boundary conditions for the shear deformable beam is summarized in Fig. 16.8.

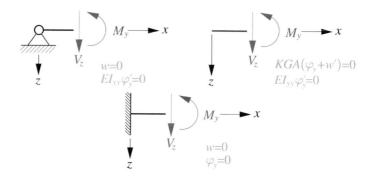

Fig. 16.8 Typical boundary conditions for the Timoshenko beam

16.7 The Force Method

16.7.1 Determination of Displacements

When considering shear-deformable beams, the transverse shear force components
of the virtual complementary internal work $\delta \bar{W}_i$ (cf. Chap. 12, Eq. (12.9)) are no
longer negligible which affects the determination of displacements and the analysis
of statically indeterminate systems as discussed in Chap. 12. We start the considerations with the determination of displacements by means of the force method and
consider again the beam of Fig. 12.3 which is shown again here for the sake of clarity
(Fig. 16.9). The deflection w_{max} in the center of the beam at the position $x = \frac{l}{2}$ is to
be determined with the help of the force method.

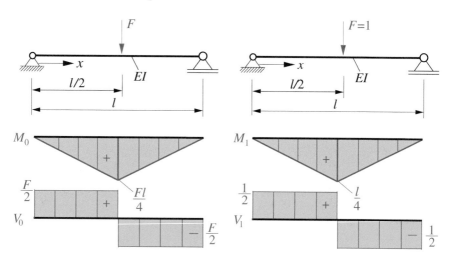

Fig. 16.9 Beam under point load F (top left), beam under virtual force $F = 1$ (top right), associated
moment and shear force diagrams (bottom)

Currently, the deflection w_{max} results from the superposition of the moment and shear force diagrams as follows:

$$w_{max} = \int_0^l \frac{M_0 M_1}{E I_{yy}} dx + \int_0^l \frac{V_0 V_1}{KGA} dx. \tag{16.62}$$

Evaluation yields:

$$w_{max} = \frac{Fl^3}{48 E I_{yy}} + \frac{Fl}{4KGA}. \tag{16.63}$$

The comparison with the result (12.29) shows that in contrast to the Euler-Bernoulli beam theory now the transverse shear term $\frac{Fl}{4KGA}$ is added.

It is of interest at this point to investigate the influence of the shear deformation component in more detail. The result (16.63) can be transformed as follows:

$$w_{max} = \frac{Fl^3}{48 E I_{yy}} \left[1 + \frac{12 E I_{yy}}{KGAl^2} \right] = w_{EB} \left[1 + \frac{w_S}{w_{EB}} \right]. \tag{16.64}$$

We assume here a rectangular cross section (width b, height h) such that:

$$KA = \frac{5}{6} bh, \quad I_{yy} = \frac{bh^3}{12}. \tag{16.65}$$

Then the result for the maximum deflection is:

$$w_{max} = w_{EB} \left[1 + \frac{6}{5} \left(\frac{h}{l} \right)^2 \frac{E}{G} \right]. \tag{16.66}$$

Thus, for the rectangular cross-section, there are two quotients that significantly control the influence of the shear deformation. These are the quotient $\frac{E}{G}$ and the ratio $\frac{l}{h}$. Assuming an isotropic material with $\nu = 0.3$, the ratio $\frac{E}{G}$ is $\frac{E}{G} = \frac{13}{5}$. Then:

$$w_{max} = w_{EB} \left[1 + 3.12 \left(\frac{h}{l} \right)^2 \right]. \tag{16.67}$$

Evaluating this expression for $\frac{h}{l} = \frac{1}{10}$, we obtain:

$$w_{max} = 1.03 w_{EB}, \tag{16.68}$$

i.e. a negligible influence of the shear deformation on the maximum deflection.

On the other hand, if a fiber composite material with $\frac{E}{G} = 15$ is given, then:

$$w_{max} = w_{EB} \left[1 + 18 \left(\frac{h}{l} \right)^2 \right]. \tag{16.69}$$

With $\frac{h}{l} = \frac{1}{10}$ we obtain:

$$w_{\max} = 1.18 w_{EB}. \tag{16.70}$$

Accordingly, the shear deformation amounts to 18% of the bending deformation, which is no longer a negligible proportion.

16.7.2 Statically Indeterminate Systems

We revisit the example of Fig. 12.27 where we want to assume for simplicity that both fields have identical lengths l, bending stiffnesses EI and shear stiffness KGA and are loaded by the uniform line load $q_z = q_0$ (Fig. 16.10, top). We consider the support moment above the middle support as the statically indeterminate quantity X_1. The statically determinate system is defined accordingly by a moment joint at this point. The moment and shear force lines due to the given load and as a result of $X_1 = 1$ are also shown in Fig. 16.10. The following compatibility equation is used to determine the moment X_1:

$$\delta_{10} + X_1 \delta_{11} = 0, \tag{16.71}$$

where:

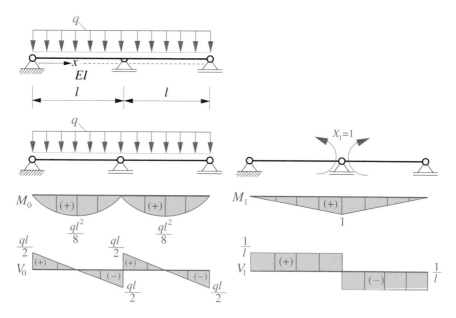

Fig. 16.10 Given static system (top); moment lines M_0, M_1 and transverse shear force lines V_0, V_1 at the statically determinate system

$$\delta_{10} = \int \frac{M_0 M_1}{EI_{yy}} dx + \int \frac{V_0 V_1}{KGA} dx = \frac{ql^3}{12EI_{yy}},$$

$$\delta_{11} = \int \frac{M_1 M_1}{EI_{yy}} dx + \int \frac{V_1 V_1}{KGA} dx = \frac{2l}{3EI_{yy}} + \frac{2}{KGAl}. \qquad (16.72)$$

Thus, the support moment X_1 follows from (16.71) as:

$$X_1 = -\frac{\delta_{10}}{\delta_{11}} = -\frac{q_0 l^2}{8} \frac{1}{1 + \dfrac{3EI_{yy}}{KGAl^2}}. \qquad (16.73)$$

A comparison with the solution (12.132) shows that currently a portion from the shear deformation has to be taken into account, which will effectively provide for a decrease of the supporting moment and thus for an increase of bending moments between the supports. Again, the influence of the ratios $\frac{E}{G}$ and $\frac{I_{yy}}{KA}$ is evident.

16.8 The Ritz Method

In this section we want to derive a general formulation of the Ritz method for the Timoshenko beam. Let us consider a beam of length l, which is loaded by the arbitrary but continuously distributed line load $q = q(x)$. The beam has the stiffnesses EI_{yy} and KGA, which are arbitrarily but continuously distributed over the beam length. Boundary conditions are assumed to be given at the points $x = 0$ and $x = l$ but are not specified further at this point.

The starting point of the considerations here is again the total elastic potential $\Pi = \Pi_i + \Pi_a$ as follows:

$$\Pi_i = \frac{1}{2} \int_0^l \left[EI_{yy} \left(\varphi_y' \right)^2 + KGA \left(\varphi_y^2 + 2w' \varphi_y + \left(w' \right)^2 \right) \right] dx,$$

$$\Pi_a = -\int_0^l qw\, dx. \qquad (16.74)$$

We now use approximations of the following form for the two degrees of freedom $w(x)$ and $\varphi_y(x)$ at hand:

$$w(x) \simeq W(x) = \sum_{i=1}^m C_i W_i(x), \quad \varphi_y(x) \simeq \Phi(x) = \sum_{i=1}^n D_i \Phi_i(x). \qquad (16.75)$$

Here $W_i(x)$ and $\Phi_i(x)$ are admissible shape functions which have to satisfy the given geometric boundary conditions. The quantities m and n are the respective degrees of the series expansions.

Substituting (16.75) into (16.74) gives the following representation for the total elastic potential Π:

$$\Pi = \frac{1}{2}\int_0^l E I_{yy}\sum_{i=1}^n\sum_{j=1}^n D_i D_j \Phi_i'\Phi_j' \mathrm{d}x + \frac{1}{2}\int_0^l KGA\sum_{i=1}^n\sum_{j=1}^n D_i D_j \Phi_i\Phi_j \mathrm{d}x$$

$$+ \int_0^l KGA\sum_{i=1}^m\sum_{j=1}^n C_i D_j W_i'\Phi_j \mathrm{d}x + \frac{1}{2}\int_0^l KGA\sum_{i=1}^m\sum_{j=1}^m C_i C_j W_i'W_j' \mathrm{d}x$$

$$- \int_0^l q\sum_{i=1}^m C_i W_i \mathrm{d}x. \tag{16.76}$$

Evaluating the Ritz equations

$$\frac{\partial \Pi}{\partial C_i} = 0, \quad \frac{\partial \Pi}{\partial D_j} = 0 \tag{16.77}$$

with $i = 1, 2, ..., m$ and $j = 1, 2, ..., n$ finally results in a linear equation system of the following form for the determination of the constants C_i and D_j:

$$\underline{\underline{K}}\,\underline{C} = \underline{F}, \tag{16.78}$$

or

$$\begin{bmatrix} \underline{\underline{K}}_1 & \underline{\underline{K}}_3 \\ \underline{\underline{K}}_3^T & \underline{\underline{K}}_2 \end{bmatrix} \begin{pmatrix} C_1 \\ C_2 \\ \vdots \\ C_m \\ D_1 \\ D_2 \\ \vdots \\ D_n \end{pmatrix} = \begin{pmatrix} F_1 \\ F_2 \\ \vdots \\ F_m \\ 0 \\ 0 \\ \vdots \\ 0 \end{pmatrix}. \tag{16.79}$$

Herein, the quantities F_i are the resultants of the applied line load $q(x)$ as follows:

$$F_i = \int_0^l q_i W_i \mathrm{d}x. \tag{16.80}$$

The submatrices $\underline{\underline{K}}_1, \underline{\underline{K}}_2, \underline{\underline{K}}_3$ of the stiffness matrix $\underline{\underline{K}}$ can be given as:

$$\underline{\underline{K}}_1 = \begin{bmatrix} K_{11}^1 & K_{12}^1 & K_{13}^1 & \cdots & K_{1m}^1 \\ K_{12}^1 & K_{22}^1 & K_{23}^1 & \cdots & K_{2m}^1 \\ K_{13}^1 & K_{23}^1 & K_{33}^1 & \cdots & K_{3m}^1 \\ \vdots & \vdots & \vdots & \ddots & \vdots \\ K_{1m}^1 & K_{2m}^1 & K_{3m}^1 & \cdots & K_{mm}^1 \end{bmatrix},$$

$$\underline{\underline{K}}_2 = \begin{bmatrix} K^2_{11} & K^2_{12} & K^2_{13} & \cdots & K^2_{1n} \\ K^2_{12} & K^2_{22} & K^2_{23} & \cdots & K^2_{2n} \\ K^2_{13} & K^2_{23} & K^2_{33} & \cdots & K^2_{3n} \\ \vdots & \vdots & \vdots & \ddots & \vdots \\ K^2_{1n} & K^2_{2n} & K^2_{3n} & \cdots & K^2_{nn} \end{bmatrix},$$

$$\underline{\underline{K}}_3 = \begin{bmatrix} K^3_{11} & K^3_{12} & K^3_{13} & \cdots & K^3_{1n} \\ K^3_{21} & K^3_{22} & K^3_{23} & \cdots & K^3_{2n} \\ K^3_{31} & K^3_{32} & K^3_{33} & \cdots & K^3_{3n} \\ \vdots & \vdots & \vdots & \ddots & \vdots \\ K^3_{m1} & K^3_{m2} & K^3_{m3} & \cdots & K^3_{mn} \end{bmatrix}. \tag{16.81}$$

Note that the two submatrices $\underline{\underline{K}}_1$ and $\underline{\underline{K}}_2$ are symmetric with m and n rows and columns, respectively, whereas the submatrix $\underline{\underline{K}}_3$ with m rows and n columns generally has no symmetry properties. The entries of the submatrices are:

$$K^1_{ij} = \int_0^l K G A W_i' W_j' \mathrm{d}x,$$

$$K^2_{ij} = \int_0^l E I_{yy} \Phi_i' \Phi_j' \mathrm{d}x + \int_0^l K G A \Phi_i \Phi_j \mathrm{d}x,$$

$$K^3_{ij} = \int_0^l K G A W_i' \Phi_j \mathrm{d}x. \tag{16.82}$$

16.9 Finite Beam Element

In this subsection we want to derive a general finite element formulation for the Timoshenko beam. For this purpose, we consider the finite beam element shown in Fig. 16.11. The element has the length l_e and a total of m equidistant nodes. The nodes divide the element into sections of lengths $\frac{l_e}{m-1}$. The nodes $(1, e)$, $(2, e)$, $(3, e)$, ..., $(m-1, e)$, (m, e) have the degrees of freedom $w_{1,e}, w_{2,e}, w_{3,e}, ..., w_{m-1,e}, w_{m,e}$ and $\varphi_{1,e}, \varphi_{2,e}, \varphi_{3,e}, ..., \varphi_{m-1,e}, \varphi_{m,e}$ as indicated. Let the stiffnesses $E I_e$ and $K G A_e$ be constant along the element length, the element load consists of the constant line load q_e. The local longitudinal axis s_e is introduced as reference axis.

Starting point of the considerations is, analogous to the Ritz method, the elastic total potential $\Pi = \Pi_i + \Pi_a$ of the finite element:

$$\Pi_i = \frac{1}{2} \int_0^{l_e} \left[E I_e \left(\varphi_e' \right)^2 + K G A_e \left(\varphi_e^2 + 2 w_e' \varphi_e + \left(w_e' \right)^2 \right) \right] \mathrm{d}s_e,$$

$$\Pi_a = - \int_0^{l_e} q_e w_e \mathrm{d}s_e, \tag{16.83}$$

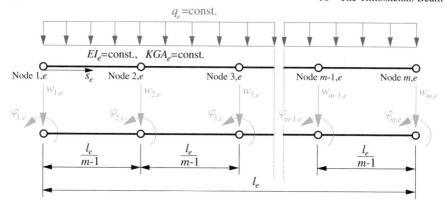

Fig. 16.11 Finite element with m nodes for the Timoshenko beam

where $w_e(s_e)$ and $\varphi_e(s_e)$ are the element deflection and the element bending angle.

For $w_e(s_e)$ and $\varphi_e(s_e)$, similar approaches of the following form are now made for an $m-$noded finite element:

$$w_e(s_e) = w_{1,e}N_{1,e}(s_e) + w_{2,e}N_{2,e}(s_e) + \ldots + w_{m,e}N_{m,e}(s_e) = \sum_{i=1}^{m} w_{i,e}N_{i,e}(s_e),$$

$$\varphi_e(s_e) = \varphi_{1,e}N_{1,e}(s_e) + \varphi_{2,e}N_{2,e}(s_e) + \ldots + \varphi_{m,e}N_{m,e}(s_e) = \sum_{i=1}^{m} \varphi_{i,e}N_{i,e}(s_e). \quad (16.84)$$

Here, $N_{1,e}(s_e)$, $N_{2,e}(s_e)$, $N_{3,e}(s_e)$,...,$N_{m,e}(s_e)$ are the m shape functions belonging to the respective element nodes which are usually formulated in the form of polynomials. It is required that the shape function of node k, e takes the value 1 in node k, e and the value zero in all other nodes.

If we consider an element with $m = 2$ nodes, then the shape functions $N_{1,e}$ and $N_{2,e}$ can be formulated as follows:

$$N_{1,e}(s_e) = 1 - \frac{s_e}{l_e}, \quad N_{2,e}(s_e) = \frac{s_e}{l_e}. \quad (16.85)$$

In the case of $m = 3$ nodes, the shape functions $N_{1,e}$, $N_{2,e}$ and $N_{3,e}$ read:

$$N_{1,e}(s_e) = 1 - 3\frac{s_e}{l_e} + 2\frac{s_e^2}{l_e^2}, \quad N_{2,e}(s_e) = 4\left(\frac{s_e}{l_e} - \frac{s_e^2}{l_e^2}\right), \quad N_{3,e}(s_e) = -\frac{s_e}{l_e} + 2\frac{s_e^2}{l_e^2}. \quad (16.86)$$

For $m = 4$ nodes, the shape functions $N_{1,e}$, $N_{2,e}$, $N_{3,e}$ and $N_{4,e}$ can be formulated in terms of cubic polynomials:

$$N_{1,e}(s_e) = 1 - \frac{11}{2}\frac{s_e}{l_e} + 9\left(\frac{s_e}{l_e}\right)^2 - \frac{9}{2}\left(\frac{s_e}{l_e}\right)^3,$$

$$N_{2,e}(s_e) = 9\frac{s_e}{l_e} - \frac{45}{2}\left(\frac{s_e}{l_e}\right)^2 + \frac{27}{2}\left(\frac{s_e}{l_e}\right)^3,$$

$$N_{3,e}(s_e) = -\frac{9}{2}\frac{s_e}{l_e} + 18\left(\frac{s_e}{l_e}\right)^2 - \frac{27}{2}\left(\frac{s_e}{l_e}\right)^3,$$

$$N_{4,e}(s_e) = \frac{s_e}{l_e} - \frac{9}{2}\left(\frac{s_e}{l_e}\right)^2 + \frac{9}{2}\left(\frac{s_e}{l_e}\right)^3. \tag{16.87}$$

Substituting the general approaches (16.84) into the potential formulations (16.83), we obtain:

$$\Pi = \frac{1}{2}\int_0^{l_e} EI_e \sum_{i=1}^m \sum_{j=1}^m \varphi_{i,e}\varphi_{j,e}N'_{i,e}N'_{j,e}\mathrm{d}s_e + \frac{1}{2}\int_0^{l_e} KGA_e \sum_{i=1}^m \sum_{j=1}^m \varphi_{i,e}\varphi_{j,e}N_{i,e}N_{j,e}\mathrm{d}s_e$$

$$+ \int_0^{l_e} KGA_e \sum_{i=1}^m \sum_{j=1}^m w_{i,e}\varphi_{j,e}N'_{i,e}N_{j,e}\mathrm{d}s_e + \frac{1}{2}\int_0^{l_e} KGA_e \sum_{i=1}^m \sum_{j=1}^m w_{i,e}w_{j,e}N'_{i,e}N'_{j,e}\mathrm{d}s_e$$

$$- \int_0^{l_e} q_e \sum_{i=1}^m w_{i,e}N_{i,e}(s_e)\mathrm{d}s_e. \tag{16.88}$$

By evaluating the Ritz equations

$$\frac{\partial \Pi}{\partial w_{i,e}} = 0, \quad \frac{\partial \Pi}{\partial \varphi_{i,e}} = 0 \tag{16.89}$$

(with $i = 1, 2, ..., m$) we obtain the following linear equation system:

$$\underline{\underline{K}}_e \underline{v}_e = \underline{F}_e, \tag{16.90}$$

or:

$$\begin{bmatrix} \underline{\underline{K}}_{1,e} & \underline{\underline{K}}_{3,e} \\ \underline{\underline{K}}^T_{3,e} & \underline{\underline{K}}_{2,e} \end{bmatrix} \begin{pmatrix} w_{1,e} \\ w_{2,e} \\ \vdots \\ w_{m,e} \\ \varphi_{1,e} \\ \varphi_{2,e} \\ \vdots \\ \varphi_{m,e} \end{pmatrix} = \begin{pmatrix} F_{1,e} \\ F_{2,e} \\ \vdots \\ F_{m,e} \\ 0 \\ 0 \\ \vdots \\ 0 \end{pmatrix}. \tag{16.91}$$

The quantities $F_{i,e}$ are the resultants of the line load q_e:

$$F_{i,e} = \int_0^l q_e N_i \mathrm{d}s_e. \tag{16.92}$$

The element stiffness matrix $\underline{\underline{K}}_e$ is composed of the submatrices $\underline{\underline{K}}_{1,e}$, $\underline{\underline{K}}_{2,e}$, $\underline{\underline{K}}_{3,e}$ (each m rows and m columns) defined as follows:

$$
\underline{\underline{K}}_{1,e} =
\begin{bmatrix}
K^1_{11,e} & K^1_{12,e} & K^1_{13,e} & \cdots & K^1_{1m,e} \\
K^1_{12,e} & K^1_{22,e} & K^1_{23,e} & \cdots & K^1_{2m,e} \\
K^1_{13,e} & K^1_{23,e} & K^1_{33,e} & \cdots & K^1_{3m,e} \\
\vdots & \vdots & \vdots & \ddots & \vdots \\
K^1_{1m,e} & K^1_{2m,e} & K^1_{3m,e} & \cdots & K^1_{mm,e}
\end{bmatrix},
$$

$$
\underline{\underline{K}}_{2,e} =
\begin{bmatrix}
K^2_{11,e} & K^2_{12,e} & K^2_{13,e} & \cdots & K^2_{1m,e} \\
K^2_{12,e} & K^2_{22,e} & K^2_{23,e} & \cdots & K^2_{2m,e} \\
K^2_{13,e} & K^2_{23,e} & K^2_{33,e} & \cdots & K^2_{3m,e} \\
\vdots & \vdots & \vdots & \ddots & \vdots \\
K^2_{1m,e} & K^2_{2m,e} & K^2_{3m,e} & \cdots & K^2_{mm,e}
\end{bmatrix},
$$

$$
\underline{\underline{K}}_{3,e} =
\begin{bmatrix}
K^3_{11,e} & K^3_{12,e} & K^3_{13,e} & \cdots & K^3_{1m,e} \\
K^3_{21,e} & K^3_{22,e} & K^3_{23,e} & \cdots & K^3_{2m,e} \\
K^3_{31,e} & K^3_{32,e} & K^3_{33,e} & \cdots & K^3_{3m,e} \\
\vdots & \vdots & \vdots & \ddots & \vdots \\
K^3_{m1,e} & K^3_{m2,e} & K^3_{m3,e} & \cdots & K^3_{mm,e}
\end{bmatrix}. \tag{16.93}
$$

As already noted for the Ritz method, also in the case of the present finite element formulation the two submatrices $\underline{\underline{K}}_{1,e}$ and $\underline{\underline{K}}_{2,e}$ are symmetric, while the submatrix $\underline{\underline{K}}_{3,e}$ is not. The entries of the submatrices can be given as follows:

$$
K^1_{ij,e} = \int_0^{l_e} K G A_e N'_i N'_j \mathrm{d}s_e,
$$

$$
K^2_{ij,e} = \int_0^{l_e} E I_e N'_i N'_j \mathrm{d}s_e + \int_0^l K G A_e N_i N_j \mathrm{d}s_e,
$$

$$
K^3_{ij,e} = \int_0^{l_e} K G A_e N'_i N_j \mathrm{d}s_e. \tag{16.94}
$$

Equation (16.90) constitutes the stiffness relation for an $m-$noded finite element in the framework of the Timoshenko beam theory. Regarding the assembly of the element contributions to a global system of equations and the general performance of a finite element analysis we refer to Chap. 14.

References

Altenbach H, Altenbach J, Rikards R (1996) Einführung in die Mechanik der Laminat- und Sandwichtragwerke. Deutscher Verlag der Grundstoffindustrie, Stuttgart, Germany

Dieker S, Reimerdes HG (1992) Elementare Festigkeitslehre im Leichtbau. Donat Verlag, Bremen, Germany

Linke M, Nast E (2015) Festigkeitslehre für den Leichtbau. Springer Berlin et al, Germany

Reddy JN (2004) Mechanics of laminated composite plates and shells, 2nd edn. CRC Press Boca Raton et al, USA

Schürmann H (2015) Konstruktiver Leichtbau I. Technische Universität Darmstadt, Germany, Department of mechanical engineering

Chapter 17
Hybrid Beams

17.1 Introduction

Until now, all explanations concerning rods and beams were based on the assumption that we are considering homogeneous structures in which the properties do not change over the cross-sectional area. However, there are a number of examples in engineering applications where this is not the case and where more advantageous properties of a structure can be achieved by the selective arrangement of different materials at certain points in the cross-section. A selection of such structures, referred to as hybrid beams in all that follows, is shown in Fig. 17.1.

In Fig. 17.1, top left, a hybrid beam structure is shown which exhibits different materials in the flanges and in the web. A special case of such a beam is shown in Fig. 17.1, top right. This structure consists of a reinforced concrete plate supported by a steel beam. Such structures, which we will not consider further at this point, are also typically referred to as hybrid structures in the context of steel construction (Petersen 1997).

A special case of a hybrid beam structure that is extremely relevant for lightweight construction and design is the so-called sandwich construction, as shown qualitatively in Fig. 17.1, bottom left. Sandwich structures have already been addressed in Chap. 1. We will discuss the mechanics of sandwich components in more detail in a subsequent volume of this book series, and this class of structural elements will be briefly discussed in terms of computational principles in the following Chap. 18. Finally, laminate structures should be mentioned (Fig. 17.1, bottom right), which consist of an arbitrary number N of layers of, for example, high-performance fiber composite materials. The mechanics of laminate rods and beams are also discussed in Chap. 18.

A very basic introduction to the mechanics of hybrid rod and beam structures can be found in Gross et al. (2017), much more in-depth explanations are given in Mahnken (2019) or Kollár and Springer (2003). A review paper was presented by Mittelstedt in Mittelstedt (2020), which, in addition to the stability behavior of thin-

© The Author(s), under exclusive license to Springer Nature Switzerland AG 2021   611
C. Mittelstedt, *Structural Mechanics in Lightweight Engineering*,
https://doi.org/10.1007/978-3-030-75193-7_17

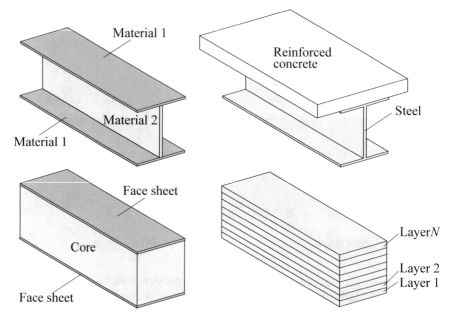

Fig. 17.1 Hybrid beams

walled laminated beams, also discusses the corresponding modeling in the context of higher-order theories.

The present chapter is devoted to the analysis of hybrid rods and beams, where we want to consider here such cross sections where individual segments may be made of different materials (see Fig. 17.2) and the material properties are constant within the n individual segments. We restrict ourselves here, as in all previous chapters, to the analysis of thin-walled structures, so that we can orient all considerations to the skeleton line of the thin-walled profiles under consideration. This, of course, also applies to the distribution of material properties.

The assumptions of the Euler-Bernoulli beam theory apply here, as they have already been discussed in detail in Chap. 5. The remarks of the present chapter shall be limited to rods and beams under normal force and bending as well as transverse

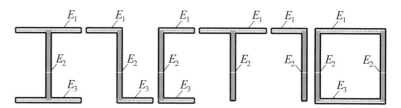

Fig. 17.2 Thin-walled hybrid beams

shear forces, and a short discussion of St. Venant's torsion is given. The corresponding Chaps. 5, 6, 7 are accordingly prerequisites for the understanding of the present chapter.

17.2 Beams Under Normal Forces and Bending Moments

17.2.1 Basic Equations for an Arbitrary Reference System

Consider a thin-walled hybrid cross-section as shown in Fig. 17.2. An arbitrary coordinate system $\bar{x}$, $\bar{y}$, $\bar{z}$ is defined for the following considerations (see also Fig. 5.5). The Euler-Bernoulli beam theory (see Chap. 5) is assumed. The starting point of the considerations is the displacement field of the Euler-Bernoulli beam (cf. Eq. (5.6) as well as the results obtained from Fig. 5.6). The longitudinal displacement $\bar{u}_P$ of an arbitrary point P at the location $\bar{y}_p$ and $\bar{z}_P$ is:

$$\bar{u}_P = \bar{u} - \bar{z}_P \bar{w}' - \bar{y}_p \bar{v}'. \tag{17.1}$$

Moreover, as already assumed with Eqs. (5.7) and (5.8):

$$\bar{v}_P = \bar{v}, \quad \bar{w}_P = \bar{w}. \tag{17.2}$$

Since these relations hold for any point of the cross-section, we will drop the subscript P from here on for the sake of readability. The strain $\varepsilon_{\bar{x}\bar{x}}$ results from (17.1) as follows:

$$\varepsilon_{\bar{x}\bar{x}} = \bar{u}' - \bar{z}\bar{w}'' - \bar{y}\bar{v}''. \tag{17.3}$$

From this, the normal stress $\sigma_{\bar{x}\bar{x},i}$ in each segment i of the hybrid cross-section can be determined as:

$$\sigma_{\bar{x}\bar{x},i} = E_i \varepsilon_{\bar{x}\bar{x}} = E_i \left(\bar{u}' - \bar{z}\bar{w}'' - \bar{y}\bar{v}'' \right), \tag{17.4}$$

with E_i being the modulus of elasticity of segment i.

As already shown in Chap. 5, Eq. (5.11), we calculate the normal force and bending moments $N_{\bar{x}}$, $M_{\bar{y}}$, $M_{\bar{z}}$ as follows:

$$N_{\bar{x}} = \int_A \sigma_{\bar{x}\bar{x}} dA, \quad M_{\bar{y}} = \int_A \sigma_{\bar{x}\bar{x}} \bar{z} dA, \quad M_{\bar{z}} = -\int_A \sigma_{\bar{x}\bar{x}} \bar{y} dA. \tag{17.5}$$

For the normal force $N_{\bar{x}}$, the integral over the cross-sectional area A can be divided into a sum over all subsegments:

$$N_{\bar{x}} = \int_A \sigma_{\bar{x}\bar{x}} dA = \sum_{i=1}^n \int_{A_i} \sigma_{\bar{x}\bar{x},i} dA_i. \tag{17.6}$$

Inserting (17.4) yields:

$$N_{\bar{x}} = \sum_{i=1}^{n} \int_{A_i} E_i \left(\bar{u}' - \bar{z}\bar{w}'' - \bar{y}\bar{v}'' \right) dA_i$$

$$= \sum_{i=1}^{n} \int_{A_i} E_i \bar{u}' dA_i - \sum_{i=1}^{n} \int_{A_i} E_i \bar{z}\bar{w}'' dA_i - \sum_{i=1}^{n} \int_{A_i} E_i \bar{y}\bar{v}'' dA_i. \quad (17.7)$$

The modulus of elasticity E_i is assumed to be constant within segment i. Moreover, the displacement quantities are not functions of the cross-section coordinates and can be drawn in front of the sum signs so that:

$$N_{\bar{x}} = \bar{u}' \sum_{i=1}^{n} E_i \int_{A_i} dA_i - \bar{w}'' \sum_{i=1}^{n} E_i \int_{A_i} \bar{z} dA_i - \bar{v}'' \sum_{i=1}^{n} E_i \int_{A_i} \bar{y} dA_i. \quad (17.8)$$

With the area integrals

$$A_i = \int_{A_i} dA_i, \quad S_{\bar{y}i} = \int_{A_i} \bar{z} dA_i, \quad S_{\bar{z}i} = \int_{A_i} \bar{y} dA_i \qquad (17.9)$$

we obtain:

$$N_{\bar{x}} = \sum_{i=1}^{n} E_i A_i \bar{u}' - \sum_{i=1}^{n} E_i S_{\bar{y}i} \bar{w}'' - \sum_{i=1}^{n} E_i S_{\bar{z}i} \bar{v}''. \qquad (17.10)$$

We introduce the following quantities at this point:

$$(EA)_{eff} = \sum_{i=1}^{n} E_i A_i,$$

$$\left(ES_{\bar{y}}\right)_{eff} = \sum_{i=1}^{n} E_i S_{\bar{y}i},$$

$$(ES_{\bar{z}})_{eff} = \sum_{i=1}^{n} E_i S_{\bar{z}i}. \qquad (17.11)$$

These can be interpreted as the effective stiffnesses of the hybrid cross-section. Thus, for the normal force $N_{\bar{x}}$ we obtain:

$$N_{\bar{x}} = (EA)_{eff} \, \bar{u}' - \left(ES_{\bar{y}}\right)_{eff} \bar{w}'' - (ES_{\bar{z}})_{eff} \, \bar{v}''. \qquad (17.12)$$

The comparison with the first equation in (5.16) shows a very obvious similarity. The difference in the case of the hybrid cross-section is the use of effective stiffness quantities.

We can proceed analogously for the bending moment $M_{\bar{y}}$:

$$
\begin{aligned}
M_{\bar{y}} &= \int_A \sigma_{\bar{x}\bar{x}}\bar{z}\,\mathrm{d}A = \sum_{i=1}^n \int_{A_i} \sigma_{\bar{x}\bar{x},i}\bar{z}\,\mathrm{d}A_i \\
&= \sum_{i=1}^n \int_{A_i} E_i\left(\bar{z}\bar{u}' - \bar{z}^2\bar{w}'' - \bar{y}\bar{z}\bar{v}''\right)\mathrm{d}A_i \\
&= \sum_{i=1}^n E_i\bar{u}' \int_{A_i} \bar{z}\,\mathrm{d}A_i - \sum_{i=1}^n E_i\bar{w}'' \int_{A_i} \bar{z}^2\,\mathrm{d}A_i - \sum_{i=1}^n E_i\bar{v}'' \int_{A_i} \bar{y}\bar{z}\,\mathrm{d}A_i.
\end{aligned}
\tag{17.13}
$$

With the area integrals

$$
I_{\bar{y}\bar{y}i} = \int_{A_i} \bar{z}^2\,\mathrm{d}A_i, \quad I_{\bar{y}\bar{z}i} = \int_{A_i} \bar{y}\bar{z}\,\mathrm{d}A_i
\tag{17.14}
$$

we obtain:

$$
M_{\bar{y}} = \sum_{i=1}^n E_i S_{\bar{y}i}\bar{u}' - \sum_{i=1}^n E_i I_{\bar{y}\bar{y}i}\bar{w}'' - \sum_{i=1}^n E_i I_{\bar{y}\bar{z}i}\bar{v}''.
\tag{17.15}
$$

With the effective stiffnesses

$$
\left(EI_{\bar{y}\bar{y}}\right)_{eff} = \sum_{i=1}^n E_i I_{\bar{y}\bar{y}i}, \quad \left(EI_{\bar{y}\bar{z}}\right)_{eff} = \sum_{i=1}^n E_i I_{\bar{y}\bar{z}i}
\tag{17.16}
$$

the following expression results:

$$
M_{\bar{y}} = \left(ES_{\bar{y}}\right)_{eff}\bar{u}' - \left(EI_{\bar{y}\bar{y}}\right)_{eff}\bar{w}'' - \left(EI_{\bar{y}\bar{z}}\right)_{eff}\bar{v}''.
\tag{17.17}
$$

An analogous expression can be derived for the bending moment $M_{\bar{z}}$. In summary, the constitutive law of the hybrid cross-section under normal force and bending moments is as follows:

$$
\begin{aligned}
N_{\bar{x}} &= (EA)_{eff}\,\bar{u}' - \left(ES_{\bar{y}}\right)_{eff}\bar{w}'' - \left(ES_{\bar{z}}\right)_{eff}\bar{v}'', \\
M_{\bar{y}} &= \left(ES_{\bar{y}}\right)_{eff}\bar{u}' - \left(EI_{\bar{y}\bar{y}}\right)_{eff}\bar{w}'' - \left(EI_{\bar{y}\bar{z}}\right)_{eff}\bar{v}'', \\
-M_{\bar{z}} &= \left(ES_{\bar{z}}\right)_{eff}\bar{u}' - \left(EI_{\bar{y}\bar{z}}\right)_{eff}\bar{w}'' - \left(EI_{\bar{z}\bar{z}}\right)_{eff}\bar{v}''.
\end{aligned}
\tag{17.18}
$$

Apparently, this set of equations is identical to (5.16), where currently effective stiffness quantities of the hybrid cross-section are used instead of the known extensional and bending stiffnesses. In a vector-matrix notation, we obtain:

$$
\begin{pmatrix} N_{\tilde{x}} \\ M_{\tilde{y}} \\ -M_{\tilde{z}} \end{pmatrix} = \begin{bmatrix} (EA)_{eff} & (ES_{\bar{y}})_{eff} & (ES_{\bar{z}})_{eff} \\ (ES_{\bar{y}})_{eff} & (EI_{\bar{y}\bar{y}})_{eff} & (EI_{\bar{y}\bar{z}})_{eff} \\ (ES_{\bar{z}})_{eff} & (EI_{\bar{y}\bar{z}})_{eff} & (EI_{\bar{z}\bar{z}})_{eff} \end{bmatrix} \begin{pmatrix} \bar{u}' \\ -\bar{w}'' \\ -\bar{v}'' \end{pmatrix}. \tag{17.19}
$$

17.2.2 First Cross-Sectional Normalization: Elastic Center of Gravity S

Analogous to the explanations in Chap. 5 we carry out the first cross-sectional normalization, i.e. we refer all considerations to a selected point, namely the so-called ideal or elastic center of gravity S_{el}. The expression of the ideal or elastic center of gravity S_{el} indicates that, in contrast to the center of gravity S of the homogeneous cross-section, it cannot be a purely geometrical quantity, but rather that the stiffnesses E_i of the individual segments of the hybrid cross-section must also play a role. For the sake of better readability, however, the indexing S_{el} is dispensed with in the following and the simpler designation S is used, although it is always assumed in the further course of this chapter that this is an ideal quantity and thus a quantity that depends on both the cross-section geometry and the stiffness distribution.

The reference system is shifted to the elastic center of gravity S of the cross-section (see also Fig. 5.11), which is located at $(\bar{y} = \bar{y}_S, \bar{z} = \bar{z}_S)$. Let the reference system with respect to the elastic center of gravity be $\hat{x}, \hat{y}, \hat{z}$, and let the force and displacement quantities be denoted as $N_{\hat{x}}, M_{\hat{y}}, M_{\hat{z}}$ and as $\hat{u}, \hat{v}, \hat{w}$, respectively. The relation between $\bar{x}, \bar{y}, \bar{z}$ and $\hat{x}, \hat{y}, \hat{z}$ is then as follows (cf. also (5.29)):

$$
\hat{x} = \bar{x}, \quad \hat{y} = \bar{y} - \bar{y}_S, \quad \hat{z} = \bar{z} - \bar{z}_S. \tag{17.20}
$$

The criterion for finding the elastic center of gravity is the vanishing of the two effective stiffnesses $(ES_{\hat{y}})_{eff}$ and $(ES_{\hat{z}})_{eff}$. From the requirement $(ES_{\hat{y}})_{eff} = 0$ it follows:

$$
\begin{aligned}
(ES_{\hat{y}})_{eff} &= \sum_{i=1}^{n} E_i \int_{A_i} \hat{z} \, dA_i = \sum_{i=1}^{n} E_i \int_{A_i} (\bar{z} - \bar{z}_S) \, dA_i \\
&= \sum_{i=1}^{n} E_i \int_{A_i} \bar{z} \, dA_i - \sum_{i=1}^{n} E_i \bar{z}_S \int_{A_i} dA_i = \sum_{i=1}^{n} E_i S_{\bar{y}i} - \bar{z}_S \sum_{i=1}^{n} E_i A_i \\
&= (ES_{\bar{y}})_{eff} - \bar{z}_S (EA)_{eff} = 0.
\end{aligned} \tag{17.21}
$$

This gives the coordinate $\bar{z}_S$ of the elastic center of gravity as:

$$
\bar{z}_S = \frac{\displaystyle\sum_{i=1}^{n} E_i \int_{A_i} \bar{z} \, dA_i}{\displaystyle\sum_{i=1}^{n} E_i \int_{A_i} dA_i} = \frac{(ES_{\bar{y}})_{eff}}{(EA)_{eff}}. \tag{17.22}
$$

Again, the comparison with the corresponding formula for the homogeneous cross section (5.31) is useful. An analogous expression can be found for the coordinate $\bar{y}_s$ from the requirement $(ES_{\hat{z}})_{eff} = 0$. We obtain:

$$\bar{y}_s = \frac{\displaystyle\sum_{i=1}^{n} E_i \int_{A_i} \bar{y}\,dA_i}{\displaystyle\sum_{i=1}^{n} E_i \int_{A_i} dA_i} = \frac{(ES_{\hat{z}})_{eff}}{(EA)_{eff}}. \qquad (17.23)$$

For illustration, consider the T-section of Fig. 17.3, i.e. a cross-section with a flange (width b, wall thickness t, modulus of elasticity E_1) and a web (height h, wall thickness t, modulus of elasticity E_2). The reference system $\bar{y}$ and $\bar{z}$ is introduced as shown, the position of the elastic center of gravity S is to be determined.

For symmetry reasons, $\bar{y}_s = 0$ must hold for the present cross-section. We obtain the coordinate $\bar{z}_s$ from (17.22) as follows:

$$\bar{z}_s = \frac{h}{2}\frac{1}{\left(1 + \dfrac{E_1 b}{E_2 h}\right)}. \qquad (17.24)$$

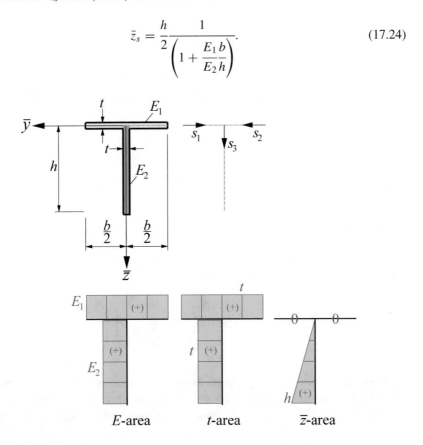

Fig. 17.3 Hybrid T-section

Here two limit cases are of interest. If $E_1 \gg E_2$ is valid, the position of the elastic center of gravity is $\bar{z}_s = 0$. The much stiffer flange thus dominates the position of the center of gravity compared to the much more compliant web. If, on the other hand, the reverse case $E_1 \ll E_2$ is present, then $\bar{z}_s = \frac{h}{2}$ follows, i.e. the T-section under consideration behaves like a narrow rectangular cross-section of height h.

We now also consider the effective bending stiffness $\left(E I_{\hat{y}\hat{y}}\right)_{eff}$, which is obtained as follows:

$$
\begin{aligned}
\left(E I_{\hat{y}\hat{y}}\right)_{eff} &= \sum_{i=1}^{n} E_i \int_{A_i} \hat{z}^2 \mathrm{d}A_i = \sum_{i=1}^{n} E_i \int_{A_i} (\bar{z} - \bar{z}_s)^2 \, \mathrm{d}A_i \\
&= \sum_{i=1}^{n} E_i \int_{A_i} \left(\bar{z}^2 - 2\bar{z}_s \bar{z} + \bar{z}_s^2\right) \mathrm{d}A_i \\
&= \sum_{i=1}^{n} E_i \int_{A_i} \bar{z}^2 \mathrm{d}A_i - 2\sum_{i=1}^{n} E_i \bar{z}_s \int_{A_i} \bar{z} \mathrm{d}A_i + \sum_{i=1}^{n} E_i \bar{z}_s^2 \int_{A_i} \mathrm{d}A_i \\
&= \sum_{i=1}^{n} \left(E_i I_{\bar{y}\bar{y}i} - 2E_i S_{\bar{y}i}\bar{z}_s + E_i A_i \bar{z}_s^2\right) = \sum_{i=1}^{n} \left(E_i I_{\bar{y}\bar{y}i} - 2E_i A_i \bar{z}_s^2 + E_i A_i \bar{z}_s^2\right) \\
&= \sum_{i=1}^{n} \left(E_i I_{\bar{y}\bar{y}i} - E_i A_i \bar{z}_s^2\right) = \left(E I_{\bar{y}\bar{y}}\right)_{eff} - (EA)_{eff}\, \bar{z}_s^2. \quad (17.25)
\end{aligned}
$$

Analogous expressions can be derived for the effective stiffnesses $(E I_{\hat{z}\hat{z}})_{eff}$ and $\left(E I_{\hat{y}\hat{z}}\right)_{eff}$. In summary, we obtain:

$$
\begin{aligned}
\left(E I_{\hat{y}\hat{y}}\right)_{eff} &= \left(E I_{\bar{y}\bar{y}}\right)_{eff} - (EA)_{eff}\, \bar{z}_s^2, \\
(E I_{\hat{z}\hat{z}})_{eff} &= (E I_{\bar{z}\bar{z}})_{eff} - (EA)_{eff}\, \bar{y}_s^2, \\
\left(E I_{\hat{y}\hat{z}}\right)_{eff} &= \left(E I_{\bar{y}\bar{z}}\right)_{eff} - (EA)_{eff}\, \bar{y}_s \bar{z}_s. \quad (17.26)
\end{aligned}
$$

This is Steiner's theorem, formulated for a hybrid beam cross-section (cf. Eq. (5.34)).

By reference to the ideal or elastic center of gravity of the cross-section, the constitutive law (17.19) takes the following form:

$$
\begin{pmatrix} N_{\hat{x}} \\ M_{\hat{y}} \\ -M_{\hat{z}} \end{pmatrix} = \begin{bmatrix} (EA)_{eff} & 0 & 0 \\ 0 & \left(E I_{\hat{y}\hat{y}}\right)_{eff} & \left(E I_{\hat{y}\hat{z}}\right)_{eff} \\ 0 & \left(E I_{\hat{y}\hat{z}}\right)_{eff} & (E I_{\hat{z}\hat{z}})_{eff} \end{bmatrix} \begin{pmatrix} \hat{u}' \\ -\hat{w}'' \\ -\hat{v}'' \end{pmatrix}. \quad (17.27)
$$

Obviously, there is a decoupling of extension and bending, whereas the two bending deformations are still coupled via the effective deviation stiffness $\left(E I_{\hat{y}\hat{z}}\right)_{eff}$. The effective behavior is thus analogous to the homogeneous beam (cf. Eq. (5.42)). The complete decoupling of the system of equations (17.27) is also achieved for the hybrid beam by reference to the principal axes, which is the subject of the following subsection.

17.2.3 Second Cross-Sectional Normalization: Principal Axes

We now determine the principal axis system x, y, z of the hybrid beam which results from the rotation of the reference system $\hat{y}$, $\hat{z}$ around the principal axis angle φ_0 about the longitudinal axis $\hat{x}$ and which is characterized by a vanishing deviation stiffness $\left(E I_{\hat{y}\hat{z}}\right)_{eff}$. In the principal axis system, moreover, the two effective bending stiffnesses become extremal, analogous to the homogeneous cross section.

The coordinate system $\hat{x}$, $\hat{y}$, $\hat{z}$ is now rotated by an arbitrary angle φ into a new reference system x, y, z (cf. Fig. 5.14). The corresponding transformation formulas are (cf. Eq. (5.48)):

$$y = \hat{y} \cos \varphi + \hat{z} \sin \varphi, \quad z = -\hat{y} \sin \varphi + \hat{z} \cos \varphi. \tag{17.28}$$

For the deviation stiffness $\left(E I_{yz}\right)_{eff}$ we then obtain:

$$
\begin{aligned}
\left(E I_{yz}\right)_{eff} &= \sum_{i=1}^{n} E_i \int_{A_i} yz \, \mathrm{d}A_i \\
&= \sum_{i=1}^{n} E_i \int_{A_i} \left(\hat{y} \cos \varphi + \hat{z} \sin \varphi\right) \left(-\hat{y} \sin \varphi + \hat{z} \cos \varphi\right) \mathrm{d}A_i \\
&= \sum_{i=1}^{n} \left[-E_i \sin \varphi \cos \varphi \int_{A_i} \hat{y}^2 \mathrm{d}A_i + E_i \left(\cos^2 \varphi - \sin^2 \varphi\right) \int_{A_i} \hat{y}\hat{z} \, \mathrm{d}A_i \right. \\
&\qquad \left. + E_i \sin \varphi \cos \varphi \int_{A_i} \hat{z}^2 \mathrm{d}A_i \right] \\
&= \sum_{i=1}^{n} \left[E_i \sin \varphi \cos \varphi \left(I_{\hat{y}\hat{y}i} - I_{\hat{z}\hat{z}i}\right) + E_i I_{\hat{y}\hat{z}i} \left(\cos^2 \varphi - \sin^2 \varphi\right) \right] \\
&= \left(\left(E I_{\hat{y}\hat{y}}\right)_{eff} - \left(E I_{\hat{z}\hat{z}}\right)_{eff} \right) \sin \varphi \cos \varphi + \left(E I_{\hat{y}\hat{z}}\right)_{eff} \left(\cos^2 \varphi - \sin^2 \varphi\right).
\end{aligned}
\tag{17.29}
$$

With $\cos^2 \varphi = \frac{1}{2}(1 + \cos 2\varphi)$ and $\sin^2 \varphi = \frac{1}{2}(1 - \cos 2\varphi)$ we then have:

$$\left(E I_{yz}\right)_{eff} = \frac{1}{2} \left(\left(E I_{\hat{y}\hat{y}}\right)_{eff} - \left(E I_{\hat{z}\hat{z}}\right)_{eff} \right) \sin 2\varphi + \left(E I_{\hat{y}\hat{z}}\right)_{eff} \cos 2\varphi. \tag{17.30}$$

This is the transformation equation for the deviation stiffness during the transition from the reference system $\hat{x}$, $\hat{y}$, $\hat{z}$ to the new reference system x, y, z by rotation about the fixed axis $\hat{x} = x$ by the angle φ. Analogous transformation relations can be given for the effective bending stiffnesses, they read:

$$\left(EI_{yy}\right)_{eff} = \frac{1}{2}\left(\left(EI_{\hat{y}\hat{y}}\right)_{eff} + \left(EI_{\hat{z}\hat{z}}\right)_{eff}\right)$$
$$+ \frac{1}{2}\left(\left(EI_{\hat{y}\hat{y}}\right)_{eff} - \left(EI_{\hat{z}\hat{z}}\right)_{eff}\right)\cos 2\varphi - \left(EI_{\hat{y}\hat{z}}\right)_{eff}\sin 2\varphi,$$
$$\left(EI_{zz}\right)_{eff} = \frac{1}{2}\left(\left(EI_{\hat{y}\hat{y}}\right)_{eff} + \left(EI_{\hat{z}\hat{z}}\right)_{eff}\right)$$
$$- \frac{1}{2}\left(\left(EI_{\hat{y}\hat{y}}\right)_{eff} - \left(EI_{\hat{z}\hat{z}}\right)_{eff}\right)\cos 2\varphi + \left(EI_{\hat{y}\hat{z}}\right)_{eff}\sin 2\varphi. \quad (17.31)$$

The comparison of (17.30) and (17.31) with (5.54) shows the similarity of these equations.

The angle $\varphi = \varphi_0$ under which the effective deviation stiffness $\left(EI_{yz}\right)_{eff}$ vanishes follows from the requirement $\left(EI_{yz}\right)_{eff} = 0$. This results in (cf. Eq. (5.53)):

$$\tan 2\varphi_0 = \frac{2\left(EI_{\hat{y}\hat{z}}\right)_{eff}}{\left(EI_{\hat{z}\hat{z}}\right)_{eff} - \left(EI_{\hat{y}\hat{y}}\right)_{eff}}. \quad (17.32)$$

Analogous to Eq. (5.56), the principal values can also be given for the effective bending stiffnesses:

$$EI_{1,2} = \frac{\left(EI_{\hat{y}\hat{y}}\right)_{eff} + \left(EI_{\hat{z}\hat{z}}\right)_{eff}}{2} \pm \sqrt{\left(\frac{\left(EI_{\hat{z}\hat{z}}\right)_{eff} - \left(EI_{\hat{y}\hat{y}}\right)_{eff}}{2}\right)^2 + \left(EI_{\hat{y}\hat{z}}\right)_{eff}^2}.$$
$$(17.33)$$

Due to the second cross-sectional normalization the constitutive law (17.27) changes into the following form:

$$\begin{pmatrix} N \\ M_y \\ -M_z \end{pmatrix} = \begin{bmatrix} (EA)_{eff} & 0 & 0 \\ 0 & \left(EI_{yy}\right)_{eff} & 0 \\ 0 & 0 & \left(EI_{zz}\right)_{eff} \end{bmatrix} \begin{pmatrix} u' \\ -w'' \\ -v'' \end{pmatrix}. \quad (17.34)$$

Solving for the displacements yields:

$$\begin{pmatrix} u' \\ w'' \\ v'' \end{pmatrix} = \begin{bmatrix} \dfrac{1}{(EA)_{eff}} & 0 & 0 \\ 0 & -\dfrac{1}{\left(EI_{yy}\right)_{eff}} & 0 \\ 0 & 0 & \dfrac{1}{\left(EI_{zz}\right)_{eff}} \end{bmatrix} \begin{pmatrix} N \\ M_y \\ M_z \end{pmatrix}. \quad (17.35)$$

From (17.4), the normal stress $\sigma_{xx,i}$ in each segment i can then be calculated as:

$$\sigma_{xx,i} = E_i \left(u' - zw'' - yv'' \right)$$

$$= E_i \left(\frac{N}{(EA)_{eff}} + \frac{M_y z}{(EI_{yy})_{eff}} - \frac{M_z y}{(EI_{zz})_{eff}} \right)$$

$$= \frac{E_i N}{(EA)_{eff}} + \frac{E_i M_y z}{(EI_{yy})_{eff}} - \frac{E_i M_z y}{(EI_{zz})_{eff}}. \tag{17.36}$$

This equation is identical to Eq. (5.65) with the exception of the effective stiffnesses to be applied here and the multiplication by the stiffness E_i of segment i. Apparently, the normal stresses of a hybrid cross-section are divided according to the ratio of stiffnesses to each other, i.e., segments with high stiffnesses attract stresses, whereas less stiff segments exhibit lower stresses. The transformation of the bending moments to the principal axis system can be performed according to Chap. 5, Eq. (5.59).

The constitutive law (17.35) shows that the relationship between displacements and internal forces and moments corresponds exactly to the relationships already discussed in detail in Chap. 5. This means that the calculation of deformations of hybrid bars and beams can be done exactly as in the case of homogeneous bars and beams while applying the corresponding effective stiffnesses.

17.2.4 Selected Basic Cases

We illustrate the analysis equations derived so far at the example of the double symmetric hybrid I-cross section of Fig. 17.4. The cross section has the web height h and the flange width $2b$. The flange has the modulus of elasticity E_1 and the wall thickness t_1, the web has the properties E_2 and t_2. The reference system $\bar{y}, \bar{z}$ is introduced as indicated.

We start the analysis with the calculation of the effective extensional stiffness $(EA)_{eff}$. With (17.9) and (17.11) we obtain:

$$(EA)_{eff} = \sum_{i=1}^{n} E_i \int_{A_i} dA_i = \sum_{i=1}^{n} E_i \int_{0}^{l_i} 1 \cdot t_i ds_i = 4E_1 b t_1 + E_2 t_2 h. \tag{17.37}$$

For the special case $E_1 = E_2 = E$ and $t_1 = t_2 = t$ we have:

$$(EA)_{eff} = Et \left(4b + h \right). \tag{17.38}$$

This corresponds to the extensional stiffness of the homogeneous I-beam (see also (5.21)).

Furthermore, $(ES_{\bar{y}})_{eff}$ and $(ES_{\bar{z}})_{eff}$ are calculated according to (17.9) and (17.11):

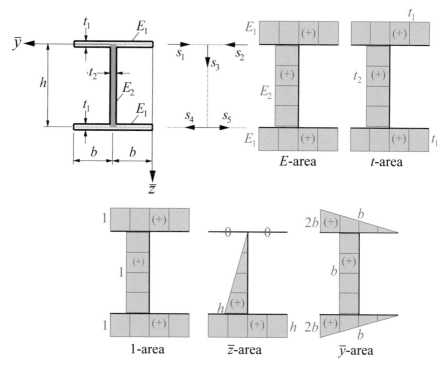

Fig. 17.4 Double symmetric hybrid I-section

$$\left(ES_{\bar{y}}\right)_{eff} = \sum_{i=1}^{n} E_i \int_{A_i} \bar{z}\, dA_i = \sum_{i=1}^{n} E_i \int_{0}^{l_i} \bar{z}t\, ds_i = 2E_1 bht_1 + \frac{1}{2}E_2 h^2 t_2,$$

$$\left(ES_{\bar{z}}\right)_{eff} = \sum_{i=1}^{n} E_i \int_{A_i} \bar{y}\, dA_i = \sum_{i=1}^{n} E_i \int_{0}^{l_i} \bar{y}t\, ds_i = 4E_1 b^2 t_1 + E_2 bht_2. \qquad (17.39)$$

For $E_1 = E_2 = E$ and $t_1 = t_2 = t$ it follows (see also Eq. (5.25)):

$$\left(ES_{\bar{y}}\right)_{eff} = Eht\left(2b + \frac{h}{2}\right), \quad \left(ES_{\bar{z}}\right)_{eff} = Ebt\left(4b + h\right). \qquad (17.40)$$

The coordinates of the elastic center of gravity S are determined with (17.22) and (17.23) as follows:

$$\bar{y}_s = \frac{\left(ES_{\bar{z}}\right)_{eff}}{\left(EA\right)_{eff}} = b, \quad \bar{z}_s = \frac{\left(ES_{\bar{y}}\right)_{eff}}{\left(EA\right)_{eff}} = \frac{h}{2}. \qquad (17.41)$$

The effective bending stiffnesses $\left(EI_{\bar{y}\bar{y}}\right)_{eff}$ and $\left(EI_{\bar{z}\bar{z}}\right)_{eff}$ as well as the effective deviation stiffness follow as:

$$\left(EI_{\bar{y}\bar{y}}\right)_{eff} = \sum_{i=1}^{n} E_i \int_{A_i} \bar{z}^2 dA_i = \sum_{i=1}^{n} E_i \int_0^{l_i} \bar{z}^2 t_i ds_i = 2E_1 bh^2 t_1 + \frac{1}{3} E_2 h^3 t_2,$$

$$\left(EI_{\bar{z}\bar{z}}\right)_{eff} = \sum_{i=1}^{n} E_i \int_{A_i} \bar{y}^2 dA_i = \sum_{i=1}^{n} E_i \int_0^{l_i} \bar{y}^2 t_i ds_i = \frac{16}{3} E_1 b^3 t_1 + E_2 b^2 h t_2,$$

$$\left(EI_{\bar{y}\bar{z}}\right)_{eff} = \sum_{i=1}^{n} E_i \int_{A_i} \bar{y}\bar{z} dA_i = \sum_{i=1}^{n} E_i \int_0^{l_i} \bar{y}\bar{z} t_i ds_i = 2E_1 t_1 b^2 h + \frac{1}{2} E_2 bh^2 t_2. \quad (17.42)$$

In the case of a homogeneous cross-section of constant wall thickness with $E_1 = E_2 = E$ and $t_1 = t_2 = t$ we obtain:

$$\left(EI_{\bar{y}\bar{y}}\right)_{eff} = Eh^2 t \left(\frac{h}{3} + 2b\right),$$

$$\left(EI_{\bar{z}\bar{z}}\right)_{eff} = Eb^2 t \left(\frac{16}{3} b + h\right),$$

$$\left(EI_{\bar{y}\bar{z}}\right)_{eff} = Ebht \left(\frac{h}{2} + 2b\right). \quad (17.43)$$

The comparison with Eq. (5.26) shows the obvious similarity.

With the coordinates y_s and z_s according to (17.41) and the determined effective stiffnesses (17.38) and (17.42), the first cross-sectional normalization can be performed. Using Steiner's theorem (17.26), the following effective stiffnesses are obtained with respect to the reference system $\hat{y}$, $\hat{z}$:

$$\left(EI_{\hat{y}\hat{y}}\right)_{eff} = \left(EI_{\bar{y}\bar{y}}\right)_{eff} - (EA)_{eff} \bar{z}_s^2 = E_1 bh^2 t_1 + \frac{1}{12} E_2 h^3 t_2,$$

$$\left(EI_{\hat{z}\hat{z}}\right)_{eff} = \left(EI_{\bar{z}\bar{z}}\right)_{eff} - (EA)_{eff} \bar{y}_s^2 = \frac{4}{3} E_1 b^3 t_1,$$

$$\left(EI_{\hat{y}\hat{z}}\right)_{eff} = \left(EI_{\bar{y}\bar{z}}\right)_{eff} - (EA)_{eff} \bar{y}_s \bar{z}_s = 0. \quad (17.44)$$

The deviation stiffness $\left(EI_{\hat{y}\hat{z}}\right)_{eff}$ vanishes at this point, so that $\hat{y}$, $\hat{z}$ are already the principal axes y, z of the cross section and the effective bending stiffnesses are already the principal stiffnesses. It is also easy to show that the result (17.44) for the special case $E_1 = E_2 = E$ and $t_1 = t_2 = t$ transitions to the bending stiffnesses of the homogeneous cross-section (cf. Eq. (5.36)).

In the following, we want to consider the selected I-section under different load cases. First, we consider a normal force $N = F$ (Fig. 17.5) acting in the elastic center of gravity. For the sake of simplicity, we set the wall thicknesses in web and flange equal in the following, so that $t_1 = t_2 = t$ applies. We also consider the case $h = 2b$. The effective stiffness (17.37) then changes into the following expression:

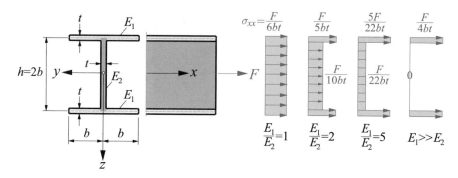

Fig. 17.5 Normal stress distribution σ_{xx} on the double symmetric hybrid I-cross-section due to the normal force $N = F$ for different ratios $\frac{E_1}{E_2}$

$$(EA)_{eff} = 2bt \, (2E_1 + E_2) \,. \tag{17.45}$$

For the normal stress σ_{xxi} in segment i of the cross section, it then follows from (17.36):

$$\sigma_{xx,i} = \frac{E_i N}{(EA)_{eff}} = \frac{E_i F}{2bt \, (2E_1 + E_2)} \,. \tag{17.46}$$

Three different ratios of the two elastic moduli are considered here, namely $\frac{E_1}{E_2} = 1$, $\frac{E_1}{E_2} = 2$, $\frac{E_1}{E_2} = 5$, and as a limit case $E_1 \gg E_2$. Fig. 17.5 shows the resulting stress distributions over the hybrid I-cross-section under consideration. It can be clearly seen that the stress $\sigma_{xx,1}$ continues to increase as the ratio $\frac{E_1}{E_2}$ increases, whereas the value for $\sigma_{xx,2}$ continues to decrease accordingly. Accordingly, for the limit case $E_1 \gg E_2$, $\sigma_{xx,2}$ tends towards zero and the entire load is carried exclusively by the two flanges. This example thus shows in a clear way that in a hybrid cross-section the stiffer segments attract stresses and participate disproportionately in the load transfer than is the case for more compliant segments. This is also evident from Fig. 17.6 where the two stresses $\sigma_{xx,1}$ and $\sigma_{xx,2}$ are shown as functions of $\frac{E_1}{E_2}$. It turns out that the stress distribution in the present case arises exactly in the ratio of the stiffnesses, as can be seen, for example, from the ratio $\frac{E_1}{E_2} = 5$: The stress $\sigma_{xx,1}$ here is exactly five times the stress $\sigma_{xx,2}$.

A simple check can be performed by taking advantage of the fact that the sum of the resultants of the stresses in the individual segments must correspond exactly to the acting normal force $N = F$. For this purpose, we first form the partial internal forces N_1 and N_2 (in Fig. 17.7 again for the case $\frac{E_1}{E_2} = 5$) as:

$$N_1 = \sigma_{xx,1} A_1 = \frac{5F}{22bt} \cdot 2b \cdot t = \frac{10F}{22} \,,$$
$$N_2 = \sigma_{xx,2} A_2 = \frac{F}{22bt} \cdot 2b \cdot t = \frac{2F}{22} \,. \tag{17.47}$$

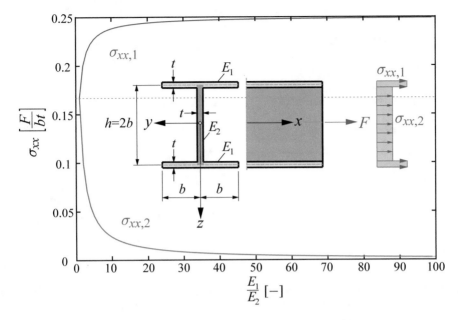

Fig. 17.6 Normal stresses $\sigma_{xx,1}$ and $\sigma_{xx,2}$ on the double symmetric hybrid I-cross-section due to normal force $N = F$ as functions of $\frac{E_1}{E_2}$

Fig. 17.7 Resultants of the normal stresses $\sigma_{xx,1}$ and $\sigma_{xx,2}$ on the double symmetric hybrid I-cross-section due to the normal force $N = F$ for $\frac{E_1}{E_2} = 5$

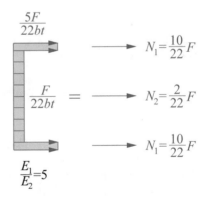

Accordingly for the present example the following must hold:

$$2N_1 + N_2 = F. \qquad (17.48)$$

It can be easily shown that this requirement is satisfied by (17.47).

As another load case we consider uniaxial bending with $M_y = M_0$ (Fig. 17.8) on the hybrid I-cross-section. The calculation of the normal stress $\sigma_{xx,i}$ then results with (17.36):

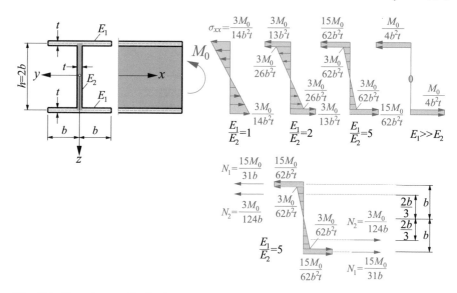

Fig. 17.8 Normal stress distribution σ_{xx} on the double symmetric hybrid I-cross-section for different ratios $\frac{E_1}{E_2}$ due to uniaxial bending $M_y = M_0$ (top), resultants of the normal stresses $\sigma_{xx,1}$ and $\sigma_{xx,2}$ for $\frac{E_1}{E_2} = 5$ (bottom)

$$\sigma_{xx,i} = \frac{E_i M_0}{\left(EI_{yy}\right)_{eff}} z, \qquad (17.49)$$

where currently the following effective bending stiffness $\left(EI_{yy}\right)_{eff}$ applies with $t_1 = t_2 = t$ and $h = 2b$:

$$\left(EI_{yy}\right)_{eff} = 4E_1 tb^3 + \frac{2}{3}E_2 tb^3. \qquad (17.50)$$

The distribution of the normal stress $\sigma_{xx,i}$ is also shown in Fig. 17.8 for different ratios $\frac{E_1}{E_2}$, where for simplicity it was assumed here that the normal stress $\sigma_{xx,1}$ is distributed constantly over the flange thickness. Obviously, the normal stress in the flanges increases as the ratio $\frac{E_1}{E_2}$ increases, again with a distribution proportional to the ratio of the stiffnesses. In the limiting case $E_1 \gg E_2$, the entire bending load is carried by the two flanges, the stresses in the web are negligible.

A simple analysis check can be performed again by first calculating resulting forces from the stress components of the individual segments. These must be in equilibrium with the acting bending moment $M_y = M_0$ (shown in Fig. 17.8, bottom, for the case $\frac{E_1}{E_2} = 5$). For the forces N_1 and N_2 we obtain:

$$
\begin{aligned}
N_1 &= \frac{15M_0}{62b^2 t} \cdot 2bt = \frac{15M_0}{31b}, \\
N_2 &= \frac{1}{2} \cdot \frac{3M_0}{62b^2 t} \cdot bt = \frac{3M_0}{124b}.
\end{aligned}
\qquad (17.51)
$$

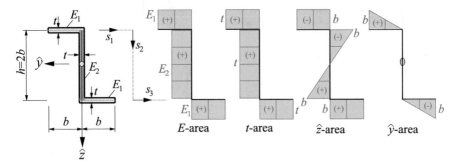

Fig. 17.9 Hybrid Z-cross-section

The according partial moments result as:

$$M_1 = N_1 h = 2N_1 b = \frac{30 M_0}{31},$$

$$M_2 = \frac{2}{3} N_2 h = \frac{4}{3} N_2 b = \frac{M_0}{31}. \qquad (17.52)$$

The sum $M_1 + M_2$ necessarily results in the applied bending moment M_0 to ensure equilibrium. Based on the amounts of M_1 and M_2, it can be seen that the significant part of the bending moment is carried by the two flanges of the hybrid I-section.

As another basic example we consider the Z-section of Fig. 17.9 which is a point-symmetric cross-section (flange: width b, wall thickness t, modulus of elasticity E_1; web: height $h = 2b$, wall thickness t, modulus of elasticity E_2), and the principal axis angle φ_0 is sought as a function of the ratio $\frac{E_1}{E_2}$.

The first cross-sectional normalization is not required explicitly here; the position of the elastic center of gravity S can be deduced from the point symmetry of the given cross-section. The effective stiffnesses related to the centroid coordinates $\hat{y}, \hat{z}$ can be calculated as:

$$\left(EI_{\hat{y}\hat{y}} \right)_{eff} = \sum_{i=1}^{2} E_i \int_0^{l_i} \hat{z}^2 t_i \, ds_i = 2b^3 t \left(E_1 + \frac{E_2}{3} \right),$$

$$\left(EI_{\hat{z}\hat{z}} \right)_{eff} = \sum_{i=1}^{2} E_i \int_0^{l_i} \hat{y}^2 t_i \, ds_i = \frac{2}{3} E_1 b^3 t,$$

$$\left(EI_{\hat{y}\hat{z}} \right)_{eff} = \sum_{i=1}^{2} E_i \int_0^{l_i} \hat{y}\hat{z} t_i \, ds_i = -E_1 b^3 t. \qquad (17.53)$$

The principal axis angle then follows with (17.32) as:

$$\tan 2\varphi_0 = \frac{2 \left(EI_{\hat{y}\hat{z}} \right)_{eff}}{\left(EI_{\hat{z}\hat{z}} \right)_{eff} - \left(EI_{\hat{y}\hat{y}} \right)_{eff}} = \frac{3 \dfrac{E_1}{E_2}}{1 + 2 \dfrac{E_1}{E_2}}. \qquad (17.54)$$

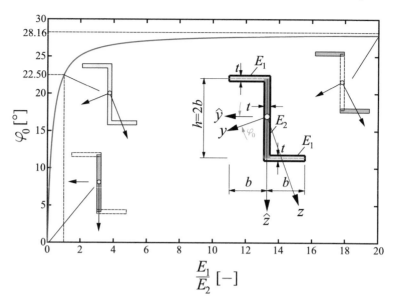

Fig. 17.10 Principal axis angle φ_0 of the hybrid Z-cross-section as a function of $\frac{E_1}{E_2}$

The variation of φ_0 as a function of $\frac{E_1}{E_2}$ is shown in Fig. 17.10. Of interest here are the two limit cases $\frac{E_1}{E_2} = 0$ and $E_1 >> E_2$. In the first case, the principal axis angle φ_0 is $\varphi_0 = 0°$, in which case the cross-section acts like a simple thin-walled rectangular section with no flanges. For the second case, on the other hand, the cross-section acts like two flanges connected by a compliant web. Accordingly, the curve $\varphi\left(\frac{E_1}{E_2}\right)$ tends towards the limit $\varphi_0 = 28.16°$.

17.3 Beams Under Transverse Shear Forces

17.3.1 Shear Flow in Open Cross-Sections

The present section is devoted to the calculation of shear flows and shear stresses due to transverse shear forces in open and closed thin-walled hybrid cross-sections. Analogous to the explanations contained in Chap. 6, the challenge arises here that the shear stresses cannot be derived from any constitutive equations due to the assumptions of the Euler-Bernoulli beam theory, so that we have to procure the shear stresses due to transverse shear forces from a post-calculation also in the case of hybrid beams. We assume here the existence of a principal axis system x, y, z.

We again consider the free-body diagram of Fig. 6.1, assuming here that the infinitesimal sectional element $dx ds_i$ shown there was cut out of segment i of the hybrid beam under consideration. Analogously to Eq. (6.5) we then obtain:

$$\frac{dn_{xx,i}}{dx} dx \, ds_i + \frac{dT_{s,i}}{ds_i} ds_i \, dx = 0. \tag{17.55}$$

Herein, $n_{xx,i}$ and $T_{s,i}$ are the normal force flow and the shear flow of segment i. Division by dx and integration with respect to s gives:

$$\int_{s_{A,i}}^{s_i} \frac{dT_{s,i}}{ds_i} ds_i = T_{s,i}(s_i) - T_{s,i}(s_i = s_{A,i}) = -\int_{s_{A,i}}^{s_i} \frac{dn_{xx,i}}{dx} ds_i, \tag{17.56}$$

where $s_i = s_{A,i}$ represents the starting point of the coordinate s_i of segment i. From (17.36), the following expression can be derived for the normal force flow $n_{xx,i}$ with the wall thickness t_i of the segment i:

$$n_{xx,i} = \sigma_{xx,i} t_i = \left(\frac{E_i N}{(EA)_{eff}} + \frac{E_i M_y z}{(EI_{yy})_{eff}} - \frac{E_i M_z y}{(EI_{zz})_{eff}} \right) t_i. \tag{17.57}$$

Assuming a constant normal force N along the longitudinal axis x of the beam, the first derivative with respect to x is then given as:

$$\frac{dn_{xx,i}}{dx} = n'_{xx,i} = \left(\frac{E_i M'_y z}{(EI_{yy})_{eff}} - \frac{E_i M'_z y}{(EI_{zz})_{eff}} \right) t_i, \tag{17.58}$$

or with $M'_y = V_z$ and $M'_z = -V_y$:

$$n'_{xx,i} = \left(\frac{E_i V_z z}{(EI_{yy})_{eff}} + \frac{E_i V_y y}{(EI_{zz})_{eff}} \right) t_i. \tag{17.59}$$

Thus, from (17.56):

$$T_{s,i}(s_i) - T_{s,i}(s_i = s_{A,i}) = -\int_{s_{A,i}}^{s_i} \left(\frac{E_i V_z z}{(EI_{yy})_{eff}} + \frac{E_i V_y y}{(EI_{zz})_{eff}} \right) t_i \, ds_i$$

$$= -\frac{E_i V_z}{(EI_{yy})_{eff}} \int_{s_{A,i}}^{s_i} z t_i \, ds_i - \frac{E_i V_y}{(EI_{zz})_{eff}} \int_{s_{A,i}}^{s_i} y t_i \, ds_i. \tag{17.60}$$

With the effective stiffnesses

$$(ES_{y,i})_{eff} = E_i \int_{s_{A,i}}^{s_i} z t_i \, ds_i, \quad (ES_{z,i})_{eff} = E_i \int_{s_{A,i}}^{s_i} y t_i \, ds_i \tag{17.61}$$

we obtain:

$$T_{s,i}(s_i) - T_{s,i}(s_i = s_{A,i}) = -\frac{V_z \left(ES_{y,i}\right)_{eff}}{\left(EI_{yy}\right)_{eff}} - \frac{V_y \left(ES_{z,i}\right)_{eff}}{\left(EI_{zz}\right)_{eff}}. \quad (17.62)$$

The comparison with Eq. (6.13) shows the obvious similarity of the resultant equation for the shear flow.

The determination of the effective stiffness $\left(ES_{y,i}\right)_{eff}$ is illustrated at the example of the double symmetric I-cross-section shown in Fig. 17.11. For the individual ordinates we obtain:

$$\left(ES_{y,1}\right)_{eff} (s_1 = 0) = \left(ES_{y,2}\right)_{eff} (s_2 = 0) = 0,$$

$$\left(ES_{y,1}\right)_{eff} (s_1 = b) = \left(ES_{y,2}\right)_{eff} (s_2 = b) = -\frac{1}{2}E_1 bht_1,$$

$$\left(ES_{y,3}\right)_{eff} (s_3 = 0) = \left(ES_{y,3}\right)_{eff} (s_3 = h) = -E_1 bht_1,$$

$$\left(ES_{y,3}\right)_{eff} (s_3 = \frac{h}{2}) = -E_1 bht_1 - \frac{1}{8}E_2 h^2 t_2,$$

$$\left(ES_{y,4}\right)_{eff} (s_4 = 0) = \left(ES_{y,5}\right)_{eff} (s_5 = 0) = 0,$$

$$\left(ES_{y,4}\right)_{eff} (s_4 = b) = \left(ES_{y,5}\right)_{eff} (s_5 = b) = \frac{1}{2}E_1 bht_1. \quad (17.63)$$

A graphical representation is given in Fig. 17.11, right.

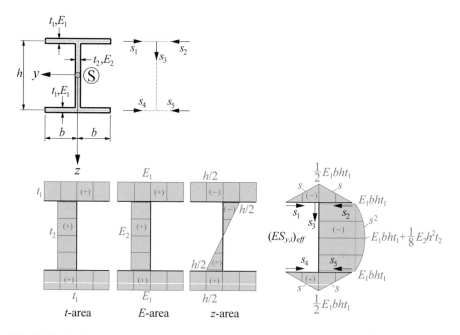

Fig. 17.11 Double symmetric hybrid I-cross-section (top), distribution of $\left(ES_{y,i}\right)_{eff}$ (bottom)

A parameter study is performed for the case $t_1 = t_2 = t$ and $h = 2b$ for the load case V_z. For the cases $E_1 = \alpha E_2$ and $E_2 = E$ with $\alpha = 1$, $\alpha = 2$, $\alpha = 5$ as well as for the limit case $E_1 \gg E_2$, the distributions of the effective stiffness $\left(ES_{y,i} \right)_{eff}$ as well as the resulting shear flows $T_s(s_i)$ are shown in Fig. 17.12. With increasing ratio $\frac{E_1}{E_2}$ the shear flow in the flange increases until for the limit case $E_1 \gg E_2$ the value $\frac{V_z}{4b}$ is reached. At the same time, the value for T_s at the transition point between flange and web increases equally, whereas the influence of E_2 decreases and consequently the maximum value at the point $s_3 = b$ decreases from $\frac{15V_z}{28b}$ for $E_1 = E_2$ to $\frac{V_z}{2b}$ for $E_1 \gg E_2$. Moreover, for the limit case $E_1 \gg E_2$, an almost constant distribution of $T_s(s_3)$ results in the web due to the negligible influence of E_2.

A simple test can be performed by checking whether the resultant F of $T(s_3)$ agrees with the applied transverse shear force V_z. This is briefly confirmed here for the case $E_1 = 2E_2$:

$$F = \frac{6}{13} \frac{V_z}{b} \cdot 2b + \frac{2}{3} \cdot 2b \cdot \frac{3}{52} \frac{V_z}{b} = V_z. \tag{17.64}$$

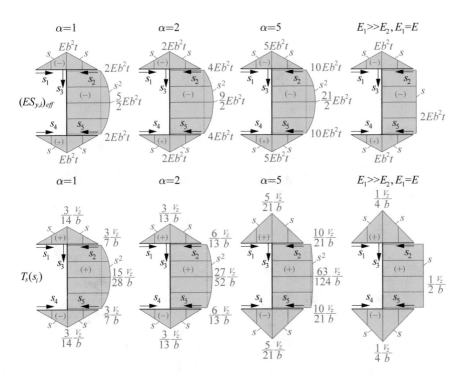

Fig. 17.12 Effective stiffness $\left(ES_{y,i} \right)_{eff}$ (top) and shear flow distribution $T_s(s_i)$ (bottom) on the double symmetric hybrid I-section for the cases $E_1 = \alpha E_2$ and $E_2 = E$ with $\alpha = 1$, $\alpha = 2$, $\alpha = 5$ as well as for the limit case $E_1 \gg E_2$

17.3.2 Shear Flow in Closed Cross-Sections

We now extend our considerations to closed cross-sections and consider the box cross-section of Fig. 17.13. Let the given cross-section consist of the two flanges (width b, elastic modulus E_1, shear modulus G_1, wall thickness t_1) and the two webs (height $h = b$, elastic modulus E_2, shear modulus G_2, wall thickness t_2) as shown and be subjected to the shear force V_z. The effective bending stiffness $(EI_{yy})_{eff}$ can be given for this cross section as:

$$(EI_{yy})_{eff} = \frac{1}{2}E_1 b^3 t_1 + \frac{1}{6}E_2 b^3 t_2. \tag{17.65}$$

To determine the shear flow at the closed cross-section, we perform a statically indeterminate calculation and open the cross-section at an arbitrary point (Fig. 17.14). At this now opened cross-section, the shear flow $T_{s,0}$ can then be determined.

The further course of the calculation is similar to what has already been explained in Chap. 6 on the cross section of Fig. 6.9. Due to the shear flow $T_{s,0}$ there will be a displacement discontinuity at the cutting edge which of course cannot occur in the closed cross-section (cf. also Eq. (6.40)):

$$\Delta u_{10} = \oint \frac{T_{s0}(s)}{Gt(s)} ds. \tag{17.66}$$

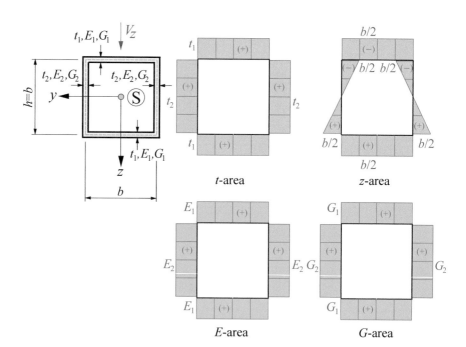

Fig. 17.13 Closed hybrid cross-section

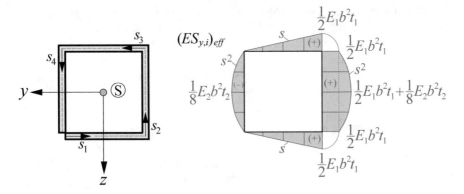

Fig. 17.14 Effective stiffness $\left(ES_{y,i}\right)_{eff}$ (s_i) at the opened hybrid box cross section

It is useful at this point to carry out the integration segment by segment. For the displacement $u_1(s_1 = b)$ one obtains (see also Fig. 6.8):

$$u_1(s_1 = b) = \int_0^b du_1 = \int_0^b \gamma_{xs,1} ds_1 = \int_0^b \frac{T_{s0,1}}{G_1 t_1} ds_1 = \frac{1}{G_1} \int_0^b \frac{T_{s0,1}}{t_1} ds_1. \quad (17.67)$$

With $T_{s0,1} = -\dfrac{V_z \left(ES_{y,1}\right)_{eff}}{\left(EI_{yy}\right)_{eff}}$ we obtain:

$$u_1(s_1 = b) = -\frac{V_z}{G_1 \left(EI_{yy}\right)_{eff}} \int_0^b \frac{\left(ES_{y,1}\right)_{eff}}{t_1} ds_1. \quad (17.68)$$

For the displacement $u_2(s_2 = b)$ we then achieve:

$$u_2(s_2 = b) = u_1(s_1 = b) - \frac{V_z}{G_2 \left(EI_{yy}\right)_{eff}} \int_0^b \frac{\left(ES_{y,2}\right)_{eff}}{t_2} ds_2. \quad (17.69)$$

We can proceed for all further displacements in an analogous manner. The displacement discontinuity Δu_{10} at the intersection edge then results from the sum of the individual contributions u_i:

$$\Delta u_{10} = -\frac{V_z}{\left(EI_{yy}\right)_{eff}} \sum_{i=1}^{n} \frac{1}{G_i} \int_0^{l_i} \frac{\left(ES_{y,i}\right)_{eff}}{t_i} ds_i. \quad (17.70)$$

For the present example we obtain:

$$\Delta u_{10} = -\frac{V_z}{2\left(EI_{yy}\right)_{eff}} E_1 b^3 t_1 \left(\frac{1}{G_1 t_1} + \frac{1}{G_2 t_2}\right). \quad (17.71)$$

In the case of a homogeneous cross section with $E_1 = E_2 = E$ and $G_1 = G_2 = G$, this gives the value $\Delta u_{10} = -\frac{3V_z}{2Gt}$ which agrees with the result (6.48).

Similarly, the displacement discontinuity Δu_{11} due to a unit shear flow $T_{s1} = 1$ can be determined as:

$$\Delta u_{11} = \sum_{i=1}^{n} \frac{1}{G_i} \int_0^{l_i} \frac{\mathrm{d}s_i}{t_i}. \tag{17.72}$$

For the given example:

$$\Delta u_{11} = 2b \left(\frac{1}{G_1 t_1} + \frac{1}{G_2 t_2} \right). \tag{17.73}$$

For the case $G_1 = G_2 = G$ we get $\Delta u_{11} = \frac{4b}{Gt}$ which agrees with (6.48).
From the compatibility requirement

$$\Delta u_{10} + T_{s1} \Delta u_{11} = 0 \tag{17.74}$$

(cf. Eq. (6.42)), the statically indeterminate shear flow T_{s1} can be determined as:

$$T_{s1} = \frac{V_z}{2b} \frac{1}{1 + \frac{1}{3} \frac{E_2 t_2}{E_1 t_1}}. \tag{17.75}$$

For the special case $E_1 = E_2 = E$ and $t_1 = t_2 = t$ we obtain $T_{s1} = \frac{3V_z}{8b}$ which agrees with (6.49).

The final shear flow $T_{s,i}$ is given as the sum of

$$T_{s,i} = T_{s0,i} + T_{s1} \tag{17.76}$$

and is shown in Fig. 17.15.

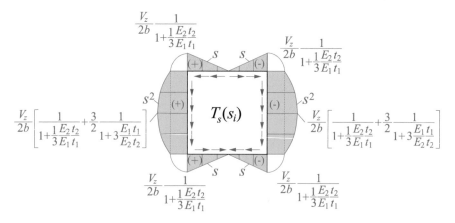

Fig. 17.15 Final shear flow $T_{s,i}$ at the hybrid box cross-section

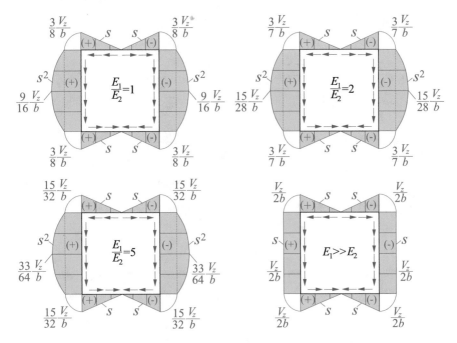

Fig. 17.16 Shear flow $T_{s,i}$ at the hybrid box cross-section for $E_1 = E_2$, $E_1 = 2E_2$, $E_1 = 5E_2$ and $E_1 >> E_2$

A parameter study is performed here for the case $t_1 = t_2 = t$ and the stiffness ratios $E_1 = E_2$, $E_1 = 2E_2$, $E_1 = 5E_2$ as well as the limit case $E_1 >> E_2$. The resulting shear flows in these cases are shown in Fig. 17.16. The influence of the variation of the ratio $\frac{E_1}{E_2}$ is similar to what has already been observed for the open cross-section of Fig. 17.12. Therefore, a renewed discussion can be omitted at this point.

If a closed cross-section with a total of n cells is considered, then the statically indeterminate calculation is analogous to the explanations of Chap. 6, based on the compatibility equations (6.56). Further explanations on this topic are not given here.

17.3.3 Elastic Shear Center

Analogous to the explanations of Chap. 6, an ideal or elastic shear center M can also be determined for thin-walled hybrid cross sections. This can be done following the example of Fig. 6.23 using the hybrid C-section of Fig. 17.17, which is subjected to a shear force V_z and consists of two flanges (E_1, t_1, b) and a web (E_2, t_2, h). The reference system $\bar{y}, \bar{z}$ is positioned as indicated. Due to the simple symmetry of the cross section $\bar{z}_s = 0$ can be concluded immediately. For the coordinate $\bar{y}_s$ of the elastic center of gravity we obtain:

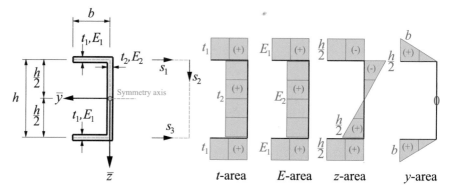

Fig. 17.17 Hybrid C-cross-section

$$\bar{y}_s = \frac{\left(ES_{\bar{y}}\right)_{eff}}{(EA)_{eff}} = \frac{\sum\limits_{i=1}^{3} E_i \int_{A_i} \bar{y}\, dA_i}{\sum\limits_{i=1}^{3} E_i \int_{A_i} dA_i} = \frac{b}{2 + \dfrac{E_2\, h\, t_2}{E_1\, b\, t_1}}. \tag{17.77}$$

For the case $E_1 = E_2 = E$, $t_1 = t_2 = t$, $h = 2b$, this gives $\bar{y}_s = \frac{b}{4}$ which is in agreement with the result (6.85).

Let us simplify the present example in all that follows for the special case $t_1 = t_2 = t$ and $h = 2b$ so that:

$$\bar{y}_s = \frac{b}{2\left(1 + \dfrac{E_2}{E_1}\right)} = \frac{b\, E_1}{2\, E_2}\, \frac{1}{1 + \dfrac{E_1}{E_2}}. \tag{17.78}$$

To determine the shear flow $T_{s,i}(s_i)$, first the effective stiffness

$$\left(ES_{y,i}\right)_{eff} = E_i \int_{S_{A,i}}^{s_i} z t_i\, ds_i \tag{17.79}$$

is determined. Its distribution across the Z-cross-section is shown in Fig. 17.18, left. The effective bending stiffness $\left(EI_{yy}\right)_{eff}$ follows as:

$$\left(EI_{yy}\right)_{eff} = \sum\limits_{i=1}^{3} E_i \int_{S_{A,i}}^{s_i} z^2\, dA_i = 2b^3 t\left(E_1 + \frac{1}{3}E_2\right). \tag{17.80}$$

The shear flow $T_{s,i}(s_i)$ is shown in Fig. 17.18, middle. The results of Fig. 17.18 correspond to those of the example of Fig. 6.25 for the case of the homogeneous cross-section.

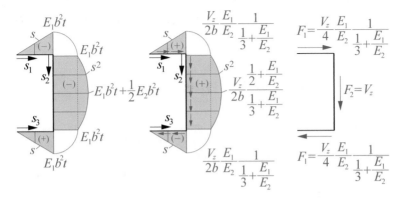

Fig. 17.18   Effective stiffness $(ES_{y,i})_{eff}$ (s_i) (left), shear flow $T_{s,i}$ (s_i) (middle) and resulting forces (right) on the hybrid C-section

Analogous to the explanations for Fig. 6.26, we determine the resulting forces F_1 and F_2 in flanges and web for the present hybrid C-section (Fig. 17.18, right). The resulting torsional moment M_x is obtained analogously to (6.90) as:

$$M_x = -2 \cdot F_1 \cdot b - F_2 \cdot \bar{y}_s = -\frac{V_z b}{2} \frac{E_1}{E_2} \left(\frac{1}{\frac{1}{3} + \frac{E_1}{E_2}} + \frac{1}{1 + \frac{E_1}{E_2}} \right). \qquad (17.81)$$

Finally, from the moment balance $M_x = V_z y_M$, the location of the elastic shear center M for the present C-section follows as:

$$y_M = -\frac{b}{2} \frac{E_1}{E_2} \left(\frac{1}{\frac{1}{3} + \frac{E_1}{E_2}} + \frac{1}{1 + \frac{E_1}{E_2}} \right). \qquad (17.82)$$

For the case $E_1 = E_2 = E$ this gives $y_M = -\frac{5b}{8}$ which agrees with the result (6.92).

A generally applicable formulation for the coordinates of the elastic shear center M for thin-walled hybrid cross sections can be developed analogously to (6.102) and (6.103) by replacing the unit shear flows T_y and T_z as well as the moments of inertia I_{yy} and I_{zz} by the corresponding effective stiffness quantities. Following Fig. 6.28 the following applies:

$$z_M = z_D + \frac{1}{(EI_{zz})_{eff}} \oint (ES_z)_{eff} \, r_{tD} ds,$$

$$y_M = y_D - \frac{1}{\left(EI_{yy}\right)_{eff}} \oint \left(ES_y\right)_{eff} r_{tD} ds. \qquad (17.83)$$

17.4 Torsion of Thin-Walled Hybrid Beams

In this section we consider the determination of the effective torsional stiffness $(GI_T)_{eff}$ of open and closed thin-walled hybrid beams. We first consider the single symmetric open I-section of Fig. 17.19, left. For the segment i (here $i = 1, 2, 3$) of the hybrid cross section the following relation between the twist ϑ_i' and the partial moment $M_{x,i}$ holds:

$$\vartheta_i' = \frac{M_{x,i}}{G_i I_{T,i}}. \qquad (17.84)$$

Herein, G_i and $I_{T,i}$ are the shear modulus and the torsional moment of inertia of segment i, respectively. Since here, as in the case of the homogeneous cross section (Chap. 7), the equality of the partial twists ϑ_i' with the twist ϑ' of the whole cross section is required, the proportional torsional moment $M_{x,i}$ in segment i can also be written as:

$$M_{x,i} = G_i I_{T,i} \vartheta'. \qquad (17.85)$$

At the same time, the constitutive law for the total cross section applies as follows:

$$M_x = (GI_T)_{eff} \, \vartheta', \qquad (17.86)$$

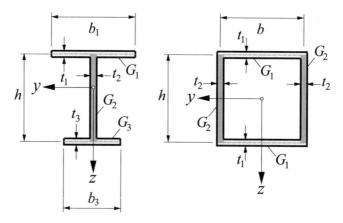

Fig. 17.19 Single symmetric hybrid I-cross-section (left), double symmetric hybrid box cross-section (right)

where $(GI_T)_{eff}$ is the effective torsional stiffness of the hybrid beam.

The total torsional moment M_x acting on the cross-section is composed of the shares of the individual segments, i.e. in the case of a cross-section consisting of n segments:

$$M_x = \sum_{i=1}^{n} M_{x,i}. \tag{17.87}$$

Using (17.85) and (17.86) after cancelling out the twist ϑ', we obtain the effective torsional stiffness $(GI_T)_{eff}$ as the sum of the torsional stiffnesses of the cross-section segments:

$$(GI_T)_{eff} = \sum_{i=1}^{n} G_i I_{T,i}. \tag{17.88}$$

In the case of the I-cross-section of Fig. 17.19, left, we obtain:

$$I_{T,1} = \frac{1}{3}b_1 t_1^3, \quad I_{T,2} = \frac{1}{3}ht_2^3, \quad I_{T,3} = \frac{1}{3}b_3 t_3^3. \tag{17.89}$$

For the effective torsional stiffness $(GI_T)_{eff}$ then follows:

$$(GI_T)_{eff} = \frac{1}{3}\left(G_1 b_1 t_1^3 + G_2 ht_2^3 + G_3 b_3 t_3^3\right). \tag{17.90}$$

The stress determination at such a hybrid cross section results in segment i following the explanations in Chap. 7:

$$\max \tau_i = \frac{M_{x,i} G_i t_i}{(GI_T)_{eff}}. \tag{17.91}$$

The determination of the effective torsional stiffness $(GI_T)_{eff}$ for a closed hybrid cross-section is also briefly discussed here using the cross section of Fig. 17.19, right. We use the second Bredt formula (7.54) to determine the effective torsional stiffness $(GI_T)_{eff}$ as follows:

$$(GI_T)_{eff} = \frac{4A_m^2}{\oint \dfrac{ds}{G(s)t(s)}}. \tag{17.92}$$

Using the example of Fig. 17.19, right, we obtain:

$$(GI_T)_{eff} = \frac{4A_m^2}{\dfrac{2b}{G_1 t_1} + \dfrac{2h}{G_2 t_2}}. \tag{17.93}$$

References

Gross D, Hauger W, Schröder J, Wall WA (2017) Technische Mechanik 2: Elastostatik, Thirteenth.
  Springer Vieweg, Wiesbaden, Germany

Kollár LP, Springer GS (2003) Mechanics of composite structures. Cambridge University Press,
  Cambridge, UK

Mahnken R (2019): *Lehrbuch der Technischen Mechanik - Band 2: Elastostatik: Mit einer Ein-führung in Hybridstrukturen*, Second edition, Springer Vieweg, Wiesbaden, Germany

Mittelstedt C (2020) Buckling and postbuckling of thin-walled composite laminated beams - a
  review of engineering analysis methods. Applied Mechanics Reviews 72:020802

Petersen C (1997) Stahlbau, Third edition, Vieweg. Braunschweig et al, Germany

Chapter 18
Laminated and Sandwich Beams

18.1 Classical Laminate Theory

A laminated beam (Fig. 17.1, bottom right, as well as Fig. 1.7) is a structural element consisting of an arbitrary number N of individual layers, where each individual layer can have an arbitrary thickness and arbitrary elastic properties. In this section, we will turn to the question of how the constitutive law for a laminated beam can be derived for an arbitrary sequence of layers based on the analysis methods outlined in Chap. 17. The nomenclature used here is shown in Fig. 18.1. The laminated beam consists of N layers, where the layer k has the elastic modulus E_k. The layer k is bounded by the coordinates z_{k-1} and z_k. The laminated beam under consideration has a constant width b and the thickness h and is divided into two halves of equal thickness at each point x by the so-called laminate middle plane.

The theoretical framework we will use in this section is the so-called Classical Laminate Theory which is the generalization of the beam theory of Euler and Bernoulli (Chap. 5) to layered structures. In particular, we assume that the assumptions stated in Chap. 5 apply here as well, so that we can directly reuse and adapt equations (17.1)–(17.5) of the hybrid beam for our purposes at this point. Detailed explanations of the treatment of laminate structures in the context of a wide variety of theories can be found, for example, in Altenbach et al. (2018), Ambartsumyan (1970), Ashton and Whitney (1970), Becker and Gross (2002), Jones (1975), Lekhnitskii (1968), Mittelstedt and Becker (2016) or Reddy (2004).

18.1.1 Constitutive Law and Equilibrium Conditions

We consider the case of a laminated beam under normal force as well as bending in the $xz-$plane. Let the coordinate origin be fixed in the laminate middle plane as

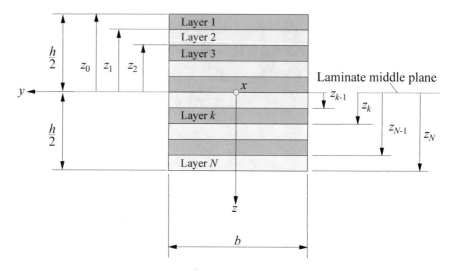

Fig. 18.1 Nomenclature for a laminated beam

shown in Fig. 18.1. The longitudinal displacement u_P of any point P at location z_P is then obtained according to Eq. (17.1) as:

$$u_P = u - z_P w'. \tag{18.1}$$

For the deflection w_P we obtain analogous to (17.2):

$$w_P = w. \tag{18.2}$$

For better readability, the subscript P is dropped from here on. We can determine the strain ε_{xx} as follows with the help of (17.3):

$$\varepsilon_{xx} = u' - z w''. \tag{18.3}$$

The normal stress $\sigma_{xx,k}$ in layer k is then obtained from (17.4) as:

$$\sigma_{xx,k} = E_k \varepsilon_{xx} = E_k \left(u' - z w'' \right). \tag{18.4}$$

The internal normal force and bending moment of the laminated beam follow with (17.5) as:

$$N = \int_A \sigma_{xx} dA, \quad M_y = \int_A \sigma_{xx} z dA. \tag{18.5}$$

We first turn to the calculation of the normal force N. The integral prescribed in (18.5) is decomposed into N partial integrals due to the layered nature of the laminate:

$$N = \sum_{k=1}^{N} \int_{A_k} \sigma_{xx,k} dA_k. \tag{18.6}$$

With $dA_k = bdz$ we obtain:

$$N = b \sum_{k=1}^{N} \int_{z_{k-1}}^{z_k} \sigma_{xx,k} dz. \tag{18.7}$$

Inserting (18.4) yields:

$$N = b \sum_{k=1}^{N} E_k \int_{z_{k-1}}^{z_k} \left(u' - zw'' \right) dz, \tag{18.8}$$

which leads to the following expression:

$$N = b \sum_{k=1}^{N} E_k \left(z_k - z_{k-1} \right) u' - \frac{b}{2} \sum_{k=1}^{N} E_k \left(z_k^2 - z_{k-1}^2 \right) w''. \tag{18.9}$$

Using the following abbreviations as they are common in Classical Laminate Theory

$$A_{11} = b \sum_{k=1}^{N} E_k \left(z_k - z_{k-1} \right),$$

$$B_{11} = \frac{b}{2} \sum_{k=1}^{N} E_k \left(z_k^2 - z_{k-1}^2 \right) \tag{18.10}$$

finally results in the following expression for the normal force N:

$$N = A_{11} u' - B_{11} w''. \tag{18.11}$$

An analogous expression can be obtained for the bending moment M_y as follows:

$$M_y = B_{11} u' - D_{11} w'', \tag{18.12}$$

with:

$$D_{11} = \frac{b}{3} \sum_{k=1}^{N} E_k \left(z_k^3 - z_{k-1}^3 \right). \tag{18.13}$$

In summary, the constitutive law for the laminated beam can be stated in a vector-matrix notation as:

$$\begin{pmatrix} N \\ M_y \end{pmatrix} = \begin{bmatrix} A_{11} & B_{11} \\ B_{11} & D_{11} \end{bmatrix} \begin{pmatrix} u' \\ -w'' \end{pmatrix}. \tag{18.14}$$

The stiffness quantities A_{11}, B_{11}, D_{11} are referred to as extensional stiffness, coupling stiffness and flexural stiffness, respectively. The constitutive law (18.14) shows that for an arbitrary laminate structure, axial and bending actions are coupled via the coupling stiffness B_{11} and can no longer be considered separately. The occurrence of a longitudinal strain u' or a curvature w'' generally causes both a normal force N and a bending moment M_y. This is connected with the notion of laminate coupling effects, and in this case specifically the so-called bending-extension coupling. This can be illustrated at a very elementary example. Consider a two-ply laminate as shown in Fig. 18.2. The stiffness quantities A_{11}, B_{11}, D_{11} are obtained in this case as:

$$A_{11} = b \sum_{k=1}^{2} E_k \left(z_k - z_{k-1}\right) = \frac{bh}{2} \left(E_1 + E_2\right),$$

$$B_{11} = \frac{b}{2} \sum_{k=1}^{2} E_k \left(z_k^2 - z_{k-1}^2\right) = \frac{bh^2}{8} \left(E_2 - E_1\right),$$

$$D_{11} = \frac{b}{3} \sum_{k=1}^{2} E_k \left(z_k^3 - z_{k-1}^3\right) = \frac{bh^3}{24} \left(E_1 + E_2\right). \tag{18.15}$$

Obviously, in this case the coupling stiffness B_{11} does not disappear, so that there is a coupling between inplane and out-of-plane action.

For the special case $E_1 = E_2 = E$ (homogeneous beam with width b and height h) it follows:

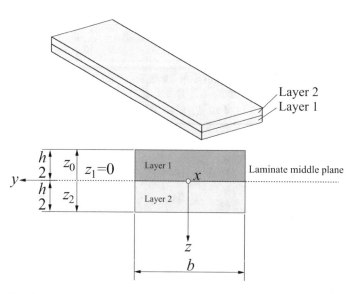

Fig. 18.2 Two-layered laminated beam

$$A_{11} = Eb \left[\frac{h}{2} - \left(-\frac{h}{2} \right) \right] = Ebh = EA,$$

$$B_{11} = E\frac{b}{2} \left[\left(\frac{h}{2} \right)^2 - \left(-\frac{h}{2} \right)^2 \right] = 0,$$

$$D_{11} = E\frac{b}{3} \left[\left(\frac{h}{2} \right)^3 - \left(-\frac{h}{2} \right)^3 \right] = E\frac{bh^3}{12} = EI_{yy}. \tag{18.16}$$

In a vector-matrix notation, the constitutive law for this special case is given as:

$$\begin{pmatrix} N \\ M_y \end{pmatrix} = E \begin{bmatrix} A & 0 \\ 0 & I_{yy} \end{bmatrix} \begin{pmatrix} u' \\ -w'' \end{pmatrix}. \tag{18.17}$$

Apparently, this corresponds to the formulation already derived in Chap. 5 with Eq. (5.62) if uniaxial bending in a principal axis system is assumed.

It can be shown that in the presence of a symmetrical laminate structure the coupling stiffness B_{11} vanishes, as can be immediately seen from (18.10). It is therefore good practice in the design of laminate structures to always use symmetrical layups and to deviate from this rule only in justified special cases.

We derive the governing equations of the laminated beam in the framework of Classical Laminate Theory from the consideration of the equilibrium conditions which we can rewrite from Chap. 5:

$$N' = -n, \quad V_z' = -q_z, \quad M_y' = V_z, \tag{18.18}$$

where we want to assume here that no longitudinal load n is applied. Using the constitutive law (18.14) then gives:

$$\left(A_{11}u' \right)' - \left(B_{11}w'' \right)' = 0,$$
$$\left(B_{11}u' \right)' - \left(D_{11}w'' \right)' = V_z,$$
$$\left(B_{11}u' \right)'' - \left(D_{11}w'' \right)'' = -q_z. \tag{18.19}$$

In the case of a prismatic beam with invariant properties along the longitudinal axis we obtain:

$$A_{11}u'' - B_{11}w''' = 0,$$
$$B_{11}u'' - D_{11}w''' = V_z,$$
$$B_{11}u''' - D_{11}w'''' = -q_z. \tag{18.20}$$

18.1.2   Calculation of Stresses

We derive an expression for the normal stress $\sigma_{xx,k}$ in layer k by inverting the constitutive law (18.14):

$$u' = \frac{D_{11}}{A_{11}D_{11} - B_{11}^2}N - \frac{B_{11}}{A_{11}D_{11} - B_{11}^2}M_y,$$

$$w'' = \frac{B_{11}}{A_{11}D_{11} - B_{11}^2}N - \frac{A_{11}}{A_{11}D_{11} - B_{11}^2}M_y, \qquad (18.21)$$

or in a vector-matrix notation:

$$\begin{pmatrix} u' \\ -w'' \end{pmatrix} = \frac{1}{A_{11}D_{11} - B_{11}^2}\begin{bmatrix} D_{11} & -B_{11} \\ -B_{11} & A_{11} \end{bmatrix}\begin{pmatrix} N \\ M_y \end{pmatrix}. \qquad (18.22)$$

The calculation of the normal stress $\sigma_{xx,k}$ in layer k is then done using (18.4) by substituting there the corresponding expressions (18.21) for u' and w''. It follows:

$$\sigma_{xx,k} = \frac{E_k}{A_{11}D_{11} - B_{11}^2}\left[D_{11}N - B_{11}M_y + z\left(A_{11}M_y - B_{11}N\right)\right]. \qquad (18.23)$$

In the case of a symmetrical laminate we obtain:

$$\sigma_{xx,k} = E_k\left(\frac{N}{A_{11}} + \frac{M_y}{D_{11}}z\right). \qquad (18.24)$$

The special case of a homogeneous beam with a rectangular cross section (width b, height h) follows from this with $A_{11} = EA$ and $D_{11} = EI_{yy}$:

$$\sigma_{xx} = \frac{N}{A} + \frac{M_y}{I_{yy}}z, \qquad (18.25)$$

which is identical to (5.65) for the case of uniaxial bending with normal force. A qualitative representation of the normal stress σ_{xx} is given in Fig. 18.3, top, for an exemplary six-layered laminate. From this it is clear that the normal stress σ_{xx} can be split into a membrane component that is constant over each layer thickness, and a bending component that is linear over each layer. As already discussed in Chap. 17 for hybrid beams, there are discontinuities in the stress profile in the interfaces between different layers. For comparison, in Fig. 18.3, bottom, the qualitative stress profile on a homogeneous beam is shown.

Analogous to Chaps. 6 and 17, we can obtain a statement about the transverse shear stresses τ_{xz} by considering the equilibrium of forces at an infinitesimal section element in a post-calculation. For this purpose, we study the infinitesimal section element of Fig. 18.4 in which we consider an element of length dx that we have cut free in the interface between layer $k - 1$ and k, so that the shear stress $\tau_{xz,k}$ of layer k is released there. The equilibrium of forces in the x−direction then gives:

$$\tau_{xz,k}dx + \sum_{i=1}^{k-1}\int_{z_{i-1}}^{z_i}\left[\left(\sigma_{xx,i} + \frac{d\sigma_{xx,i}}{dx}dx\right) - \sigma_{xx,i}\right]dz = 0, \qquad (18.26)$$

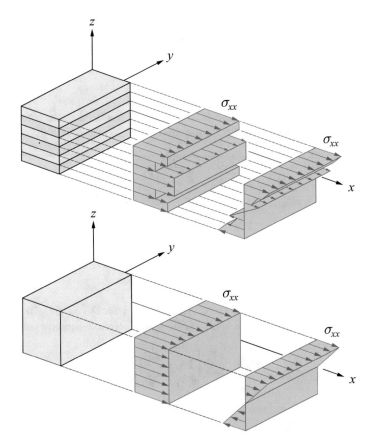

Fig. 18.3 Qualitative distribution of the normal stress σ_{xx} for an exemplary laminate (top) and a homogeneous beam (bottom)

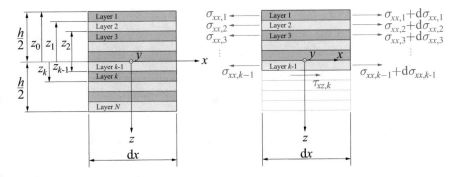

Fig. 18.4 Determination of the shear stress $\tau_{xz,k}$

which can be immediately solved for the shear stress $\tau_{xz,k}$:

$$\tau_{xz,k} = -\sum_{i=1}^{k-1} \int_{z_{i-1}}^{z_i} \frac{\mathrm{d}\sigma_{xx,i}}{\mathrm{d}x} \mathrm{d}z. \tag{18.27}$$

We consider here the special case of a symmetric laminated beam with $B_{11} = 0$. The derivation of the normal stress σ_{xx} required in (18.27) results for the case $n = 0$ considering $M_y' = V_z$:

$$\frac{\mathrm{d}\sigma_{xx,i}}{\mathrm{d}x} = \frac{E_i}{D_{11}} z V_z \tag{18.28}$$

Thus:

$$\tau_{xz,k} = -\sum_{i=1}^{k-1} \int_{z_{i-1}}^{z_i} \frac{E_i}{D_{11}} z V_z \mathrm{d}z = -\frac{V_z}{D_{11}} \sum_{i=1}^{k-1} E_i \int_{z_{i-1}}^{z_i} z \mathrm{d}z = -\frac{V_z}{2D_{11}} \sum_{i=1}^{k-1} E_i \left(z_i^2 - z_{i-1}^2 \right). \tag{18.29}$$

In the case of the homogeneous beam cross section with $D_{11} = E I_{yy}$, the formula for the rectangular cross section with width b and height h already provided by (16.30) results:

$$\tau_{xz} = \frac{3V_z}{2bh} \left[1 - 4 \left(\frac{z}{h} \right)^2 \right]. \tag{18.30}$$

18.1.3   Deformations

The starting point for calculating the deformations of a laminated beam are the differential equations (18.20). We differentiate the first equation in (18.20) once with respect to x and obtain:

$$A_{11}u''' - B_{11}w'''' = 0, \tag{18.31}$$

or

$$u''' = \frac{B_{11}}{A_{11}} w''''. \tag{18.32}$$

Substituting this expression into the second equation in (18.20) then gives:

$$\left(D_{11} - \frac{B_{11}^2}{A_{11}} \right) w'''' = q_z, \tag{18.33}$$

or with the abbreviation $\bar{D}_{11} = D_{11} - \frac{B_{11}^2}{A_{11}}$:

$$\bar{D}_{11} w'''' = q_z. \tag{18.34}$$

This provides an expression for determining the deflection w, analogous to the expression (5.101) for the Euler-Bernoulli beam, from which the deflection w can be obtained by fourfold integration:

$$\bar{D}_{11} w'''' = q_z,$$
$$\bar{D}_{11} w''' = -V_z = \int q_z \mathrm{d}x + C_1,$$
$$\bar{D}_{11} w'' = -M_y = \int \int q_z \mathrm{d}x \mathrm{d}x + C_1 x + C_2,$$
$$\bar{D}_{11} w' = \int \int \int q_z \mathrm{d}x \mathrm{d}x \mathrm{d}x + \frac{1}{2} C_1 x^2 + C_2 x + C_3,$$
$$\bar{D}_{11} w = \int \int \int \int q_z \mathrm{d}x \mathrm{d}x \mathrm{d}x \mathrm{d}x + \frac{1}{6} C_1 x^3 + \frac{1}{2} C_2 x^2 + C_3 x + C_4. \tag{18.35}$$

The constants C_1, C_2, C_3, C_4 are then adjusted to given boundary conditions.

The longitudinal displacement u can be determined accordingly from the expression (18.32) in a post-calculation.

18.1.4 Energetic Consideration

For the energetic consideration of the prismatic laminated beam the strain energy Π_i is required. It can be stated in general terms as:

$$\Pi_i = \frac{1}{2} \int_V \sigma_{xx} \varepsilon_{xx} \mathrm{d}V. \tag{18.36}$$

The volume integral can be decomposed into an area integral and an integral with respect to the longitudinal axis of the beam:

$$\Pi_i = \frac{1}{2} \int_A \int_0^l \sigma_{xx} \varepsilon_{xx} \mathrm{d}x \mathrm{d}A. \tag{18.37}$$

With $\mathrm{d}A = b\mathrm{d}z$:

$$\Pi_i = \frac{b}{2} \int_{-\frac{h}{2}}^{+\frac{h}{2}} \int_0^l \sigma_{xx} \varepsilon_{xx} \mathrm{d}x \mathrm{d}z. \tag{18.38}$$

Since the normal stress is not a continuous function of z, the integration with respect to the thickness coordinate z must be done in a layerwise fashion as follows:

$$\Pi_i = \frac{b}{2} \int_0^l \sum_{k=1}^N \int_{z_{k-1}}^{z_k} \sigma_{xx,k} \varepsilon_{xx,k} dz dx. \tag{18.39}$$

Expressing the normal stress $\sigma_{xx,k}$ by the strain $\varepsilon_{xx,k}$ we get:

$$\Pi_i = \frac{b}{2} \int_0^l \sum_{k=1}^N E_k \int_{z_{k-1}}^{z_k} \varepsilon_{xx,k}^2 dz dx = \frac{b}{2} \int_0^l \sum_{k=1}^N E_k \int_{z_{k-1}}^{z_k} \left(u' - zw'' \right)^2 dz dx. \tag{18.40}$$

After some elementary transformations this finally gives the following expression for the inner potential Π_i:

$$\Pi_i = \frac{1}{2} A_{11} \int_0^l \left(u' \right)^2 dx - B_{11} \int_0^l u' w'' dx + \frac{1}{2} D_{11} \int_0^l \left(w'' \right)^2 dx. \tag{18.41}$$

We now use the principle of the minimum of the total elastic potential $\delta \Pi = 0$ to derive both the governing differential equations and all potential combinations of boundary conditions. The requirement $\delta \Pi = \delta (\Pi_i + \Pi_a) = 0$ gives for the case of the arbitrarily but continuously distributed line load $q_z = q_z(x)$ with

$$\Pi_a = - \int_0^l q w dx \tag{18.42}$$

the following expression:

$$\int_0^l \left(-A_{11} u'' + B_{11} w''' \right) \delta u dx + \int_0^l \left(-B_{11} u''' + D_{11} w'''' - q_z \right) \delta w dx$$
$$+ \left(A_{11} u' - B_{11} w'' \right) \delta u \big|_0^l + \left(B_{11} u'' - D_{11} w''' \right) \delta w \big|_0^l + \left(-B_{11} u' + D_{11} w'' \right) \delta w' \big|_0^l = 0. \tag{18.43}$$

From the integral terms appearing here we can conclude the equilibrium conditions that we have already derived with (18.20):

$$A_{11} u'' - B_{11} w''' = 0,$$
$$B_{11} u''' - D_{11} w'''' = -q_z. \tag{18.44}$$

The boundary terms in (18.43), on the other hand, give the boundary conditions to be observed. At the points $x = 0$ and $x = l$ we have:

$$\text{Either} \quad A_{11} u' - B_{11} w'' = 0 \quad \text{or} \quad \delta u = 0. \tag{18.45}$$

With (18.11) it follows that either the normal force N vanishes at $x = 0$ and $x = l$, or that the variation δu becomes zero, which is equivalent to prescribing the displacement u with a fixed value.

Furthermore, it follows from the second boundary term in (18.43):

$$\text{Either} \quad B_{11}u'' - D_{11}w''' = 0 \quad \text{or} \quad \delta w = 0. \tag{18.46}$$

With (18.20) it follows that either the transverse shear force V_z becomes zero, or that the deflection w is specified with a fixed value.

From the last boundary term in (18.43) it follows:

$$\text{Either} \quad B_{11}u' - D_{11}w'' = 0 \quad \text{or} \quad \delta w' = 0. \tag{18.47}$$

This is equivalent to either the bending moment M_y (cf. Eq. (18.12)) vanishing, or the rotation w' being prescribed with a fixed value.

18.1.5 The Ritz Method

From the previous explanations, a formulation for the Ritz method can be derived for the laminated beam. For this purpose, we consider the prismatic laminated beam of Fig. 18.5 which has the length l and is subjected to the line load q_z. Let the properties A_{11}, B_{11}, D_{11} be given. For the given beam situation, a formulation for the Ritz method is to be developed. We start from the expressions (18.41) and (18.42) for the inner and the outer potential and use approximations for the displacements u and w as follows:

$$w(x) \simeq W(x) = \sum_{i=1}^{N} C_i W_i(x), \quad u(x) \simeq U(x) = \sum_{i=1}^{M} D_i U_i(x). \tag{18.48}$$

The shape functions $W_i(x)$ and $U_i(x)$ are required to satisfy at least the given static boundary conditions. Substituting the approaches (18.48) into (18.41) and (18.42), we get:

Fig. 18.5 Simply supported beam under line load q_z

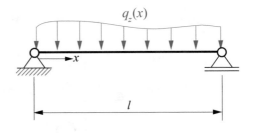

$q_z(x)$

$$\Pi_i = \frac{1}{2} A_{11} \int_0^l \sum_{i=1}^M \sum_{j=1}^M D_i D_j U_i' U_j' \mathrm{d}x$$

$$- B_{11} \int_0^l \sum_{i=1}^M \sum_{j=1}^N D_i C_j U_i' W_j'' \mathrm{d}x$$

$$+ \frac{1}{2} D_{11} \int_0^l \sum_{i=1}^N \sum_{j=1}^N C_i C_j W_i'' W_j'' \mathrm{d}x,$$

$$\Pi_a = - \int_0^l q_z \sum_{i=1}^N C_i W_i \mathrm{d}x. \tag{18.49}$$

Evaluating the Ritz equations

$$\frac{\partial \Pi}{\partial C_i} = 0, \quad \frac{\partial \Pi}{\partial D_i} = 0 \tag{18.50}$$

finally results in a linear system of equations of the following form:

$$\underline{K} \, \underline{C} = \underline{F}, \tag{18.51}$$

or

$$\begin{bmatrix} \underline{\underline{K}}_1 & \underline{\underline{K}}_3^T \\ \underline{\underline{K}}_3 & \underline{\underline{K}}_2 \end{bmatrix} \begin{pmatrix} C_1 \\ C_2 \\ \vdots \\ C_N \\ D_1 \\ D_2 \\ \vdots \\ D_M \end{pmatrix} = \begin{pmatrix} F_1 \\ F_2 \\ \vdots \\ F_N \\ 0 \\ 0 \\ \vdots \\ 0 \end{pmatrix}, \tag{18.52}$$

wherein:

$$F_i = \int_0^l q_i W_i \mathrm{d}x. \tag{18.53}$$

The submatrices $\underline{\underline{K}}_1$, $\underline{\underline{K}}_2$, $\underline{\underline{K}}_3$ of the stiffness matrix $\underline{K}$ are obtained as:

$$\underline{\underline{K}}_1 = \begin{bmatrix} K_{11}^1 & K_{12}^1 & K_{13}^1 & \cdots & K_{1N}^1 \\ K_{12}^1 & K_{22}^1 & K_{23}^1 & \cdots & K_{2N}^1 \\ K_{13}^1 & K_{23}^1 & K_{33}^1 & \cdots & K_{3N}^1 \\ \vdots & \vdots & \vdots & \ddots & \vdots \\ K_{1N}^1 & K_{2N}^1 & K_{3N}^1 & \cdots & K_{NN}^1 \end{bmatrix},$$

$$\underline{\underline{K}}_2 = \begin{bmatrix} K^2_{11} & K^2_{12} & K^2_{13} & \cdots & K^2_{1M} \\ K^2_{12} & K^2_{22} & K^2_{23} & \cdots & K^2_{2M} \\ K^2_{13} & K^2_{23} & K^2_{33} & \cdots & K^2_{3M} \\ \vdots & \vdots & \vdots & \ddots & \vdots \\ K^2_{1M} & K^2_{2M} & K^2_{3M} & \cdots & K^2_{MM} \end{bmatrix},$$

$$\underline{\underline{K}}_3 = \begin{bmatrix} K^3_{11} & K^3_{12} & K^3_{13} & \cdots & K^3_{1N} \\ K^3_{21} & K^3_{22} & K^3_{23} & \cdots & K^3_{2N} \\ K^3_{31} & K^3_{32} & K^3_{33} & \cdots & K^3_{3N} \\ \vdots & \vdots & \vdots & \ddots & \vdots \\ K^3_{M1} & K^3_{M2} & K^3_{M3} & \cdots & K^3_{MN} \end{bmatrix}, \quad (18.54)$$

where the following abbreviations were used:

$$K^1_{ij} = D_{11} \int_0^l W''_i W''_j dx,$$

$$K^2_{ij} = A_{11} \int_0^l U'_i U'_j dx,$$

$$K^3_{ij} = -B_{11} \int_0^l U'_i W''_j dx. \quad (18.55)$$

It turns out that the two submatrices $\underline{\underline{K}}_1$ and $\underline{\underline{K}}_2$ are symmetric, whereas $\underline{\underline{K}}_3$ shows no symmetry properties.

18.2 First-Order Shear Deformation Theory

18.2.1 *Constitutive Law and Equilibrium Conditions*

In many practical application cases, one will obtain sufficiently accurate and useful results with the help of Classical Laminate Theory. However, especially when using modern fiber composites which are characterized by particularly advantageous properties in their fiber direction, but also weak stiffnesses perpendicular to the fiber direction, the use of an improved theory to determine the structural response is indicated. Such an improved theory is the so-called First-Order Shear Deformation Theory (commonly abbreviated as FSDT) which is a generalization of Timoshenko's beam theory (see Chap. 16) for the analysis of laminated structures. We can thus follow up on what was already presented in Chap. 16 and assume a displacement field according to (16.1) (see also Fig. 16.2) as follows:

$$u_P = u + z_P \varphi_y, \quad w_P = w. \quad (18.56)$$

The resulting strain components are given by (16.2):

$$\varepsilon_{xx} = u' + z\varphi_y', \quad \gamma_{xz} = \varphi_y + w'. \tag{18.57}$$

In layer k we obtain the following stress components:

$$\sigma_{xx,k} = E_k \varepsilon_{xx} = E_k \left(u' + z\varphi_y' \right),$$
$$\tau_{xz,k} = G_k \gamma_{xz} = G_k \left(\varphi_y + w' \right). \tag{18.58}$$

It can be seen that the normal stress $\sigma_{xx,k}$ within layer k is linear over z and again there will be discontinuities at the interfaces between different laminate layers. The shear stress $\tau_{xz,k}$, on the other hand, is constant in each individual layer k over the entire layer thickness.

From the expressions (18.58) for the stresses, the internal forces and bending moment N, M_y, V_z can be determined as follows:

$$N = \int_A \sigma_{xx} dA, \quad M_y = \int_A \sigma_{xx} z dA, \quad V_z = \int_A \tau_{xz} dA. \tag{18.59}$$

For the determination of the normal force N with $dA = b dz$ we obtain:

$$N = b \int_{-\frac{h}{2}}^{+\frac{h}{2}} \sigma_{xx} dz. \tag{18.60}$$

Decomposing the integral into integrals over the single layer thicknesses and substituting the expression (18.58) for the normal stress $\sigma_{xx,k}$ yields:

$$N = b \sum_{k=1}^{N} E_k \int_{z_{k-1}}^{z_k} \left(u' + z\varphi_y' \right) dz. \tag{18.61}$$

Performing the integrations prescribed here yields:

$$N = A_{11} u' + B_{11} \varphi_y', \tag{18.62}$$

with the stiffnesses A_{11} and B_{11} already known from classical laminate theory.

We obtain an analogous expression for the bending moment M_y which is given here without further derivation:

$$M_y = B_{11} u' + D_{11} \varphi_y', \tag{18.63}$$

where the bending stiffness occurring here is identical to the expression already derived in the context of Classical Laminate Theory.

The transverse shear force V_z is determined from the integration of the shear stress over the cross-sectional area. Again, we use the relation $dA = b dz$, so that:

$$V_z = b \int_{-\frac{h}{2}}^{+\frac{h}{2}} \tau_{xz} dz. \tag{18.64}$$

Using (18.58), the shear stress can be expressed by the displacement quantities φ_y and w. It follows:

$$V_z = b \sum_{k=1}^{N} G_k \int_{z_{k-1}}^{z_k} \left(\varphi_y + w' \right) dz. \tag{18.65}$$

This can also be written as:

$$V_z = b \left(\varphi_y + w' \right) \sum_{k=1}^{N} G_k \left(z_{k-1} - z_k \right). \tag{18.66}$$

At this point, we introduce a new stiffness quantity that describes the shear stiffness of the laminate in the thickness direction. In accordance with the nomenclature commonly used in the context of First-Order Shear Deformation Theory, we denote this quantity as A_{55}. It is obtained as:

$$A_{55} = b \sum_{k=1}^{N} G_k \left(z_{k-1} - z_k \right). \tag{18.67}$$

Then the transverse shear force V_z results as:

$$V_z = K A_{55} \left(\varphi_y + w' \right), \tag{18.68}$$

where, as already shown in the context of Timoshenko's beam theory, a shear correction factor K is applied. The determination of this factor in the case of laminated beams will not be discussed in detail here; the interested reader is referred to the available technical literature.

The constitutive law for the laminated beam in the context of First-Order Shear Deformation Theory can be stated in a vector-matrix notation as follows:

$$\begin{pmatrix} N \\ M_y \\ V_z \end{pmatrix} = \begin{bmatrix} A_{11} & B_{11} & 0 \\ B_{11} & D_{11} & 0 \\ 0 & 0 & K A_{55} \end{bmatrix} \begin{pmatrix} u' \\ \varphi_y' \\ \varphi_y + w' \end{pmatrix}. \tag{18.69}$$

The equilibrium conditions (18.18) can also be applied here and are given again for clarity:

$$N' = -n, \quad V_z' = -q_z, \quad M_y' = V_z, \tag{18.70}$$

We assume for the moment a symmetric laminate with $B_{11} = 0$, so that with (18.69):

$$\left(A_{11}u'\right)' = -n,$$
$$\left[KA_{55}\left(\varphi_y + w'\right)\right]' = -q_z,$$
$$\left(D_{11}\varphi_y'\right)' = KA_{55}\left(\varphi_y + w'\right). \tag{18.71}$$

In the case of a prismatic beam with invariant properties over x, it follows:

$$A_{11}u'' = -n,$$
$$KA_{55}\left(\varphi_y' + w''\right) = -q_z,$$
$$D_{11}\varphi_y'' = KA_{55}\left(\varphi_y + w'\right). \tag{18.72}$$

For the case of the homogeneous beam with $A_{11} = EA$, $D_{11} = EI_{yy}$ and $KA_{55} = KGA$ this results in the equations already known from the Timoshenko beam:

$$EAu'' = -n,$$
$$KGA\left(\varphi_y' + w''\right) = -q_z,$$
$$EI_{yy}\varphi_y'' = KGA\left(\varphi_y + w'\right). \tag{18.73}$$

18.2.2   Calculation of Stresses

Expressions for the normal stress $\sigma_{xx,k}$ and the shear stress $\tau_{xz,k}$ can be obtained by inverting the constitutive law (18.69), where we assume here arbitrary unsymmetric layups. It follows:

$$\begin{pmatrix} u' \\ \varphi_y' \\ \varphi_y + w' \end{pmatrix} = \begin{bmatrix} \dfrac{D_{11}}{A_{11}D_{11} - B_{11}^2} & -\dfrac{B_{11}}{A_{11}D_{11} - B_{11}^2} & 0 \\ -\dfrac{B_{11}}{A_{11}D_{11} - B_{11}^2} & \dfrac{A_{11}}{A_{11}D_{11} - B_{11}^2} & 0 \\ & & \dfrac{1}{KA_{55}} \end{bmatrix} \begin{pmatrix} N \\ M_y \\ V_z \end{pmatrix}. \tag{18.74}$$

Substituting in the expressions (18.58) gives the following formulas for the normal stress $\sigma_{xx,k}$ and the shear stress $\tau_{xz,k}$:

$$\sigma_{xx,k} = \frac{E_k}{A_{11}D_{11} - B_{11}^2}\left[D_{11}N - B_{11}M_y + z\left(A_{11}M_y - B_{11}N\right)\right],$$
$$\tau_{xz,k} = \frac{G_k V_z}{A_{55}}. \tag{18.75}$$

Obviously, the expression for the normal stress $\sigma_{xx,k}$ is identical to (18.23).

18.2.3   Deformations

We consider the calculation of the rod and beam deformations for the case of a symmetrically laminated structure. The determination of the displacement u can be done quite simply by integrating the first equation in (18.71) and (18.72), respectively:

$$A_{11}u'' = -n,$$
$$A_{11}u' = -\int n\,dx + D_1,$$
$$A_{11}u = -\int\int n\,dx\,dx + D_1 x + D_2. \tag{18.76}$$

The two integration constants D_1 and D_2 are then fitted to given boundary conditions.

To determine the beam deflection w, we use the equilibrium conditions $V'_z = -q_z$ and $M'_y = V_z$ from which we obtain:

$$D_{11}\varphi'''_y = -q_z. \tag{18.77}$$

Moreover, it follows with $V_z = KA_{55}\left(\varphi_y + w'\right)$:

$$D_{11}\varphi''_y - KA_{55}\left(\varphi_y + w'\right) = 0. \tag{18.78}$$

We now integrate the expression

$$KA_{55}\left(\varphi'_y + w''\right) = -q_z \tag{18.79}$$

once with respect to x:

$$KA_{55}\left(\varphi_y + w'\right) = -\int q_z\,dx + C_1. \tag{18.80}$$

We insert this expression into (18.78). It follows:

$$D_{11}\varphi''_y = -\int q_z\,dx + C_1. \tag{18.81}$$

From this, φ_y can be obtained by twofold integration:

$$D_{11}\varphi'_y = -\int\int q_z\,dx\,dx + C_1 x + C_2,$$
$$D_{11}\varphi_y = -\int\int\int q_z\,dx\,dx\,dx + \frac{1}{2}C_1 x^2 + C_2 x + C_3. \tag{18.82}$$

From (18.80) we can derive the following expression for w':

$$w' = \frac{1}{KA_{55}} \left(-\int q_z \mathrm{d}x + C_1 \right) - \varphi_y. \tag{18.83}$$

Inserting (18.82) yields:

$$w' = \frac{1}{KA_{55}} \left(-\int q_z \mathrm{d}x + C_1 \right)$$
$$- \frac{1}{D_{11}} \left(-\int\int\int q_z \mathrm{d}x \mathrm{d}x \mathrm{d}x + \frac{1}{2}C_1 x^2 + C_2 x + C_3 \right). \tag{18.84}$$

Integration with respect to x finally gives the deflection w of the beam as follows:

$$w = \frac{1}{KA_{55}} \left(-\int\int q_z \mathrm{d}x \mathrm{d}x + C_1 x \right)$$
$$- \frac{1}{D_{11}} \left(-\int\int\int\int q_z \mathrm{d}x \mathrm{d}x \mathrm{d}x \mathrm{d}x + \frac{1}{6}C_1 x^3 + \frac{1}{2}C_2 x^2 + C_3 x + C_4 \right). \tag{18.85}$$

The integration constants C_1, C_2, C_3, C_4 are adjusted to given boundary conditions.

18.2.4   Energetic Consideration

For the energetic consideration of the laminated beam within the framework of First-Order Shear Deformation Theory, the strain energy Π_i is required. It can be given as:

$$\Pi_i = \frac{1}{2} \int_V \left(\sigma_{xx}\varepsilon_{xx} + \tau_{xz}\gamma_{xz} \right) \mathrm{d}V. \tag{18.86}$$

Splitting the volume integral into a line integral with respect to x and an area integral with respect to A and also using the area element $\mathrm{d}A = b\mathrm{d}z$ results in:

$$\Pi_i = \frac{b}{2} \int_0^l \int_{-\frac{h}{2}}^{+\frac{h}{2}} \left(\sigma_{xx}\varepsilon_{xx} + \tau_{xz}\gamma_{xz} \right) \mathrm{d}z\mathrm{d}x. \tag{18.87}$$

Substituting the stresses according to $\sigma_{xx,k} = E_k\varepsilon_{xx}$ and $\tau_{xz,k} = G_k\gamma_{xz}$ leads to:

$$\Pi_i = \frac{b}{2} \int_0^l \sum_{k=1}^N \int_{z_{k-1}}^{z_k} \left(E_k\varepsilon_{xx}^2 + G_k\gamma_{xz}^2 \right) \mathrm{d}z\mathrm{d}x. \tag{18.88}$$

If we express the strains by $\varepsilon_{xx} = u' + z\varphi'_y$ and $\gamma_{xz} = \varphi_y + w'$, then the following expression for the inner potential Π_i results after some transformations:

$$\Pi_i = \frac{1}{2} A_{11} \int_0^l (u')^2 \, dx + B_{11} \int_0^l u' \varphi'_y dx$$
$$+ \frac{1}{2} D_{11} \int_0^l (\varphi'_y)^2 \, dx + \frac{1}{2} K A_{55} \int_0^l (\varphi_y + w')^2 \, dx. \qquad (18.89)$$

We derive the governing differential equations as well as all potential combinations of boundary conditions using the principle of the minimum of the total elastic potential $\delta\Pi = 0$ for the case of a beam under the line load q_z (Fig. 18.5). The external potential Π_a remains unchanged compared to (18.42). The requirement for the vanishing of the first variation $\delta\Pi = \delta (\Pi_i + \Pi_a) = 0$ eventually results in:

$$-\int_0^l \left(A_{11}u'' + B_{11}\varphi''_y\right) \delta u dx - \int_0^l \left(K A_{55} \left(\varphi'_y + w''\right) + q_z\right) \delta w dx$$
$$-\int_0^l \left(B_{11}u'' + D_{11}\varphi''_y - K A_{55} \left(\varphi_y + w'\right)\right) \delta\varphi_y dx$$
$$+ \left(A_{11}u' + B_{11}\varphi'_y\right) \delta u \Big|_0^l + K A_{55} \left(\varphi_y + w'\right) \delta w \Big|_0^l$$
$$+ \left(D_{11}\varphi'_y + B_{11}u'\right) \delta\varphi_y \Big|_0^l = 0. \qquad (18.90)$$

From the three integral terms, the governing differential equations of the problem can be determined:

$$A_{11}u'' + B_{11}\varphi''_y = 0,$$
$$K A_{55} \left(\varphi'_y + w''\right) = -q_z,$$
$$B_{11}u'' + D_{11}\varphi''_y = K A_{55} \left(\varphi_y + w'\right). \qquad (18.91)$$

The boundary terms appearing in (18.90) describe the boundary conditions of the given laminated beam. A discussion of these terms is omitted here.

18.3 Sandwich Beams

18.3.1 Introduction

The common idealization approach for sandwich structures (Fig. 17.1, bottom left, as well as Fig. 18.6) assumes a clear division of tasks between the individual components, i.e. face sheets and core. The sandwich under consideration consists of two

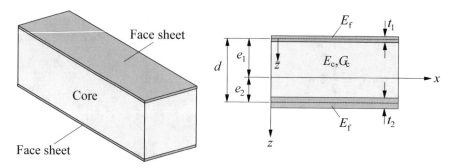

Fig. 18.6 Sandwich beam

face sheets (thicknesses t_1 and t_2), for which we want to assume, as is usually the case in lightweight engineering practice, that they are made of identical material (modulus of elasticity E_f), but may well have different thicknesses. The core has the thickness d (measured between the skeleton lines of the two facesheets). Its elastic properties are E_c and G_c. It is usually assumed that $E_f \gg E_c$ holds. The elastic center of gravity has the distance e_1 from the upper face sheet and the distance e_2 from the lower face sheet. For the later determination of the elastic center of gravity the $\bar{z}$−axis is introduced as indicated. The origin of the reference system x, z is located in the elastic center of gravity.

It is usually assumed that the core of a sandwich beam is not involved in the transfer of normal forces and bending moments (i.e. the extensional and bending stiffness are negligible) and thus is only involved in the transfer of transverse shear forces. The face sheets are assumed to be much thinner than the sandwich core so that $t_1, t_2 \ll d$. As a consequence, the bending stiffness of the face sheets is neglected so that only the Steiner components of the face sheets are included in the calculation of the sandwich bending stiffness. The bending moment $M_y = N_2 e_2 - N_1 e_1$ of the sandwich beam is introduced as a pair of forces ($N_1 = N_2 = \frac{M_y}{d}$) into the face sheets which transfer the resulting forces similar to rods. The same applies to the transfer of the normal force $N = N_1 + N_2$. As a consequence, the normal stresses occurring in the face sheets due to normal force N and bending moment M_y are assumed to be constant over the face sheet thickness. Moreover, the face sheets do not participate in the transfer of transverse shear forces. These are transferred exclusively through the core, and the resulting shear stresses are assumed to be constant over the core thickness. The core is also assumed to be incompressible in the thickness direction so that the sandwich does not experience any expansion in the thickness direction (Fig. 18.7).

Ultimately, the modeling approach described above corresponds to the application of the shear field beam model (Chap. 15) to the analysis of sandwich structures. Detailed discussions of sandwich structures can be found, e.g., in Stamm and Witte (1974), Wiedemann (2007) or Zenkert (1995).

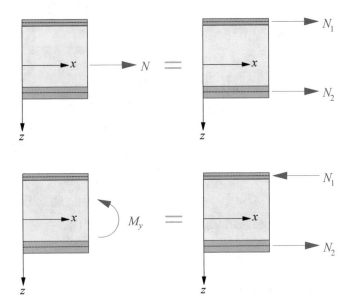

Fig. 18.7 Transfer of normal force N and bending moment M_y in a sandwich beam

18.3.2 Constitutive Law

It is necessary for the further explanations to be clear first about the position of the elastic center of gravity of the sandwich beam under consideration. We can use the calculation rules already discussed in Chap. 17 for this purpose. Since we again consider a beam with constant width b under normal force and bending moment in the $xz-$plane, only the coordinate $\bar{z}_S$ has to be determined here (cf. Eq. (17.22)):

$$\bar{z}_s = \frac{\sum\limits_{i=1}^{3} E_i \int_{A_i} \bar{z} dA_i}{\sum\limits_{i=1}^{3} E_i \int_{A_i} dA_i} = \frac{\left(ES_{\bar{y}}\right)_{eff}}{(EA)_{eff}}. \tag{18.92}$$

The effective static moment $\left(ES_{\bar{y}}\right)_{eff}$ is given as:

$$\left(ES_{\bar{y}}\right)_{eff} = E_f \cdot d \cdot b \cdot t_2 + E_c \cdot \frac{d}{2} \cdot b \cdot d. \tag{18.93}$$

Under the condition $E_f \gg E_c$ the second term can be neglected, so that:

$$\left(ES_{\bar{y}}\right)_{eff} = E_f dbt_2. \tag{18.94}$$

Fig. 18.8 Location of the
elastic center of gravity of
the sandwich beam

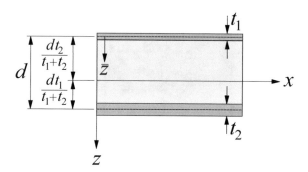

The effective extensional stiffness $(EA)_{eff}$ follows as:

$$(EA)_{eff} = E_f bt_1 + E_f bt_2 + E_c bd, \tag{18.95}$$

or after neglecting the last term:

$$(EA)_{eff} = E_f b (t_1 + t_2). \tag{18.96}$$

For the centroid coordinate $\bar{z}_S$ then follows (Fig. 18.8):

$$\bar{z}_s = \frac{dt_2}{t_1 + t_2} = e_1. \tag{18.97}$$

The lower distance $e_2 = d - e_1$ then results as:

$$e_2 = \frac{dt_1}{t_1 + t_2}. \tag{18.98}$$

In all that follows we will assume that a sandwich beam behaves like a shear-compliant beam within the framework of Timoshenko's theory (cf. Chap. 16). In particular, we assume here the validity of the displacement field (16.1), the strain field (16.2), and the equations for the stresses (16.3), which we state again here for ease of reading. The displacement field in the case of bending in the $xz-$plane for any point P at $z = z_P$ is as follows:

$$u_P = u + z_P \varphi_y,$$
$$w_P = w. \tag{18.99}$$

For the strains ε_{xx} and γ_{xz} we then obtain:

$$\varepsilon_{xx} = u' + z\varphi'_y,$$
$$\gamma_{xz} = \varphi_y + w'. \tag{18.100}$$

We obtain the stresses in segment k of the sandwich beam as:

$$\sigma_{xx,k} = E_k \varepsilon_{xx} = E_k \left(u' + z\varphi'_y \right),$$
$$\tau_{xz,k} = G_k \gamma_{xz} = G_k \left(\varphi_y + w' \right). \tag{18.101}$$

We first determine the normal force N as:

$$N = \int_A \sigma_{xx} \mathrm{d}A, \tag{18.102}$$

where we decompose the area integral into two integrals concerning the two face sheets (the core is not involved in the transfer of the normal force):

$$N = \int_{A_1} \sigma_{xx,1} \mathrm{d}A_1 + \int_{A_2} \sigma_{xx,2} \mathrm{d}A_2. \tag{18.103}$$

Inserting (18.101) yields:

$$N = E_f \int_{A_1} \left(u' + z\varphi'_y \right) \mathrm{d}A_1 + E_f \int_{A_2} \left(u' + z\varphi'_y \right) \mathrm{d}A_2$$
$$= E_f \left(A_1 + A_2 \right) u' + E_f b \left(e_2 t_2 - e_1 t_1 \right) \varphi'_y. \tag{18.104}$$

With (18.97) and (18.98), the second term occurring here results in zero, and with $A_1 = bt_1$, $A_2 = bt_2$:

$$N = E_f b \left(t_1 + t_2 \right) u'. \tag{18.105}$$

The expression $E_f b \left(t_1 + t_2 \right)$ can be interpreted as the effective extensional stiffness of the sandwich beam (see also Eq. (18.96)), so that:

$$N = (EA)_{eff} \, u'. \tag{18.106}$$

In an analogous way, we determine the bending moment M_y as:

$$M_y = \int_A \sigma_{xx} z \mathrm{d}A, \tag{18.107}$$

where we also decompose the area integral into two integrals:

$$M_y = \int_{A_1} \sigma_{xx,1} z \mathrm{d}A_1 + \int_{A_2} \sigma_{xx,2} z \mathrm{d}A_2. \tag{18.108}$$

Substitution of (18.101) leads to the following expression:

$$M_y = E_f \int_{A_1} \left(u'z + z^2\varphi'_y \right) \mathrm{d}A_1 + E_f \int_{A_2} \left(u'z + z^2\varphi'_y \right) \mathrm{d}A_2$$

$$= E_f u' \int_{A_1} z\mathrm{d}A_1 + E_f u' \int_{A_2} z\mathrm{d}A_2 + E_f \varphi'_y \int_{A_1} z^2\mathrm{d}A_1 + E_f \varphi'_y \int_{A_2} z^2\mathrm{d}A_2$$

$$= E_f u' \left(e_2 A_2 - e_1 A_1 \right) + E_f \varphi'_y \left(e_1^2 A_1 + e_2^2 A_2 \right). \tag{18.109}$$

With (18.97) and (18.98) it turns out that the first term vanishes, so that we obtain:

$$M_y = E_f d^2 b \frac{t_1 t_2}{t_1 + t_2} \varphi'_y. \tag{18.110}$$

This is the constitutive relation for the bending moment M_y with the effective bending stiffness

$$(EI)_{eff} = E_f d^2 b \frac{t_1 t_2}{t_1 + t_2}. \tag{18.111}$$

Finally, an expression for the transverse shear force V_z must be obtained, which can be determined from the shear stress τ_{xz} as follows:

$$V_z = \int_A \tau_{xz} \mathrm{d}A. \tag{18.112}$$

Since we assume that transverse shear is transferred exclusively by the core, we can also write:

$$V_z = \int_{A_c} \tau_{xz} \mathrm{d}A_c. \tag{18.113}$$

With (18.101) this takes on the following form:

$$V_z = G_c \int_{A_c} \left(\varphi_y + w' \right) \mathrm{d}A_c = G_c A_c \left(\varphi_y + w' \right) = G_c bd \left(\varphi_y + w' \right). \tag{18.114}$$

In this, the expression $G_c bd$ can be interpreted as the effective shear stiffness $(GA)_{eff}$ of the sandwich beam. The application of a shear correction factor is possible, but will not be further elaborated here.

In total, the following constitutive relations are thus available for the sandwich beam:

$$N = E_f b \left(t_1 + t_2 \right) u',$$
$$M_y = E_f d^2 b \frac{t_1 t_2}{t_1 + t_2} \varphi'_y,$$
$$V_z = G_c A_c \left(\varphi_y + w' \right). \tag{18.115}$$

The comparison with the constitutive relations of the Timoshenko beam (see Chap. 16, Eq. (16.6)) shows that these are, except for the stiffness quantities to be applied, completely identical to the present equations, so that all explanations on shear-deformable beams can be directly applied to the sandwich beams treated here. Further explanations can thus be omitted at this point.

18.3.3 Calculation of Stresses

In this section, we want to investigate the question of which stress components occur in the individual components of the sandwich beam. The normal force N is assumed to be carried exclusively by the face sheets. From Eq. (17.36), the normal stress in the face sheets can be calculated as:

$$\sigma_{xx,k} = \frac{E_k N}{(EA)_{eff}}, \tag{18.116}$$

with $k = 1$ (upper face sheet) or $k = 2$ (lower face sheet). Since we have assumed identical material for both face sheets, identical stresses result in the two face sheets:

$$\sigma_{xx,1} = \sigma_{xx,2} = \frac{E_f N}{E_f b (t_1 + t_2)} = \frac{N}{b (t_1 + t_2)}. \tag{18.117}$$

A test of this result is carried out by checking whether the resultants of the face sheet stresses in sum correspond to the normal force N. The result is:

$$N_1 = \sigma_{xx,1} A_1 = \sigma_{xx,1} b t_1 = \frac{N t_1}{t_1 + t_2},$$

$$N_2 = \sigma_{xx,2} A_2 = \sigma_{xx,2} b t_2 = \frac{N t_2}{t_1 + t_2}. \tag{18.118}$$

It can be seen that the requirement $N_1 + N_2 = N$ is satisfied here.

We further consider the normal stresses in the face sheets as a consequence of the bending moment M_y. Then, according to (17.36):

$$\sigma_{xx,k} = \frac{E_k M_y z}{(EI_{yy})_{eff}}. \tag{18.119}$$

With $(EI_{yy})_{eff}$ according to (18.111) we get:

$$\sigma_{xx,1} = -\frac{M_y}{b t_1 d}, \quad \sigma_{xx,2} = \frac{M_y}{b t_2 d}. \tag{18.120}$$

A simple test of this result can also be performed here by checking whether the resulting forces of the normal stresses $\sigma_{xx,1}$ and $\sigma_{xx,2}$, multiplied by their lever arms e_1 and e_2, respectively, correspond in sum to the acting bending moment M_y. We first calculate the two partial moments M_1 and M_2 as:

$$M_1 = \sigma_{xx,1} A_1 e_1 = \frac{M_y t_2}{t_1 + t_2},$$
$$M_2 = \sigma_{xx,2} A_2 e_2 = \frac{M_y t_1}{t_1 + t_2}. \tag{18.121}$$

It turns out that $M_1 + M_2 = M_y$ holds, which confirms the correctness of the calculations.

Finally, the shear stresses of the core must be considered. Due to the assumption of a constant shear stress distribution over the cross-sectional area of the core, these can be determined in an elementary simple way as:

$$\tau_c = \frac{V_z}{A_c} = \frac{V_z}{bd}. \tag{18.122}$$

References

Altenbach H, Altenbach J, Kissing W (2018) Mechanics of composite structural elements. 2nd edn. Springer Nature, Singapore

Ambartsumyan SA (1970) Theory of anisotropic plates. Technomic Publishing Co., Inc, Stamford, USA

Ashton JE, Whitney JM (1970) Theory of laminated plates. Technomic Publishing Co., Inc, Stamford, USA

Becker W, Gross D (2002) Mechanik elastischer Körper und Strukturen, Springer. Berlin et al, Germany

Jones RM (1975) Mechanics of composite materials. Scripta Book Co., Washington, USA

Lekhnitskii SG (1968) Anisotropic plates, Gordon and Breach. New York et al, USA

Mittelstedt C, Becker W (2016) Strukturmechanik ebener Laminate. Technische Universität Darmstadt, Germany, Studienbereich Mechanik

Reddy JN (2004) Mechanics of laminated composite plates and shells. 2nd edn, CRC Press. Boca Raton et al, USA

Stamm K, Witte H (1974) Sandwichkonstruktionen. Springer, Wien et al, Austria

Wiedemann J (2007) Leichtbau 1: Elemente. Springer. Berlin et al, Germany

Zenkert D (1995) An introduction to sandwich construction. EMAS—Engineering Materials Advisory Services Ltd., UK

Index

Printed in the United States
by Baker & Taylor Publisher Services